Developments in Primatology: Progress and Prospects

For other titles published in this series, go to
www.springer.com/series/5852

Volker Sommer • Caroline Ross
Editors

Primates of Gashaka

Socioecology and Conservation in Nigeria's Biodiversity Hotspot

Springer

Editors
Volker Sommer
University College London
Department of Anthropology
Gower Street
WC1E 6BT London
United Kingdom
v.sommer@ucl.ac.uk

Caroline Ross
Roehampton University
Department of Life Sciences
Holybourne Avenue
SW15 4JD London
United Kingdom
c.ross@roehampton.ac.uk

ISBN 978-1-4419-7402-0 e-ISBN 978-1-4419-7403-7
DOI 10.1007/978-1-4419-7403-7
Springer New York Dordrecht Heidelberg London

Library of Congress Control Number: xxxxxxxx

Printed on acid-free paper

Springer is part of Springer Science+Business Media (www.springer.com)

In memoriam
Jeremiah Adanu (1969–2004)
Mallam Salamu Waziri (c1945–2007)
Doris Nyanganji (c1976–2008)
Yakubu Ahmadu (c1967–2009)
Bobbo Buba (c1978–2010)
Ymke Warren (1970–2010)

Foreword

The importance of research for human development and the management of protected areas cannot be overemphasized. Early in the history of national parks, the primary role of research was to provide information for interpretative talks and displays, as a result of visitors' demands to learn more about the interesting phenomena they encountered. More recently, research has been directed toward collecting data in the field and providing analyses that inform and influence policy recommendations, thereby ultimately improving management effectiveness of protected areas and development of community projects.

Nigeria's Gashaka Gumti National Park embodies unique biological, physical, and cultural resources. The expanses of this wilderness, combined with a protective government policy, attract researchers from varied disciplines to this part of West Africa. Their efforts are unique, considering the challenging nature of the area, the invested time, and the remote location of the research centre. Researchers conducting studies of these primate communities deserve commendation, considering their patience, sacrifice, and commitment while trailing their subjects and collecting detailed information on their socio-ecology.

The results of such in-depth research as published here showcase the rich biodiversity potential embedded in Nigeria's largest national park. I warmly recommend this book, which adds much scientific knowledge and highlights some of Africa's hidden treasures for the world to see. It is my hope that this book will serve as a reference and source of knowledge and information to both scientists and nature lovers, particularly those in the field of biodiversity conservation and eco-tourism.

With the support and cooperation of all stakeholders, we are dedicated to the enforcement of conservation policies that aim to protect unique ecosystems such as the vast and serene Gashaka Gumti National Park for the benefit of wildlife and humans alike.

Alhaji Haruna Tanko Abubakar

Conservator General,
Nigeria National Park Service
Abuja, Nigeria, April 2010

Contents

Contributors

Jeremiah Adanu
c/o Joy Adanu, Kona Rd, near Akuka's Lodge, Jalingo, Taraba State, Nigeria
jereadanu@yahoo.com

Callistus Akosim
Department of Forestry and Wildlife Management, Federal University of Technology, PMB 2076, Yola, Adamawa State, Nigeria

Kate Arnold
School of Psychology, University of St. Andrews, St. Andrews, KY16 9JP, Scotland UK
ka11@st-andrews.ac.uk

Richard Barnwell
The Old Well, Fewster Road, Nailsworth GL6 0DH, UK
janebarnwell@btinternet.com

Jan Bauer
Research Group Evolutionary Ecology, Leibniz Institut für Zoo- und Wildtierforschung, Alfred-Kowalke-Str. 17, 10315 Berlin, Germany
Jan.Bauer.Berlin@web.de

David Bennett
Centre for Research in Evolutionary Anthropology, Roehampton University, Holybourne Avenue, London SW15 4JD, UK
dawudbennett@googlemail.com

Umaru Buba
Gashaka Primate Project, PMB 08, Serti 663001, Taraba State, Nigeria
bumami2004@yahoo.com

Elodie Ey
Research Group Human genetics and cognitive functions, Pasteur Institute, 25 rue du Docteur Roux, 75724 Paris cedex 15, France
elodie.ey@pasteur.fr

Julia Fischer
Department of Cognitive Ethology, German Primate Centre,
Kellnerweg 4, D-37077 Göttingen, Germany
fischer@cog-ethol.de

Andrew Fowler
Max-Planck-Institut für Evolutionäre Anthropology,
Deutscher Platz 6, 04103 Leipzig, Germany
andrew_fowler@eva.mpg.de

Neil Gretsky
218 16th Street, Hermosa Beach, CA 90254, USA
ngretsky@econ.ucla.edu

James P. Higham
Institute for Mind and Biology, University of Chicago, Chicago, IL 60637, USA
jhigham@uchicago.edu

Nicola Hughes
Department of Anthropology, University College London,
Gower Street, London WC1E 6BT, UK
nicolalhughes@yahoo.co.uk

Yianna Koutsioni
11 Konitsis Street, Kifissia 14561, Athens, Greece
yiannakoutsioni@hotmail.com

Ann M. Maclarnon
Centre for Research in Evolutionary Anthropology, Roehampton University,
Holybourne Avenue, London SW15 4JD, UK
a.maclarnon@roehampton.ac.uk

Aylin McNamara
Zoological Society of London, Regent's Park, London NW1 4RY, UK
aylin.mcnamara@zsl.org, aylin.mcnamara@zsl.org

Stephen Nash
Department of Anatomical Sciences, H.S.C., State University of New York,
Stony Brook, NY 11794-8081, USA
snash@ms.cc.sunysb.edu, s.nash@conservation.org

Gilbert Nyanganji
Department of Biological Sciences, University of Maiduguri,
PMB 1069, Maiduguri, Borno State, Nigeria
gilsnyanganji@yahoo.com

Sylvia Ortmann
Research Group Evolutionary Ecology, Leibniz Institute for Zoo
and Wildlife Research, Alfred-Kowalke-Str. 17, 10315 Berlin, Germany
ortmann@izw-berlin.de

Alejandra Pascual-Garrido
School of Biological Sciences, Universidad Complutense de Madrid,
Madrid 28040, Spain and J. F. Seguí 4751 1p "E" (1425), Buenos Aires, Argentina

Yvonne Pohlner
Institut für Biologie I, Albert Ludwigs Universität, Hauptstrasse 1,
79104, Freiburg, Germany
yvonne.pohlner@googlemail.com

Christian Roos
Primate Genetics Laboratory, Deutsches Primatenzentrum,
Kellnerweg 4, 37077 Göttingen, Germany
croos@dpz.eu

Norman Rosen
27 16th Street, Hermosa Beach, CA 90254, USA
normrosen@aol.com

Caroline Ross
Centre for Research in Evolutionary Anthropology,
Roehampton University, Holybourne Avenue, London SW15 4JD, UK
c.ross@roehampton.ac.uk

Caspar Schöning
Centre of Social Evolution, Department of Biology,
University of Copenhagen, Universitetsparken 15, 2100 Copenhagen/Denmark;
currently Länderinstitut für Bienenkunde, Friedrich-Engels-Str. 32,
16540 Hohen Neuendorf, Germany
caspar@zedat.fu-berlin.de

Volker Sommer
Department of Anthropology, University College London,
London WC1E 6BT, UK
v.sommer@ucl.ac.uk

Sandra Tranquilli
Department of Anthropology, University College London,
Gower Street, London WC1E 6BT, UK
sandra_tranq@hotmail.com

Ymke Warren
Research Coordinator, Takamanda-Mone Landscape Project,
Africa Programme – Wildlife Conservation Society, c/o Limbe Botanic Gardens,
PO Box 1308, Limbe, South West Province, Cameroon
anicholas@wcs.org

Dietmar Zinner
Cognitive Ethology Laboratory, Deutsches Primatenzentrum,
Kellnerweg 4, 37077 Göttingen, Germany
dzinner@gwdg.de

Klaus Zuberbühler
School of Psychology, University of St. Andrews,
St. Andrews, KY16 9JP, Scotland, UK
kz3@st-andrews.ac.uk

Bio-Sketches

Jeremiah Adanu (1969 – 2004) held a BTech (1999) and MTech (2002) in Forestry and Wildlife Management from Federal University of Technology at Yola / Nigeria, where he was also a PhD student. His research focused on fish, hippopotamus, forest guenons, olive baboons and chimpanzees. He assisted the Gashaka Gumti Project of the Nigerian Conservation Foundation (1997 – 1999) before joining and developing the Gashaka Primate Project.

Callistus Akosim is Professor at the Federal University of Technology at Yola. He and his students conduct research into the natural resources and wildlife of Nigeria.

Kate Arnold is a Research Fellow in the School of Psychology, University of St. Andrews. She conducted zoo based research on monkey social behaviour (1995 – 1996), on chimpanzee social behaviour in Uganda (1998 – 1999) and field work on monkey cognition and communication in Nigeria (since 2003).

Richard Barnwell is a conservationist and long-term officer for WWF-UK who worked mainly in West and East Africa and Oman. From its early inception as a game reserve, he cooperated closely with the Nigerian Conservation Foundation and Nigerian government at the state and federal level to create the protected area that ultimately became Gashaka Gumti National Park.

Jan Bauer works at the Tierpark Gera. Degree as Diplombiologe from the Humboldt-Universität Berlin (2006), with a thesis on digestive strategies of chimpanzees and bonobos at the Leibniz-Institut für Zoo- und Wildtierforschung, Berlin.

David Bennett obtained a masters in Human Evolution and Behaviour at University College (2003), and a PhD at Roehampton University (2009). His research investigates the interaction between people and the environment, explores perceptions of the environment and how these might be linked to actions and beliefs.

Umaru Buba obtained an MTech in Wildlife Conservation and Management from the Federal University of Technology Yola / Nigeria (2007), where he is currently completing a PhD. Project manager of the Gashaka Primate Project since 2006. Assistant lecturer at Taraba State University since 2010.

Elodie Ey completed her masters on vocal communication of prosimians at the University of Paris. She then studied vocal communication in olive baboons in Nigeria and completed her PhD at the University of Göttingen in 2008. She is now at the Pasteur Institute in Paris.

Julia Fischer heads the research group Cognitive Ethology at the German Primate Centre since 2004, and is Professor of Cognitive Ethology at the University of Göttingen / Germany. Her field work has focused on free-ranging chacma baboons in Botswana and currently on Guinea baboons in Senegal.

Andrew Fowler obtained a BA in Social Anthropology from the University of London (1987), an MSc in Computer Sciences from Middlesex University (1991), an International MBA from IAL Agenzia Formativa, Pordenone, Italy and Clemsen University USA (1993), an MSc in Human Evolution and Behaviour (1999) and - based on field work on chimpanzees in Nigeria – a PhD in Anthropology at University College London (2006). He is now studying bonobos in the Democratic Republic of Congo in conjunction with the Max-Planck-Institute für Evolutionäre Anthropologie, Leipzig / Germany.

Neil Gretsky is Professor Emeritus, Department of Mathematics, University of California, Riverside and currently Visiting Professor, Department of Economics, University of California, Los Angeles.

James P. Higham is a Post-doctoral Researcher at the Institute for Mind and Biology, University of Chicago. His research interests include life-history, sexual and social strategies and conflict, behavioural endocrinology, animal communication, and the conservation of biological diversity.

Nicola Hughes holds a BA in Archaeology and Anthropology from Keble College, Oxford University (1996) and an M.Sc. in Human Evolution and Behaviour from University College London (2003). She has a particular interest in primate ecology and conservation.

Yianna Koutsioni holds a BSc in Anthropology from the University of East London (2001) and an MSc in Human Evolution and Behaviour from University College London (2003). Part-time lecturer at University of East London of Field Methods in Urban Anthropology. Her research interests include the causes and consequences of drug use.

Ann M. Maclarnon is Professor of Evolutionary Anthropology at Roehampton University and was President of the Primatological Society of Great Britain (2006-2009). Her research interests encompass palaeoanthropology, primate biology, and recently endocrinological aspects of primate reproductive ecology, stress, and sociability.

Aylin McNamara obtained a BSc in Biology at University College London. Volunteer with the Gashaka Primate Project. Currently Climate Change Programme Coordinator at the Zoological Society of London.

Stephen Nash is scientific illustrator for Conservation International, and produces images for conservation education and biological publications. He is based at S.U.N.Y. Stony Brook, where he is a Visiting Research Associate in the Department of Anatomical Sciences, and an Adjunct Associate Professor in the Department of Art.

Gilbert Nyanganji worked as a research assistance in the Gashaka Gumti project of the Nigerian Conservation Foundation (1994 – 2006) and holds a BSc in Zoology from the Department of Biological Sciences, University of Maiduguri / Nigeria (2006).

Sylvia Ortmann is head of the Nutrition-Lab of the Leibniz Institute for Zoo and Wildlife Research. Research interests include life-history strategies and adaptation to seasonality, feeding ecology, and physiological strategies of nutrition and digestion in herbivores.

Alejandra Pascual-Garrido worked on behavioural ecology and conservation of Andean vultures, badgers, dormice, olive baboons, and chimpanzees. Field research assistant for Oxford University's Wildlife Conservation Research Unit (2002 – 2003), Roehampton University (2004) and Wildlife Conservation Society, Argentina (2005). MSc in Biological Sciences at Royal Holloway, University of London (2002). Completing PhD student at School of Biological Sciences, Universidad Complutense de Madrid.

Yvonne Pohlner studied biology at the Albert-Ludwigs-University of Freiburg / Germany; Diplomarbeit on guenon vocal communication in Nigeria (2004), in conjunction with St. Andrews University. Studies on behaviour of captive chimpanzees (2002) and animal ecology in Kenya (2005) for the University of Freiburg and on endocrinology of common marmosets at the University of Vienna (2003).

Christian Roos is senior scientist in the Primate Genetics Laboratory at the German Primate Center, focusing on molecular evolution, phylogeography, population genetics, and taxonomy. He heads the Gene Bank of Primates and is South-east Asian Deputy Regional Coordinator of the IUCN / SSC Primate Specialist Group.

Norman Rosen holds an MA in Anthropology from California State University Fullerton, a Master of Liberal Arts from the University of Southern California, and currently completes his PhD in Biological Anthropology at University College London. After a career in publishing, he developed new concepts towards Great Ape conservation, e.g., in the Conversation Breeding Specialist Group of IUCN and on the boards of BOS USA / Indonesia (Orangutan) as well as PASA (Pan African Sanctuaries Alliance).

Caroline Ross is Reader in Evolutionary Anthropology at Roehampton University, London and director of baboon studies in the Gashaka Primate Project / Nigeria. Her research interests include primate socioecology, life-history evolution, and human-wildlife conflict.

Caspar Schöning is a postdoctoral researcher in one of Germany's honey-bee research institutes. Major research interests include behavioural ecology and systematics of army ants and honey-bees and predator-prey relationships between African apes and social insects.

Volker Sommer is Professor of Evolutionary Anthropology in the University of London, Pro-Provost for the International Strategy (Africa) of University College London, and founder and director of the Gashaka Primate Project. He conducts long-term field research into the socioecology and conservation of primates, in particular langur monkeys in India (since 1981), gibbons in Thailand (since 1989), and chimpanzees in Nigeria (since 2000).

Sandra Tranquilli obtained a Diploma di Laurea at La Sapienza University, Rome (2004) based on field work on vocal behaviour of guenons in Ivory Coast. Field work on meerkats in South Africa (2005), chimpanzees in Nigeria (2006); research assistant at the Zoological Society of London (2007). Completing PhD student at the Department of Anthropology, University College London, in conjunction with the A.P.E.S. database project at the Max-Planck-Institut für Evolutionäre Anthropology, Leipzig.

Ymke Warren (1970 – 2010) was Research Coordinator for the Takamanda-Mone Landscape Project in Cameroon. She obtained an MSc in Human Evolution and Behaviour at University College London (1998) and a PhD at Roehampton University (2003). Her research interests include human-wildlife conflict, primate ecology, and conservation. She worked on the behavioural ecology of mountain gorillas, olive baboons, black mongoose, and mountain zebra.

Dietmar Zinner is senior scientist in the Cognitive Ethology Laboratory at the German Primate Center, focusing on behavioural ecology, phylogeography, and phylogenetics. Since 1995, he has conducted field studies in Africa on baboons, including Guinea baboons in Senegal.

Klaus Zuberbühler is Professor of Psychology at the University of St. Andrews. His research on primate cognition is based on field work in Ivory Coast, Uganda, Sierra Leone, Thailand, and Nigeria. He is co-director of the Tai Monkey Project/ Ivory Coast, scientific director of the Budongo Conservation Field Station, Uganda, and Deputy Director of the Living Links Centre for Human Evolution in Edinburgh/ Scotland.

Chapter 1
Exploring and Protecting West Africa's Primates: The Gashaka Primate Project in Context

Volker Sommer and Caroline Ross

Abstract The remote Gashaka region is still largely unexplored, although this area in north-eastern Nigeria is a premier wilderness where monkeys and apes still survive in large numbers. They are part of a rich assemblage of, often endemic, wildlife that has developed at this interface between the dry sub-Saharan Guinea savannah and the moist Cameroonian highlands. Primates include a large population of the rarest chimpanzee subspecies as well as colobus, guenons and baboons, which thrive here despite an unusually wet climate. The main ethnic groups – Fulani cattle herders and Hausa speaking subsistence farmers – still follow age-old traditions. Conservation challenges come in the form of human settlements in national parks, deforestation, annual bush fires, livestock grazing and hunting for bush-meat. This chapter reviews the inception and history of the *Gashaka Primate Project*, founded in the year 2000 and dedicated to research and conservation in this iconic primate habitat. The project's international network of researchers and conservationists engage in areas such as primate socioecology; genetics and phylogeography; nutritional ecology; vocal communication and cognition; ethno-botany and ethno-primatology; human subsistence strategies and conflicts with wildlife; as well as habitat surveys assessing success and failure of conservation approaches. The contributions assembled in the volume *Primates of Gashaka* aim for interdisciplinarity and comparative dimensions, across species and the African continent.

Keywords Gashaka Primate Project • Biodiversity • Nigeria • Conservation

---

V. Sommer (✉)
Department of Anthropology, University College London,
London WC1E 6BT, UK
e-mail: v.sommer@ucl.ac.uk

V. Sommer and C. Ross (eds.), *Primates of Gashaka*, Developments in Primatology: Progress and Prospects 35, DOI 10.1007/978-1-4419-7403-7_1,

The Remote Wilderness of Gashaka

The wilderness of north-eastern Nigeria is one of the least known iconic primate habitats on Earth (*Fig. 1.1*). Choruses of black-and-white colobus resonate through the woodlands, the boom of the mona monkey rolls down from thickly forested slopes, hacks of putty-nosed guenons echo through the canopy, dense riparian vegetation shakes from a pandemonium of baboon barks. And "the monkey with the drum" bangs against the buttress roots of tall emergents. Monkeys and apes still survive here in their thousands – along with a multitude of other wildlife and precious flora.

Ethnic groups with ancient traditions call the rugged landscape their home. Many of them are Fulani, living a proud pastoralist life; others are Hausa-speaking subsistence farmers. Islam is the predominant religion, and the Muslim of Nigeria's north co-exist here peacefully with Christians, who tend to dominate the south. Few roads penetrate the expanse that stretches along the border with neighbouring Cameroon. Trips to the market and visits to a doctor can take days. On foot, that is across trails, which snake through mountains and floodplains, with precarious crossings of turbulent rivers thrown in for good measure.

The area became a focus of systematic primatological studies with the founding of the *Gashaka Primate Project* in the year 2000. Our pioneering volume summarises highlights of research related to this important but hitherto largely unexplored region conducted since then. Most of the contributions report on the interactions between primate societies and their natural environment. But the times of science for science's sake are long gone – as those, who study nature, feel that they should also aim to be her guardians. Our compendium therefore also contains chapters about the challenges and prospects for conservation.

Central West Africa is increasingly recognised as a biodiversity hotspot. The richness in species of plants and animals is particularly great and levels of endemism are exceptionally high in the zone stretching from the Gulf of Guinea through the Cameroonian Highlands, with its northern limits marked by the Mambilla Plateau in Nigeria (Oates *et al.* 2004). Such hotspots seem to develop in areas characterised by high precipitation, a large altitudinal range, a moist and warm climate, and surrounding geographical barriers. But, as favourable as such habitat is for the development of a diverse flora and fauna, it is also good for humans. Consequently, there are very high human densities throughout much of this region, and most original forest cover has been cleared and converted into settlements and farmland. Hunting for bush-meat is rife, and many animal taxa, especially large mammals, have already been driven into local extinction.

The extensive destruction has left only a few areas relatively intact. One of the most important is the relatively inaccessible landscape around Gashaka in Nigeria's north-eastern state of Taraba (*Fig. 1.2*). The Muslim taboo of not eating pigs or primates has also ensured lower levels of hunting here. But it is foremost its remoteness that has so far shielded this terrain against commercial logging and other human intrusion that so often lead to the demise of natural treasures.

Fig. 1.1 Southern sector of Gashaka Gumti National Park. (**a**) View from Gashaka up the Gamgam valley towards Kwano, shrouded in mist (photo: James Higham). (**b**) Rolling hills and mountains near the Kwano field station, towards the Cameroonian border (photo: Alejandra Pascual-Garrido)

Gashaka is only a small village. A few hundred people reside here, most of them in small compounds of huts with mud-brick walls and roofs covered by grass. However, Gashaka has a long history as a regional centre of power, associated with a 19th century *jihad* when Fulani from the north invaded and converted much of the populace to Islam. Today, the provincial emirate is still named after Gashaka, as well as a local government area. The settlement also lends its name to Nigeria's

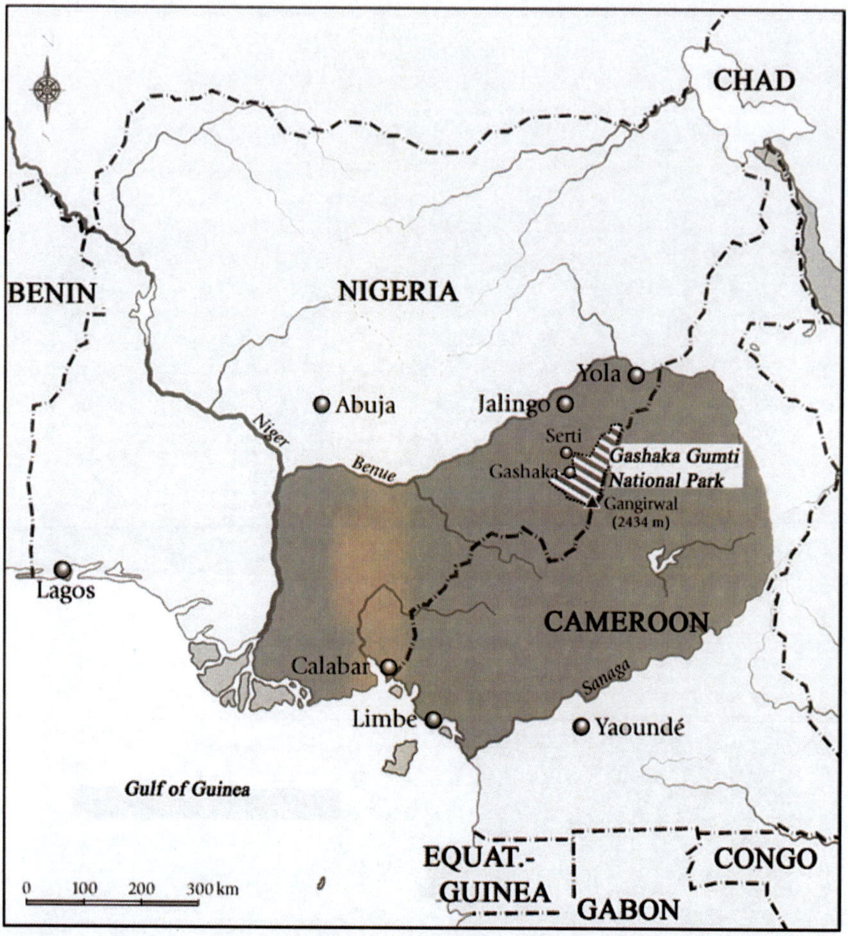

Fig. 1.2 Location of Gashaka Gumti National Park in West Africa (courtesy Maren Gumnior)

largest nature reserve: Gashaka Gumti National Park (Dunn 1999). The park (06°55′ – 08°05′ N and 11°11′ – 12°13′ E), created in 1991, covers 6731 km^2 – a vast area, roughly the size of the combined landmass of the Canary Islands.

The park's northern sector – named after the village of Gumti, and stretching far into neighbouring Adamawa state – is a flat biome of grassland with small trees, home to savannah animals such as elephants, hyenas, wild dogs, lions, roan antelopes and giant elands. The southern Gashaka sector includes a succession of lowland (< 825 m), sub-montane and montane (> 1650 m) strata, rising to 2419 m at Gangirwal, the "Mountain of Death", Nigeria's highest peak on the Chappal Wadde escarpment. Accordingly, vegetation cover is a mosaic of savannah-woodland, riverine and lowland rain forest as well as grassland and montane forests.

The habitat shelters carnivores such as civets, golden cat and leopard, ungulates such as buffalo, bushbuck, duikers, waterbuck, hartebeest, red river hogs and one of the last remaining populations of giant forest hogs. Rare fresh-water fish, otters as well as sizeable crocodiles and even a few hippopotamus thrive in the translucent rivers. More than 500 feathered species led to the designation of the park as an "important bird area".

The park's flag-ship species are diurnal primates – such as olive baboon, putty-nosed guenon, mona monkey, black-and-white colobus, tantalus monkey, patas monkey as well as grey-cheeked mangabeys, the latter recorded only a short while ago. The Gashaka region also provides refuge for the largest surviving population of *Pan troglodytes vellerosus*, the "Nigerian chimpanzee". This subspecies, only recently recognised, is the genetically most distinct (Gagneux *et al.* 2001; see Oates *et al.* 2008 for taxonomic revision to the name *P. t. ellioti*). Sadly, the subspecies is also the only one of four for which there are calls to change its Red List status from "endangered" to "critically endangered".

A Fragile Paradise

The complex patterns of diverse ecosystems and correspondingly high levels of biodiversity assign significant national and international conservation priority to Gashaka Gumti National Park. The park is also a major watershed for the Taraba, which feeds the River Benue and ultimately the River Niger. The livelihoods of millions of Nigerians downstream depend on the perennial flow of water, stored in the forests of the Gashaka mountains and slowly but steadily released. Primates in particular are important in keeping that vegetation cover alive, as, along with birds and some other mammalian species, they act as seed dispersers (Chapman 1995).

Vegetation and corresponding wildlife as well as human existence in the Gashaka area is greatly influenced by climatic features, in particular a pronounced fluctuation between wet and dry seasons (*Fig. 1.3*). For about 5 months, from mid-November until mid-April, very little rain falls, and mostly none at all. A dry dusty wind, the Harmattan, blows from the Sahara. The skies clear again once heavy downpours get going from about mid-April. Weather stations at Gashaka and Kwano (*App. 1.1*) recorded an average of 139 rainy days per year. The wettest day saw 152 mm rain. The yearly rainfall averages 1935 mm, with a range of 1683 – 2337 mm. The mean minimum temperature is 20.9°C, the coolest recorded temperature 12°C, the mean maximum 32.2°C, and the hottest day on record 43°C.

Much of West Africa is characterised by a belt of savannah landscapes, about 1500 km broad, in parallel alignment with the equator further south. Only 100 – 600 mm rain fall in the most northern savannah belt, the Sahel Zone, whereas the adjoining Sudan Zone sees up to 1000 mm of rain, with a correspondingly higher densities of trees. The Guinea zone, with up to 1500 mm of rain, borders against what was once contiguous rain forest (Reikat 2002).

The Gashaka area is thought to be part of the Guinea zone. However, its average rainfall is considerably higher, and we would therefore expect a rather extensive forest cover. The existing savannah landscape, or at least parts of it, could be a natural consequence of, for example, iron concretions in soils. However, there is little doubt that much of the current savannah is derived – a result of anthropogenic

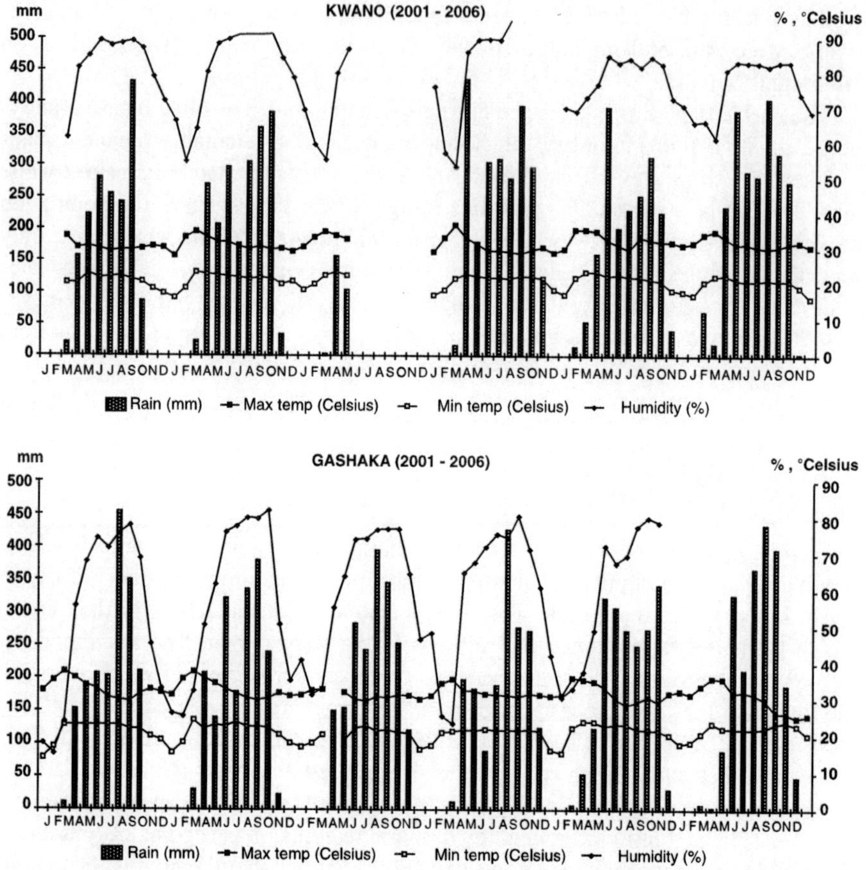

Fig. 1.3 Weather patterns at Kwano and Gashaka (2001 – 2006), indicating pronounced seasonality

influences dating back centuries, if not millennia. Most notable is the deliberate yearly burning of grass, for about 3 months from December onwards. This fire-regime prevents the natural development of semi-deciduous forest and keeps large areas as grassy woodland (Louppe *et al.* 1995). The ancient habitat thus probably resembled a mosaic of moister and drier, semi-deciduous forests, which were then gradually replaced by woodlands with trees that are fire-resistant (pyrophitic). A human-made origin must also be assumed for the grassland of the higher elevations.

The dynamics of the region, largely climate-driven, that aided the past development of a complex assemblage of vegetation cover and wildlife faces a whole catalogue of imminent threats. Logging is the smallest danger as access is difficult – although many slopes in the park's wider vicinity have been denuded by the timber industry. More and more poor migrants move from far-away into the park's buffer zone, with little means of survival other than targeting natural resources. Several illegal settlements have encroached well beyond the park boundary. Poaching is rampant in the park surroundings, and most forests have been shot empty of large mammals. Hunters, particularly coming from across the Cameroonian border, are increasingly targeting the reserve itself.

Even greater hazards come from a combination of fire and overgrazing, which lead, in turn, to erosion. The burning of grass at the advent of the dry season is meant to extend cattle pasture grounds and stimulate sprouting of new vegetation. However, the fire-regime is counterproductive to these goals, as it selects for grasses that are not only fire-resistant but also hardly palatable to cattle. Moreover, nutrients are lost, simply blown away, and as the soil becomes compacted, it is difficult even for fire-resistant savannah trees to rejuvenate. Cattle grazing reduces the vegetation cover still further. Cows trod a zigzag of tracks into the hills, creating perfect channels for rainwater to rush down. A chain of erosion is thus set in motion, ultimately causing whole hillsides to collapse in spectacular gullies. The devastating effects are particularly obvious on the Mambilla Plateau, just south of the national park. Anybody who brings up the region on *Google Earth* cannot fail to notice how the image is littered with blackish pock-marks.

One would think that the succession of erosion – burning, grazing, washing off of top-soil – would be the fate of landscape only outside the reserve. Unfortunately, this is not the case. The southern Gashaka sector of the park includes a handful of enclaves, particularly in the highlands, where tens of thousands of cattle are kept by Fulani clans. The enclaves are legal. But there is little constructive engagement of the park authorities with the Fulani, nor law enforcement to keep destructive consequences associated with cattle management restricted to these islands. More cattle are intermittently brought across the border from Cameroon. Human impact within the park is thus considerable – a shocking realisation to all those who thought that national parks are indeed sanctuaries of nature. Many areas in and around the enclaves in particular have lost more than a quarter of their vegetation cover from 1988 – 2000, as a comparison of satellite-generated maps reveals (Gumnior 2008).

It is, of course, the intention of the Nigerian government and its National Park Service to protect one of its premier and last remaining wildernesses. However, the area is vast, and resources available to the management and about 270 staff are very limited. Rangers have to confront pastoralists who graze cattle, illegal settlers and increasingly violent poachers and armed robbers. Year after year, some of these guards are losing their lives in the line of duty. Still, there can be little doubt that the park's creation made a significant difference. Now at least the option exists that natural treasures can be effectively protected for the benefit of future generations.

Until then, one cannot help but notice that the fire-regime lends an ironic note to the labelling of Gashaka as a biodiversity "hotspot".

The Gashaka Primate Project

The *Gashaka Primate Project* (GPP) has several bases. The administrative centre is the Department of Anthropology at University College London (UCL), home institution of GPP's founder and director, Volker Sommer. The co-editor of this volume and director of baboon studies at Gashaka, Caroline Ross, is based at the Centre for Research in Evolutionary Anthropology at Roehampton University in London. In Nigeria, the project maintains a 5-room research station near Gashaka village, at the edge of Gashaka Gumti National Park (GGNP), and a 16-room field station at the site of the abandoned village Kwano, inside the park, about 11 km from Gashaka. A dirt road of 35 km, suitable for 4-wheel drive vehicles, connects Gashaka with the town of Serti and the national park headquarters at Bodel. The Kwano site, for more than half of the year, can be accessed only on foot (or, for the daring, by motorcycle), and the trek requires challenging river crossings.

Volker Sommer first visited GGNP in 1999, accompanied by Isabelle Faucher, then a PhD student at the Department of Anthropology at UCL. Isabelle's field work in the Bakossi Mountains of neighbouring Cameroon had taken an all-too-familiar turn, when poachers killed her study group of drills, a priced prey for hunters, because of the monkeys' heavy weight and their unfortunate habit to escape from hunting dogs into trees – from where they can be shot down at leisure. There were rumours that primates at Gashaka were not hunted, because the indigenous Muslim population doesn't eat them. An overland trip from Lagos via Ibadan to Taraba State became stalled every so often in some lost village, or before a gushing river, while the map had proudly stated a properly tarred road. Upon arrival, the site seemed immediately promising. The park's general manager, Jarafu Mamza, was welcoming and helpful. Also, the Nigerian Conservation Foundation (NCF) maintained a project at Gashaka from 1992 – 2008; project managers Faith Ananze and Andrew Dunn volunteered invaluable advise. Salamu Waziri, a former hunter working for NCF, guided Volker Sommer and Isabelle Faucher on an initial expedition – and made it seem all too easy to find and watch wild chimpanzees. NCF also had a Nigerian student at hand, Jeremiah Adanu, who was keen to work with foreign researchers.

Thus, a decision was quickly arrived at to give it a go and establish a new research site. The underlying academic rationale was that nobody had so far investigated the fourth subspecies of chimpanzees. However, it was envisioned from the beginning to embed chimpanzee studies into a wider framework of primate socioecology – not least, because of the multiple direct and indirect interactions between sympatric monkeys and apes. It was also obvious that one would need to study the extent and impact of anthropogenic forces if one would want to contribute to habitat conservation.

Back in Britain, a first group of PhD researchers and volunteers was assembled, capitalising on the rather widespread interest in primate field work amongst anthropology students. The Nigeria National Park Service in the capital Abuja was approached with a request for permits. The then Conservator General, Alhaji Lawan B. Marguba, who had just visited the UK including UCL, drafted a memorandum of understanding that included a 10-year licence for research at Gashaka. The agreement was signed in Abuja in Jan 2001. The MoU was formulated in a highly constructive manner and removed much of the red tape that so often turns field research into a bureaucratic nightmare. GPP's mission was broadly stated as "understanding how environmental factors influence the structure of primate societies, while contributing towards the conservation of their habitats".

Various strands of investigations developed rather quickly – although the chimpanzees proved to be more elusive than envisioned. However, there was also an advantage to the difficulty with direct observations, because a wealth of detailed indirect data was assembled, on habitat ecology as well as remnants left behind by the apes such as nests, tools, and faeces. Two baboon troops were soon habituated to humans, and research into their behaviour and ecology would soon flourish.

Much energy of the first years went into improving the site's infrastructure. NCF, headquartered in Lekki / Lagos, and its partner WWF-UK provided vital logistics. NCF labour renovated a field station at Gashaka and constructed buildings at Kwano. Initial support from *in situ* conservation initiatives of Chester Zoo and the associated North of England Zoological Society fortunately developed into major core-funding for the project. This sponsorship covers GPP's running costs, such as additions to the stations and their maintenance, salaries for local field assistants, or purchase of a project vehicle. Chester Zoo also supports various other conservation and research initiatives in the Gashaka area, often designed and supervised by GPP, such as bursaries for African students, travel grants for UK students, satellite-based vegetation mapping, assistance for the Nigeria Montane Forest Project and infrastructural developments of the national park – including ranger training, boosting the eco-tourism potential, and the vital boundary demarcation (see below). Such extensive portfolio needed a name, and is thus known as the "Nigeria Biodiversity Programme of Chester Zoo".

Researchers, students and volunteers have since flocked to Gashaka and Kwano; they have to obtain their own funding, but can count on relatively modest camp fees – about 5 dollars per day – for accommodation and food in the bush they explore.

Research as Conservation

People in the Gashaka terrain live at an unhurried pace. No television reception permeates their life-style – except for a few communities along the park border, and only for the privileged few who can afford a satellite dish. Telephone landlines have never existed; cell-phone coverage reached Serti only in mid-2007, but reception is spotty and restricted to town centres. The next Internet connection is a couple of hundred kilometres away, in the state capital of Jalingo. The thousands of square kilometres inside the park will probably never be penetrated by these technological advances – and part of its serenity surely relies on the absence of such commodities. The windows to the outside world are long-wave transistor radios, which require constant and skilled twists to catch a viable frequency somehow. Centralised power-supply is non-existent, too, and whatever little light flickers through the night is nourished by generators that are loud, dirty, unreliable and expensive. Fuelling is a perennial problem, given a chronic shortage of petrol – although we are in one of the largest oil-producing countries in the world.

The lack of communication in particular is a serious challenge to effective patrols of the vast park, as messages from the interior travel only as fast as somebody walks. Operations at our field station were particularly affected by the lack of electricity for tasks such as computing, printing, deep-freezing of biological samples, food storage, recharging of battery-operated data-collection equipment – and to make day-to-day living in the bush a bit more comfortable. In any given year, more than a dozen researchers and field assistants were around. We therefore yearned for a solution to the power problem that was clean, sustainable and maintenance free – while ideally also setting an example for the potential of environmentally friendly energy.

Our project was lucky to secure exceptional help from two German institutions: the Oskar-von-Miller-Polytechnic in Kassel, where solar-engineering is taught; and the firm Bilfinger-Berger in Wiesbaden, the founder of one of the largest sub-Saharan construction companies, Julius-Berger-Nigeria (JBN), headquartered in Abuja. A team of students from the German polytechnic designed a "power-island" – a small, independent plant that supplies stable electricity around the clock, relying on a unique combination of photo-voltaic cells (solar panels), a water-driven turbine, and a computerised "heart" built into a steel-container with 30 sizeable dry-cell batteries that store the harvested power. After a year of planning, convincing local sponsors, construction and trial runs, more than 6 tons of equipment was shipped from Germany to West Africa by JBN. The construction firm had, at the same time, conceived a solution to ensure communication across the wilderness: a steel tower, crowned by an antenna and lightning rods, 24 m high, connected to a solar-powered repeater-station, plus dozens of walky-talkies and car radios.

Finally, engineers from Kassel and JBN travelled together to GGNP. In 2 weeks, they installed a power island as well as a repeater station. The tower had to be erected on a mountain cliff to secure a maximum radius of transmission. All equipment had therefore been partitioned into segments not heavier than 60 kg. Local porters, with their amazing strength, head-carried the loads several kilometres across steep slopes for assemblage on the rim of the vertical precipice. It was a thrilling moment when the team on the mountain top made first contact with the national park headquarters at the distant horizon.

Since then, protection efforts, logistics and research have been transformed and become much more efficient. Park management, ranger patrols, field station personnel, NGOs, students and doctors alike can now communicate across more than 5000 km^2, far into the Mambilla Plateau – whether they work in the headquarters, track a community of chimpanzees, or pursue poachers through the thickets of jungle. With power at the Kwano camp 24 hours a day, 365 days a year, we probably enjoy the most secure and stable supply of electricity anywhere in Nigeria, given constant black-outs even in the capital Abuja. Our bush facilities, in particular the environmentally sound "power-island", have become an attraction for visitors, officials and students who realise that advanced but simple technology can hold enormous potential.

Nevertheless, the legal guardians of the national park faced another serious challenge: the lack of any border demarcation. Cattle-grazers, settlers and hunters like to claim that they are operating outside the reserve. Our project therefore assisted the park authorities and initiated the demarcation of the Nigerian side of the border, again with full-fledged support from the "Chester Zoo Nigeria Biodiversity Programme". Decade-old maps were consulted, and experienced rangers were brought out of retirement to help clarify ambiguous portions of the park limits. Village meetings addressed encroachment problems. Then, managed by the park's technical department, chainsaws, cutlasses and diggers moved in, chopping a 6-m wide corridor through bushland that traced the border. This seemed like an unlikely conservation activity, but was indispensable to make the border clearly recognizable. A bulldozer ultimately created a motorable track across all but the most jagged parts. Workers camped in the bush. Meanwhile, truck-loads of empty oil drums were brought from afar, cut in half, and transported to the site, along with water, granite gravel, and cement. Beacons were constructed from this raw material, typically 1 m high, and in visible distance from each other. The top was sealed by heavy concrete to prevent easy destruction, and a plate with GPS coordinates inserted (*Fig. 1.4*).

The flat northern Gumti-sector proved much less of a challenge then the mountainous Gashaka sector in the south. It took more than 700 beacons to demarcate the entire border length of 225 km. The operation stretched over 4 years, but was good value for money – as 1 km of boundary cost just 115 £, roughly 200 US dollars at the time. The demarcation is highly praised by all staff of GGNP, as it makes patrols more effective, reducing uncertainty and thus conflicts with local communities. Judges also have now much clearer evidence at hand should a case come before them. A major concern of conservation initiatives is sustainability – which is hard to measure, let alone to achieve. Our project is therefore proud to be part of a lasting contribution to conservation that will stand – no matter what the fiscal situation of the park management might be.

Fig. 1.4 Demarcation of the national park border with beacons began in 2005 (photo: David Bennett)

There were fittingly festive inaugurations of the power island and radio-communication system (12 Jan 05) as well as the border demarcation (19 Mar 07). The celebrations were held in the presence of His Royal Highness, the Emir of Serti, a crowd of local and international visitors, stakeholders, and other dignitaries such as the Jauro (Chief) of Gashaka village; the Executive Chairman, Gashaka Local Government Council, Serti; the Commanding officer or Serti's 20th Mechanised Battalion, as well as high-ranking officers from the Divisional Police and State Security. Such events are an important demonstration of the fruitful collaboration between the national park and our project, showing to the wider community that foreign researchers do not just come to take, but can contribute to the long-term future of a landscape and its wildlife they learned to treasure.

There is little doubt that research is one of the best ways to achieve conservation goals – simply because researchers are jealous of their study subjects and study areas. They don't like to see them destroyed. The *Gashaka Primate Project* has therefore, from its inception, been engaged in capacity building and law enforcement on numerous fronts. Many initiatives were one-off occasions, several lasted for years, and others still go on. It is hard to know if such efforts will make a difference in the long term. At least we did what we thought was right to do.

Nature conservation works better if local stakeholders support a national park's mission – particularly if it improves living standards. Projects such as ours have collateral benefits for the local economy, as we employ up to a dozen family heads as field assistants; buy food and supplies locally; and we hire local carpenters, construction workers and commercial vehicles. Another incentive is to sell nature's beauty to tourists. Hundreds of pupils and students from secondary schools, colleges and universities flock every year into the park for field trips – where many of them see a forest for the first time. The park has also outstanding potential as a destination for foreign "eco-tourists" tired of the safari in the zebra-striped bus, who look for more individualistic experiences. Only a few dozens foreigners stray so far into the park each year – not least, because the Nigerian Immigration Service invests considerable imagination to make it all but impossible to obtain a tourist visa. Nevertheless, the park management recently upgraded a tourist camp in Serti, and also boasts one of the best guide books of any national park, produced in conjunction with NCF. Our project tries likewise to attract local and foreign tourists. We trained villagers as guides, and advertise the park via our website. Our field assistants also tagged almost 100 km of trails through some beautiful parts of forest, and it is always a highlight for visitors to accompany researchers or assistants to view primates.

We try to be proactive about community outreach. At a minimum, projects such as ours can instil a sense of pride about the wonders of nature at the doorstep of local people – not least, because foreigners are attracted to this remote place. Already in 2003, we produced a high-quality documentary with the Bavarian Television and distributed an English version – "Nigeria's Primate Paradise" – on video cassettes and DVDs. Footage of primates easily transports conservation messages, because viewers empathise with monkeys and apes and find them attractive. Project affiliates gave innumerable talks to students, school classes, rangers, in army barracks, and to visitor groups. And we should not forget the bi-annual event of a fierce football tournament between teams of GPP, NCF and GGNP. The competition draws hundreds of spectators, leads to weeks of limping, and to one party proudly holding the German-made trophy for another 2 years. Importantly, we also provide bursaries for African students, including park staff eager to further their education. African students enrolled for a university degree at the undergraduate, graduate or doctoral level will often work in tandem with a foreign student, thus picking up skills not taught in Nigeria and also imparting their important local knowledge to those who come from abroad.

All such initiatives are ultimately pointless if legislation and law enforcement are weak. We therefore work closely with the park authorities in that we identify priority areas for ranger patrols, design rotas, pay field-allowances and assist in acquisition of field equipment. The mere presence of researchers and field assistants alone is already an effective deterrent against poachers. Gunshots were regularly heard from the Kwano camp during the first years; now, there are only a few each year. Wire traps are common in the wider area – but over 9 years of field work at Kwano, we detected a single snare in those 50 km^2 where we operate.

Finally, one should not forget the intrinsic benefits of research: the generation of knowledge – on taxonomy of fauna and flora, ecology and behaviour of animals, vegetation cover and distribution of natural resources. Here, GPP has worked its way to the international forefront of research into baboon socioecology and primate cognition – and conducts the most detailed study of the fourth subspecies of chimpanzee, *P. t. vellerosus*, so far.

An International Network

The infrastructural improvements to the national park and our research stations assured the success of numerous research projects.

This is reflected in the ever-growing network of collaboration of the *Gashaka Primate Project* that spans 33 institutions in 10 countries. With respect to capacity building, it is important that 9 partner institutions are based in Nigeria itself. Involved are also a further 7 from Germany, 6 from the USA, 5 from the UK, while the Czech Republic, Ivory Coast, Denmark, Portugal, New Zealand and Spain are also represented. Individual collaborators were affiliated with 23 universities (Albert-Ludwigs-Universität Freiburg, American University Yola, California State University Fullerton, Federal University of Technology Bauchi, Federal University of Technology Yola, Gombe State University, Humboldt-Universität Berlin, Ibadan University, Oxford Brookes University, Universidad Complutense de Madrid, Universidade Tecnica de Lisboa, Universität Frankfurt, Universität Würzburg, University College London, University of Alabama at Birmingham, University of California Berkely, University of California Riverside, University of Canterbury / New Zealand, University of Chicago, University of Cocody-Abidjan, University of Copenhagen, University of Maiduguri, University of St. Andrews), 5 research institutes (Academy of Sciences of the Czech Republic Bron; Deutsches Primatenzentrum,

Göttingen; International Institut for Tropical Agriculture, Ibadan; Leibniz Institut für Zoo- und Wildtierforschung, Berlin; Max-Planck-Institut für Evolutionäre Anthropology, Leipzig), 2 zoological societies (Zoological Society of San Diego; North of England Zoological Society), 2 conservation NGOs (Nigerian Conservation Foundation, WWF-UK) plus the Nigeria National Park Service.

With this, our project is truly international (*Fig. 1.5*). Students, researchers and volunteers have come from 21 countries and 6 continents to work at Gashaka (Argentina, Austria, Cameroon, Canada, France, Germany, Greece, Italy, Ivory Coast, Japan, New Zealand, Nigeria, Portugal, Spain, Sweden, Switzerland, Taiwan, The Netherlands, United Kingdom, USA).

The project can also point to a formidable output in publications – 107 in its first 9 years (*App. 1.2*). A fair proportion is "grey literature", in form of 11 reports to donors and supporters. Moreover, the project enabled original research for both undergraduate and graduate students, leading so far to an impressive 32 dissertations – including 4 bachelor projects and 15 master's dissertations. The remainder are PhD theses, with 5 completions so far and another 8 PhD students working towards submission. More than two dozen abstracts of talks and posters – a total of 25 – testify to the participation of GPP affiliates at conferences. The hard currency of science is, of course, research articles. GPP affiliates so far produced 23 in peer-reviewed journals and edited volumes – not counting the one dozen contributions to this volume.

Interestingly, not a single research article was published during the project's first 3 years, and only one during the next 2. Output took off from the sixth year only, with 5 publications in 2006, 6 in 2007, 6 in 2008, and many more currently under revision or in press. Funding agencies such as research councils will often not appreciate the slow pace of field work, which requires the establishment of rapport with local stakeholders, the building up of infrastructure, habituation of animals of study groups to human observers, and the collection of detailed background information on the local ecology before valid interpretations are possible, particularly of data related to behaviour. Whoever asks for quick results from a project such as ours will thus be disappointed. Instead, output will only slowly gather speed – but steadily, if the project design is sustainable and holds a firm course over years.

Students and researchers also wrote a dozen popular pieces about Gashaka, while print media including dailies and magazines will often report on their own initiative. Popular dissemination of results is important – not least, because the public that contributes to donor agencies has a right to learn about research they support. A popular book about the Gashaka project entitled "*Schimpansenland. Wildes Leben in Africa*" ("Chimpanzeeland. Wild life in Africa") written by the project director and produced by renowned publishers C. H. Beck hit the shelves in German-speaking countries in February 2008 (Sommer 2008). A flurry of reviews in TV, radio, magazines and newspapers and invitations to radio and television shows helped to secure a stream of small and larger donations from individuals and institutions. The book was also a small hit amongst employees of the construction company Julius-Berger-Nigeria, whose expats are virtually all native German speakers.

The growing reputation of GPP as a major research and conservation initiative can also be measured through the number of talk invitations to the project director.

Fig. 1.5 *Gashaka Primate Project* team members. FA = local field assistant, SR = seniour researcher, DR = doctoral researcher, MS = master's student, VO = volunteer (photos: VS). (**a**) Jan 04. Back row, from l. to r.: FA Felix Vitalis, SR Kate Arnold, SR Hazel Chapman (of Nigeria Montane Forest Project), MS Yvonne Pohlner, DR Jeremiah Adanu, DR Andrew Fowler, SR Volker Sommer; front row, from l. to r.: FA Bobbo Buba, FA Nuhu Husseini, FA Ali Tappare, FA Sam Yusufu, FA Hammaunde Guruza. (**b**) Mar 06. Back row, from l. to r.: FA Isaac Timothy, FA Maigari Ahmadu, FA Ali Tappare, SR Kate Arnold, FA Halidu Ilyasu, DR David Bennett, DR Andrew Fowler, DR Sandra Tranquilli, DR Elodie Ey, FA Felix Vitalis, VO Aisha Dasgupta; front row, from l. to r.: FA Zacchariah Dollar, VO Sonia Richter, FA Hammaunde Guruza, FA Buba Bello, FA Buba Hammaselbe

The resulting tally of 44 presentations can be broken down to 1 each in the years 2000, 2001 and 2002; 2 each in 2003 and 2004; 5 in 2005; 10 in 2006; 2 in 2007; 8 in 2008; and 12 in 2009. Lectures were delivered in 10 countries, most in Germany (24) and the UK (12), but likewise Argentina, Austria, Italy, Nigeria, Portugal, Switzerland and the USA.

We try to foster an interest for future field work in Nigeria through our bi-annual "Gashaka Field Days", with presentations by students, post-docs and seniour researchers. These one-day events took place three times – at UCL in Nov 04, Roehampton in Nov 06, and again at UCL in Jan 09. All were attended by an audience of 50 – 100 participants. And each time, some attendees became so intrigued that they were soon on their way to the wilderness of Gashaka.

Contributions to *Primates of Gashaka*

This volume reflects the multi-faceted activities of the *Gashaka Primate Project* over the first decade since its inception. The contributions, quite independent from this particular publication, are testimony of the rich research potential the Gashaka site holds, of its biodiversity as well its cultural diversity – but also of the dangers that threaten this unique ecotome.

Aspects of several contributions have been previously described, scattered throughout journals and edited volumes. We explicitly requested summaries of these results, along with general introductions to the research topic. This seemed necessary, as the chapters range from botany to socioecology to nutritional analyses, from molecular studies to acoustic analyses, from social anthropology to conservation biology. Moreover, although all contributions are related to research in Gashaka Gumti National Park or the wider region, each can be read independently. Thus, we consciously allowed for minor repetitions, mainly with respect to descriptions of the study site. This editorial approach is meant to make the volume widely accessible – to an audience of primatologists and conservationists alike, but chiefly also to readers in primate habitat countries such as Nigeria.

Chapter 2 (Barnwell): "*To save a wilderness. Creation and development of Gashaka Gumti National Park.*" One could not have wished for a more fitting opening than this personal account. Richard Barnwell has been involved in nature conservation since the 1960s. He was one of the first conservationists to recognise the importance and potential of the wider Gashaka region and became a driving force behind attempts to protect it. Richard Barnwell retells the story that ultimately led to the establishment of the vast reserve that is now Gashaka Gumti National Park. Barnwell argues with authority about the need to engage with the enclave communities, so that they can become part of the conservation enterprise as was once envisioned. But, as Barnwell has seen many failures of "soft" community-development approaches in conservation, he also advocates that park rangers should be trained and equipped as a paramilitary force. Hence they may protect a place that is special to him – as it is to its people and wildlife – for future generations.

Chapter 3 (Adanu, Sommer & Fowler): "*Hunters, fire, cattle. Conservation challenges in eastern Nigeria, with special reference to chimpanzees.*" The contribution is an update on the current conservation situation in the wider Taraba region. The historic range of the fourth subspecies of chimpanzee, *P. t. vellerosus* extended roughly between the Niger River in Nigeria and the Sanaga River in Cameroon. Most original habitat is lost, and the remote Taraba region of north-eastern Nigeria with its Gashaka Gumti National Park considered the stronghold of this taxon. The authors explored the region's conservation prospects for large mammals, and in particular chimpanzees, through foot surveys that lasted 2 months. They found that even remote forests not yet destroyed are often "hunted empty". The park is the big exception. A contiguous population of about 1000 chimpanzees survives here, mainly along the rugged axis Gashaka – Kwano – Yakuba – Chappal Wade. Nevertheless, the park's natural habitat is likewise shrinking, especially in and around the enclaves of settled pastoralists, due to cattle grazing, bush-burning, hunting and a lack of law enforcement.

Chapter 4 (Nyanganji, Fowler, McNamara & Sommer): "*Monkeys and apes as animals and humans. Ethno-primatology in Nigeria's Taraba region.*" The contribution extends the scope of the foot surveys, as it explores local attitudes towards primates in southern Taraba through extensive appraisals via questionnaires. The questionnaires revealed striking differences between the Cameroon Border Area (intense agriculture; influenced by Christian culture; cross-border trade; remote from the national park) and the Park Support Zone (more pastoralist activities; largely influenced by Muslim culture; greater awareness of protected areas). As a result, there is much more hunting near Cameroon, with a consequently dramatic decrease in sightings of primates. Contrary to monkeys, chimpanzees are credited in widespread folklore with human-like qualities. This entails certain taboos, including Islamic prescriptions, associated with eating of ape meat. On the other hand, chimpanzees are at times targeted by hunters precisely because of their human-like character. Their body parts provide powerful ingredients for medicine, and killing a chimpanzee can bestow special honours. In any case, traditional restrictions are increasingly breaking down, because of an influx of immigrants from non-Muslim areas and a commercialisation of the bush-meat trade.

Chapter 5 (Koutsioni & Sommer): "*The bush as pharmacy and supermarket. Plant use by human and non-human primates at Gashaka.*" Neither monkeys nor apes or humans could survive in the Gashaka area if they would not exploit the natural resources of savannah-woodland and forests. The study established a database on the usage of plants as food, medicine and implements, assembling information on more than 300 species. For this, previous unpublished reports were supplemented with original survey data and data accumulated by primate researchers. Four (at first glance) disparate consumer groups were compared – humans, domestic animals, baboons, chimpanzees. This reflects a conscious effort to work towards a merger of ethno-botany and zoo-botany with its emerging sub-discipline of animal self-medication. The authors report on two case studies of plant use by non-human primates with obvious medicinal effects. The consumption of African black plum was found to have a contraceptive effect on baboons, with the potential benefit of reducing mortality during the rainy season, whereas the swallowing of intact leaves of a coarse herbaceous plant by chimpanzees led to the expulsion of parasitic worms. The authors reflect on potential co-evolutionary processes that lead to the preference for certain plant families and plant parts. A comparison with a compendium of medicinal plants in Nigeria revealed that many taxa of the Gashaka area are not yet included. Future work should also engage with traditional concepts of how to classify plants, and explore plant properties in more detail as this might affect their usage as nutrition, for treatments, or as equipment.

Chapter 6 (Bennett & Ross): "*Fulani of the highlands. Costs and benefits of living in national park enclaves*." Exploitation of park resources is especially pronounced by those Fulani pastoralists who had settled in the highlands and were not evicted when the national park was created. Instead, they were allowed to raise cattle in designated enclaves, and it was envisioned that they would assist in protection of the surrounding reserve. The contribution is related to this unique conservation experiment. Living in the park is on the one hand perceived as beneficial by the Fulani because livestock prospers and family and cows are safe. On the other hand, difficult access to their settlements and prohibitions on land use increasingly forces Fulani into farming – although they originally chose to live in the highlands because it is a good place to rear cattle. Consequently, crops are often damaged by wild animals such as baboons. The authors stress that an appreciation of local knowledge systems could mitigate human-animal conflicts as well as assist to better integrate the Fulani into the park management.

Chapter 7 (Zinner, Buba, Nash & Roos): "*Pan-African voyagers. Phylo-geography of baboons*". This contribution focuses on the taxonomy of the most widespread primate of the Gashaka region – the baboon, one of Africa's ubiquitous monkeys. Experts have long disagreed on how many morphotypes are to be distinguished, and if these represent species or subspecies. The authors obtained an exhaustive collection of faecal samples from many populations across Africa, including, for the first time, from Nigeria. They extracted mitochondrial DNA to reconstruct the phylogenetic relationships of baboons and propose to recognise at least 6 species (chacma, Kinda, yellow, olive, Guinea, hamadryas). The evolutionary history of baboons seems to be even more complicated than previously thought, as it is difficult to distinguish monophyletic clades. The detection of several para- and polyphyletic clades suggests multiple phases of fragmentation, isolation, hybridisation and introgression. These processes, were likely triggered by multiple cycles of expansion and retreat of savannah biomes during late Pliocene and Pleistocene glacial and inter-glacial periods.

Chapter 8 (Warren, Higham, MacLarnon & Ross): "*Crop-raiding and commensalism in olive baboons. The costs and benefits of living with humans*". The chapter explores how different ecologies influence the life-history of the two main study groups of olive baboons at the Gashaka site. One troop, Gamgam, is crop-raiding, while the other, Kwano, exclusively feeds on wild foods – thus providing the setting for a natural experiment. Data collected over 8 years indicate that crop-raiding provides both energetic and reproductive advantages. For example, the Gamgam troop spent less time travelling and feeding and more time resting and socialising, had shorter inter-birth intervals and lower infant mortality rates. There are also costs to crop-raiding, such as the risk of being chased or even getting killed by farmers. But these are outweighed by the benefits of increased access to high quality foods, reduced susceptibility to pathogen loads, and a consequently increased reproductive output. Baboons may be one of the few primate species to benefit from anthropogenic alterations of habitats, although often to the detriment of their human neighbours.

Chapter 9 (Ross, Warren, MacLarnon & Higham): *"How different are Gashaka's baboons? Populations of forests and open country compared"*. Previously published socioecological and life-history models, primarily based on data from open-country baboons, do not always give good predictions for the socioecology of forest baboons. That is a major conclusion of this chapter, which compares the forest-living Gashaka baboons with their better studied open-country relatives in East and South Africa. High rainfall and relatively low predation pressure at Gashaka are major factors that produce this difference. The rain is linked to high productivity and this, in turn, creates a diet that is diverse, high in fruit but low in subterranean foods and leaves. But living in a wet habitat also comes with stresses that lead to a need for wild-feeding mothers to invest heavily in their offspring, resulting in a relatively long inter-birth interval. The combination of high rainfall, low predation and availability of fruit all year round allows small group sizes and a relatively small home-range size.

Chapter 10 (Ey & Fischer): "*Keeping in contact. Flexibility in calls of olive baboons*". The diversity of acoustic signals is remarkable between, but also within primate taxa. Phylogenetic relatedness, social structure and habitat may all affect the usage and structure of vocal signals. Given a multitude of grunts, barks and calls, baboon groups are ideal to investigate the interplay between senders and recipients of auditory signals. The chapter explores how habitat influences usage and structure of contact calls, focusing on troops in Nigeria and including a comparison with populations in Uganda and Botswana. All populations ranged through various habitats, and the physics of call production can be expected to reflect this. In more closed forests, one would expect higher call rates, longer calls and lower frequencies. Indeed, grunts were generally longer in closed habitat, but bark rate did not vary significantly. This suggests that other factors, such as the emotional state of the caller, the context, and group dispersal, might also play a role.

Chapter 11 (Arnold, Pohlner & Zuberbühler): "*Not words but meanings? Alarm calling behaviour in a forest guenon*". A further investigation of monkey vocalisation relates explicitly to the evolution of language. Human speech is based on rule-governed assemblage of morphemes into more complex vocal expressions. The authors were able to establish an interesting analogy with calls of free-ranging putty-nosed monkeys at Gashaka, setting up elaborate experiments with play-backs and predator models. Putty-nosed monkey males combine the loud alarm calls "hack" and "pyow", into different call series, depending on external events. "Pyow"-series are a common response to leopards, while "hacks" or "hacks" followed by "pyows" are regularly given to crowned eagles. One important finding relates to a further sequence of about 1 – 4 "pyows", that is followed by 1 – 4 "hacks", as this doesn't relate to predators but predicts a group progression. The authors conclude that, in this primate, meaning is encoded by call sequences, not individual calls – a claim that has far-reaching consequences for our understanding of how language evolved. Many animals such as birds and primates are limited by small vocal repertoires, and this constraint may have favoured the evolution of such combinatorial signaling.

Chapter 12 (Sommer, Bauer, Fowler & Ortmann): "*Patriarchal chimpanzees, matriarchal bonobos. Potential ecological causes of a Pan dichotomy*". The chapter sets the tone for the last three contributions, which all relate to *Pan troglodytes vellerosus*. Their starting point is the realisation that chimpanzees and bonobos, despite being so closely related, have different species' psychologies. Violent male aggression and female-female competition is typical for chimpanzees, whereas bonobo females form coalitions and often dominate males. This dichotomy is also of interest for the understanding of sex-differential power-asymmetries and violent inter-group conflict in contemporary human societies. A comparison between feeding ecologies of chimpanzees at Gashaka / Nigeria and bonobos at Salonga / Democratic Republic of Congo predicts that female gregariousness will be impeded if environmental quality is lower. Indeed, the chimpanzee habitat is found to be poorer (more seasonal diet, more anti-feedants, lack of high-protein herbs), and chimpanzees invest more time in the removal of seeds from fruit and in digestion. Their nest-groups sizes (a proxy for sociality) thus vary strongly with food availability – whereas bonobos are much less affected. These findings support the hypothesis that ecological differences are at the heart of the dichotomy of sociality in *Pan*. Nonetheless, the conclusion is based on only two populations. Future studies will have to address the considerable intra-specific variability processes and environmental parameters, in particular for chimpanzees.

Chapter 13 (Fowler, Pascual-Garrido, Buba, Tranquilli, Akosim, Schöning & Sommer): "*Panthropology of the fourth chimpanzee. A contribution to cultural primatology*". Biological anthropologists define culture broadly as "socially transmitted behaviour". This mechanism produces considerable intra-specific variability, as exemplified by human "cultural diversity". Chimpanzee behaviour varies likewise between populations – although virtually nothing was known about *P. t. vellerosus* – until research began at Gashaka. The contribution addresses the question of intra-specific variation via an analyses of the subsistence technology of Nigerian chimpanzees. The central question is, if and how the material culture of chimpanzees at Gashaka reflects environmental constraints or arbitrary "cultural variants". Toolkits are found to be quite elaborate, probably because the chimpanzees need to rely on extractive foraging – given that their habitat is so poor (at least in comparison to bonobos). Tools exploit the colonies of social insects (bees, ants), but strangely, termites are ignored. Many expressions of material culture seem to environmentally influenced, such as that tools are longer during the dry season, when insects retreat deeper into their nests. Other traits may represent arbitrary cultural variation. For example, hard-shelled nuts are not hammered open with tools, unlike elsewhere in West Africa. Given such differences, extinctions of local populations of chimpanzees do not only entail a loss of biodiversity, but also of cultural diversity.

Chapter 14 (Hughes, Rosen, Gretsky & Sommer): "*Will the Nigeria-Cameroon chimpanzee go extinct? Models derived from intake rates of ape sanctuaries*". The final chapter tries to predict the future of *P. t. vellerosus*. Assessments of reductions of wild ape populations are notoriously unreliable since they are mostly based on indirect evidence such as surveys of markets and interviews with hunters. The authors employ an alternative measure, and calculate annual loss from the wild through intake rates of sanctuaries in Africa, which currently hold close to 1000 gorillas, chimpanzees and bonobos. They relate arrivals into sanctuaries to the proportions of infants in wild groups, hunting strategies, and the likelihood that captured babies make it to a sanctuary. About 3000 – 5000 *P. t. vellerosus* survive in the wild. The limited geographical range of this subspecies renders it likely that all orphans who make it to a sanctuary will end up in one of the four institutions that exist in Cameroon and Nigeria. The intake rates are so high that the future seems bleak. VORTEX – a population viability analysis tool – predicts extinction of this subspecies in the wild in the next 20 – 50 years. Thus, sanctuaries will become increasingly important conservation tools, if only because they may soon harbour more apes than survive in the wild.

These last results, depressing as they may be, link back to where the volume started: to Gashaka Gumti National Park, a wilderness that not only holds the key to the survival of the fourth chimpanzee subspecies, but is also a haven for myriads of other taxa of plants and animals.

Those Who Made It Happen

We are deeply indebted to the many individuals who, over the years, contributed to the success of the *Gashaka Primate Project* (*App. 1.3*). We single out a few, those, who as members of various organisations helped beyond the call of duty, and those with that special kind of enthusiasm without whom long-term projects could never get underway and persist:

- Isabelle Faucher, wildlife biologist and gifted artist, introduced Volker Sommer to Nigeria and Gashaka Gumti National Park.
- Andrew Dunn, Deputy Manager of the Nigerian Conservation Foundation project at Gashaka-Gumti, provided crucial advice during the initial years.
- Richard Barnwell of UK's World-wide-Fund-for-Nature supported our initiative in a place for which he had a special love throughout his career as a conservationist.
- Alhaji Lawan B. Marguba, visionary first Conservator General of Nigeria's national parks, had the wisdom to encourage long-term research in the Gashaka-Gumti reserve by furnishing GPP with a 10-year permit.
- Jarafu Mamza, Modu Sherif and Okeyoyin Agboola embody the unflinching support of the national park service as successive heads of GGNP (with an interesting evolution of titles from "General Manager" via "Director" to "Conservator of Parks").
- Andrew Fowler and Ymke Warren, untiring field workers got the chimpanzee and baboon studies off the ground.

- Jeremiah Adanu was our crucial link to Nigerian mentality, both academically and as a friend.
- Hammaunde Guruza of Gashaka village, our Head field assistant, displayed unrivalled loyalty and expertise in the bush.
- Callistus Akosim, professor at Yola University's Department of Wildlife Management, sent the best of his students into the bush and proved to be a reliable partner in supervision, aiming to raise a new generation of Nigerian scientists.
- Roger Wilkinson, Head of Field Programmes and Research at Chester Zoo, our project's core funding organisation, reviewed numerous funding applications professionally and with sympathy – not to mention unforgettable miles of *in situ* trekking.
- Horst Hoppe of the Oskar-von-Miller-Polytechnic in Kassel / Germany, assembled a team of highly motivated engineering students who fitted our research station with the miraculous "power-island".
- Volker "Papa" Gallitz of Julius-Berger mobilised many departments of the mighty Nigerian construction firm to support conservation of a last wilderness with solid technology.
- Umaru Buba worked with GPP since his university career began to become its first Project Manager, renowed for honesty and reliability.

Looking Towards the Future

The *Gashaka Primate Project* has grown into one of the largest research and conservation activities in West Africa. At present, it keeps going on the initiative of the editors of this volume and their academic home institutions. An alternative legal and logistic structure will have to be found, if the project is to continue in perpetuity. The first prerequisite will be legal security, the second high-calibre scientific management, the third reliable management on the ground, and the fourth assurance of long-term core-funding. We are currently pondering ideas such as: turning the project into an NGO; integrating it with other research activities (such as the national park's research department); making it the field station of a regional university (such as the newly founded Taraba State University); handing the management over to a long-established nature conservancy (such as WCS); or convincing a world-class zoo (such as Chester) to dedicate itself to long-term in-situ conservation.

One would probably also have to think about a name-change, given that research topics have long developed into areas other than orthodox primate research. An obvious candidate would be "Gashaka Biodiversity Project". Our logo was designed with this idea in mind and a change from GPP to "GBP" would be easy.

The appearance of this volume marks the 10th anniversary of the *Gashaka Primate Project*. Let's hope that "the monkey with the drum"will join in the celebrations.

Literature

Chapman, C. (1995). Primate seed dispersal: coevolution and conservation implications. *Evolutionary Anthropology* 20: 74–82.

Dunn, A. (1999). *Gashaka Gumti National Park: a Guidebook.* Lagos: Gashaka Gumti National Park, NCF / WWF-UK.

Gagneux, P., Gonder, M. K., Goldberg, T. L. & Morin, P. A. (2001). Gene flow in wild chimpanzee populations: what genetic data tell us about chimpanzee movement over space and time. *Philosophical Transactions of the Royal Society B Biological Sciences* 356: 889–897.

Gumnior, M. (2008). *Satellite-based Habitat Mapping of Gashaka Gumti National Park (GGNP), Nigeria.* Report to North of England Zoological Society.

Louppe, D., Oattara, N. & Coulibaly, A. (1995). The effects of brush fires on vegetation: the Audrville fire plots after 60 years. *Commonwealth Forestry Review* 74: 288–292.

Oates, J. F., Bergl, R. A. & Linder, J. M. (2004). *Africa's Gulf of Guinea Forests: Biodiversity Patterns and Conservation Priorities.* (Advances in Applied Biodiversity Science 6). Washington, DC: Conservation International.

Oates, J. F., Groves, C. & Jenkins, P. D. (2008). The type locality of *Pan troglodytes vellerosus* (Gray, 1862), and implications for the nomenclature of West African chimpanzees. *Primates* 50: 78–80.

Reikat, A. (ed.) (2002). *Leben in Westafrika* Deutsche Forschungsgemeinschaft: Sonderforschungsbereich 268 (Kulturentwicklung und Sprachgeschichte im Naturraum Westafrikanische Savanne).

Sommer, V. (2008). *Schimpansenland. Wildes Leben in Afrika.* Munich: C. H. Beck.

Appendix 1. Climate in Study Areas at Kwano and Gashaka

Summary of Weather Parameters, 2000 – 2008

Data from weather stations of NCF (Gashaka 2000 – 2004) and GPP (Gashaka 2005 – 2008; Kwano 2001 – 2008). Mean humidity measured at Gashaka at 16:00, but at 19:00 at Kwano. Total rainfall adjusted for 4/17 individual years (indicated by asterisks) by adding averages from other years to compensate for months without data (Gashaka in 2006: Nov, Dec; 2007: Jan, Feb, Mar; Kwano in 2001: Mar; 2006: Mar, Apr).

GASHAKA

Mean humidity (%)	JAN	FEB	MAR	APR	MAY	JUN	JUL	AUG	SEP	OCT	NOV	DEC	MEAN
2000	-	-	-	-	-	71	76	70	72	76	51	36	64
2001	18	15	24	56	68	74	72	76	78	69	46	33	52
2002	26	26	32	51	62	76	78	80	80	82	51	35	57
2003	41	31	33	55	64	74	74	77	77	77	65	47	60
2004	49	26	24	65	68	72	76	75	81	71	61	42	59
2005	31	33	38	49	72	68	70	78	80	79	-	-	60
2006	-	-	-	-	-	-	-	-	-	-	-	-	-
2007	-	-	-	-	-	-	-	-	-	-	-	-	-
2008	-	-	36	40	66	69	-	-	-	-	-	-	53
MEAN	33	26	31	53	67	72	74	76	78	76	55	39	58

Rain days / month (%)	JAN	FEB	MAR	APR	MAY	JUN	JUL	AUG	SEP	OCT	NOV	DEC	MEAN
2000	0	0	0	57	48	63	58	71	60	58	0	0	35
2001	0	0	6	40	58	57	71	55	70	52	0	0	34
2002	0	0	6	43	52	67	58	65	67	58	7	0	35
2003	0	0	3	33	35	73	58	68	80	77	23	0	38
2004	0	0	3	50	65	30	65	58	57	52	30	0	34
2005	0	7	6	30	55	67	65	52	73	65	7	0	35
2006	0	4	6	13	55	63	61	55	70	65	-	-	39
2007	-	-	-	47	52	53	55	55	73	71	16	0	47
2008	0	0	3	33	-	-	78	55	72	29	-	-	34
MEAN	0	1	4	38	52	59	63	59	69	58	12	0	35

Wettest day (mm rain)	JAN	FEB	MAR	APR	MAY	JUN	JUL	AUG	SEP	OCT	NOV	DEC	MAX
2000	-	0	0	21	26	69	99	57	53	78	0	0	99
2001	0	0	9	31	32	38	64	144	53	53	0	0	144
2002	0	0	22	32	23	43	29	54	58	36	18	0	58
2003	0	0	1	43	62	36	39	89	75	46	56	0	89
2004	0	0	12	41	30	29	31	152	34	70	54	0	152
2005	0	6	41	42	103	40	39	52	50	53	23	0	103
2006	0	8	3	36	95	47	74	83	114	40	-	-	114
2007	-	-	-	34	50	26	33	61	51	55	23	0	61
2008	0	0	8	13	-	-	75	45	60	31	-	-	75
MAX	0	8	41	43	103	69	99	152	114	78	56	0	152

Rain (mm)	JAN	FEB	MAR	APR	MAY	JUN	JUL	AUG	SEP	OCT	NOV	DEC	TOTAL
2000	-	0	0	89	145	215	446	399	280	302	0	0	1875
2001	0	0	10	153	190	207	203	453	349	211	0	0	1777
2002	0	0	31	209	140	321	178	335	379	240	23	0	1856
2003	0	0	1	150	155	284	243	395	345	253	121	0	1946
2004	0	0	12	198	178	90	190	426	277	272	125	0	1767
2005	0	8	55	124	323	307	273	250	275	342	31	0	1989
2006	0	8	4	90	327	212	367	435	397	189	-	-	2081 *
2007	-	-	-	106	251	178	204	296	329	438	63	0	1882 *
2008	0	0	8	79	-	-	153	281	363	208	-	-	-
MEAN	0	2	15	133	214	227	251	363	333	273	52	0	1897

KWANO

Mean humidity (%)	JAN	FEB	MAR	APR	MAY	JUN	JUL	AUG	SEP	OCT	NOV	DEC	MEAN
2001	-	-	62	82	85	89	88	89	89	87	79	72	82
2002	67	55	64	81	88	90	91	91	91	91	84	79	81
2003	70	60	56	80	87	-	-	-	-	-	-	-	70
2004	76	58	54	86	90	90	90	95	99	-	-	-	82
2005	70	69	74	77	85	83	84	82	85	83	73	71	78
2006	66	67	-	-	84	83	83	82	83	83	74	69	-
2007	62	52	57	78	81	82	83	82	84	83	80	70	-
2008	66	51	56	75	80	83	81	81	82	78	71	69	73
MEAN	68	59	60	80	85	86	86	86	87	84	77	72	78

Rain days / month (%)	JAN	FEB	MAR	APR	MAY	JUN	JUL	AUG	SEP	OCT	NOV	DEC	MEAN
2001	0	0	0	53	68	73	74	68	73	45	0	0	38
2002	0	0	10	50	61	63	55	84	57	74	10	0	39
2003	0	0	6	47	42	-	-	-	-	-	-	-	19
2004	0	0	13	63	77	60	52	58	70	48	33	0	40
2005	0	11	29	40	84	77	58	61	87	71	7	0	44
2006	0	21	-	-	87	87	68	74	87	84	7	0	51
2007	0	4	13	60	68	60	68	68	70	55	23	0	41
2008	0	0	3	60	77	63	71	68	53	26	0	0	35
MEAN	0	4	11	53	71	69	64	69	71	58	11	0	40

Wettest day (mm rain)	JAN	FEB	MAR	APR	MAY	JUN	JUL	AUG	SEP	OCT	NOV	DEC	MAX
2001	0	0	0	51	45	45	32	65	122	23	0	0	122
2002	0	0	15	70	48	41	55	50	65	49	14	0	70
2003	0	0	2	65	21	-	-	-	-	-	-	-	65
2004	0	0	7	70	30	104	63	93	63	90	22	0	104
2005	0	11	14	44	59	45	39	41	40	30	26	0	59
2006	0	37	-	-	47	41	34	82	91	35	3	0	91
2007	0	8	9	40	61	26	82	67	68	40	27	0	82
2008	0	0	8	47	37	51	125	25	64	76	0	0	125
MAX	0	37	15	70	61	104	125	93	122	90	27	0	125

Rain (mm)	JAN	FEB	MAR	APR	MAY	JUN	JUL	AUG	SEP	OCT	NOV	DEC	TOTAL
2001	0	0	-	157	222	280	255	242	432	86	0	0	1683 *
2002	0	0	23	270	207	297	178	306	359	384	34	0	2056
2003	0	0	3	157	104	-	-	-	-	-	-	-	-
2004	0	0	16	436	179	305	310	279	394	297	122	0	2337
2005	0	15	54	164	392	259	230	253	314	226	40	0	1945
2006	0	70	-	-	387	291	283	406	319	274	3	0	2279 *
2007	0	8	17	240	231	194	345	186	274	205	87	0	1786
2008	0	0	8	184	296	266	437	169	233	134	0	0	1726
MEAN	0	12	20	229	252	270	291	263	332	229	41	0	1973

GASHAKA

Coolest day (°C)	JAN	FEB	MAR	APR	MAY	JUN	JUL	AUG	SEP	OCT	NOV	DEC	MIN
2000	-	17	16	21	22	21	21	20	21	21	16	16	16
2001	13	14	19	20	20	22	21	20	21	19	17	13	13
2002	13	14	19	16	21	20	21	21	21	21	17	14	13
2003	13	13	17	-	16	19	19	19	20	19	18	13	13
2004	15	19	19	19	16	20	21	20	20	21	16	14	14
2005	14	19	22	22	21	22	20	19	21	20	17	16	14
2006	16	16	22	18	21	21	21	21	22	22	18	17	16
2007	12	12	18	18	21	20	20	21	20	20	17	13	12
2008	13	13	18	20	21	21	21	21	21	18	20	-	13
MIN	12	12	16	16	16	19	19	19	20	18	16	13	12

Hottest day (°C)	JAN	FEB	MAR	APR	MAY	JUN	JUL	AUG	SEP	OCT	NOV	DEC	MAX
2000	-	39	41	40	38	35	35	33	34	33	34	32	41
2001	35	39	40	39	37	36	35	34	33	34	34	34	40
2002	34	39	39	42	36	36	35	33	33	33	36	34	42
2003	34	39	36	-	34	36	36	36	33	33	32	32	39
2004	33	42	41	37	36	34	33	33	34	33	34	33	42
2005	36	40	39	40	38	35	36	34	33	33	35	34	40
2006	34	38	38	40	34	34	34	33	34	28	28	28	40
2007	37	38	40	39	36	35	33	32	32	32	32	34	40
2008	34	39	41	42	-	35	34	34	34	37	33	-	42
MAX	37	42	41	42	38	36	36	36	34	37	36	34	42

Mean max. temp. (°C)	JAN	FEB	MAR	APR	MAY	JUN	JUL	AUG	SEP	OCT	NOV	DEC	MEAN
2000	-	36	37	36	35	32	31	30	31	31	33	31	33
2001	32	35	38	36	34	33	31	30	30	32	33	32	33
2002	31	36	38	36	35	33	32	30	30	30	32	31	33
2003	31	33	33	-	32	30	30	31	31	31	31	30	31
2004	31	35	36	33	33	32	31	31	31	32	31	31	32
2005	31	36	36	35	34	31	29	29	31	29	32	32	32
2006	32	34	36	36	32	32	32	30	27	26	25	26	31
2007	32	35	37	35	32	31	30	30	30	31	31	31	32
2008	32	35	37	37	-	-	-	-	-	-	-	-	35
MEAN	32	35	36	35	33	32	31	30	30	30	31	31	32.5

Mean min. temp. (°C)	JAN	FEB	MAR	APR	MAY	JUN	JUL	AUG	SEP	OCT	NOV	DEC	MEAN
2000	-	19	21	24	23	22	23	22	22	22	20	14	21
2001	14	17	23	23	23	23	23	23	22	22	20	19	21
2002	16	18	24	22	23	23	24	23	23	22	21	18	21
2003	17	18	20	-	19	22	23	21	22	21	21	16	20
2004	18	21	21	22	22	22	22	22	22	22	22	16	21
2005	16	22	24	24	23	23	23	22	22	22	20	18	22
2006	19	21	24	22	22	22	22	22	24	24	23	20	22
2007	14	17	22	23	22	22	22	22	22	21	20	16	20
2008	15	15	21	23	23	22	22	22	22	20	21	-	21
MEAN	16	19	22	23	22	22	23	22	22	22	21	17	21.0

KWANO

Coolest day (°C)	JAN	FEB	MAR	APR	MAY	JUN	JUL	AUG	SEP	OCT	NOV	DEC	MIN
2001	-	-	18	17	17	21	21	20	20	18	16	16	16
2002	14	16	22	20	21	20	21	20	20	20	17	15	14
2003	14	17	20	22	21	-	-	-	-	-	-	-	14
2004	16	16	20	21	21	20	21	21	20	21	19	16	16
2005	14	19	22	22	21	21	20	20	20	20	16	16	14
2006	17	19	-	-	20	20	20	21	19	20	15	14	14
2007	14	16	18	21	21	18	20	20	20	20	17	14	14
2008	14	16	20	21	20	20	20	20	19	20	17	15	14
MIN	14	16	18	17	17	18	20	20	19	18	15	14	14

Hottest day (°C)	JAN	FEB	MAR	APR	MAY	JUN	JUL	AUG	SEP	OCT	NOV	DEC	MAX
2001	-	-	43	35	35	34	32	34	33	33	32	35	43
2002	31	39	37	39	34	36	36	34	35	34	32	32	39
2003	35	37	38	38	36	-	-	-	-	-	-	-	38
2004	31	36	40	37	35	33	31	34	32	32	33	31	40
2005	31	36	40	37	35	33	31	34	32	32	33	31	40
2006	36	40	39	39	35	34	33	34	34	33	35	34	40
2007	35	37	-	-	36	35	33	32	33	33	33	33	37
2008	29	38	41	37	34	33	32	32	33	35	34	33	41
MAX	36	40	43	39	36	36	36	34	35	35	35	35	43

Mean max. temp. (°C)	JAN	FEB	MAR	APR	MAY	JUN	JUL	AUG	SEP	OCT	NOV	DEC	MEAN
2001	-	-	34	31	31	30	30	30	30	30	31	31	31
2002	28	34	35	34	32	32	31	30	31	30	30	30	31
2003	31	34	35	34	33	-	-	-	-	-	-	-	33
2004	29	33	37	33	32	30	30	30	29	30	31	29	31
2005	30	36	36	35	33	31	30	31	32	31	32	31	32
2006	32	34	-	-	32	32	31	31	31	32	32	31	32
2007	29	35	37	34	32	31	30	30	31	32	32	32	32
2008	30	34	37	33	32	31	31	32	31	32	32	33	32
MEAN	30	34	36	33	32	31	30	31	31	31	32	31	31.9

Mean min. temp. (°C)	JAN	FEB	MAR	APR	MAY	JUN	JUL	AUG	SEP	OCT	NOV	DEC	MEAN
2001	-	-	21	21	23	22	22	22	22	21	19	18	21
2002	16	19	24	23	23	22	22	22	22	22	20	21	21
2003	19	20	23	24	23	-	-	-	-	-	-	-	22
2004	17	19	22	23	22	22	22	22	23	22	22	19	21
2005	17	22	24	24	23	23	22	22	22	21	19	18	21
2006	17	21	-	-	22	21	21	21	21	21	19	16	20
2007	15	19	21	22	22	21	21	21	21	21	21	16	20
2008	17	17	22	22	22	21	21	21	21	21	21	19	20
MEAN	17	20	22	23	22	22	22	22	22	21	20	18	20.9

Rainfall Pattern

Examples for the year 2004. "x" indicates drizzle.

RAINFALL (mm) AT KWANO (2004)

Day	JAN	FEB	MAR	APR	MAY	JUN	JUL	AUG	SEP	OCT	NOV	DEC
1				x		4		23	45		6	
2			1	31				26	25		10	
3			7		4			4		27	14	
4			4				21	8		18	18	
5				17	14		37	1	35	2		
6				12	9	67	x		1	30		
7			4	60	4	2			17		18	
8				20	9	x	32	93	4		22	
9				x	10	1			2		2	
10				x	3	3		8	3	2	12	
11				23	5			7	63	25		
12				3						26		
13				x	8	8				27		
14				18	x	36	63				11	
15				x	30	19						
16				32	4		24	18				
17				x	8	3	14	1	4	13		
18				8	6		2	9	35	10		
19				x	6		47	28	21	15		
20				39	24	6	x			10		
21				5	9	7	x		26	1	9	
22				20		2	41		2			
23				26		22		37	23			
24				6	13			8	24	90		
25				15			10	1	21			
26				70			3		15			
27				22	4	22	8		22			
28				10			9		2	1		
29				x	4	104	x	4	4			
30					9			1				
31								3				
Sum			16	436	179	305	310	279	394	297	122	
Total												2337

RAINFALL (mm) AT GASHAKA (2004)

Day	JAN	FEB	MAR	APR	MAY	JUN	JUL	AUG	SEP	OCT	NOV	DEC
1								30				
2						1	2	1	10	1		
3			12		30	16		22		1		
4					10	9	8		24	70	11	
5					25		31	14	34	1	7	
6				3			1	36		27	54	
7				6	10	6	1	152		1	6	
8				41		1			31	37	34	
9				17		x			x	4	1	
10					1	1	28					
11					11	6	1	31		6		
12					6			1	33	19		
13						21	18		2		7	
14					7				3	38		
15				9	4				8			
16				x	21		4	9	13		1	
17							16	10	7	12		
18				2	5		1					
19					12			10				
20				37	6		4		12	27	5	
21							24			2		
22				4	17		22	12	25			
23				2	1			41	19			
24				23	1		3	3	27	3		
25				5			1	1		25		
26				3	2			2	9			
27				25	2		5					
28				19			x					
29				4			7		11			
30					4	29	13	46	11			
31					2		4	8				
Sum			12	198	177	90	190	426	277	272	125	
Total												1870

Appendix 2. Publications of the Gashaka Primate Project

Total

n = 107, reference date: 01 Jan 09. Listed chronological within the following categories: 2 books, 23 research articles in journals and edited volumes, 10 unpublished reports, 4 PhD theses (completed), 9 PhD theses (expecting completion), 15 masters dissertations and diplomarbeiten, 4 undergraduate dissertations, 25 abstracts, 11 popular writing, 3 documentaries.

Books

Sommer, Volker (2008). *Schimpansenland. Wildes Leben in Afrika*. Munich: C. H. Beck. 251 pp, 8 plates. ["Chimpanzeeland. Wild life in Africa"].

Sommer, Volker & Caroline Ross (eds.) (2011). *Primates of Gashaka. Socioecology and Conservation in Nigeria's Biodiversity Hotspot*. Springer: New York.

Research Articles in Journals & Edited Volumes

Sommer, Volker; Jeremiah Adanu, Isabelle Faucher & Andrew Fowler (2004). The Nigerian chimpanzee (*Pan troglodytes vellerosus*) at Gashaka: Two years of habituation efforts. *Folia Primatologica* 75: 295–316.

Arnold, Kate & Klaus Zuberbühler (2006) The alarm calling system of adult male putty-nosed monkeys, *Cercopithecus nictitans martini*. *Animal Behaviour* 72: 643–653.

Arnold, Kate & Klaus Zuberbühler (2006). Alarm calls and organised imperatives in male putty-nosed monkeys. In Cangelosi, A., Smith, A. D. M., & Smith, K. (eds.), *The Evolution of Language* (Proc. 6th International Conference [Evolang]. Singapore: World Scientific Publishing Company, pp. 389–390.

Arnold, Kate & Klaus Zuberbühler (2006). Semantic combinations in primate calls. *Nature* 441: 303.

Hohmann, Gottfried; Andrew Fowler, Volker Sommer & Sylvia Ortmann (2006). Frugivory and gregariousness of Salonga bonobos and Gashaka chimpanzees: the abundance and nutritional quality of fruit. In: Hohmann, G., Robbins, M. M. & Boesch, C. (eds.), *Feeding Ecology in Apes and Other Primates*. Cambridge: Cambridge University Press, pp. 123–159.

Weyher, Anna H.; Caroline Ross & Stuart Semple (2006). A comparison of gastrointestinal parasites in a crop raiding and a wild foraging troop of olive baboons (*Papio cynocephalus anubis*) in Nigeria. *International Journal of Primatology* 27: 1519–1534.

Schöning, Caspar & Mark Moffett (2007). Driver ants invading a termite nest - why do the most catholic predators of all seldom take this abundant prey? *Biotropica* 39: 663–667.

Fowler, Andrew; Yianna Koutsioni & Volker Sommer (2007). Leaf-swallowing in Nigerian chimpanzees. Assumed evidence for self-medication. *Primates* 48: 73–76.

Schöning, Caspar; Darren Ellis, Andrew Fowler & Volker Sommer (2007). Army ant prey availability and consumption by chimpanzees at Gashaka (Nigeria). *Journal of Zoology* 271: 125–133.

Fowler, Andrew & Volker Sommer (2007). Subsistence technology in Nigerian chimpanzees. *International Journal of Primatology* 28: 997–1023.

Warren Ymke; Bobbo Buba & Caroline Ross (2007). Patterns of crop-raiding by wild and domestic animals near Gashaka Gumti National Park Nigeria. *International Journal of Pest Management* 53: 207–216.

Higham, James P.; Caroline Ross, Ymke Warren, Michael Heistermann & Ann M. MacLarnon (2007). Reduced reproductive function in wild olive baboons (*Papio hamadryas anubis*) related to natural consumption of the African black plum (*Vitex doniana*). *Hormones and Behavior* 52: 384–390.

Higham, James P.; Michael Heistermann, Caroline Ross, Stuart Semple & Ann MacLarnon (2008). The timing of ovulation with respect to sexual swelling detumescence in wild olive baboons. *Primates* 49: 295–299.

Higham, James P.; Ann MacLarnon, Caroline Ross, Michael Heistermann & Stuart Semple (2008). Baboon sexual swellings: information content of size and color. *Hormones and Behavior* 53: 452–462.

Higham, James P.; Stuart Semple, Ann MacLarnon, Michael Heistermann & Caroline Ross (2008). Female reproductive signaling, and male mating behavior, in the olive baboon. *Hormones and Behavior* 55: 60–67.

Higham, James P. & David M. Bennett (2008). Perspectives on wildlife, and wildlife consumption, in Eastern Nigeria. *Gorilla Journal* 36: 11–13.

Wilkinson, Roger (2008). Some recent records of birds from Gashaka Gumti National Park and Ngel Nyaki, Nigeria, and the Gotel Mountains, Cameroon. *Malimbus* 30: 156–164.

Arnold, Kate & Klaus Zuberbühler (2008). Meaningful call combinations in a non-human primate. *Current Biology* 18: R202–203.

Higham, James P.; Ymke Warren, Jeremiah Adanu, Buba N. Umaru, Ann M. MacLarnon, Volker Sommer & Caroline Ross (2009). Life on the edge: Life-history of olive baboons at Gashaka-Gumti National Park, Nigeria. *American Journal of Primatology* 71: 293–304.

Higham, James P.; Ann MacLarnon, Michael Heistermann, Caroline Ross, & Stuart Semple (2009). Self-directed behaviour and faecal glucocorticoids are not correlated in wild female baboons (*Papio hamadryas anubis*). *Stress* 12: 526-532.

Ey, Elodie; Charlotte Rahn, Kurt Hammerschmidt & Julia Fischer (2009). Wild female olive baboons adapt their grunt vocalizations to environmental conditions. *Ethology* 115: 493-503.

Unpublished Reports

Sommer, Volker (ed.) (2000, 2001, 2002, 2003, 2004, 2005, 2006, 2007, 2008). *Gashaka Primate Project: Annual Report*. London: University College London.

Fowler, Andrew; Gilbert Nyanganji, Aylin McNamara, Jeremiah Adanu & Volker Sommer (2006). *Conservation Status of Primates in Eastern Nigeria's Taraba State With Special Reference to Chimpanzees* (Pan troglodytes vellerosus). Report to WWF-UK.

Gumnior, Maren (2008). *Satellite-based Habitat Mapping of Gashaka Gumti National Park (GGNP), Nigeria*. Report to North of England Zoological Society.

PhD Theses (completed)

Warren, Ymke (2003). *Olive Baboons* (Papio cynocephalus anubis): *Behaviour, Ecology and Human Conflict in Gashaka Gumti National Park, Nigeria*. PhD Thesis, Roehampton University, London.

Fowler, Andrew (2006). *Socio-ecology of Nigerian Chimpanzees at Gashaka*. PhD Thesis; Department of Anthropology, University College London.

Higham, James (2006) *The Reproductive Ecology of Female Olive Baboons* (Papio hamadryas anubis) *at Gashaka-Gumti National Park, Nigeria*. PhD thesis, Roehampton University, London.

Ey, Elody (2008). *Influences of Ecological Factors on Vocal Communication in Wild Olive Baboons*. PhD thesis, Georg-August Universität Göttingen, Germany.

Bennett, David (2009). *Human-Wildlife Conflict: A Case Study of the Fulani at Gashaka Gumti National Park, Nigeria*. Roehampton University, London.

PhD Theses (expecting completion; working titles)

Adanu, Jeremiah † (2003–2005). *Socio-ecology of Baboons at Kwano in Gashaka Gumti National Park*. Federal University of Technology Yola, Nigeria.

MacGregor Inglis, David (since 2006). *Vocal and Gestural Communication in Olive Baboons*. Roehampton University, London.

Pascual-Garrido, Alejandra (since 2007). *Insectivory in Nigerian Chimpanzees* (Pan troglodytes vellerosus): *Habitat Ecology and Harvesting Strategies*. Universidad Complutense de Madrid.

Buba, Umaru (since 2007). *Habitat Ecology of Nigerian Chimpanzees at Gashaka Gumti National Park*. Federal University of Technology Yola, Nigeria.

Alberts, Nienke (since 2008). *Social Networking in Wild Baboons at Gashaka Gumti National Park, Nigeria*. Roehampton University, London.

Weaver, Daniel (since 2008). *Where is Biodiversity Located? Spatial and Temporal Patterns from a Nigerian Tropical Forest*. Roehampton University, London.

Lodge, Emily (since 2008). *Feeding Ecology of Olive Baboons*. Roehampton University, London.

Auckland, Kate (since 2009). *Population Genetics of Nigerian Chimpanzees*. Department of Anthropology, University of Cambridge.

Harvey, Suzanne (since 2009). *Audience Effects in Olive Baboons. A Field Study in Nigeria*. Department of Anthropology, University College London.

Jesus, Gonçalo (since 2008). *The Influence of Habitat Productivity on Private Gregariousvess and Reproduction at gashaka / Nigeria*. Departmet of Anthropology, University College London.

Masters Dissertations and Diplomarbeiten

Adanu, Jeremiah (2003). *Socio-ecology of Forest Monkeys at Kwano in Gashaka Gumti National Park*. MTech dissertation, Wildlife Conservation & Management, Federal University of Technology Yola, Nigeria

Hughes, Nicola (2003). *Great Ape Sanctuaries in Nigeria and Cameroon: a Window into the Bushmeat-Trade*. MSc dissertation, Masters in Human Evolution and Behaviour, Department of Anthropology, University College London

Koutsioni, Yianna (2003). *Merging Ethnobotany and Zoopharmacognosy: A Case Study of Humans and Non-Human Primates at Gashaka / Nigeria*. MSc dissertation, Masters in Human Evolution and Behaviour, Department of Anthropology, University College London

Weyher, Anne (2004). *A Comparison of Gastrointestinal Parasites in a Crop-Raiding and a Forest-Dwelling Troop of Olive Baboons* (Papio cynocephalus anubis) *in Gashaka Gumti National Park, Nigeria*. MRes dissertation in Primatology, Roehampton University, London

Ellis, Darren (2005). *Seasonal Variation in Availability and Consumption of Army Ants by Nigerian Chimpanzees*. MSc dissertation, Masters in Human Evolution and Behaviour, Department of Anthropology, University College London, England

Pohlner, Yvonne (2005). *Ruf-Muster arborealer Affen in Gashaka, Nigeria*. Diplomarbeit, Institut für Zoologie, Universität Freiburg, Germany

Walravens, Fionnuala B. C. (2005). *Fission-fusion Sociality in Olive Baboons* (Papio cynocephalus anubis) *at Gashaka Gumti National Park, Nigeria*. MRes dissertation in Primatology, Roehampton University, London

Bauer, Jan (2006).Springer Science+Business Media, LLC *Vergleichende Faecesuntersuchungen bei*Pan paniscus,Pan troglodytes vellerosus *und* Pan troglodytes verus. Diplomarbeit, Humboldt-Universität Berlin, Germany

Buba, Umaru (2007). *An Investigation of the Behaviour of Pregnant and Lactating Baboons* (Papio anubis) *at Gashaka Gumti National Park, Nigeria*. MTech dissertation in Wildlife Conservation & Management, Federal University of Technology Yola, Nigeria

Couto, Sarah (2006). *Activity and Association Pattern of Wild Olive Baboons at Gashaka/Nigeria: Spatio-Temporal Variation in Relation to Resource Availability*. MSc dissertation, Masters in Human Evolution and Behaviour, Department of Anthropology, University College London

Ihuba, Jerome (2007). *Frugivore Communities Within Ngel Nyaki Forest Reserve and the Influence of Forest Fragmentation on These Communities*. MTech dissertation in Wildlife Conservation & Management, Federal University of Technology Yola, Nigeria

Price, Tabitha K. (2007). *Information Content of Male Putty-Nosed Monkey* (Cercopithecus nictitans) *Vocalisations*. MRes dissertation in Primatology, Roehampton University, London

Goffe, Adeelia S. (2007). *Examining Social Relationships and Stress Levels of Wild Yearling and Juvenile Baboons* (Papio hamadryas anubis) *in Gashaka Gumti National Park*. MRes dissertation in Primatology, Roehampton University, London

Ellis, Jacklyn J. (2008). *The Social and Behavioural Correlates of Stress in Wild Male Olive Baboons* (Papio hamadryas anubis) *in Gashaka-Gumti National Park, Nigeria*. MRes dissertation in Primatology, Roehampton University, London

Buhl, Jacqueline S. (2008). *Influence on Play Behaviour in Infant and Juvenile Olive Baboons* (Papio hamadryas anubis) *in Gashaka-Gumti National Park*. MRes dissertation in Primatology, Roehampton University, London

Gordon, Gemma (2009). *Male Consortship in Wild Olive Baboons. A Study at Gashaka-Gumti National Park, Nigeria*. MSc dissertation, Masters in Human Evolution and Behaviour, Department of Anthropology, University College London

Undergraduate Dissertations

Rawle, Elaine (2002). *Crop Raiding and Its Implication for Conservation in and Around Gashaka-Gumti National Park*. Bachelor dissertation, Department of Anthropology, Oxford Brookes University

Buba, Umaru (2004). *Nest-building Behaviour in Chimpanzees at Gashaka-Gumti National Park.* Bachelor dissertation, Department of Forestry & Wildlife Management, Federal University of Technology Yola, Nigeria

Allon, Oliver (2009). *Army Ant Harvesting in Nigerian Chimpanzees: Predator-Prey Ecology and Implications for Human Evolution.* Bachelor dissertation, Human Sciences/Department of Anthropology, University College London

Nodza, George (2009). *Phenology Transects in a Long-term Primate Research Site.* Bachelor dissertation, Federal University of Technology Bauchi, Nigeria

Abstracts (Conference Talks or Posters)

Ross, Caroline; Catherine M. Hill & Ymke Warren (2004). Friends or foes? Farmers' attitudes towards primate crop raiders. (Abstracts, UFAW International Symposium "Science in the Service of Animal Welfare", 02–04 Apr 03, Edinburgh) *Animal Welfare* 13: 254

Sommer, Volker; Andrew Fowler & Jeremiah Adanu (2004). The Nigerian chimpanzee (*Pan troglodytes vellerosus*) at Gashaka: Two years of habituation efforts. (Abstracts, 8. Kongress, Gesellschaft für Primatologie, 02–05 Oct 03, Leipzig) *Folia Primatologica* 75

Ross, Caroline & Ymke Warren (2004). West African baboons: the influence of forest living and food-enhancement on activity budgets. (Abstracts, 8. Kongress, Gesellschaft für Primatologie, 02–05 Oct 03, Leipzig) *Folia Primatologica* 75: 328

Warren, Ymke & Caroline Ross (2004). Baboon crop raiding: pre-adaptation and success. (Abstracts, 8th Congress, Gesellschaft für Primatologie, 02–05 Oct 03, Leipzig) *Folia Primatologica* 75: 425

Arnold, Kate & Klaus Zuberbühler (2005) The alarm call system of free-ranging putty-nosed monkeys. (Abstract, Budapest, XXIX Int. Ethological Conference, Budapest) Advances in *Ethology (Supplements to Ethology)*: 274

Fowler, Andrew & Volker Sommer (2005). Material culture in Nigerian chimpanzees: a contribution to cultural primatology. (Abstracts, 1. Kongress, Europäische Vereinigung für Primatologie, 09–12 Aug 05, Göttingen) *Primate Report* 72: 34f

Higham, James, Caroline Ross, Ymke Warren, Ann MacLarnon, Volker Sommer & Jeremiah Adanu (2005). 4 years of rainforest baboons at Gashaka. (1. Kongress, Europäische Vereinigung für Primatologie, 09–12 Aug 05, Göttingen) *Primate Report* 72: 46f

Higham, James; Caroline Ross, Ymke Warren, Ann MacLarnon, Volker Sommer & Jeremiah Adanu (2005). Four years of rainforest baboons at Gashaka. (Abstract, Primate Society of Great Britain Winter Meeting, 09 Dec 05, London) *Primate Eye* 87: 29–30

Hill, Catherine M. & Caroline Ross (2006). Symposium: conflict and conservation: can non-human primates and farmers live together? (Abstracts, 21st Congress, International Primatological Society, Jul 06, Entebbe, Uganda) *International Journal of Primatology* 27: 394

Ross, Caroline & Catherine M. Hill (2006). Non-human primates as pests: variation in primate species and human cultures. (Abstracts, 21st Congress, International Primatological Society, Jul 06, Entebbe, Uganda) *International Journal of Primatology* 27: 398

Ross, Caroline & Ymke Warren (2006). Primate and other mammalian maize pests in Gashaka, Nigeria. (Abstracts, 21st Congress, International Primatological Society, Jul 06, Entebbe, Uganda) *International Journal of Primatology* 27: 403.

Bennett, David M. & Caroline Ross (2006). Crop-raiding on the Fulani of Gashaka Gumti National Park: The impact of animal-human conflict on the political ecology of a settled pastoral community. (Abstracts, 21st Congress, International Primatological Society, Jul 06, Entebbe, Uganda) *International Journal of Primatology* 27: 404

Higham, James P. (2007). The reproductive ecology of female olive baboons (*Papio hamadryas anubis*) at Gashaka-Gumti National Park, Nigeria. (PhD Abstracts) *Primate Eye* 91: 35f

Higham, James P.; Stuart Semple, Ann MacLarnon, Michael Heistermann & Caroline Ross. (2007). Baboon sexual swellings in a multiple-signal framework. (Abstracts, Primate Society of Great Britain Spring Meeting, 17–18 Apr 07, Durham) *Primate Eye* 92: 7–8

Higham, James P.; Michael Heistermann, Caroline Ross, Stuart Semple & Ann MacLarnon (2007). The relationship between ovulation and maximal swelling in wild baboons. (Abstracts, Primate Society of Great Britain Spring Meeting, 17–18 Apr 07, Durham) *Primate Eye* 92: 15

Higham, James P.; Caroline Ross, Ymke Warren, Michael Heistermann & Ann MacLarnon (2008). Reduced reproductive function in wild baboons (*Papio hamadryas anubis*) related to natural consumption of plant compounds. (Abstracts, European Federation of Primatology, 03 – 07 Sep 07, Prague) *Folia Primatologica* 79: 338–339

MacLarnon, Ann M.; Stuart Semple, Kathryn Shutt, Michael Heistermann & James P. Higham (2008). Sociality and stress in macaques and baboons. (Abstracts, European Federation of Primatology, 03 – 07 Sep 07, Prague) *Folia Primatologica* 79: 355–355

Ey, Elodie; Kurt Hammerschmidt, Dietmar Zinner, Julia Fischer (2007). Influences of environmental factors on vocal communication in baboons. (Abstracts, European Federation of Primatology, 03 – 07 Sep 07, Prague) *Folia Primatologica* 79: 327

Buba, Umaru N.; James P. Higham, Ann M. MacLarnon, Ymke Warren & Caroline Ross (2008). The natural history and behaviour of olive baboons at Gashaka, Nigeria. (Poster *abstracts* No. 139, XXII Congress, *The International Primatological Society*, 03 – 08 Aug 08, Edinburgh, UK)

Inglis, David MacGregor; Caroline Ross & Stuart Semple (2008). Multi-modal communication in olive baboons (*Papio hamadryas anubis*). (Poster *abstracts* No. 168, XXII Congress, *The International Primatological Society*, 03 – 08 Aug 08, Edinburgh, UK)

Price, Tabitha K.; Kate Arnold, Klaus Zuberbühler & Stuart Semple (2008). Vocal responses of male putty-nosed monkeys reflect a response-urgency rather than a referential calling system. (*Abstracts* No. 341, XXII Congress, *The International Primatological Society*, 03 – 08 Aug 08, Edinburgh, UK)

Fowler, Andrew; Umaru Buba & Volker Sommer (2008). Nest-building ecology of *Pan troglodytes vellerosus* at Gashaka, Nigeria. (*Abstracts* No. 533, XXII Congress, *The International Primatological Society*, 03 – 08 Aug 08, Edinburgh, UK)

Arnold, Kate & Klaus Zuberbühler (2008). A meaningful call combination in a forest guenon. (*Abstracts* No. 680, XXII Congress, *The International Primatological Society*, 03 – 08 Aug 08, Edinburgh, UK)

Ey, Elody; Charlotte Rahn, Kurt Hammerschmidt, Dietmar Zinner, Julia Fischer (2008). Do olive baboons modify their vocalizations according to their environment? (*Abstracts* No. 817, XXII Congress, *The International Primatological Society*, 03 – 08 Aug 08, Edinburgh, UK)

Higham James P.; Caroline Ross, Michael Heistermann & Ann M. MacLarnon (2008). Diet and reproductive function in wild olive baboons (*Papio hamadryas anubis*) at Gashaka-Gumti national park, Nigeria. (*Abstracts* No. 844, XXII Congress, *The International Primatological Society*, 03 – 08 Aug 08, Edinburgh, UK)

Hughes, Nicola; Norm Rosen, Neil Gretsky & Volker Sommer (2009). Extinction risk of *Pan troglodytes vellerosus*. Predictions based on intake rates of chimpanzee sanctuaries. (*Abstracts*, 11. Meeting, Gesellschaft für Primatologie, 24 – 26 Feb 09, Hannover), p. 31

Bauer, Jan; Sylvia Ortmann, Heidrun Barleben, Gottfried Hohmann, Volker Sommer, Andrew Fowler, Heribert Hofer, Andreas Elephandt (2009). Comparative structural analysis of faeces from *Pan paniscus* and *Pan troglodytes* (*Abstracts*, 11. Meeting, Gesellschaft für Primatologie, 24 – 26 Feb 09, Hannover), p. 61

Pascual-Garrido, Alejandra; Oliver Allon, Caspar Schöning & Volker Sommer (2009) (*Abstracts*, 2nd Iberian Primatological Conference, 01 – 03 Oct 09, Madrid) *Folia Primatologica* 80: 402

Popular Writing

Adanu, Jeremiah (2000). Screaming chimpanzees. *WWF-UK Primate Report* (Summer)

Sommer, Volker (2000). Bald sind wir allein. Die Menschenaffen sterben – ein Verwandtschaftsbesuch in Nigeria. *Die Welt (Literarische Welt)* 09 Sep 00: 3

Fowler, Andrew (2001). Chimpanzees of Nigeria. *WWF-UK Primate Report* (Autumn)

Sommer, Volker (2001). The monkey with the drum. Inside Nigeria's primate paradise. *Nigeria Parks*: 22–25 (Reprint, The Week, Nigeria: 15, 2002)

Sommer, Volker (2002). Primate paradise. Chester Zoo: *Zoo Life* 10: 18–19

Warren, Ymke (2002). More to baboons than meets the eye. *WWF-UK Primate Report* (Summer)

Warren, Ymke (2002). Baboons: Gashaka's most wanted. *WWF-UK Primate Report*

Sommer, Volker (2003). Bruder Affe. Titel-Essay für "Wir Affen. Der Mensch und seine Verwandten", *Folio, Neue Zuercher Zeitung* (August): 14–18

Sommer, Volker (2003). Geistlose Affen oder äffische Geistwesen? Eine Exkursion durch die mentale Welt unserer Mitprimaten. Pp. 112–136 in: Alexander Becker, C. Mehr, H. H. Nau, G. Reuter & Dagmar Stegmüller (eds.), *Gene, Meme und Gehirne. Geist und Gesellschaft als Natur*. Frankfurt / M.: Suhrkamp

Garcia, Susana (2008). Memória da passagem por Gashaka. Pp. 213–222 in: Frias, S. (ed.), *Etnografia & Emoções*. Lisboa: Instituto Superior de Ciências Sociais e Políticas. [Portuguese]

Marina Aizen (2007). Una monada. Revista *Viva*, diario *Clarin*, 11 Nov, pp. 103–106 (story about Alejandra Pascual-Garrido in Sunday-Magazine of leading national Argentinian newspaper) [Spanish]

Documentaries

"*Der Affenwald von Gashaka*" (2003). German Television, ARD / Bayerischer Rundfunk; 45 min documentary produced by Felix Heidinger; first screening (with many repeats): 30 Jul 03

"*Nigeria's Primate Paradise*" (2003). German Television, ARD / Bayerischer Rundfunk; 25 min documentary produced by Felix Heidinger & Volker Sommer; for distribution since Aug 03

"*Culture in Chimpanzees*" (2006). Display within the exhibition "Evolution", Deutsches Hygienemuseum, Dresden, Germany, Sep 05 – Feb 06; Florence, Italy, mid 2006, Kerkrade (Netherlands) 2009

Appendix 3. Affiliates and Supporters of the Gashaka Primate Project

Core Funding

Chester Zoo Nigeria Biodiversity Programme (since 2002, including contributions to "African Student Fund"; head of Field Programmes & Research: Roger Wilkinson; Conservation & Education Director: Mark Pilgrim; Director General: Gordon McGregor Reid; www.chesterzoo.org)

Cooperation and Logistic Support

Nigeria National Park Service (Conservator Generals: Alhaji Lawan B. Marguba, Haruna Tanko Abubakar; Conservator of Parks: Jarafu Mamza, Modu Sherif, Okeyoyin Agboola; Contact officers: Abubakar A. Kaigama, Inwon Fingesi, Yoanna Saidu)

Gashaka Gumti National Park (all staff on the ground and in the bush; in particular Head of Department of Park Protection and Conservation: Mohammed Umar; Range Officers Filinga: George Karngong, Pepeh Kamaya, Mwajim Mbaya; Technical department: Joseph Kehinde Lawan, Babangida Kawuji; carpenter Aminu "Kano" Ibrahim Tanko; rangers: Anthony David, Saidu Isa Masabere, Dahiru Lawal; the Jauro and villagers of Gashaka)

Nigerian Conservation Foundation (executive directors: Muhtari Aminu-Kano, Emmanuel Obot; Gashaka Gumti Project Managers: Tirimisiyu Ashimi, Abdullahi N. Zabudum, Faith Ananze; Abba Leuro, Andrew Dunn, Aaron Nicholas, Doris Nyanganji †, Stephen Gawaisa, Salamu Waziri †)

WWF-UK (Richard Barnwell, Helen Gardiner, Mark Wright, Callum Rankine)

Leventis Foundation / Fauna & Flora International (Phil Hall, Tasso Leventis, Barbara Privett)

Federal University of Technology Yola (Callistus Akosim)

Development Corporation Ireland (Russell Hanley, Abuja)

Nigeria Montane Forest Project (Hazel Chapman)

International Institut for Tropical Agriculture, Ibadan (Richard Markham, Isabelle Faucher, Taiwo Oviasuyi)

University College London (Department of Anthropology Administrator Alena Kocourek)

Infrastructural Development, in Particular "Power-Island" at Kwano Field Station and Radio Communication Throughout Gashaka Gumti National Park

Oskar-von-Miller Schule (Horst Hoppe, Karsten Reichart, Markus Engelhardt, Timo König, Martin Münstedt, Stefan Werner, as well as Boris Kraft Kai von Schumann, Meik Mitter, Martin Wegner, Kadir Dönmez, Thomas Muth)

Deutsche EXIDE GmbH

Bilfinger-Berger, Wiesbaden and Julius-Berger Nigeria, Abuja (Hans Wittmann, Volker Gallitz sen., Reinhard Schmidt-Tuexen, Eleonore Schlueter, Martina Lobenstein, Reinhard Grimpo, Herbert Kloepper, Volker Gallitz jun., Werner Fettinger, Katja Rosenke, and the "Berger Ladies Charity" headed by Irma Hermannsdoerfer)

Stadtwerke Kassel AG

Jürgen Volmar, dentistry, Espenau and Holzhausen am Reinhardswald

SMA Kassel (Günther Häckl, Günther Cramer, Werner Reitze, Patrick Mark)

Companies in and near Kassel / Germany. ABB Stotzkontakt (Michael Hueck), CL Bergmann (Thomas Siebert), Digital Copyblitz (Adrian Meilinger, Manfred Köhler), EAM Energie AG, Firma Mitter, L&S Landwehr & Schulz (Waldemar Boguslawski), Firma Rittal, Rotary Club Kassel (Norbert Greib), Rotary Klub Hofgeismar-Wolfhagen (Wolfgang Annecke), SAB, Solara (Frank Heise), Wago Kontakttechnik (Dieter Muxfeldt)

Individual supporters. Renate and Volker Storch, Armin and Waltraud Sommer, Günter Haaf, Hark Bohm, Susanne Hassenkamp

Webmaster

Peter Schauer, Department of Archaeology, University College London

Volunteers

Claire Bracebridge (England), Julia Greenslade (England), Judith Bovensiepen (Germany), Jennifer Rogan (Scotland), Asako Saegusa (Japan), Alejandra Pascual-Garrido (Argentina), Maria Susana Garcia (Portugal), Helen Cross (England), Aylin McNamara (England), Aisha Dasgupta (England), Sonja Richter (Austria), Vincent Thuet (France), Daphné Kerhoas-Essens (France), Nienke Alberts (Netherlands), Lauren Adele Holt (England), James Vernon (England), Anna Hopkins (England), James Tallant (UK), Ruth Sharpe (UK), Tom Pollard (England), Hsueh Chen (Taiwan).

GPP Field Assistants and Camp Management

Head FA: Hammaunde Guruza; Senior FA: Felix Vitalis, Halidu Ilyasu (Dan-Azumi); FA: Bobbo Buba †, Ali Tappare, Maigari Ahmadu, Buba Bello, Zacchariah Dollar, Isaac Timothy, Maikanti Hassan, Yakubu Ahmadu †, Buba Hammaselbe, Haruna Ahmadu, Agnes Bulus, Nuhu Husseini, Musa Basago, Sam Yusufu, Mahmoda Dabo

Contact

Gashaka Primate Project, PMB 08, Serti 663001, Taraba State, Nigeria

GPP c/o Volker Sommer, Department of Anthropology, University College London, Gower Street, London WC1E 6BT, v.sommer@ucl.ac.uk

www.ucl.ac.uk/gashaka

Chapter 2
To Save a Wilderness: The Creation and Development of Gashaka Gumti National Park, Nigeria

Richard Barnwell

Abstract The chapter reconstructs the process that led to the creation of Nigeria's largest national park at Gashaka Gumti from the initial inception as a chain of game sanctuaries. Its development is closely linked with political processes in the 1960s after Nigeria gained independence and the involvement of local communities. The contribution reiterates the need to revive and expand this early approach to conservation.

Keywords National park • Conservation policy

Early Efforts of Protection

Gashaka Gumti National Park is situated in the rugged mountains of north-eastern Nigeria, a sparsely populated tract of wild country that has had a long history of being the haunt of a wealth of wildlife. The forests were known to harbour many species of primates, including putty-nosed, mona and colobus monkeys and chimpanzees. Leopards also roamed through these forests and were hunted for their skins. The woodlands were home to herds of hartebeest, roan antelope, kob and water buck and the favourite quarry of the hunters, the buffalo, were found everywhere. Pastoralists who had moved into the mountains with their herds of cattle swiftly discovered that they had to guard them vigilantly against the numerous lions and hyenas. This wilderness was a renowned source of ivory in the 19th and early 20th centuries and as recently as 1949 elephants were still being hunted in these mountains. The last surviving animals were seen heading northwards towards the River Benue in 1952.

R. Barnwell (✉)
World Wide Fund for Nature – UK, Goldalming, UK (1990 – 2006)
e-mail: janebarnwell@btinternet.com

V. Sommer and C. Ross (eds.), *Primates of Gashaka*, Developments in Primatology: Progress and Prospects 35, DOI 10.1007/978-1-4419-7403-7_2,

As elsewhere on this globe, exploitation of this wilderness gathered speed, however slow, until, from the mid-20th century onwards, its very survival as a harbour for wildlife came under threat. In 1966, a land systems survey of this mountainous countryside was commissioned by the Government of Northern Nigeria, resulting in the publication of a thorough study by Mike Bawden and Paul Tuley (Bawden & Tuley 1966). This work drew attention to the urgent need to protect the watersheds in these forested mountains that supplied the River Taraba, one of the largest tributaries of the River Benue, with a reliable, year-round supply of water. In 1969 another study by a wildlife expert from the Food and Agriculture Organization (FAO) of the United Nations recommended that a game reserve or national park be created to protect the local wildlife resources. These reports were commissioned at a time when the management of Nigeria's natural resources was high on the agenda of the recently independent nation and significant financial resources were being invested in a range of conservation programmes throughout the country. In 1970 the Forestry Department of what was then the Government of the North-East State began the task of putting the findings of these two studies into action. Alan Fox, a forestry officer with considerable experience of the adjoining Mambilla Plateau, started the arduous job of trekking on foot around the entire area, meeting the local people and discussing the options for conservation. He spoke both Hausa and Fulani, he was a skillful negotiator and he was greatly respected by the local people. He was also a prodigious, indefatigable walker. Many years later, one could still meet elderly villagers who remembered working with "Mista Fok"! Indeed, Alan Fox laid the foundations for the future national park.

In 1972, the Government of the North-East State, in collaboration with the local governments and traditional chiefs of Ganye and Gashaka Gumti divisions, created three adjoining game sanctuaries (Serti, Gashaka and Gumti) in which hunting was no longer to be permitted. The agreement of the traditional chiefs to the creation of the sanctuaries was a fundamental requirement for the success of the process. In those days it was the voice of the traditional chief (the Lamido), supported by the sub-chief (the Jauro) and the village headman (the Mai'ungwa), which most accurately reflected the opinions of the local people. This was traditional democracy at work, with all the people, both men and women, having immediate access to a well-understood system of grassroots local government. It was then the responsibility of this traditional structure to interface with the local and state government officials in the implementation of development services, law and order etc. The responsibility for establishing and managing these sanctuaries was given to the newly created Wildlife Unit, part of the Forestry Division, Ministry of Agriculture and Natural resources, with headquarters in Yola and a base camp in the quiet, dusty town of Serti.

"Mista Fok" initially sent three key people, two Nigerians and one Canadian to Serti to initiate the development of the three sanctuaries. Peter Bubur Thiliza was a young forestry assistant and Mallam Musa Gassol (a Nigerian ex-army officer) was a game assistant. The assistant game warden was Barry Snider, a volunteer with CUSO (Canadian University Services Overseas). This small team set out to learn what was in this vast, road-less wilderness, walking many miles through the mountains, staying in remote villages and spreading the conservation message. The first 12 game guards were recruited from the local villages. They were all ex-hunters and they were the people who really knew this rough and rugged country, its people and its wildlife and, most importantly, they knew how to move through it in safety. They were all well renowned throughout the area and their reputations as respected hunters immediately made them effective "educators" and "ambassadors", who were listened to by the villagers. One of these guards, Hamajidda Gashaka, was a key individual from a long line of hunters from the village of Gashaka. His elder brother Ahmadu was a village headman and a renowned hunter, his three other brothers were all hunters and their father and grandfather had both been elephant hunters. Hamajidda guided the team over the mountains, across the rivers and through the forested valleys, remembering invisible paths that he had followed with his father many years before.

Work was started to build a base camp at Serti and to construct a road through the bush to Gashaka village. By September 1972 the initial team had been joined by three more assistant game wardens. Dick Menefy and Al Soziak were Canadian wildlife ecologists sent out by CUSO. They initiated the first surveys of the sanctuaries' wildlife and one of their early findings was that chimpanzees were present in the area.

This is where I became part of the story. I was posted to Nigeria through VSO (Voluntary Service Overseas), a UK-based international development charity. I had initially been selected for the placement, because I was a wildlife ecologist. However, since the new construction work was considered to be a top priority and I had experience of working in the building trade during my university vacations, I was given the job of resident engineer. In fact, to this date I am known throughout the area as the "enginiya".

Little did I know that this initial placement would have deep impacts on my personal life, too. I had just constructed a house at Gashaka, not in the traditional style of using mud-bricks, but from solid stone (which, four decades later, is still referred to as "the stone house"). A reason to celebrate, surely – all the more, as Christmas was approaching. Invitations went out to other VSOs in the wider area of north-eastern Nigeria. It was on this occasion, fateful in hindsight, that I met my future wife. From Gashaka, I took her horse-back riding to the glorious highlands, and this certainly helped to bring some fortunate development about. With Jane and our soon-to-be-born son, I share many happy memories of this very special place.

A local contractor called Sarkin Tasha – "the king of the lorry park" – put up the first buildings in the base camp at Serti. His work was so shoddy that it was decided to carry out all further construction with direct labour, i.e., workers employed directly by the Wildlife Unit. This early decision fostered a very strong spirit of self-reliance within the people developing the sanctuaries, so that if a road needed to be built then they just got on and built it and if something needed fixing then they fixed it. To survive and succeed in a remote area like Gashaka Gumti, self-reliance was the first requirement for success. A team of very competent masons, carpenters and labourers was rapidly brought together, supervised by the excellent foreman Mohamed Sani, and some remarkably effective headmen including Yakubu Dikko,

Bako Baiboko and Garba Company (who had served with the British West Africa force in Burma during World War II). A bulldozer was available but it was so unreliable and so thirsty for fuel that it was parked up most of the time. Consequently, much of the road building was done by manual labour. All the workers were very tough local men who thought nothing of living in the bush for days on end, excavating the hard laterite soil and stumping out the trees. Within a short time, the Wildlife Unit was the largest employer in the area and much-needed cash was being taken home to the remotest villages.

By mid-1973, Barry Snider and Al Soziak had returned to Canada and Dick Menefy had moved to the Mambilla Plateau, leaving the rest of the team to continue with the work, with me now appointed as the reserve's warden. In the same year, following negotiations between the State Government and the traditional rulers, the area covered by the three sanctuaries was formally proposed as the Gashaka Gumti Game Park Forest Reserve, covering some 6700 km^2. (For comparison: This is quite a bit larger than either of two world-famous national parks in the USA, the Everglades National Park and the Grand-Canyon National Park.) Within this area hunting and fishing were banned, forests were protected and people were not allowed to open up new farmland. It was at this early stage that an agreement was made with the traditional rulers to allow pastoralists and farmers to continue living on the high plateau in the proposed reserve, within demarcated enclaves. Life in the enclaves continued more or less unchanged, with the Fulani pastoralists raising herds of healthy cattle in these tsetse fly-free highlands. A ban on hunting and the strict protection of the montane forests were the only restrictions to land use. A few small hamlets were scattered inside the new reserve but most were soon deserted as their inhabitants voluntarily moved closer to Serti or to the Mambilla Plateau. One large community in the Sukare valley – between Gashaka village and the Gumti enclave – remained in residence, pending future negotiations with the State Government.

Over the next few years the development of the reserve continued apace. An all-weather road of about 40 km was constructed to Gashaka village where the reserve's headquarters was then sited and many miles of motorable tracks were built into and around the reserve, including the northern sector near the village of Toungo. The protection of the wildlife was a top priority, with considerable energy being expended on ensuring that all the surrounding villages and the communities within the enclaves were fully aware of the ban of hunting, fishing and the destruction of forests. However, despite the awareness campaigns, illegal exploitation of wildlife continued in many places and a number of determined poachers gained great notoriety as they strived to slip through the game guards clutches. One of most notorious villains was an old "gobirawa" hunter from Sokoto in north-western Nigeria, called Bagobri "mai yasa tara", translated as Bagobri "with the nine fingers". He had lost a finger many years before when his musket had exploded. Bagobri eluded the game guards for several years until he finally came to a gruesome end. He had wounded a buffalo with a musket shot and before he could fire his second musket, the buffalo caught him and literally shredded his body! One of his assistants brought the pieces of body back to Serti in a sack. Other hunters from

Sokoto also caused serious problems within the reserve until six of them were drowned when their canoe overturned in a raging river during the rainy season.

The reserve's force of game guards (*Fig. 2.1*), which reached a total of 80 men in 1979, successfully arrested many of the trespassers. The guards were posted in teams of three to four men in the villages around the boundary of the reserve and up in the enclaves. From these small, strategically positioned, widely dispersed bases they were able to patrol on foot into the remotest corners. The successful management of these guards required the game / forest assistants and me to regularly visit them and their patrol areas. It was difficult for the guards to fabricate their patrol activities when they knew that their senior officers were acquainted with their beat areas and were not ignorant of what was happening on the ground. The guards were relatively well paid during the 1970s and those jobs were much sought after. Those who failed to work diligently were swiftly disciplined. Most of the older game guards were ex-hunters but in later years government policy required the employment of educated school-leavers as game guards. Many good young people were subsequently employed but these new recruits were never able to match the deep wilderness lore of the older ex-hunters.

Ultimately, two officers from the Forestry Division (Faith Ananze and Stephen Gawaisa) joined the staff of the reserve in the mid-1970s and for many years contributed greatly to the development programme. A visible increase in wildlife populations throughout the reserve was recorded. At Selbe, a village in the highland enclaves, people used to gather in the market place during the early evening to watch the buffaloes grazing freely on the surrounding hillsides. Certain parts of the reserve were recognised as being more important habitats for some species of animals than

Fig. 2.1 Rangers of the newly created Gashaka Gumti Game Park Forest Reserve at Serti in 1976 (photo: RB)

other locations. Small groups of roan antelope were recorded in the northern hills of the reserve, but nowhere else. Large herds of kob were the most common animal in the open country near the village of Gumti. Small groups of buffalo were found throughout the reserve but large herds of these big beasts – 10 – 60 animals – were associated with the woodlands and forests of the Yakuba plains and around the headwaters of the Ngiti and Gamgam rivers. Hartebeest were always seen along the entrance road to Gashaka village in the area called Mayo Kpaa. Leopards, lions and wild dogs were recorded in many parts of the reserve. Chimpanzee were occasionally seen and frequently heard across a wide range of the southern sector of Gashaka Gumti. Even the crocodiles staged a come-back! The game guards associated specific, deep, dark pools in the reserve's many rivers as being the haunts of ancient crocodiles that their grandfathers had known. They insisted that these old reptiles were still alive and that they had become very clever at hiding themselves. Indeed, after a few years of protection from hunters and fishermen, very large crocodiles began to be spotted in these deep pools, where they would gorge themselves with often gigantic fish (*Fig. 2.2*), corroborating the stories of the guards and showing the effectiveness of their anti-poaching patrols. The first checklist of the reserve's rich

Fig. 2.2 Richard Barnwell with a 35-kg Nile perch, (legally) caught during the dry season of 1978 in the River Kam, one of the reserve's rivers renowned for its wealth of fish (photo: Jane Barnwell)

birdlife was compiled in the mid-1970s by wildlife officer Philip Hall, a conservationist who – as of 2009 – is still living in Nigeria and working as a conservationist. The increase in wildlife was also illustrated by the intensified damage to the crops of the villagers who lived in and around the reserve. In response to complaints, the game guards and I were obliged to shoot problem animals, including tantalus monkeys, baboons, warthogs, bush pigs and porcupines. Primatologists will be aghast to read that I became very proficient at hunting tantalus monkeys in the farmland around Gashaka village, using two dogs to corner the monkeys in a tree and a shotgun to bring them down to earth! We killed scores of these monkeys, thereby saving the maize crops of many farmers and earning their gratitude. Baboons were far more difficult to hunt. If I tried using the dogs to corner them, the big male baboons would turn on the dogs and send them to flight before I could get close enough to shoot. But when I borrowed a rifle from a friend, the tables were turned on the baboons!

Outsiders began to visit the reserve. Many of them were expatriates resident in Nigeria and looking for a quiet holiday in beautiful surroundings. Some were Nigerian guests of the Lamido of Gashaka who enjoyed coming into the reserve to see the wildlife and to witness the progress of the developments. A few self-catering chalets were constructed at the Gashaka camp, from where energetic people were escorted by the game guards on walking trips into the wild interior. Horses maintained up on the Mambilla Plateau could be used by visitors seeking to climb Nigeria's highest mountain, Chappal Waddi, the "Mountain of Death".

The all-important task of demarcating the exact boundaries of the reserve was initiated in 1977. This time-consuming task required lengthy negotiations with the Local Government officials, the traditional rulers and the local communities, to decide the approximate locations of the boundaries. In the southern Gashaka sector of the park, I led a demarcation team made up of representatives of the Local Government, traditional rulers and local communities on the exhausting process of trekking hundreds of kilometres along the exact line of the boundaries, constructing stone beacons at key points and marking these beacons on 1 : 50000 scale maps. A second demarcation team worked in the northern, Gumti sector of the park. Needless to say, lengthy discussions took place before every individual in the demarcation party was happy – or too tired to argue any further – with the position of each individual beacon! An exhausting and time-consuming task indeed! In addition to the external boundary, the boundaries of the highland enclaves inside the reserve also had to be demarcated. The demarcation task was still being carried out in 1979 and even this back-breaking work was not the final product required by a fully legal demarcation. The final product would require a survey party to carry out a detailed chain and compass survey of the boundary lines, draw an accurate map of these boundaries and replace the stone beacons with concrete pillars. This final task was never completed for the enclaves, although the entire length of the Nigerian border of the park has meanwhile been marked with beacons and through a motorable track, thanks to efforts of the outreach programme of Chester Zoo from 2005 onwards.

In mid-1979 my family and I left Nigeria and a new reserve warden took over. This man was unable to live in the remoteness of the Gashaka base camp and complained that the warden's house was haunted. He moved back to the town of Serti taking with him many of the staff. A young game assistant, Kwada Drambi, was left to hold the fort in the wilderness. The delicate network intended to protect the precious Gashaka Gumti area began to unravel again.

From Fragile Reserve to Full-Fledged Park

In the early 1980s, Nigeria's booming economy started to falter and government budgets were slashed, including the budgets for new conservation initiatives such as that at Gashaka Gumti. The intensive job of maintaining the roads was neglected, buildings were not looked after and vehicles were kept running less through technology than by the sheer willpower of their determined drivers. The game guards continued to be employed, but lacking strong leadership their patrolling was no longer effective.

It was then, in 1983 – 84, that a catastrophe hit the park: the pan-African rinderpest. Thousands of ungulates were wiped out, with herds of buffalo most severely decimated. When the disease had passed through, the surviving remnants of the wildlife were assailed by a new wave of poaching that the demoralised guards were quite incapable of confronting. As a consequence, extensive tracts of precious montane forests were destroyed by illegal farmers and the rivers were plundered for their fish – with the help of nets, dynamite and poison.

In 1986, the Nigerian Conservation Foundation (NCF) – one of the pioneering NGOs in West Africa, dedicated to the preservation of nature and community development – was commissioned by the Forestry Division of what was now Gongola State to produce a management plan for the reserve. The field work was undertaken by an American biologist, Arthur Green, who commented on the scarcity of the wildlife. The management plan never got beyond a draft document. At about the same time, a Ghanaian conservationist worked in the reserve and recommended that, for welfare reasons, the game guards should be withdrawn from their remote patrol posts and assembled in a few towns around the boundary of the reserve. The reserve management unfortunately implemented this disastrous recommendation, leaving much of Gashaka Gumti wide open to poachers, illegal farmers and illegal pastoralists. By the end of the 1980s, the reserve was well on the way to becoming yet another forgotten, failed conservation area.

While Gashaka Gumti waned towards a slow death, the World Wide Fund for Nature-UK (WWF-UK) and its Nigerian partner NCF were working with the Cross River State Government to develop a new conservation area in south-eastern Nigeria, the Cross River National Park. I had started working for WWF-UK in 1990 and on my first visit to Nigeria that year, to review the progress with the Cross River work, I was asked by the NCF to make a quick visit to Gashaka Gumti and assess the prospects for rescuing the failing reserve. I borrowed a landrover and driver from the Cross River project and made a long, hot and dusty journey to Serti, where we found Faith Ananze in the old Wildlife Unit base camp. The road into the park was no longer passable and so we hiked the 30-odd km along a traditional bush-path to Gashaka. Here, I spent one night camping in my old "stone-house", before trekking back to Serti the next day. It was somewhat strange to be literally walking back in time, but really wonderful to be speaking Hausa again with friends and colleagues from way back. After meeting the Lamido in Serti, Faith and I reached the conclusion that the State Government could be supported in bringing the reserve back from the brink. I made a second trip to Nigeria later that year and developed a 5-year project proposal. Financial support was secured from WWF-UK and one of its main funding partners, the British Government's Department for International Development (DFID). A life line had been thrown.

At the same time, the NCF proposed to the Federal Government of Nigeria that Gashaka Gumti should be designated a national park, under new legislation that was in the process of being formulated. Eleven main reasons for the designation of Gashaka Gumti National Park were identified:

- Protection of important watersheds;
- Conservation of biodiversity;
- Preservation of genetic diversity;
- Protection of rare species;
- Protection of rare montane habitats;
- Prevention of erosion;
- Stabilisation of climate;
- Protection of protein resources, both fish and wild herbivores;
- Maintenance of traditional livelihoods;
- Demonstration of successful enclave management;
- Development of ecotourism.

And at last, the new Gashaka Gumti project was launched in 1991. It was managed by the NCF, working in partnership with the Gongola State Government's Wildlife Unit. The project manager employed by the NCF was Faith Ananze (on secondment from Wildlife Unit) who knew the area and its people extremely well. He swiftly assembled a team of reliable workers including drivers, carpenters, masons, labourers, headmen and a mechanic – all the essential people without whom no field project can ever begin. Many had worked in the reserve in earlier years but had been laid off when government funding had run low. Two tractors and trailers and two landrovers were purchased and with the labour force thus mobilised, the work began to rebuild the all-weather road to Gashaka and to renovate the camps at Gashaka and Serti. Money immediately began to trickle into the villages once again through wage packets of the workers.

The protection of the reserve also required urgent attention. About 40 game guards living in the towns around the reserve (permanent staff of the State Government) were still receiving basic wages from the Wildlife Unit, although, for several years, few patrols into the reserve had been undertaken. The guards preferred the easy option of trying to catch hunters as they brought their bush-meat into the towns or travelled along the main roads in the area. The guards could not really be blamed for this lazy way of working since they had not been issued with new uniforms or equipment for many years nor had they received any financial incentives to undertake long distance patrols. After negotiations with the State Government, the management of the game guards and their anti-poaching activities were brought under the control of the NCF project manager; while they continued to receive their basic wages from the Wildlife Unit, their field allowances and fresh uniforms and equipment were provided by the new project. Patrols were sent out and poachers started to be apprehended and prosecuted in the court in Serti. The basic work of re-energising a protected area was making good progress through a comfortable partnership between the NCF / WWF project and the State Government.

In 1991 the Federal Government of Nigeria created the National Park Service (NPS). Along with 5 other conservation areas in the country, Gashaka Gumti was declared a national park under the National Parks Decree Number 36 of 1991.

The NPS appointed a general manager to the new Gashaka Gumti National Park (GGNP) and started the task of establishing a temporary park headquarters in Serti early in 1992, recruiting staff and purchasing vehicles and equipment. Jarafu Mamza, the general manager, had worked for the Forestry Division of Gongola State for many years and therefore already had some knowledge of the area, and its potential and challenges. The game guards who were still serving under the State Government were subjected to a selection process, with the successful candidates appointed as park rangers. Additional rangers and senior staff were recruited from elsewhere in the state and further afield in the country.

With the new national park authorities now taking direct control of the management of the park, the terms of reference for the NCF / WWF project (which had been de facto managing the former reserve) had to be renegotiated. This turned out to be a rather tortuous process, to which were added the not unexpected conflicts over who should have authority over what. As it is natural in an undertaking of such scale, inevitable strains appeared in the relationship between the NCF / WWF project and the general management of the national park, and the Conservator General of the NPS in Abuja, Alhaji Lawan Marguba, was frequently required to sort out problems and make decisions one way or the other.

Ultimately, the negotiations resulted in the park management taking direct responsibility for: park protection activities; political relationships with the State and Local Governments; establishing a new park headquarters at the village of Bodel near Serti; demarcating the park boundaries; resettling illegal farmers; and developing ecotourism. The NCF / WWF project was given the responsibility for: re-building the network of roads and tracks within the park; renovating the camps at Gashaka and Toungo; supporting park protection activities through the provision of equipment, training and the construction of ranger posts; initiating a biological research programme; launching an environmental education programme; initiating development activities within the communities living in and around the park; and preparing a park management plan. These remained the core activities of the NCF / WWF project for the next 15 years.

The biological research programme was initiated by the UK biologist Andrew Dunn in September 1992, when he was contracted by WWF-UK for an initial period of 3 months to carry out surveys of fauna in the forests and to lay the foundations for a comprehensive research programme. Andrew had not been to Nigeria before, but he had worked for 2 years in the forests of Liberia, thus proving that he could master the rigors of West Africa. In Gashaka he teamed up with Salamu Waziri, a renowned former hunter who had worked with me in the park during the 1970s. Salamu's extraordinary knowledge of this wild country enabled Andrew and a succession of other biologists to work safely and productively. This knowledge was the foundation of the park's research programme. Andrew's report of January 1993, on *The Large Mammals of Gashaka-Gumti National Park* (Dunn 1993) was the first formal biological report produced in the park. It set an intensive research programme in motion and it has since been followed by a long list of exciting biological and socio-economic work. Andrew Dunn worked as the WWF-UK / NCF conservation adviser in Gashaka Gumti from 1993 until 2000, contributing immensely to the development of the park through his diligent field work, his networking skills, his remarkably succinct writing skills and his inspiration to the young Nigerians working in the park.

These skills brought members of the Chester Zoo / North of England Zoological Society to Gashaka in 1994, initially in the person of its Director, Gordon MacGregor Reid, who carried out the very first survey of the rich and diverse fish life that is found in the park's many rivers. He was followed in 1995 by Nick Ellerton and Caroline Harcourt who briefly surveyed the park's primates, including the chimpanzees. Interest in these apes was picked up by one of the zoo's keepers, Steve Hogarth, who spent several months in the park establishing initial estimates of their population and their distribution. He was very popular with all the local people, who appreciated his strength and his great sense of humour. They nicknamed him "bature manga" – the big European. Sadly, for personal reasons, he had to return to the UK earlier than expected, leaving the chimpanzee work to be continued by a very tough, young American woman, Keri Foster. The faunal research work was complemented by vegetation surveys undertaken by Dr. Aderopo Akinsoji from the University of Lagos. The universities at Maiduguri and Yola were also engaged in the research programme.

At the same time, a number of masters (MSc) students from the Department of Anthropology at University College London (UCL), undertook socio-economic surveys in the highland enclaves, following earlier contacts that Andrew Dunn had made with the Department of Anthropology at UCL. Incidentally, Volker Sommer, a newly appointed Professor of Evolutionary Anthropology with a particular interest in primate socioecology at the same department at UCL, learned about the thriving chimpanzee population around Gashaka. He made an initial visit to the park in 1999. This swiftly led to his development of a proposal to start a primate research project in Gashaka Gumti. The first UK students interested in primates duly arrived in Nigeria to launch the *Gashaka Primate Project* (GPP) at the beginning of 2000. Since then, GPP has gone from strength to strength. Working under a mandate from the National Park Service, with financial support from the Chester Zoo and other donors and with logistical assistance from the NCF / WWF project, it has established a wide-ranging and exciting research programme and contributed greatly to the protection and development of the park.

In addition to initiating and guiding the NCF / WWF research programme, Andrew Dunn wrote the park's first, very attractive guide book, illustrated by a Lagos-based artist, George Ejebare (Dunn 1999). Andrew was also commissioned to write the park's first management plan. This was a prodigious undertaking that required extensive consultations with individuals and communities within and surrounding the park, with government officials, with the National Park Service and the management of Gashaka Gumti (NPS / NCF / WWF 1998). All existing literature about the park and the vicinity was carefully reviewed and the 5-year plan produced, covering the period 1998 – 2002, intended to serve as the baseline for the future planning of the park's development.

The management plan established 10 main objectives for the park's effective management:

- Protect the forested water catchment area of the River Taraba;
- Maintain the diversity of ecosystems, species, genetic varieties and ecological processes;
- Manage enclave areas through conservation partnerships formed with local people designed to safeguard long-term conservation goals and traditional livelihoods;
- Promote sustainable rural development in a support zone surrounding the national park intended to improve the living standard of local people;
- Encourage sustainable land-use practices within the enclaves and in a support zone surrounding the park boundary;
- Promote ecotourism in an area of considerable untapped tourism potential;
- Develop an environmental education and interpretation programme;
- Conserve all cultural, archaeological and historical features of special interest;
- Encourage all forms of scientific research that are compatible with the park's management objectives;
- Encourage the involvement of NGOs in the protection and development of the national park.

More than a decade on, these main objectives are still serving to guide the management of the park, with a range of partners working under the overall control of the park authorities in Bodel and the NPS in Abuja.

Future Challenges

The park has slowly but steadily "grown up" with a succession of leaders posted to this site by the NPS – first as "General Manager", then as "Director". Both Jarafu Ulam Mamza (1991 – 2003) and Modu Sherif (2003 – 2006) have left their mark in taking the management and infrastructural development of the reserve forward. Currently, the park is under the leadership of Dr. Okeyoyin Agboola who now carries the title "Conservator of Park". Because his name is rather difficult to master, even for Nigerians, he prefers to be known as "Mr. George". He is all the more suitable for this appointment, as he was one of first federal officers to work in the park, a hardworking, dedicated conservationist who, by trekking all over the hills, valleys and forests, had developed a genuine interest in this wilderness already decades ago.

The future development of Gashaka Gumti will require the park management to ensure that the park is effectively protected. Protection has to be the top priority, a very difficult task to undertake in mountainous terrain. The rangers will need to be professionally led in the field by their officers, properly taught and adequately equipped. They must be fully armed and trained as a paramilitary force. Anything short of this basic requirement will render them highly vulnerable to the increasingly well-armed poachers. In the past, illegal pilferers of wildlife were armed with spears, bows and arrows and locally made muzzle-loaders. These days, poachers in Nigeria are using shotguns, hunting rifles and in some cases, automatic military rifles and they have no qualms about using these weapons on the rangers if they are challenged. Indeed, two park rangers were killed by poachers as recently as July 2008 in the Gumti area. A new quarterly newsletter of the national park, *The Chimp*, describes the depressing incident in its September 2008 issue as follows: "On that fateful day, the Park Rangers who were six in number all armed with gun except one, came across the camp of the poachers, arrested the porter and handcuffed him. While the Park Rangers were still deciding on the possible area of ambush, the poachers emerged from the bush and immediately opened fire on the team. In the process, two Rangers were killed while one was severely wounded."

An additional and relatively new threat are the armed robbers who have operated in Gashaka Gumti in recent years and have become a serious threat to the local people, park rangers, researchers and tourists. These robbers have sometimes been associated with poachers, and it is difficult to distinguish one group of criminals from another. Sadly enough, in early 2007, robbers stopped my friend Salamu Waziri when he rode on a motorcycle back to his village of Gashaka. When Salamu resisted the threats and demands, he was shot and bled to death – in the bush he had loved and respected so much.

Given these circumstances, it has become unjustifiable and irresponsible to send poorly armed rangers out to confront such well-armed law-breakers. Throughout most of Africa's parks and reserves, rangers have had to be increasingly better armed and trained in order to apprehend increasingly determined and violent poachers and criminals. I have worked with park rangers right across Africa, from Nigeria to Cameroon, Kenya, Rwanda and Tanzania and I have no doubt that the Nigerian rangers are the most poorly armed. Without doubt, it is high time for Nigeria's park rangers to be properly equipped as an effective paramilitary force.

Rangers should be stationed in and around the park in secure, strategically positioned, well maintained ranger posts from where they will undertake regular patrols that will cover their areas of responsibility. A number of posts have been built in recent years, but more will be needed in the future. The park management will need to maintain the access roads and motorable tracks within and around the park, but the rugged nature of the terrain means that the bulk of rangers' work will always have to be carried out on foot. Day-to-day protection measures will be enhanced by the regular maintenance of the park boundaries, an exercise in that the *Gashaka Primate Project* and Chester Zoo have recently assisted the park management. There is also need to keep up the extensive infrastructure of offices, housing, workshops, radio-communications network and tourist facilities. A working fleet of vehicles is of paramount importance, which in turn necessitates a properly equipped workshop and a team of professional mechanics led by an experienced engineer.

Illegal settlers continue to reside within the park, particularly along the southern boundary with the Mambilla Plateau. These trespassers destroy valuable forests and hunt illegally within the park. Unlawful settlers should be swiftly evicted from the park, a difficult task that will require full political support from the State and Local Governments and the traditional rulers. The situation of the villages in the Sukare valley needs to be urgently sorted out. They were supposed to have been resettled outside the park many years ago, but it would appear that the exercise failed so far.

Of critical importance is the status of the enclaves. They lie within the heart of the park, and any breakdown in their management affects the conservation of the entire area. Why were these enclaves established? Over the years, I have often been asked this question by a succession of park managers, NGOs and researchers. The answer is that, back in 1971–1973, when first protection plans were drafted, the traditional rulers did not agree to the Fulani pastoralists being moved from the highlands. The Fulani and their valuable cattle were too important to the local economy. Thus, despite other opinions, the eviction of the Fulani was never part of the agenda! The only alternative to the establishment of the enclaves would have been to completely excise the cattle-rearing highlands from the conservation area, thus creating a game park-forest reserve that did not include the highlands within its boundaries. This alternative would have been far worse for conservation than the enclaves have proved to be, despite all their problems. The establishment of the enclaves within the overall conservation area gave the park management a degree of authority and control inside the enclaves. If the highlands had been left outside the conservation area, then the management would have had no authority at all within these highlands, whereas now, hunting and habitat destruction are illegal in the enclaves. But, this authority has to be effectively exercised. During the 1970s, game guards stationed within the enclave communities undertook constant anti-poaching patrols and successfully arrested numerous hunters. The impact of this work was plain for all to see – plenty of visible wildlife! However, during the past two decades, there has not been enough visible presence of the management authorities in the enclaves and so the inhabitants have been able to operate under virtually no controls. They should be encouraged and supported to work as partners in conservation, while people who persistently carry out illegal activities should be required to leave the area. Political support will again be required.

Gashaka Gumti has great tourism potential, with its wonderful sceneries, its rich biodiversity and its extensive tracts of wild, unspoiled wilderness. Small numbers of intrepid visitors have ventured into this precious landscape since the 1970s and have enjoyed exciting experiences. The park has a limited network of motorable tracks that permit people to view the wildlife and the habitats from the comfort of a vehicle, in the classic "safari" style of wildlife tourism. However, Gashaka Gumti needs to be explored on foot in order to see the best of the park and to have an authentic wilderness experience. There is an extensive network of trails and footpaths through the reserve and some of the park's personnel and the local people have experience of guiding visitors through the forests, across the rivers and up onto the high mountains. It is a very beautiful place, as beautiful as any other conservation area that I have seen in Africa, and it richly deserves to be shown to a wider audience. However, Nigeria ranks very low on the priority lists of most tourists, even the more adventurous ones, and a great improvement in Nigeria's international image will be required before globe-trotting tourists will regularly venture to its remaining natural treasures. In the meantime, the focus should be on attracting the domestic tourism that already exists within the nation, with Gashaka Gumti's unique assets being brought to the attention of Nigerians themselves.

Environmental education and community development work within the support zone around the park can and should be undertaken primarily by NGOs with proven experience of this type of work. Good contacts were established between the park and local schools many years ago, through the assistance of the NCF / WWF project. School visits into the forests and savannah-woodlands were organised, environmental literature was distributed and pupils were engaged in essay-writing competitions. These foundations need to be further developed. The park management would be a partner in this work, although its core responsibilities should be the activities within the park.

Moreover, the park authorities need to develop their own research expertise and retain an overview of the long-term research requirements of Gashaka Gumti. Nevertheless, specific research programmes should continue to be undertaken by partner organisations with greater experience of this type of work, such as University College London and Nigerian universities. The *Gashaka Primate Project* should be encouraged to continue its valuable programmes for many more years into the future, to expand these programmes as necessary and to further develop its role as the senior research organisation in the park – with the ultimate goal of creating a Gashaka Gumti research institute. The park's unique combination of diverse habitats makes it one of the most exciting places in Africa for enthusiastic researchers. A most important lesson to learn in African conservation is that the parks with active research programmes have always managed to survive the shocks of collapsing budgets, management crises, poaching onslaughts and vanishing tourists. Active research programmes therefore keep conservation areas on the agenda and in the headlines.

Bernhard and Michael Grzimek's research work and their famous Oscar-winning film and book *Serengeti Shall Not Die* (Grzimek & Grzimek 1959) placed Tanzania's Serengeti National Park at the top of Africa's conservation agenda, where it remains to this day, as one of the wildlife wonders of the world. On the other side of Africa, Gashaka Gumti National Park also has its own unique qualities. This marvel, too, deserves to be retained firmly on an international conservation agenda – and never allowed to die!

Literature

Bawden, M. G. & Tuley, P. (1966). *The Land Resources of Southern Sardauna and Southern Adamawa Provinces, Northern Nigeria* (Land Resource Study No. 2). Tolworth, Surrey: Directorate of Overseas Surveys.

Dunn, A. (1993). *The Large Mammals of Gashaka-Gumti National Park, Nigeria: Line Transect Surveys from Forest and Savannah.* Report to WWF-UK, NCF & Federal Ministry of Agriculture, Water Resources and Rural Development / Nigeria.

Dunn, A. (1999). *Gashaka Gumti National Park: a Guidebook.* Lagos: Gashaka Gumti National Park, NCF / WWF-UK.

Grzimek, B. & Grzimek, M. (1959). *Serengeti darf nicht sterben. 367.000 Tiere suchen einen Staat.* ("Serengeti shall not die"). Frankfurt: Ullstein.

NPS / NCF / WWF (1998). *Gashaka Gumti National Park: A Management Plan for Developing the Park and its Support Zone.* Nigeria National Park Service, Nigerian Conservation Foundation, World Wide Fund for Nature (UK).

Chapter 3
Hunters, Fire, Cattle: Conservation Challenges in Eastern Nigeria, with Special Reference to Chimpanzees

Jeremiah Adanu, Volker Sommer, and Andrew Fowler

Abstract The historic range of the recently recognised fourth subspecies of chimpanzee, *Pan troglodytes vellerosus*, extended roughly between the Niger River in Nigeria and the Sanaga River in Cameroon. Most original habitat is lost, but a stronghold remains in the remote Taraba region of north-eastern Nigeria, in particular Gashaka Gumti National Park (GGNP). We explored the region's conservation prospects for large mammals, and in particular chimpanzees, through foot surveys, for a total of 59 days. No evidence for chimpanzees was found in the Shebshi Mountains, which represent the potential northern distribution, but it is unclear if their historic range ever extended so far. The Cameroon border area and the Fali mountains that demarcate the southern distribution, and the vicinity of GGNP including the Mambilla Plateau, have experienced considerable deterioration of habitat. Surviving forests are often practically devoid of large mammals, as poaching and hunting are rampant, while patrolling and protective measures are all but absent. Forest destruction is progressive due to fire damage from seasonal bush burning and cattle grazing. It is thus unlikely that large chimpanzee populations survive outside GGNP. Whatever populations of chimpanzees may have been found there in the past have now either disappeared, are on the brink of extinction or reduced to small remnant groups. On the more positive side is the realisation that GGNP seems to fulfil its elementary function of providing a haven for threatened wildlife, including charismatic megafauna such as the chimpanzee. A contiguous population of about 1000 chimpanzees survives in the Gashaka sector of the park, in the axis Gashaka – Kwano – Yakuba – Chappal Wade (Nigeria's highest peak). Nevertheless, it is the remoteness and large extent of the park, rather than effective protection measures, that have so far ensured the existence of the apes. In fact, the park's chimpanzee habitat is also shrinking, particularly in and around the enclaves of settled pastoralists, due to cattle grazing, bush-burning, hunting and a lack of patrols. Community-based conservation approaches seem to have made little progress, and protective measures

A. Fowler (✉)
Max-Planck-Institute for Evolutionary Anthropology, Leipzig 04103, Germany
e-mail: andrew_fowler@eva.mpg.de

V. Sommer and C. Ross (eds.), *Primates of Gashaka*, Developments in Primatology: Progress and Prospects 35, DOI 10.1007/978-1-4419-7403-7_3,

emanating from research activities are too localised. There is an urgent need for law enforcement, at least in the medium term, as conservation measures will otherwise be meaningless, given that little would be left to protect.

Keywords Conservation • Primates • Bush-meat • Nigeria

Introduction

Chimpanzees are the closest living relatives of humans and are found from Senegal in West Africa to Angola south of the equator and throughout Central Africa up to the East African Rift Valley (reviews in Goodall 1986, Heltne & Marquardt 1989, Wrangham *et al.* 1994, Boesch & Boesch-Achermann 2000, Reynolds 2005; see also Sommer *et al.* this volume [Ch. 12]). However, the survival of these apes – as well as that of many other non-human primate species – is increasingly threatened by habitat destruction, disease, civil strife and a trade in bush-meat (reviews in Ammann *et al.* 2003, Peterson & Ammann 2003, Kormos *et al.* 2003, Caldecott & Miles 2005; see also Hughes *et al.* this volume [Ch. 14]).

It is believed that concerted efforts by governments, researchers, conservation agencies and local communities are needed to prevent further demise, although the effectiveness of such measures is disputed (Oates 1994). In any case, conservation efforts demand knowledge of basic distribution and population densities as well as of the land-use characteristics of human stakeholders in primate habitat, including an understanding of socio-economics, cultural practice and local attitudes towards nature preservation (Poulsen & Clark 2004, Redmond 2005). With such information to hand, it is, for example, possible to decide on whether to attempt to conserve species across their range, or to focus efforts on bolstering those areas where they are already legally protected (SGA / IUCN / PSG 2005).

Recent genetic studies based on hairs sampled from nests built by wild chimpanzees have revealed the existence of a fourth subspecies of chimpanzee: *Pan troglodytes vellerosus* (Gray 1862) (*Fig. 3.1*), the "Nigerian chimpanzee" (Gonder *et al.* 1997, Kormos *et al.* 2003) or "Nigeria-Cameroon chimpanzee" (Inskipp 2005), which is found in Nigeria and adjacent Cameroon. (Note: The scientific name might need revision to *Pan troglodytes ellioti* [Matschie 1914]; Oates *et al.* 2008).

Numbers of *P. t. vellerosus* surviving in the wild are estimated to be 3000-5000 in Cameroon and 2000 – 3000 in Nigeria (Kormos *et al.* 2003, Ngalla *et al.* 2005, Inskipp 2005, McManus 2005, Forboseh *et al.* 2007), although such figures have to be treated with caution and may well be overly optimistic (see Hughes *et al.* this volume [Ch. 14]). The historic range of *P. t. vellerosus* (*Fig. 3.2*) in Cameroon is believed to extend up to the river Sanaga (Ngalla *et al.* 2005), whereas the range extent is unclear on the Nigerian side. We assume here that it is demarcated by the Benue River and Niger River, although others maintain that these streams have not been a substantial boundary to gene exchange (Gonder *et al.* 1997, McManus

Fig. 3.1 *Pan troglodytes vellerosus*, fourth subspecies of chimpanzees: Mother and infant in Gashaka Gumti National Park (photo: AF)

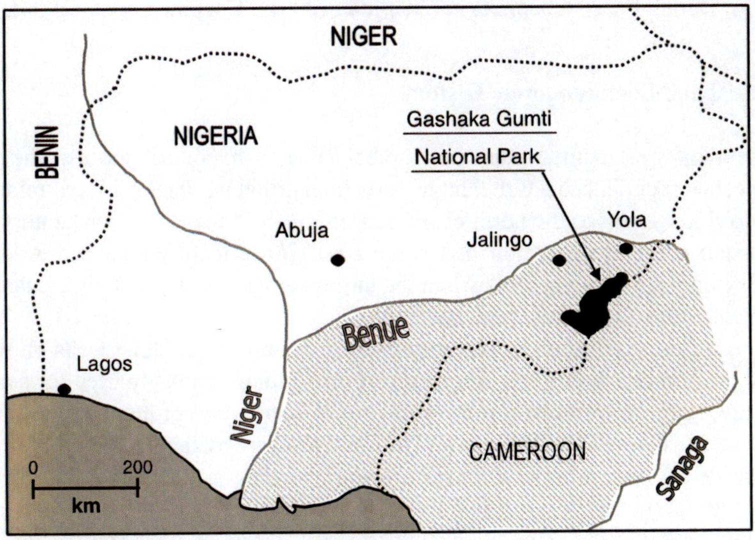

Fig. 3.2 Approximate historical distribution of *P. t. vellerosus*, shown as a hatched area

2005). Small populations (which may belong to *P. t. vellerosus*, or to the Western chimpanzee, *P. t. verus*) survive in Nigeria west of the Niger / Benue divide. Larger populations are believed to survive in the Oban and Okwango division of Cross River National Park and Afi River Forest Reserve (McManus 2005). The stronghold of *P. t. vellerosus* is assumed to lie in the mountains of the southern section of Taraba State in north-eastern Nigeria, in particular the large *Gashaka Gumti National Park* (GGNP; Dunn 1999). This region thus represents a high-priority population for conservation efforts by SGA / IUCN / PSG (2005).

Long-term research on chimpanzees is conducted by the *Gashaka Primate Project* (GPP) in the central section of GGNP (Sommer *et al.* 2004; Sommer & Ross this volume [Ch. 1]). As part of this study, we surveyed areas of historic, current or potential chimpanzee habitat in the wider Taraba area. With this, we aimed to (a) gain a better understanding of the current conservation situation against the backdrop of anthropogenic influence and (b) assess the potential of current and future protection measures in southern Taraba State.

Materials and Methods

The Taraba Region

Data were collected in eastern Nigeria in the states of Taraba and Adamawa that border Cameroon. GGNP stretches over large parts of both states. The following brief account of the study region's political and natural history paraphrases the works and summaries found in Bawden & Tuley (1966), NPS / NCF / WWF (1998), Dunn (1999), Chapman & Chapman (2001), Chapman *et al.* (2004).

Political and Socioeconomic History

The Taraba area, named after the Taraba River, a major tributary to the larger Benue, has experienced a turbulent history. Inter-tribal warfare and slave-raiding by powerful kingdoms to the north characterised much of the early 19th century, until European colonial expansion had some pacifying effect. Taraba and Adamawa States came into being in 1991 when the larger Gongola state, initially created after independence from British rule, was split.

The region is relatively sparsely populated. Remnants of stone walls on remote hill tops, grinding stones and fragments of clay tubing, probably used for smelting iron ore, suggests a formerly more widespread distribution of human communities. It is likely that human settlements and movements were markedly affected by violent upheavals and insect-borne diseases, notably sleeping sickness (trypanosomiasis), elephantiasis (filariasis), river-blindness (onchocerciasis) and, above all, malaria.

The Gashaka area in particular experienced waves of immigration by groups such as the Jukun, Chamba, Bata and Fulani. The Jukun were powerful pagans, with influence up to Kano for several centuries. Their power was eventually curbed by the spread of the Fulani. Another wave involved the Chamba, who were driven out of what is now Cameroon by the Bata. The Bata themselves invaded the area later, until they were gradually displaced westwards by the Islamic Fulani. These pastoralists first entered Nigeria around the year 1300. Their numbers became substantial in the 1800s with the declaration of a Jihad. This Holy War, largely fought by horse power, conquered many of local pagan tribes, although those in the hills avoided subjugation. Finally, a sub-Emirate was established over the region in the name of the Emir of Yola (now in Adamawa State), spreading to the Mambilla Plateau and Baissa, of which Gashaka became the headquarters. Germany invaded in the late 1800s, but first limited her activities to expeditions. Consolidated administration was not enforced until the eve of World War I when the historical link with Yola was severed, and Gashaka made a separate division. The Germans created a network of roads and footpaths, several of which remain in use to date. After their defeat, the area was mandated to the British. They returned control to the Yola Emirate, with Gashaka established as the region's administrative centre. By 1940, the area had been divided into two chiefdoms, Gashaka (administered from Serti) and Mambilla (administered from Gembu), which were ruled by descendants of the former Gashaka village chiefs. The current Emir or Lamdo, who has ruled since 1965, also holds the office of Chairman of GGNP that was declared in 1991.

Road-building has been undertaken since the 1960s, most importantly the Beli – Serti – Gembu road. It connects the state capital Jalingo with the Taraba valley, from where it branches into the Serti plain before ascending to the Mambilla highlands. As a result, by 1978, while the population of the Gashaka region generally had remained static at around 1 person / km^2, that of Serti had increased as agriculturally based settlements were attracted. Agriculture is still the economic basis for 80 % of local inhabitants. Maize and guinea-corn represent the most common cereals, with some rice and yam cultivated. Avocado, banana, cola and mango are the most important fruits. Fertile land is abundant, but a lack of access and transportation means that most agriculture remains subsistence oriented.

Ethnic and linguistic diversity typify the modern villages, where Fulani is used as the primary lingua franca but Hausa and English are the media of instruction in schools. English is rarely spoken by women, and few girls attend school beyond primary level. Educational and medical resources in villages are generally basic at best.

Islam, brought by the Fulani, first displaced the local animist religions, and Christianity became another powerful force with the European colonial expansion. Larger towns, such as Serti, tend to be a mixture of both Christian and Muslim, with some elements of animism surviving.

Physical Environment and Habitat Types

Taraba comprises a varied physical environment, including plains at 120 – 450 m above sea level and escarpments above 2000 m. Physiographic regions consist of the flat plains of the Benue valley to the north-west, and the Adamawa highlands to the south and east. The highland region comprises the Alantika Mountains in the north-east and the Shebshi Mountains in the centre. In the south lie the Gotel Mountains, which include Chappal Wade, Nigeria's highest elevation, and the Mambilla Plateau, approximately 3000 km^2 of predominantly open grassland.

Soils are heterogeneous, based on geological structures from the Lower Palaeozoic to Pre-Cambrian era, and tend to be shallow, stony skeletal, with a high iron content. Boulders and rock exposures intersperse the area. Soils derived from volcanic rocks tend toward higher nutrient status and richness, which affects natural vegetation growth, as well as suitability for agriculture. The region is highly susceptible to erosion after removal of natural vegetation, over-grazing and the effects of fire.

Main vegetation types include: savannah-woodland (*Fig. 3.3a*); lowland rain forest (*Fig. 3.3b, c*); grassland and montane grassland (*Fig. 3.3d*); and remnants of montane rain forest. The year is divided into the dry season from about December to March, and the wet season from about April to November (*Fig. 3.4*). Rainfall approximates an average of 1200 mm in the north and 3000 mm in the south of Taraba.

Fig. 3.3 Chimpanzee habitat in Taraba State / Nigeria. (**a**) Savannah-woodland in GGNP at the onset of the rainy season (photo: James Higham). (**b**) Lowland rain forest in the forest reserve of Zedah, Cameroonian Border – an "empty" forest, as indicated by the lack of animal tracks in the river bed. (**c**) Intact lowland rain forest in the Gamgam headwater area, GGNP, leading up to the Chappal Wade escarpment in the background. (**d**) Hillsides degraded by cattle grazing and burning near the enclave of Mayo Sabere in GGNP (photos: VS)

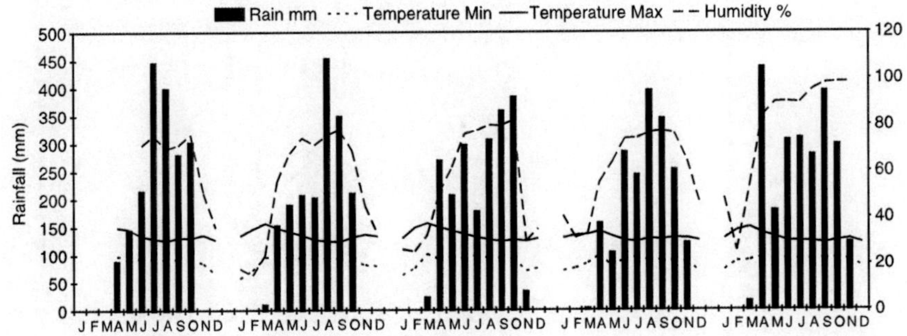

Fig. 3.4 Alternation of wet and dry season in Taraba (based on data collected 2001 – 2005 at GPP station at Gashaka, GGNP)

Gashaka Gumti National Park

GGNP is, at about 6700 km^2, the largest of the 7 national parks currently existing in Nigeria, and is divided into the flat Gumti sector in the north and the hilly and mountainous Gashaka sector in the south. Habitat types represent the diversity of Taraba as a whole, including Guinea savannah-woodland, riverine and gallery forest, lowland rain forest, montane forest and montane grassland (Akinsoji 1996). The park is at the northern edge of Africa's Gulf of Guinea forests, considered a hotspot of biodiversity (Oates *et al.* 2004). The composition of savannah-woodland has been greatly altered by the yearly burning of grasses to facilitate cattle grazing and clearing of forests for agriculture. Montane forests have been even more affected, as it was here that the most extensive grasslands were created. The national park, within its boundaries, contains several high-altitude enclaves occupied by Fulani who with their cattle escaped the disease carrying tse-tse flies prevalent in low lying regions (Bennett & Ross this volume [Ch. 6]).

Surveys

Habitats in Taraba State support a wide variety of large animals, as exemplified by the portfolio of taxa encountered in the vicinity of the GPP field station at Gashaka-Kwano (*Tab. 3.1*). This includes mammals such as carnivores (African civets, golden cats, leopards, hyenas, lions, wild dogs), ungulates (red river hogs, giant forest hogs, African buffalo, bushbuck, red-flanked duiker, yellow-backed duiker, waterbuck), rodents (crested porcupine), and aardvarks (Dunn 1999). The area is also well known for its diurnal primates including chimpanzees, baboons, putty-nosed monkeys, mona monkeys, tantalus monkeys, black-and-white colobus monkeys and patas monkeys. Gorillas have been rumoured to occur, too (White 1990). No evidence has so far been discovered to support this claim, which, however, is not completely unreasonable, as remnant populations of *Gorilla gorilla diehli* are found not so far to the south-west (Caldecott & Miles 2005).

Table 3.1 Large animals in the Gashaka-Kwano area of Gashaka Gumti National Park (2000 – 2007)

Taxon	Common name	Latin name	Hausa name	Sightings (a)
Primates	Nigerian chimpanzee	*Pan troglodytes vellerosus*	Biri mai ganga	C
	Olive baboon	*Papio anubis*	Gogo	C
	Putty-nosed monkey	*Cercopithecus nictitans*	Bakinbiri	C
	Mona monkey	*Cercopithecus mona*	Gimchiki	C
	Tantalus monkey	*Cercopithecus aethiops*	Kirikaa	O
	Patas monkey	*Erythrocebus patas*	Jambiri	(b)
	Guereza, black-and-white colobus	*Colobus guereza*	Biri mai roro	C
	Grey-cheeked mangabey	*Lophocebus albigena*		R (c)
	Potto	*Perodicticus potto*		R
Pangolins	Tree pangolin	*Phataginus tricuspis*	Dankunya	R
Rodents	Derby's flying-squirrel	*Anomalurus derbianus*	Kurege	R
	Gambian sun-squirrel	*Heliosciurus gambianus*		C
	Brush-tailed porcupine	*Atherurus africanus*	Beguwa	O
	Crested porcupine	*Hystrix cristata*		O
	Gambian giant rat	*Cricetomys gambianus*		Probably present
Carnivores	Spot-necked otter	*Lutra maculicollis*	Karenruwa	R
	African civet	*Civetticus civetta*	Tunkun juda	O
	Leopard	*Panthera pardus*	Damisa	R (also heard, scratch marks)
	Golden cat	*Felis aurata*	Kawondamisa	R
	Wild dog (b)	*Lycaon pictus*	Kerkeshi/Karendaji	R
	Lion (b)	*Panthera leo*	Zaki	R (seen at Yakuba, 15 km away)
	Spotted hyena (b)	*Crocuta crocuta*	Kura	R (heard at Selbe, 15 km away)
Aardvarks	Aardvark	*Orycteropus afer*	Dubganya	C (burrows)

Ungulates	Rock hyrax	*Procavia ruficeps*	Agwada	O
	Red river hog	*Potamochoerus porcus*	Jan alde	O
	Warthog	*Phacochoerus africanus*	Alden daji/Alhanzir	O
	Giant forest hog	*Hylochoerus meinertzhageni*	Bakin alde	O
	African buffalo	*Syncerus caffer*	Bauna	O
	Bushbuck	*Tragelaphus scriptus*	Mazo	O
	Red-flanked duiker	*Cephalophus rufilatus*	Makurna	O
	Yellow-backed duiker	*Cephalophus silvicultor*	Boka	O
	Waterbuck	*Kobus ellipsiprymnus*	Gwombaza	O
	Hartebeest	*Alcelaphus buselaphus*	Kanki	O
Reptiles	Nile monitor lizard	*Varanus niloticus*	Guza	R
	Nile crocodile (b)	*Crocodylus niloticus*	Kada	R
Birds	(d)			

(a) Actual sightings by fieldworkers only; C = common, O = occasional, R = rare
(b) Not at Kwano but in surrounding area (< 20 km distance)
(c) Since 2005 (K. Arnold pers. comm.; V. Sommer pers. obs.)
(d) 431+ species identified in Gashaka Gumti National Park>

We conducted foot surveys and utilised information from previous reports and interviews with locals and rangers to assess the conservation status of wildlife in southern Taraba. Surveys drew on a pool of Nigerian and non-Nigerian personnel affiliated with GPP, the Nigerian National Park Service and the Nigerian Conservation Foundation (NCF), all based in the Gashaka area. Surveys covered areas considered of strategic importance with respect to chimpanzee distribution, including sections of the Cameroon border area, the park vicinity, and GGNP itself. The Shebshi mountains in northern Taraba / southern Adamawa were also visited as they might, at least in historic times, have constituted the northern-most distribution of the apes.

Records of wildlife were based on assessments of the presence or absence of certain taxa. For chimpanzees, we noted sightings, calls and nests. We did not attempt to estimate densities as figures derived from nest counts along transects, decay rates of nests and nest group sizes (e.g., Wrogemann 1992) are often questionable. One main problem is the mosaic nature of the habitat, which makes extrapolation from one patch of habitat to another unreliable. Still, our surveys are likely to indicate whether or not a given area contains a healthy population of chimpanzees. Similarly, we recorded sightings, tracks, dung, calls and animal shelters, which indicate the presence of other large mammals (e.g., monkeys, ungulates, carnivores). Anthropogenic activities such as hunting (wire traps, discarded ammunition, remnants of on-site butchering, people carrying guns), burning, cattle-grazing and settlements were likewise noted.

Surveys were restricted to the dry seasons (Oct – Apr) of 2002 – 2007. These expeditions often covered difficult and remote terrain and were subject to idiosyncratic influences such as changing weather conditions, available expertise, transport, manpower, fatigue and illness. Detailed narratives, as provided in the *Appendix*, are intended to relay a flavour of logistic challenges, including local attitudes, as such factors strongly impact on conservation measures, current or future.

Results

Foot surveys of potential chimpanzee habitat in the wider Taraba region consisted of 10 parts, executed on 59 days from 2002 – 2007 (*Fig. 3.5*).

Potential Northern Distribution: Shebshi Mountains

Dakka (4 days, 2004). The Shebshi mountains – shared between Taraba and Adamawa states – are sometimes considered to represent the northern edge of the range the Nigerian chimpanzee (McManus 2005: Map 16.15). The remote town of Dakka was chosen as the western entry gate to the Shebshi range. Here, an extensive bush-meat market was evident. Forests were reduced to burnt scrub, with small patches of bush around streams. Evidence for large wild mammals was absent. The chief of hunters in a small village had never seen chimpanzees despite a life-time of hunting. Thus, chimpanzees appear to be absent from the wider area.

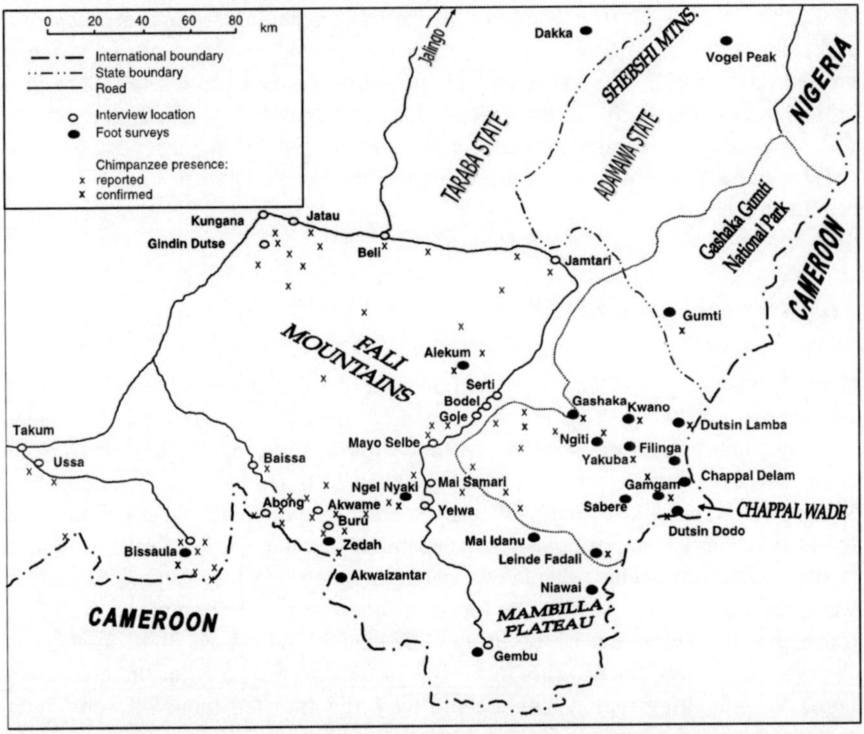

Fig. 3.5 Chimpanzee distribution in Taraba and Adamawa state: Confirmed presence (foot surveys) and reported presence (former reports; ranger reports; interviews by Nyanganji *et al.* this volume [Ch. 4])

Vogel Peak (4 days, 2004). Gurum Pawu below the Vogel Peak (1680 m; located in Adamawa state) was chosen as the eastern entry gate to the Shebshi range. Heavy cultivation and cattle grazing was evident up the slopes. No large wild animals were encountered in the rugged, hilly terrain, although the peak itself was not reached. People were neither carrying guns nor were shots heard – but there may well not have been anything there left to shoot at. The survey was short, but the presence of chimpanzees seems rather unlikely given that the local Chamba tribe sports fierce hunters.

Southern Distribution: Cameroon Border

Bissaula (4 days, 2004). Around this town near Cameroon, guns were openly carried in the forest, and local guides were ready to kill any animal found. The forest was relatively intact, although heavily burnt. No large wild mammals were seen, but a few chimpanzee nests were located.

Zedah (4 days, 2004). Human cultivation, particularly of palm trees, was extensive, although rain forest remained, particularly near rivers. Guns were openly carried in the forest. The extended beds of a larger river were virtually devoid of animal tracks (*Fig. 3.3b*). Little evidence for large mammals was found, except for the vocalisations of mona monkeys, and finds of 2 old chimpanzee nests. Thus, despite an NCF community forest project at Buru, the Zedah forest has been virtually hunted empty.

Vicinity of the National Park

Ngel Nyaki (4 days, 2002). This State Forest Reserve is an isolated 5-km^2 fragment of montane forest on the edge of the largely deforested and eroded Mambilla Plateau. The grassland around the reserve was found trampled by cattle. According to the chief ranger, the buffalo population had been halved in the last 5 years by poachers. Wire snares were removed during the survey and blood of an animal slaughtered was found. The basic composition of large mammal populations is similar to that of nearby GGNP and included ungulates, carnivores, rodents, and monkeys. Chimpanzees were heard and seen, and 38 nests found in 6 nest groups with a mean size of 6.3 (range 1 – 21). Thus, there are probably more than 30 individuals in the forest.

Alekum (4 days, 2004). This forest lies in the eastern fringe of the Fali mountains, an undulating region with steep, rocky and forested mountain tops, intersected by valleys, about a 5-h walk from the road near Serti town. The foot-path followed small rivers and passed several poor subsistence farming communities towards a pass behind which the water flows towards the centre of the Fali mountains. Men on the foot-path carried guns and also bows and arrows. Gunshots were heard. The area, until perhaps only a decade ago, must have been prime habitat for chimpanzee and other wildlife. However, the forest is now open to rampant destruction by fire, cattle grazing and hunting. Evidence for large wild mammals was consequently restricted to sightings of a civet, tracks of a bushbuck and dung of a red-flanked duiker. No monkey except one baboon was seen. No chimpanzee sightings or calls were recorded, but 18 nests of varying ages found.

Gashaka Gumti National Park (GGNP)

Chappal Wade to Mambilla Plateau (6 days, 2005). A trail leads from Gashaka-Kwano south-east through areas rich in wildlife, including forests and savannah-woodland (*Fig. 3.3a*) as well as valleys of the Ngiti and Gamgam rivers, before reaching Sabere. The surrounding hills of this park enclave have been entirely grazed-out (*Fig. 3.3d*), with few forested patches remaining, often comprising non-native species. This condition prevailed for 2 days of hiking to the Chappal Wade escarpment, which peaks at 2420 m. Here, huts of cattle-tenders abound, as do large herds of cows. The montane vegetation has been all but eradicated by fire and unchecked cattle grazing (see Chapman *et al.* 2004). Chimpanzees were heard and seen at various points along the stretch from Kwano to the Gamgam, and calls heard from the rocky outcrop of Dutsin Dodo in forests below the ridge of the escarpment. The return trip via the park border at Njawai and towards Ngoroje revealed that this part of the Mambilla Plateau, which must once have been blanketed by montane forest, is now deforested, barren and pockmarked by gigantic erosion gullies.

Gamgam headwaters (13 days, 2005 – 2006). The west-facing slopes below the Chappal Wade escarpment constitute the least explored part of the national park. No ranger or local could be identified who had previously ventured into this densely forested area (*Fig. 3.3c*). We entered through water-courses and gorges up from the Gamgam River. The forests became denser closer to the steep ridges of the escarpment, until signs of human activity such as cattle grazing, farming and poaching ceased. Much wildlife was encountered, including tracks of leopard. Chimpanzees were heard and seen, as well as many nests found.

Eastern enclaves Chappal Delam – Filinga (9 days, 2007). The remote highland enclaves towards the Cameroonian border contain vast stretches of grassland. Human population density is low, but herds of cattle number in the thousands. Access to hundreds of square kilometres of pasture seems to be controlled by just a few Fulani clans. More grassland is slowly but steadily being created, through fire and grazing, often far beyond the original demarcation of the enclaves (see also Chapman *et al.* 2004). The de-facto continuous expansion of activities, theoretically confined to enclave areas, comes with a virtually unchecked elimination of wildlife. Chimpanzees were not found to utilise the open grassland.

Gashaka-Kwano (continuous presence of GPP researchers since 2000). This site is exemplary for the diversity of wildlife once native to Taraba as a whole. A community of chimpanzees lives in the surrounding forests (Sommer *et al.* 2004). The site also harbours 6 species of monkey, i.e., olive baboon, tantalus monkey, black-and-white colobus, mona and putty-nosed monkeys. Many other species of large mammals are also common, such as antelopes and pigs as well as carnivores including leopards and golden cats. Gunshots were sometimes heard from 2000 –2003, but the frequency decreased markedly once the settlements of Gidan Hitla and Tonga were abandoned in 2004. Wire traps are hardly ever found. Evidence for poaching of primates is very sparse, although a poachers camp in the study area was raided by GPP personnel in 2005 after at least 13 baboons and 2 black-and-white colobus monkeys had been killed.

Discussion

Our study aimed to explore the current situation of, and prospects for, the conservation of primates in Nigeria's north-eastern Taraba area, not least because this region is the stronghold of the rarest chimpanzee subspecies, *P. t. vellerosus*. For this, we conducted *foot surveys* of historic and current chimpanzee habitat.

Trends of Chimpanzee Distribution

Our habitat surveys on foot and a compilation of published and unpublished reports (*Tab. 3.2*) were used to detect patterns in the chimpanzee distribution of the Taraba region (cf. *Fig. 3.5*).

Shebshi Mountains and Northern Distribution

We were unable to confirm the assumption that chimpanzees occur as far north as the Shebshi mountains, but more survey work is needed. If this area was indeed included in the historical range, then it is likely that chimpanzees are extinct here. The current northern distribution is assumed to follow the line Gumti – Jamtari – Beli – Jatau. North of this line is mostly savannah-woodland, rather unsuitable for chimpanzees, whereas the Fali mountains south of the line provide better habitat.

Cameroon Border and Southern Distribution

These populations are either in imminent danger of extinction (Bissaula, Zedah) or have already disappeared because of forest destruction.

Park Vicinity

Areas west of Serti (Alekum) still harbour chimpanzees but survival for much longer into the future is unlikely, given the hunting pressure. Survey work still needs to be carried out in the interior of the Fali Mountains as the available information is too scarce. Chimpanzees have been continuously reported from Ngel Nyaki since 1973. The population in this biodiverse montane forest fragment is certainly small and perhaps isolated. However, it has good chances of survival, if research and recently stepped-up conservation efforts are to be continued, such as those associated with the establishment of the Nigeria Montane Forest Project (NMFP; Chapman *et al.* 2004). The NMFP estimated that the forest fragment harboured 12 weaned individuals in 2005 – 2006 (Beck & Chapman 2008). The vast Mambilla Plateau towards the south-east (triangle Yelwa – Njawai – Gembu) is completely deforested, and at least a third of the area is severely eroded. It is safe to assume that no chimpanzees survive in this part of their former range.

Gashaka Gumti National Park

Chimpanzee records stretch as far back as the 1970s, and numerous surveys have since confirmed their existence in numerous locations. The northern distribution

Table 3.2 Chimpanzee occurrences in north-eastern Nigeria: Previous reports, recent foot surveys and future prospects

				Habitat				Chimpanzees	
Source	Survey date	Area	Location	Type	Status	Anthropogenic influence	Large mammals	Evidence	Population status
SHEBSHI MOUNTAINS									
FS	2004		Vogel Peak: Gurum Pawu	S	FR?	C+++, H+++,	–	–	PE?
FS	2004		Dakka	S, G	FR	C+++, H+++	–	–	PE?
CAMEROON BORDER AREA									
IN	2004		Ussa, 3 locations				AM+, TM++	R	
4	1989		Bissaula	L	R			H	
FS	2004					H+++, V++	–	N+	IP?, SU
IN	2004		Bissaula, 6 locations				AM +++, TM++	R	
IN	2004		Baissa				AM+++, TM+++	R	
IN	2004		Abong, 5 locations				AM+, TM+	R	
IN	2004		Akwame, 2 locations				AM+, TM+	R	
IN	2004		Buru, 3 locations				AM+++, TM++	R	
FS	2002		Zedah	L	R	H+++, V+, P	–, +	N+	IP?, SU
6	1970s		Akwaizantar	L	U	V, H+	AM, UN, CA	S	
RR	1996							H++, N+++, S+	
7	2002					C+++ (destroyed)		–	PE
IN	2005		Gembu				TM+	–	PE
PARK SUPPORT ZONE									
IN	2005		Gidin Dutse				–	R	
IN	2005		Kungana, 8 locations				AM+, TM++	R	
IN	2005		Beli, 5 locations				AM++, TM+++	R	
IN	2005		Jamtari, 3 locations				AM++, TM++	R	
RR	1998		Serti: Abadogo		U	V		S	IP?, SU
FS	2004		Alekum	L, W	U	V, H+++	–, +	N+++	IP?, SU

(continued)

Table 3.2 (continued)

Source	Survey date	Area	Location	Habitat: Type	Habitat: Status	Habitat: Anthropogenic influence	Large mammals	Chimpanzees: Evidence	Chimpanzees: Population status
IN	2004		Bodel				AM+, TM+	R	
IN	2004		Mayo Selbe , 3 locations				AM++, TM++	R	
5	1973		Ngel Nyaki	M, G	R		AM, UN	S	
6	1970s							S	
9	1989							present	
19	1990					G, H, V	AM	N++	
15	1994						AM, CA, UN	N++	
17	1997						AM, TM	N, H?	
FS	2002					G++, H++, V	AM, UN	H, N+++, S+	
7	2002					G, H++, V	AM, TM, UN	N, S+	
NMFP	2003 –2006					G++, H++, V	AM, TM, UN	continuous H, N, S	IP?, SL
IN	2005						AM+++, TM+++	R	
17	1997		Mai Samari	W	U			N+	
IN	2005						TM	R	
RR	1980		Mai Idanu: Mayo Taya	?	U	C+++		S	
FS	2005		Mai Idanu	E	U		–	–	PE
GASHAKA GUMTI NATIONAL PARK									
13	1996		Gumti	W, L	E	H++	AM, TM, UN	H, N+++, S+	
1	1998							N++	
11	1997							N+++	
14	2002							S++	CP?, SL
10	1992		Gumti: Mayo Yum	L, W	P		AM, TM, UN, CA	H+++	
13	1996					H+++	AM, TM, UN,	N+++	
11	1997							N+++	

12	1998						N+	
1	1998						N+++	CP?, SL
14	2002	Goje to Mayo Kam	L, W	R, P		AM, TM, UN, CA	S++	CP?, SL
10	1992	Gashaka: Bakashi / Gamgam	L, W			AM, TM, UN	H+++, N+++, S?	
15	1994	Gashaka: Boloko river	L, W		H+++		N++	
15	1994	Gashaka: Mayo Sannere	L, W			TM, UN, CA	N+++	
15	1994	Gashaka: Curmin Gamagi	L, W			AM, TM, UN, CA	N+	
13	1996	Gashaka: Mayo Kpa'a	L, W				H, N+++	
11	1997						N+	
17	1997	Gashaka: Mayo Garandi	L, W			AM, TM, UN	N+++	
12	1998	Gashaka North–East					H+	
RR	2000s	Gashaka: Zaria Road					S	
RR	2001	Gashaka: German fort					H	
18	2000–2003	Gashaka: near Village		R			S++	
GPP	2004–2006						continuous H, N	CP, SL
10	1992	Kwano	L, W, G	P		AM, TM, UN	H, N+, S+	
15	1994				H+		H, N+++, S++	
16	1997						H, N, S	
11	1997						N+++	
17	1997				H+	AM, TM, UN	H, N+++, S+	
12	1998						H, N++	
11	1998						N+++, S+++	
1	1998						N++	

(continued)

Table 3.2 (continued)

				Habitat				Chimpanzees	
Source	Survey date	Area	Location	Type	Status	Anthropogenic influence	Large mammals	Evidence	Population status
GPP	Since 1999					H+	AM+++, TM++, UN+++, CA+	continuous H+++, N+++, S++	CP, SL
10	1992		Ngiti: Gamgam confluence	L			AM, TM, UN	H, N+++, S++	
15	1994			L	NP	H+		H, N+++	
11	1998							S+	CP, SL
11	1997		Yakuba: Transect	L, W	P			N++	
11	1997							N+++	
12	1998							S+	
RR	2002							S+	CP, SL
RR	2000		Yakuba – Gamgam Road	L, W	P			H	
RR	2000		Yakuba – Filinga	L, G	P	C+++		H	
RR	2001							H++	
RR	?		Ngiti: Hamman Kankadu	M				H	CP, SL
FS	2005		Sabere: Mayo Tapare	L, G	P, E			N+++, S+	CP, SL
6	1970s		Leinde Fadali	M	P, E			S	
9	1989							present	
11	1997							N+++	
RR	1998?					C++, V, H?		H, S+++	
7	2002							H, N++	
RR	2003					C++, H?		H	IP?
3	1970s		Chappal Wade / Dutsin Dodo	M, G	P		AM, UN	present	
6	1974–5							S	

9	1989						present	
7	2002						H, N	
FS	2005				C+, G+++, H+		H, N	CP, SL
FS	2005	Gamgam headwaters	M	P			H++, N++, S+	CP, SL
RR	>2000	Chappal Hendu – Selbe	G, M	E			S	
11	1998	Kogitan	F, V	P			N++	
6, 2	1970s	Chappal Hendu	M, G	P			present	
RR	1980s						present	
9	1989						–	
10	1992					AM, TM, UN,	H+++, N+++, S+++	
11	1997						N+++	
8	2002						–	
6	1970s	Chappal Hendu: Dutsin Lamba	M, G	P			S	
9	1989						present	
8	2002						present	IP?</TB>

Sources. 1 = Adanu 1998; 2 = Barnwell 1993a; 3 = Barnwell 1993b, cit. in NPS / NCF / WWF 1998; 4 = Barnwell *pers. comm.*, cit. in NPS / NCF / WWF 1998; 5 = C.F.A. Onochie *pers. comm.* to J. Chapman, cit. in Chapman & Chapman 2001; 6 = Chapman & Chapman 2001; 7 = Chapman *et al.* 2003 (published in part as Chapman *et al.* 2003); 8 = Chapman *et al.* 2004; 9 = Dowsett & Dowsett-Lemaire 1989; 10 = Dunn 1993; 11 = Foster 1998a; 12 = Foster 1998b; 13 = Gawaisa 1997; 14 = Gawaisa 2002; 15 = Harcourt & Ellerton 1995; 16 = Hogarth 1997; 17 = Ristau & Waziri 1997 ; 18 = Warren 2003; 19 = White 1990; FS = Foot survey involving authors; RR = Ranger report (personnel of GGNP, Nigerian Conservation Foundation; compiled by AF and VS); NMFP = Nigeria Montane Forest Project, founded 2002; director: H. Chapman; GPP = Gashaka Primate Project, founded 2000, director: V. Sommer; IN = interviews, see Nyanganji *et al.* this volume [Ch. 3]. *Habitat type*: L = lowland forest, M = montane forest, W = savannah-woodland, S = scrub, G = grassland, E = eroded landscape. *Habitat status*: U = unprotected, R = forest reserve, P = national park, E = enclave. *Anthropogenic influence*: C = forest cutting, G = cattle grazing, H = hunting, S = settlements, V = village, farmland, P = plantations. *Large mammals*: AM = arboreal monkeys (putty-nosed monkey, mona monkey, black-and-white colobus), TM = terrestrial monkeys (baboon, tantalus monkey, patas monkey); UN = ungulates; CA = carnivores; missing entries indicate missing data, not necessarily absence of mammals. *Chimpanzee evidence*: H = calls heard, N = nests, S = seen, R = reported by local interviewed for this survey, present = based on source in far left column. *Chimpanzee population status*: IP = isolated population, CP = contiguous population, PE = population extinct; SL = survival for more than 10 yrs likely, SU = survival over next 10 yrs unlikely. *Quantifiers*: – = none / nothing; + = few (1 – 4) / little; ++ = moderate (5 – 9) / medium; +++ = many (10+) / much.

reaches to the Gumti area, which borders on the extensive woodland-savannah and grassland areas of the park's northern sector. To the west, chimpanzees occur right up to the park boundary close to the Serti – Mayo Selbe road and beyond, into the Fali Mountain range. Chimpanzees are also found in the surroundings of various enclaves in small isolated forests such as Leinde Fadali near Sabere and Dutsin Lamba near Chappal Hendu. These populations appear to be vulnerable, particularly as inhabitants of the enclaves constantly encroach upon the national park, creating more and more grassland through burning and cattle grazing. Least explored and influenced by anthropogenic alterations are the forests leading up to Chappal Wadde from its western side. One should not exclude the possibility that gorillas survive in this area of about 200 km^2, as their historic range in Cameroon extended certainly as close as 100 km to the south, if not nearer. The idea that species of large mammals might have escaped discovery till now is not unrealistic, as it was only from 2000 onwards that lone specimens of grey-cheeked mangabeys were noticed in the forests around Kwano (I. Faucher, K. Arnold *pers. comm.*, VS *pers. obs.*). This area is also a most important reservoir for *P. t. vellerosus*, as the vast majority of chimpanzees are found in the interior of the park's southern half, with a most likely contiguous population found along the axis Gashaka – Ngiti – Yakuba – Gamgam headwaters – Chappal Wade (roughly 50 km × 15 km = 750 km^2). Data from the Gashaka-Kwano community suggest a population density of 1.3 chimpanzees / km^2 (Sommer *et al.* 2004). Thus, we assume at this stage that the park harbours a minimum of 1000 apes.

A satellite image of the central section of GGNP around Gashaka (*Fig. 3.6*) illustrates the challenges to the future survival of chimpanzees and other wildlife. Chimpanzee distribution is contiguous in the remaining forests along the axis Gashaka – Kwano – Mayo Sabere and the Gamgam headwaters leading up to Chappal Wade (lower right corner). However, widespread deforestation can be recognised in the area around the town of Serti, the enclave areas north-east of Kwano leading to Filinga, around the enclave of Mayo Sabere leading on to Mai Idanu, a settlement on the eroded northern Mambilla Plateau outside the park. The unprotected Cameroonian side of the Chappal Wade Mountains is completely deforested.

Hunting, Burning, Grazing: Main Threats to Wildlife

Habitat loss is often accompanied by an overkill of wildlife, "as local people scour the forests within walking distance of their villages in search of game or marketable animal products. Later, as the agricultural frontier expands into the forest, the most suitable sites are cleared first, resulting in a fragmented landscape. As more settlers arrive […], the forest melts away and eventually only small, biologically unviable fragments remain, typically in deep ravines and on ridge tops. The fragments then experience cascades of secondary extinctions, frequently exacerbated by the invasion of nonforest and exotic species from adjoining disturbed areas" (Terborgh & van Schaik 1997: 26). Our specific findings should be viewed against this general backdrop

Fig. 3.6 Remote sensing scene of the central section of GGNP around Gashaka (courtesy Maren Gumnior, Universities of Frankfurt and Gombe)

(see also Cowlishaw & Dunbar 2000, Caldecott & Miles 2005). Similarly, the decline of wildlife populations and consequent reduction of biodiversity in Taraba can be related to three principal forces of destruction: (a) hunting for bush-meat and the pet trade; (b) fire damage to forests; (c) cattle grazing. In Taraba, they dwarf the more outright destruction of habitat due to logging as low-lying areas have already been exploited and the remaining terrain is too rugged for large-scale timber extraction (Bawden & Tuley 1966, Chapman & Chapman 2001).

Hunting is largely driven by subsistence needs and monetary profit made from the sale of bush-meat. The trade in primate babies as pets should be seen as a corollary, not the prime motivation behind hunting (see Nyanganji *et al.* this volume [Ch. 4], Hughes *et al.* this volume [Ch. 14]). Hunting reduces wildlife, but, as a secondary consequence, also prevents the rejuvenation of the vegetation cover, as primates such as chimpanzees, baboons and other forest monkeys are important dispersers of large seeds (Chapman 1995, Kunz & Linsenmaier 2008). Of course, hunting is not restricted to primates. In fact, our surveys indicate that hunters shoot anything that they can, although larger-bodied animals are preferred. There are certain taboos fostered by the Islamic religion against killing primates from which monkeys and apes benefit, but these restrictions are being weakened by a growing influx of migrant populations (Nyanganji *et al.* this volume [Ch. 4]). Thus, we share the concern that grew out of a recent assessment of changes over 30 years to the montane forests of Taraba: "In the long- or medium-term the forests described here are in jeopardy of becoming 'empty forests'" (Chapman *et al.* 2004: 7).

The second danger is fire. Most seasonal tropical forests and savannah-woodlands have been shaped by anthropogenic fires (Goldammer & Crutzen 1993). They are a regular occurrence in Taraba during the dry season from November till March (Dunn 1994, NPS / NCF / WWF 1998). Fires are set with the intention of improving pasture by encouraging the growth of new grasses to feed livestock during the dry season, clear and fertilise the land for cultivation, flush out prey for hunting or destroy parasites such as tse-tse flies (Dyer *et al.* 2001). Seasonal burning is also encouraged by the national park authorities with the explicit goal of preventing major natural burning from damaging the forest when it occurs late in the dry season (Gyar 1999). However, long-term effects of burning are adverse. The top-soil is degraded because organic materials are depleted and microorganisms are decimated. The surface hardens, which increases run-off once the rains set in. This, in turn, leads to erosion by water and wind as well as flooding and silting of lower lying areas (Egunjobi 1979). In Taraba, burning also encourages the dominance of a single grass species, *Sporobolus africanus*, which is not very palatable to livestock (NPS / NCF / WWF 1998: 7–13). Fires late in the dry season gradually destroy the forest edge, occasionally penetrate the forest interior and thus prevent forest regeneration. Taraba's current mosaic habitat is clearly the result of such a fire-regime, which has increased the proportion of savannah-woodland against lowland and riperian forest and created vast areas of montane grassland at the expense of formerly existing biodiverse montane forests (Chapman *et al.* 2004). Chimpanzees avoid open grassland, and while they make use of savannah-woodland, they prefer forest as these provide more fruit and suitable nesting sites. Thus, the fire-regime in Taraba does have a severe negative impact on the arboreal monkeys and chimpanzees.

The negative consequences of bush-burning are exacerbated by pastoralist practices. Many communities in Taraba depend on cattle-keeping – most intensely on the Mambilla Plateau. The dominant pastoralists are the Fulani, some of whom have given up their nomadic or semi-nomadic lifestyle, instead settling in villages (Barnwell this volume [Ch. 2], Bennett & Ross this volume [Ch. 6]). Intense cattle herding leads to trampling of the top-soil and overgrazing with the resulting dangers of erosion (Chapman & Chapman 2001). This is a severe problem throughout Taraba, but particularly in the national park. Fulani pastoralists regularly drive large herds through the protected area and cattle-herders from Cameroon use the grasslands on the Chappal Wade escarpment. The enclaves are a grazing ground not only for at least 10000 – 15000 cattle of the inhabitants, but also livestock from Fulani from outside the enclaves (Dunn 1994, NPS / NCF / WWF 1998). The borders of the enclaves are not respected, and more grassland is constantly created within the national park proper.

The combination of fire damage and cattle-grazing has devastating effects on the landscape. This is most obvious on the Mambilla Plateau. Once blanketed by montane forest, at least one third of its 3000 km^2 have been reduced to a barren landscape with virtually no trees left. During the dry season, often not even goats or sheep will find enough pasture, and gigantic erosion gullies dot the landscape.

Factors Aiding Conservation

Community-based Approaches

Many institutions advocate nature conservation on the basis that "the key to protecting any cherished landscape lies within those communities that call it home" (Dunn 1999: 58). In this spirit, organisations such as the Nigerian National Park Service, NCF, the Nigeria Montane Forest Project at Ngel Nyaki, GPP and Taraba State Broadcasting have been promoting infrastructural development in the health and educational sectors as well as community forestry. Such programmes are active in the park's enclaves and support zone but also near the Cameroonian border where NCF maintains a long-standing community forestry project at Buru / Zedah (NCF 2003). The aim of these activities is to reduce the level of hunting and boost conservation-oriented attitudes.

However, given the vast extent of the Taraba region and park vicinity, such efforts are necessarily patchy, often low-key, and measures of success are hard to come by. Surely, such outreach is not yet effective in replacing deeply engrained traditional practices of environmental exploitation on a wider scale, despite the fact that practices such as deforestation, over-grazing and non-sustainable hunting and fishing are clearly detrimental to communities in the long run. Moreover, new challenges come with the constant waves of immigration into the region.

There is particular need to comment on the community-based approach that came with the creation of enclaves within the national park. A number of Fulani settlements developed after 1950 when it became clear that the highlands were relatively free from tse-tse flies, thus providing an ideal grazing ground for cattle (NPS / NCF / WWF 1998: 7-12). The national park's guide book spells out the philosophy behind the policy not to evict the resident pastoralists once the park was created: "It was decided that the essential needs of the resident pastoral people and the demands of wildlife conservation could both be accommodated within the same protected area. At Gashaka-Gumti, enclaves are managed jointly by the National Park Service and local people in such a way as to safeguard long-term conservation interests and local livelihoods. [..] These pastoral and farming communities are now among the keenest park supporters and assist in its protection by reporting the presence of poachers and other offenders to park rangers. [..] Gashaka Gumti National Park represents a pioneering experiment in multi-purpose land use in Nigeria" (Dunn 1999: 58; see also NPS / NCF / WWF 1998: Section 7, and Dunn 1995).

In hindsight, this well-meaning experiment seems to have largely failed as little of the described vision became reality. Poaching from within the enclaves is common and much forest in the wider diameter of the enclaves has been destroyed by fire and cattle-grazing. Cooperation with the National Park Service is close to non-existent (see also Chapman *et al.* 2004). The reasons for this is are complex (Bennett & Ross this volume [Ch. 6]), but include the desire of powerful Fulani to maintain as large herds of cattle as possible and the rather spotty efforts of park authorities to communicate with the settled pastoralists. This harsh judgement is not shared by everybody who cares about the future of GGNP (see Barnwell this volume [Ch. 2], Bennett & Ross this volume [Ch. 6]). However, there can be little doubt that a community-based approach does not in itself guarantee the success of conservation policies (Oates 1995, 1999, Terborgh 1999).

Protected Areas and Law Enforcement

An often disparaged alternative to the community-based, incentive-driven approach to nature conservation is that of a stricter enforcement of existing laws up to the extent that protected areas are viewed as fortresses (reviews in Kramer *et al.* 1997).

In Taraba, such enforcement is absent in most locations. State-run forest reserves in particular are little more than "paper-parks" (Peterson & Ammann 2003). Again, numerous factors are to be blamed for this. Forest guards and park rangers are underpaid, often lack gear, and thus tend not to leave their stations for patrol. The Nigerian government, while barely paying salaries to its national park staff, did not provide an operational budget for years. Encounters with law-breakers are often settled through bribes than in a court of law, notwithstanding the fact, that court-cases against poachers and illegal settlers *are* brought by the National Park Service.

Having said all this, our survey results tend to suggest that the presence of a national park does make a difference, although the evidence is not unequivocal. Firstly, while, for example, the forests of Alekum were found to be all but devoid of wildlife, the park nearby still harbours many large animals. Thus, it cannot be denied that the park, however incomplete, provides more effective protection than a state forest reserve. Secondly, a questionnaire survey showed that people in the vicinity of the park had a greater awareness than inhabitants of the area near the Cameroonian border that Nigeria has laws protecting wildlife (Nyanganji *et al.* this volume [Ch. 4]).

To turn the tide, it would be necessary for the central government and the Taraba state government to dispense considerably more resources towards nature conservation. However, even such measures will not be effective as long as the widespread corruption is not tackled, which persistently chokes the development of the Nigerian nation (Smith 2006). Until then, biodiversity will continue to decline in Taraba State.

Research

Naturalistic studies of primate communities have proven to be an effective conservation tool (see Harcourt & Stewart 2007). Notable examples are the mountain gorillas in Rwanda (Fossey 1983) and the chimpanzees of Gombe in Tanzania (Goodall 1986) as well as Budongo in Uganda (Reynolds 2005) that almost certainly would not exist today were it not for continuous efforts of researchers to protect their study subjects. A major factor in these protection efforts are the selfish motivations of researchers to save from extinction what they want to explore, and – almost paradoxically – that they are normally not embedded in the cultural, sociopolitical and economic background at their study site. This allows them to be less corruptible and implement "imported" protective measures such as patrols, law enforcement, employment of locals, training of native students, and development of eco-tourism.

The fragments of montane forest at Ngel Nyaki could potentially capitalise on such conditions. Since 2002, this reserve has received increasing attention with the foundation of the Nigeria Montane Forest Project (see above; Chapman *et al.* 2004) and with Yelwa as site of a new NCF project. The small size and the surrounding grassland may allow for some effective protection measures such as the control of poaching. There is thus reason for optimism for the wildlife in this isolated patch of montane forest. A similar approach, albeit on a much larger geographical scale, is taken by GPP (Sommer & Ross this volume [Ch. 1]). Situated in the midst of what is the remaining prime habitat of chimpanzees left in the Taraba region, the project has tried to add Gashaka-Kwano to the list of sites where nature conservation benefits from the presence of researchers (Sommer 2008). There is little doubt that GPP research activity has been beneficial for the Gashaka-Kwano area. However, this success is localised, and it is likely that the external forces of destruction will exert more and more pressure even on the core area of the park's southern sector.

Prospects

A mere 1 % of Taraba State is still under forest vegetation (Chapman & Chapman 2001), which makes the remaining fragments all the more important as havens for threatened flora and fauna. The national park in particular has, despite all shortcomings, contributed positively to biodiversity conservation. However, it is hard to assess whether the park's existence will simply delay the ultimate eradication of wildlife or if this refuge provides a prospect for long-term survival. The existence of the enclaves is clearly one of the most serious problems. The national park will certainly be further degraded if they remain (cf. *Fig. 3.6*). In the light of this, one might be tempted to "argue for resettling the local people and cattle outside the park" (Chapman *et al.* 2004: 8). However, the eviction of Fulani settlers would inevitably open up the highlands for even more rampant poaching operations and cattle grazing from the Cameroonian side. Park protection would therefore gain nothing unless this operation would include a substantial and sustained upgrade of protection measures on the part of the Nigerian National Park Service. A major step in increasing the effectiveness of law enforcement was the demarcation of the Nigerian side of GGNP over its entire length of more than 200 km, through beacons and a motorable track, an initiative of GPP was funded by the outreach programme of Chester Zoo and the North of England Zoological society from 2005–2008 (www.chesterzoo.org / WhatWeDo; accessed 12 Sep 08).

In any case, the destructive forces in operation are too strong and too complex to allow for simple recommendations on how to protect Taraba's wildlife. Nevertheless, it would seem to be all but impossible to make a difference via a community-based approach, given rapid population growth and the dynamics of immigration to the area. In the current situation, law-enforcement and fortress conservation must surely have priority. Otherwise, there will be nothing left to protect – even if alternative strategies should one day become effective.

In light of this, one might perhaps want to resort to a rather existentialist philosophy, which advocates that there is dignity in resisting the demise, even if prospects seem dim (Sommer 2009). This echoes the deeply sceptical tone of Joseph Conrad's famous tale *Heart of Darkness* where he laments the exploitation of Africa's grand biomes; with vivid vision did Conrad anticipate the slow destruction of the treasures of forests and wildlife, given human nature and the forces of "civilisation" – a seemingly unstoppable obliteration that we see in full swing today. Of course, everybody is contributing to this, the readers of this chapter as well as the writers and those about whom they are writing. Still, the realisation reverberates with a "vibrating note of revolt" (Conrad 2008 [1899]: 179). Perhaps it is the chimpanzees, beating their drums.

Acknowledgements Permissions were granted by the Nigeria National Park Service for GGNP, and for other regions within Taraba State by the State Forestry Department and Ministry of Environment and Solid Minerals, Jalingo. The State Ministry of Environment, Yola (officer Steven Gawaisa) gave permission for visits to the Vogel Peak area in Adamawa State. Thanks are due to WWF-UK, who initiated the survey by part-sponsoring it, and to Leventis Ltd. & Pro-Natura International for commissioning the survey of Ngel Nyaki. Maren Gumnior, Department of Geography, Universities of Frankfurt / Germany and Gombe State / Nigeria, kindly provided the satellite image. JA was part-supported by Primate Conservation, AF by The Leakey Foundation International, VS by the Graduate School and the FSHS Dean's Fund of UCL. The Nigeria Biodiversity Programme of Chester Zoo provided the running costs of the *Gashaka Primate Project*.

Literature

Adanu, J. (1998). *Distribution and Abundance of Chimpanzee (Pan troglodytes troglodytes) in Gashaka Gumti National Park, Nigeria.* UG dissertation, Department of Forestry and Wildlife Management, Federal University of Technology Yola.

Akinsoji, A. (1996). *Vegetation Types and Ethnobotanical Studies of Gashaka Gumti National Park, Nigeria.* Report in conjunction with Department of Biological Sciences, University of Lagos / Nigeria, to NCF & WWF-UK.

Ammann, K., Rose, A. L., Mittermeier, R. A., Langrand, O., Ampadu-Agyei, O. & Butynski, T. M. (2003). *Consuming Nature. A Photo Essay on African Rain Forest Exploitation.* Palos Verdes, CA: Altisima.

Barnwell, R. (1993a). *Personal Notes: Some Additions and Revisions to the 1988 Draft Management Plan by Arthur Green* (cited in: NPS / NCF / WWF 1998).

Barnwell, R. (1993b). WWF / NCF wildlife conservation initiatives in Nigeria. In *Wildlife Conservation in West Africa* (Proceedings, Biennial Symposium Nigerian Field Society UK Branch & WWF-UK). Kew: Royal Botanical Gardens.

Bawden, M. G. & Tuley, P. (1966). *The Land Resources of Southern Sardauna and Southern Adamawa Provinces, Northern Nigeria* (Land Resource Study No. 2). Tolworth, Surrey: Directorate of Overseas Surveys.

Beck, J. & Chapman, H. (2008). A population estimate of the endangered chimpanzee *Pan troglodytes vellerosus* in a Nigerian montane forest: implications for conservation. *Oryx* 42: 448–451.

Boesch, C. & Boesch-Achermann, H. (2000). *The Chimpanzees of the Taï Forest. Behavioural Ecology and Evolution.* Oxford: Oxford University Press.

Caldecott, J. & Miles, L. (ed.) (2005). *World Atlas of Great Apes and Their Conservation.* Los Angeles: University of California Press.

Chapman, C. A. (1995). Primate seed dispersal: coevolution and conservation implications. *Evolutionary Anthropology* 20: 74–82.

Chapman, J. D. & Chapman, H. M. (2001). *The Forest Flora of Taraba and Adamawa States, Nigeria. An Ecological Account and Plant Species Checklist.* New Zealand: University of Canterbury.

Chapman, C. A., Chapman, L. J., Rode, K. D., Hauck, E. M. & McDowell, L. R. (2003). Variation in the nutritional value of primate foods: among trees, time periods, and areas. *International Journal of Primatology* 24: 317–333.

Chapman, H. M., Olson, S. M. & Trumm, D. (2004). An assessment of changes in the montane forests of Taraba State, Nigeria, over the past 30 years. *Oryx* 38: 1–9.

Conrad, J. (2008 [1899]). *Heart of Darkness* (edited & introduced by Watts, C.; Oxford World's Classics). Oxford: Oxford University Press.

Cowlishaw, G. & Dunbar, R. (2000). *Primate Conservation Biology.* Chicago: The University of Chicago Press.

Dowsett, R. J. & Dowsett-Lemaire, F. (1989). Larger mammals observed in the Gotel Mts and on the Mambilla Plateau, eastern Nigeria. *Turaco Research Report* 1: 38–41.

Dunn, A. (1993). *The Large Mammals of Gashaka-Gumti National Park, Nigeria: Line Transect Surveys from Forest and Savannah.* Report to WWF-UK, NCF & Federal Ministry of Agriculture, Water Resources and Rural Development, Nigeria.

Dunn, A. (1994). *Empowerment or Eviction? The Highland Enclaves of Tale, Shirgu, Hendu, Nyumti and Filinga, and Their Future Management.* Report to WWF-UK, NCF & Federal Ministry of Agriculture, Water Resources and Rural Development, Nigeria.

Dunn, A. (1995). Gashaka Gumti, Nigeria – From Game Reserve to National Park. *RDFN Paper* 18.

Dyer, R., Jacklyn, P., Partridge, I., Russell-Smith, J. & Williams, D. (eds.) (2001). *Savannah Burning. Understanding and Using Fire in Northern Australia. Vol. 57.* Natural Heritage Trust.

Egunjobi, J. K. (1979). An appraisal of savanna burning problem. In Ajaji, S. S. & Hallstead L. B. (eds.), *Wildlife Management in Savannah Woodland*. London: Taylor and Francis Ltd., pp. 160–165.

Forboseh, P. F., Eno-Nku, M. & Sunderland, T. C. H. (2007). Priority setting for conservation in South-West Cameroon based on large mammal surveys. *Oryx* 41: 255–262.

Fossey, D. (1983). *Gorillas in the Mist*. Boston: Houghton Mifflin.

Foster, K. L. (1998a). *Chimp Photo Project Summary – Gashaka Gumti National Park, June – August 1998*. Report to NCF.

Foster, K. L. (1998b). *Censusing Chimpanzees in Gashaka-Gumti National Park, Taraba and Adamawa States, Nigeria*. Report to WWF-UK, NCF & Federal Ministry of Agriculture, Water Resources and Rural Development / Nigeria.

Fruth, B. & Hohmann, G. (1994). Comparative analyses of nest building behavior in bonobos (*Pan paniscus*) and chimpanzees (*Pan troglodytes*). In Wrangham, R. W., McGrew, W. C., de Waal, F. B. M. & Heltne, P. G. (eds.), *Chimpanzee Cultures*. Cambridge, MA: Harvard University Press, pp. 109–128.

Gawaisa, S. G. (1997). *Chimpanzees of Gashaka Gumti National Park. More Line Transects into Savannah Habitat*. Student Industrial Work Experience Scheme Report, Department of Forestry and Wildlife Management, Federal University of Technology Yola.

Gawaisa, S. G. (2002). *Annual Participatory Road Transect Monitoring of Large Mammals in Gashaka Gumti National Park, Nigeria (Mayo Kpa'a, Goje and Gumti)*. Report to NCF & WWF-UK.

Goldammer, J. G. & Crutzen, P. J. (1993). Fire in the environment: scientific rationale and summary of results of the Dahlem Workshop. In Crutzen, P. J. & Goldammer, J. G. (eds.), *Fire in the Environment: the Ecological, Atmospherics, and Climatic Importance of Vegetation Fires*. John Wiley and Sons Ltd., pp. 1–13.

Gonder, M. K., Oates, J. F., Disotell, T. R., Forstner, M. R. J., Morales, J. C. & Melnick, D. J. (1997). A new West African chimpanzee subspecies? *Nature* 338: 337.

Goodall, J. (1986). *The Chimpanzees of Gombe: Patterns of Behaviour*. Cambridge, MA: Harvard University Press.

Gyar, D. T. (1999). *Habitat Management for Nature Conservation: Fire, Salt-lick and Water Regime*. Nigeria Parks, May: 12–14.

Harcourt, A. H. & Stewart, K. J. (2007). *Gorilla Society: Conflict, Compromise, and Cooperation Between the Sexes*. Chicago: The University of Chicago Press.

Harcourt, C. S. & Ellerton, N. G. (1995). *A Brief Look at the Primates of Gashaka Gumti National Park, Nigeria*. Report to NCF & WWF-UK.

Heltne, P. G. & Marquardt, L. A. (eds.) (1989). *Understanding Chimpanzees*. Cambridge, MA: Harvard University Press.

Hogarth, S. A. (1997). *The Distribution and Abundance of the Chimpanzees (Pan troglodytes troglodytes) of the Lowland Gallery Forest of Gashaka-Gumti National Park Nigeria: A Preliminary Report to the Gashaka Sector*. Report to WWF-UK, NCF & Federal Ministry of Agriculture, Water Resources and Rural Development / Nigeria.

Inskipp, T. (2005). Chimpanzee (*Pan troglodytes*). In Caldecott, J. & Miles, L. (eds.), *World Atlas of Great Apes and Their Conservation*. Los Angeles: University of California Press, pp. 53–81.

Kormos, R., Boesch, C., Bakarr, M. I. & Butynski, T. M. (eds.) (2003). *West African Chimpanzees. Status Survey and Conservation Action Plan*. Gland & Cambridge UK: IUCN / SSC Primate Specialist Group.

Kramer, R., van Schaik, C. & Johnson, J. (1997). *Last Stand. Protected Areas and the Defense of Tropical Biodiversity*. Oxford: Oxford University Press.

Kunz, B. K. & Linsenmair, K. E. (2008). Seed dispersal by olive baboons in the Comoé National Park, Ivory Coast. *Folia Primatologia* 71: 233–234.

McManus, E. (2005). Federal Republic of Nigeria. In Caldecott, J. & Miles, L. (eds.), *World Atlas of Great Apes and Their Conservation*. Los Angeles: University of California Press, pp. 378–383.

NCF [Nigerian Conservation Foundation] (2003). *Annual Report*. Lagos: Academy Press.

Ngalla, P. T., Miles, L. & Caldecott, J. (2005). Republic of Cameroon. In Caldecott, J. & Miles, L. (eds.), *World Atlas of Great Apes and Their Conservation*. Los Angeles: University of California Press, pp. 305–314.

NPS / NCF / WWF (1998). *Gashaka Gumti National Park: A Management Plan for Developing the Park and its Support Zone*. Nigeria National Park Service, Nigerian Conservation Foundation, World Wide Fund for Nature (UK).

Oates, J. F. (1994). Africa's primates in 1992: conservation issues and options. *American Journal of Primatology* 34: 61–71.

Oates, J. F. (1995). The dangers of conservation by rural development – a case study from the forests of Nigeria. *Oryx* 29: 115–122.

Oates, J. F., Bergl, R. A. & Linder, J. M. (2004). *Africa's Gulf of Guinea Forests: Biodiversity Patterns and Conservation Priorities*. (Advances in Applied Biodiversity Science 6). Washington, DC: Conservation International.

Oates, J. F., Groves, C. & Jenkins, P. D. (2008). The type locality of *Pan troglodytes vellerosus* (Gray, 1862), and implications for the nomenclature of West African chimpanzees. *Primates* 50: 78–80.

Peterson, D. & Ammann, K. (2003). *Eating Apes*. California: University of California Press.

Poulsen, J. R. & Clark, C. J. (2004). Densities, distribution and seasonal movements of gorillas and chimpanzees in swamp forest in Northern Congo. *International Journal of Primatology* 25: 285–306.

Redmond, I. (2005). Where are the great apes and whose job is it to save them? In Caldecott, J. & Miles, L. (eds.), *World Atlas of Great Apes and Their Conservation*. Los Angeles: University of California Press, pp. 286–292.

Reynolds, V. (2005). *The Chimpanzees of the Budongo Forest*. Oxford: Oxford University Press.

Ristau, C. A. & Waziri, S. (1997). *Notes on a Trip to Gashaka Gumti National Park and Ngel Nyaki, Nigeria (01–19 July 1997)*. Report to WWF-UK & NCF.

SGA / IUCN / PSG [Section on Great Apes, IUCN Primate Specialist Group] (2005). Minutes, 1st meeting, *Joining Efforts for Great Ape Conservation*. April 17–19, Chicago / USA.

Sommer, V. (2008). *Schimpansenland. Wildes Leben in Afrika*. Munich: C. H. Beck.

Sommer, V. (2009). The anthropologist as a primatologist. Mental journeys of a fieldworker. In MacClancy, J. & Fuentes, A. (eds.), *Fieldwork: Applying Social Anthropological Insights to its Practice by Biological Anthropologists and Primatologists*. Cambridge: Cambridge University Press.

Sommer, V., Adanu, J., Faucher, I. & Fowler, A. (2004). The Nigerian Chimpanzee (*Pan troglodytes vellerosus*) at Gashaka: Two years of habituation efforts. *Folia Primatologica* 75: 295–316.

Terborgh, J. (1999). *Requiem for Nature*. Washington: Island Press.

Terborgh, J. & van Schaik, C. (1997). Minimizing species loss: The imperative of protection. In Kramer, R., van Schaik, C. & Johnson, J. (eds.), *Last Stand. Protected Areas and the Defense of Tropical Biodiversity*. Oxford: Oxford University Press, pp. 15–35.

Warren, Y. (2003). *Olive Baboons (Papio cynocephalus anubis): Behaviour, Ecology And Human Conflict In Gashaka Gumti National Park, Nigeria*. PhD thesis, Roehampton University.

White, D. (1990). *Report on a Visit to Ngel Nyaki Forest Reserve, Mambila Plateau (07 – 11 April 1990)*. Report to NCF & WWF-UK.

Wrangham, R. W., McGrew, W. C., de Waal, F. B. M. & Heltne, P. G. (eds.) (1994). *Chimpanzee Cultures*. Cambridge, MA: Harvard University Press.

Wrogemann, D. (1992). *Wild Chimpanzees in Lope, Gabon. Census Methods and Habitat Use*. PhD thesis, Universität Bremen.

Appendix. Narratives of Foot Surveys of Chimpanzee Habitats in the Taraba Region

The *Gashaka Primate Project* (GPP) conducted foot surveys to assess the current distribution and conservation status of chimpanzees and other wildlife first hand. These surveys supplemented previous reports, and covered areas not previously visited by naturalists. Moreover, oral reports from locals, wildlife officers and guides may not always be correct or may be outdated. The following descriptions are intended to relay a flavour of the encountered survey logistics and local attitudes, as such factors strongly impact on conservation measures, current or future.

Starting point of all surveys was Gashaka-Kwano, GPP's long-term research site. Kwano lies in the centre of the Gashaka sector of Gashaka Gumti National Park (GGNP). The forests and savannah-woodland that surround this site are exemplary for the biodiversity that is native to the wider Taraba region. The survey consisted of 10 parts, executed on 59 days from 2002 – 2007. Narratives were sketched out by AF and VS and altered in consultation with other survey members. The style is biased towards observations which non-Nigerians might make.

Cameroon Border (Southern Distribution)

Bissaula

Survey rationale, date and team

Former reports had stated the presence of chimpanzees around Bissaula. This town lies close to the border with Cameroon, which predicts a heavy hunting presence, and no taboos on the killing of primates. The survey was conducted from 04 – 07 Mar 04 by AF and JA with 2 local guides.

Survey narrative

Public transport was used from Jalingo to Takum, where we boarded a truck to Bissaula on a very bad road. Progress was slow. We took up local lodging shortly before midnight and struggled with large and inquisitive rats throughout the night. We registered at the local immigration office, where reception was initially harsh, but ultimately helpful, especially because we know the Head of Immigration in Gembu, who turned out to be a friend of this officer. With advice from the local forest officer, we hired two guides to lead us for a day's trip to the forest 6 hours walk away, into hilly terrain. We encountered 2 separate men carrying guns and 3 men separately collecting "ogbono" in large quantities, i.e., fruit of the "bush mango" (*Irvingia gabonensis*), which yield aromatic nuts. Much fire damage is evident in the forest and all grass was burnt. The forest seemed similar to the Gashaka area, with trees including *Pycnanthes angolensis, Pseudospondias microcarpa, Uapaca togoensis, Crossopteryx febrifugia, Piliostigma thonningi* and *Khaya senegalensis*. We spent the second night in the police station because we did not feel secure in the hostel after our presence had become known. We slept on the floor of a room next to the cells, where various prisoners were incarcerated for crimes unknown. The local population showed a lot of interest in us, and crowds gathered outside any dwelling we visited. Public transport was used to return to Jalingo.

Large mammals

No mammal was observed, not even a squirrel. One of our guides attempted to call some baboons for us by imitating them, which he says often works when he wants to shoot them, but nothing responded. He would unshoulder his gun at the slightest sound and was ready to kill any animal that moved, even lizards.

Chimpanzee distribution

On a rugged corner on top of a hill, we found 4 chimpanzee nests. No clean water was to be found, so we drank from a stagnant smelly pool. This gives some idea of the relative difficulty of the area, and may explain why the chimpanzees have survived here. Our guides lamented that they were not present when the nests had been built, so that they might have killed the chimpanzees. They told us that chimpanzees take refuge further up slopes inaccessible to humans, venturing down occasionally, hence the presence of these nests. The guides sometimes try to follow the chimpanzees further up, but it is difficult and not always successful. We tried to climb higher, but the terrain was too rugged. The lack of drinking water was a major constraint. However, chimpanzees do not need to drink everyday, so they could venture down for short forays, although this would seem a risky under-taking. This may also explains the absence of chimpanzee vocalisations, since the apes can rapidly learn to repress their usual ebullience in the proximity of humans who mean them harm (Sept 1992).

Conservation prospects

What little wildlife remains, will most probably not do so for long. Whether the small remnant population of chimpanzees could recover in any significant way is doubtful.

Zedah / Buru

Survey Rationale, Date and Team

Locals in this southern tip of the Fali mountains are predominantly Muslim (who do not typically eat primates) but Christian hunters from across the nearby

Cameroon also operate here. The presence of an NCF Community Forestry Project at Buru made the site interesting as a comparison with areas without direct conservation measures. The survey was conducted from 25 – 29 Jan 04 by AF, JA, VS, GPP field assistant H. Guruza and local guide Anthony.

Survey Narrative

We travelled from Serti to Buru in a land-rover supplied by NCF Gashaka, driven by Garba Mungo, thus circumventing the Fali mountain range that is not crossed by any motorable road. From Beli, the road is mostly not tarred, although the bridges, dating back to the 1950s, are mostly intact. Many surrounding hills have been clear-cut by earlier logging operations. Trade relations with Cameroon are obvious from Baissa onwards, and the non-Nigerian team members' passports were subjected to close scrutiny at a border control station. On our arrival at Buru, we introduced ourselves to the recently posted NCF Project Officer Tirimisiyu Ashimi, who was known to JA and AF from his previous position at NCF headquarters in Lagos. In the evening, NCF station showed an action-video to the local population, and a sizeable crowd of mostly youngsters gathered around the generator-powered video-recorder. That evening we attempted to locate the village chief, but he was unavailable, working on his farm. Next morning, the chief visited us in the NCF residence, and it was not a happy encounter. He claimed we were, "hiding behind NCF" to avoid paying him the required "fee" for entering the community forest. The discussion became heated at times, but as we had a permit from the state forestry department, we ultimately decided to enter the forest in any case, if not exactly with the chief's blessing. We were accompanied by a guide known to AF from Gashaka the previous year when a team from NCF Buru visited for ranger training. Locals lectured us about the extent of the forest, although their measurements were not congruent. The primary school teacher said: "It takes you 4 days to walk through", whereas our guide insisted that "You can walk in this forest for 4 weeks and not come out". In any case, we walked for about 3 h, on a well used path, passing several farms. Cultivation was extensive, with many palm trees present. Nevertheless, there was good standing forest, much of it secondary growth, particularly near rivers. We established a camp beside the foot-path as running water was nowhere else to be found. This also prevented us from venturing further. We met two single young men, both carrying guns. They were polite and shy, explaining without prompting in one case that they were not hunting, although they clearly were.

Large Mammals

No large mammals were seen, except for one potential glimpse of a mona monkey; 2 faint vocalisations of these monkeys were heard. Burrows of Gambian rats were recorded, but those of aardvarks were absent, which is peculiar, given their high abundance in GGNP. Old buffalo tracks, 2 probable diggings by red river hog and fresh red-flanked duiker faeces complete the indirect evidence of large mammals. The extended beds of a larger river were virtually devoid of animal tracks.

Chimpanzee Distribution

The primary school teacher insisted that chimpanzees can sometimes be heard "crying" from the NCF office. In the forest, despite much searching and reassurance of the local guide (who had previously hunted both primates and ungulates here), only 2 old chimpanzee nests were found. Our guide said he knew a place where chimpanzees nest continuously. This is certainly unlikely, given typical chimpanzee ranging patterns, and it was therefore no surprise that the guide failed to locate the area.

Conservation Prospects

GPP field assistant H. Guruza had visited Buru in 1996 with K. Gonder in the course of her work in establishing *P. t. vellerosus* as a fourth distinct sub-species (Gonder *et al.* 1997). They walked to Akoforo village near Buru, 4 h from our camp. People there had told that they eat chimpanzees, and in neighbouring Sabon Gida he saw a baby chimpanzee for sale. The nearby forest of Akwaizantar has since been destroyed (Chapman *et al.* 2004), and conditions in the Buru forests certainly have also not improved. The road from Beli to Takum has been widened and tarred since 2006, thus providing easy access to the entire north-eastern flank of the Fali mountains. This easy vehicular access will without doubt increase the pressure on remaining wildlife. Since the implementation of its community forest programme, NCF has formed a "Community Council of Buru" and a "Hunters' Association" (NCF 2003). Our local guide was careful to explain that while most of the men there, including himself, had hunted chimpanzees in the past, that this had stopped since NCF arrived. Other locals also adamantly expressed their conservation-mindedness, using phrases such as "NCF taught us that there are endangered species. We do not eat monkeys any more". It is hard to refrain from the cynical comment that this is accurate, because there are hardly any monkeys left. The NCF project seems to intend to give locals some role in administering the use of the forest, and a bee-keeping initiative was also mentioned. People also gratefully acknowledge construction of the primary school, but lament the lack of desks and chairs as the children either squat on the floor or use logs. The project impact will likely be very localised and probably achieve little to enhance sustainable forest management in the wider area. In any case, the Zedah forest, while some portions of large standing trees remain, has been virtually hunted empty. Conservation initiatives are unlikely to lead to a recovery. Instead, it seems only a question of time before the few remaining chimpanzees are killed.

Shebshi Mountains (Potential Northern Distribution)

Dakka

Survey Rationale, Date and Team

The Shebshi mountains – shared between Taraba and Adamawa states – might represent the limit of the range of *P. t. vellerosus* to the north (cf. McManus 2005: Map 16.15). Dakka was chosen as the western entry gate to the Shebshi range. The survey was conducted from 07 – 10 Feb 04 by AF and JA.

Survey Narrative

From Jalingo to Dakka it took 13 h of public transport, in the back of a pick-up truck and on motorbikes. The road from Jalingo to Sulkani, although tarred, is holed. Progress on the untarred road beyond was even slower, with long detours around collapsed bridges. We were warned of occasional armed robberies. The landscape is barren and dusty. Rain is said to be sporadic but severe when it comes. It was a surprise when we entered the large, orderly town of Dakka. People are mostly Chamba, with some Fulani, who predominantly farm and hunt. A Danish missionary woman called Hajia Jummai runs a clinic. We found accommodation with the local National Youth Corps. They were mostly from Lagos and Ibadan, and a place like Dakka provided a solid culture shock for them. The young women particularly complain of being ignored and ill-treated by senior staff at the school where they teach. The local food is extremely bad. We met the chief in his sizeable palace, which awaits completion in grey concrete. He wanted us to bring animals from the national park of Yankari to boost tourism of which there is currently none. He was benevolent but not really interested after he understood our mission. The local forest officer thought we wanted a permit to remove animals from the area. He was agreeable for a price and led us to believe that it has happened before. An extensive bush-meat market was evident. Local militia check vehicles for meat and other forest products. However, the purpose is not to protect wildlife per se, but to obtain money from offenders who have no licence – although we were not sure if this was an officially sanctioned system or something locally imposed. People also carry meat on motorbikes at night to avoid check points.

Large Mammals

Only small walks were conducted in the surroundings of Dakka. The forest is heavily affected by fire and reduced to burnt scrub, with small patches of bush around streams. No wild mammals were seen. No baboons were heard.

Chimpanzee Distribution

The chief of hunters in a small village reported that he has never seen chimpanzees, despite a lifetime of hunting in this bush. He said he knows of them from the Gashaka area, and recommended that we go there.

Conservation Prospects

One would need to trek deeper into the Shebshi mountains to ascertain the status of wildlife. However, at least the vicinity of Dakka is devoid of wildlife. Chimpanzees in particular appear to be absent from the wider area. It is unlikely that conservation measures of any kind would reach and be enforced in this remote location.

Vogel Peak

Survey Rationale, Date and Team

The area below the Vogel Peak (1680 m) was chosen as the eastern entry gate to the Shebshi range that might represent the northern limit of *P. t. vellerosus* (cf. McManus 2005: Map 16.15). This survey, the only one to include a location in Adamawa State, was conducted from 05 – 08 May 04 by AF, JA and local guide Ado.

Survey Narrative

We hired a taxi from Jalingo for the whole period. In Ganye, we met the local forest officer who directed us to Gurum Pawu, the last village at the foot of the hills. Here, we met the local second class chief, Alhaji Alyu, who told us that he had visited Vogel Peak once, that it was a 4-h walk, and that chimpanzees were present there. However, he clearly mistook baboons for chimpanzees, as he told us they could readily be seen on the farms. In fact, we heard baboons barking near to the village. Many farms were protected by robust wooden fences against animals, indicating that crop-raiding is a problem. The chief helped us to find a capable guide, Ado, who set out with us at 05:00 the next morning. It was cold and a strong breeze blew against us throughout. The area was heavily cultivated, particularly with Yam, and used extensively for grazing. We met many herders moving their cattle up the steep tracks leading towards Vogel Peak. We ascended through rugged, hilly terrain for several hours but did not reach any large tracts of forest. The vegetation was stunted and montane, dissimilar to that of the Gashaka area, and included trees of *Piliostigma thonningi*, *Vitex doniana* and thorny bushes. Our guide then told us that Vogel Peak was up to 12 h distant, not the 4 h originally stated, that chimpanzees were there, but that he didn't know exactly where. With no source of running water along the footpaths, we decided to return. We spent the night in Gurum Pawu – not enjoying the bad taste of the local food -, and the next day returned via Yola to Jalingo.

Large Mammals

No large mammals were encountered

Chimpanzee Distribution

We cannot ascertain whether or not chimpanzees are present around Vogel peak, as our survey was incomplete, but it seems unlikely.

Conservation Prospects

Stephen Gawaisa, forestry official at the State Ministry of Environment, Yola, and an experienced conservationist and former employee of NCF Gashaka, was openly sceptical about the prospects of finding chimpanzees in this area. He pointed out that locals are mostly Chamba, known to be fierce hunters. On site, we were not able to discern the level of hunting, but we neither saw people carrying guns nor did we hear shots, although there may well not have been anything there left to shoot at. Heavy cultivation and cattle grazing up the slopes of Vogel peak do not allow much optimism that significant wildlife can be found there. A more extensive survey would require porters and plenty of food and water.

Park Support Zone

Ngel Nyaki

Survey Rationale, Date and Team

Ngel Nyaki, on the edge of the largely deforested Mambilla Plateau, is an isolated 5-km^2 patch of plant species rich montane forest, with several endangered tree species including *Entandrophragma angolense*, *Lovoa trichilioides*, *Milletia conraui* and *Prunus africana*. Because of its biodiversity, the forest has been subject to various past surveys (e.g., Dowsett & Dowsett-Lemaire 1989, Chapman & Chapman 2001). The presence of chimpanzees has been continuously reported, and Ngel Nyaki was (mistakenly) even thought to harbour gorillas (White 1990). We surveyed from 25 to 28 Feb 02, involving team members AF, JA, field assistants H. Guruza (GPP), S. Waziri (NCF) and local guides including Misa Zubaru.

Survey Narrative

An NCF vehicle transported the team from Gashaka via Serti about 55 km to Mai Samari. Soon afterwards we turned into the dusty, red dirt road leading to

Yelwa, a small village 9 km from the forest reserve. We met with chief ranger Hamidu Njeke, recruited some porters and walked for about 1 1 / 2 hours to the edge of the forest where we set up camp for the next 3 nights. Human encroachment was obvious. The grassland around the reserve is trampled by cattle, which grazed at least into the edge of the forest. Large-scale logging was absent, but locals exploit the forest for products such as fire-wood and honey. Hunting, also illegal, clearly occurred. The chief ranger was bitter that poaching could not be controlled due to a shortage of manpower. There were, for example, 36 buffalo in 1996 but now only 16 were left. Fresh blood of an animal slaughtered by poachers was found during the survey, and 3 wire snares were removed.

Large Mammals

The basic composition of large mammal populations at Ngel Nyaki is similar to that found in nearby GGNP (*Tab. 3.1*). Our records of sightings, faeces, tracks, hearings and guide reports together suggest that none of the species reported during earlier surveys have gone extinct – although densities for some favourite hunting species such as buffalo may certainly have declined. The absence of some taxa such as hartebeest, yellow-backed and red-flanked duiker may be due to the relatively short survey period. Curiously, White (1990) produced a report entitled "*Gorilla gorilla mambilla*" in which he strongly asserts the existence of gorillas at Ngel Nyaki, based on observations of nests of 1.5 m width and built at heights of 4 – 11 m. However, tree-nests of gorillas, while occurring (Fruth & Hohmann 1994) are rare, and there is no reason to assume that these structures were built by anything else than chimpanzees.

Chimpanzee Distribution

Chimpanzees were seen on occasion by our local guides over the years. We found faeces and tracks of chimpanzees, heard them calling on 4 occasions, saw 1 individual and recorded 38 nests that belonged to 6 groups with an average size of 6.3 (range 1 – 21). While some day and night-nests might have been conflated in the largest count of 21, there are probably more than 30 individuals in the forest. The otherwise knowledgeable local guide Misa told that female chimpanzees sometimes leave their babies in the nest during the day and that some males do not construct or sleep in nests. He also reported that while chimpanzees do not hunt mammals, they do scavenge. These reports are almost certainly baseless, which warned us once more against taking the word of locals for granted without corroborating facts. Poaching of chimpanzees reportedly occurred either 3 or 20 years in the past. Of course, this might not be reliable either, as our informants may well have been themselves involved in this hunting, and may therefore have downplayed such incidences.

Conservation Prospects

Ngel Nyaki harbours a diverse set of fauna and flora, including a sizeable, if isolated, population of chimpanzees. The area has been receiving increasing attention since 2002 when Hazel Chapman from Canterbury University / New Zealand, initiated long-term research via the foundation of the "Nigeria Montane Forest Project" (Chapman *et al.* 2004, Beck & Chapman 2008). A research station at the forest edge opened in early 2006. In addition, Yelwa was selected by NCF in 2005 as project headquarter for a UK-funded Darwin-initiative on "Participatory Management of Priority Biodiversity Sites in Taraba State, Nigeria" that also focuses on Ngel Nyaki. This patch of montane forest and its wildlife is likely to benefit from this attention.

Alekum

Survey Rationale, Date and Team

Alekum forest lies about 5 h walk from Serti town, off the main road between Gembu on the Mambilla Plateau and Beli. Its location should be representative of the eastern fringe of the Fali mountains, an undulating region with steep, rocky and forested mountain tops, intersected by valleys with small settlements. The area is a State Forest Reserve. Alekum was chosen for survey due to its proximity to, but exclusion from, the national park. This should allow a near enough direct comparison between forests within and without the park, and give an indication of the effectiveness of this conservation landmark. The survey took place from 18 – 21 Jan 04, and included AF, VS, JA and GPP field assistant H. Guruza.

Survey Narrative

We were dropped by NCF land-rover at the last point accessible to vehicles on the road behind Serti, near the village of Abadogo. Then it was half a day's walk through farms with guinea corn, maize, cassava and yam along a well-used dirt track. We came across several Fulani herder shelters. Cow tracks for grazing were ubiquitous, and all grass was burnt. In contrast to walking within the national park, there was a noticeable lack of rubbish on the footpath, such as paper wrappers and the ubiquitous small plastic bags. People are very poor and seem not to have any possessions, so biscuits etc. are probably not common. Particularly the inhabitants of the village of Tapare, about half way along our route, seemed to exist in an environment of scramble subsistence. It was surprising to see such elevated levels of poverty just behind Serti, with hardly any clothes, no shoes and very skinny dogs. We followed small rivers until we crossed a pass behind which the water flows towards the centre of the Fali mountains and established camp at a small creek. The following people were met on the foot-path: man with palm wine, local medical attendant, 1 hunter with bow and arrows, 1 man with a gun 3 men with guns in a party of 4. Thus, half of all people met carried guns, some of whom made efforts to avoid us by entering the bush. Gunshots were heard on the evenings of 18 and 19 Feb.

Large Mammals

From Abadogo to Alekum forest, no mammal except one baboon was seen in an area of heavy farming. Even outside the farmed areas, the forest was silent. Throughout the 2 days, 2 baboon calls were heard, but not a single other monkey. Civet faeces were recorded, and one civet was seen at night. Tracks of bushbuck and red-flanked duiker were recorded, in addition to dung of red-flanked duiker.

Chimpanzee Distribution

No sightings or calls were recorded, but 18 nests of varying ages were found. The difficulty is knowing whether these nests represent a small group or even pair, frequently using a limited area, or a larger group passing through a much wider area.

Conservation Prospects

The area, until perhaps only a decade ago, must have clearly been prime habitat for chimpanzees and other wildlife. However, even outside the valley beds with their farming activity, the area is now open to rampant destruction by cattle grazing and hunting. Incidentally, a letter was shown to JA by Tapara villagers because they could not read it; it was from the Ministry of Environment and Minerals in Jalingo, dated 2003, stating that the ministry wants to collaborate with NCF in conservation efforts and that villagers should cooperate. This was copied to various settlements around the area, but without indications of follow up action. We were unable to ascertain if hunting was legal, in other words, whether the area is inside the "buffer zone" of the park or not. It is very unlikely that protection measures would come into effect before the area is completely devoid of wildlife and most of its forest cover. Villagers in Tapare said they eat monkeys but not chimpanzees "because they are like humans". Interestingly, these were indeed the only mammal we saw obvious signs of in what is most probably a remnant of an isolated population.

Gashaka Gumti National Park

Chappal Wade to Mambilla Plateau

Survey Rationale, Date and Team

Gangirwal is Nigeria's highest peak, rising to 2420 m within the Chappal Wade escarpment, and straddling the international border with Cameroon. The unique montane flora and the mountain range's vegetation cover has been decimated over the last decades by fire and cattle grazing (Chapman *et al.*, 2003). The trip intended to explore some of the remaining forests on the way from the Gashaka-Kwano area of GGNP towards Chappal Wade and assess potential threats posed by illegal activities emanating from the park enclaves and Cameroon itself. The survey from 12 – 17 Feb 05 included AF, VS, guide Saidu Isa Sabere, researchers J. Higham, D. Ellis, visitor Klaus Meister and 6 porters including camp manager Anthony David.

Survey Narrative

A potentially motorable dirt road leads from the GPP field station at site of the abandoned former village of Kwano to the site of the likewise deserted village of Yakuba. This track has deteriorated in recent years, necessitating a 4-h walk. At Yakuba, one crosses the river Ngiti to continue on a well established foot-path towards the river Gamgam, where we set up camp. The area is mostly grassland and savannah-woodland, with small patches of riverine forest. The perfectly clear rivers contain abundant fish and are populated by many bird species. A 4-h walk brought us the next morning to Mayo Sabere, a sizeable enclave containing substantial cattle and extensive farms. Inhabitants are supposed to limit their activities to the enclave area itself, but thick cattle tracks start many kilometres before the settlement. We camped beside a small river, where local produce was offered to us, including wild honey. The enclave surroundings have been entirely grazed-out, and only a few patches of trees remain, primarily non-native species. A further walk, mostly along the international border, brought us in the late afternoon to Jauro Hammasale, a remote Fulani settlement already in Cameroon. The village was replete with inquisitive children, good hospitality and some French speakers. The area is mountainous and denuded. It is difficult indeed to imagine the forests that must once have stood there. The next morning, it was a fairly gentle climb to Chappal Wade, through heavily grazed land. Occasional isolated Fulani homesteads were encountered, mostly deserted as the occupants were tending their herds. The vegetation became coarse and montane on the slopes, rising from short grassy meadows. We camped for one night on a meadow beside a small river before ascending to the top. The international border with Cameroon runs along the summit that is a series of barren ridges rather than a single peak. Extensive over-grazing abounds, and large herds of cattle of often enormous proportions are frequently met, as well as small pastoralist dwellings. All this is perfectly illegal but is executed in broad daylight due to a complete lack of law enforcement. The terrain is spectacular and the views suitably epic, as befitting Nigeria's highest peak. Upon our descent, we visited the vista near the spectacular rocky outcrop Dutsin Dodo, from where one has a grand view into extremely steep slopes covered by rather extensive sub-montane and montane forest. We returned via Jauro Hammasale for a final night half way towards the ranger post at Njawai. The next morning we walked along the southern border of the park, where numerous large compounds including surrounding farmland have been set up inside the protected area. Rangers were found to guard their post in Njawai, where we were picked up at noon by an NCF vehicle. A trip of about 5 h over curvy and precarious dirt roads towards Ngoroje followed. This trip took us from east to west across the northern rim of the Mambilla Plateau. The terrain in this area looks like a moonscape. The montane forest that must have previously stretched from horizon to horizon has been replaced with an utterly barren landscape, over-grazed and with huge erosion gullies that have often swallowed whole slopes. This part of the Mambilla Plateau clearly demonstrates what is likely to happen to the whole southern part of the national park. The scale of this human-made destruction in an area well over 1000 km^2 was simply mind-boggling. We traversed a dirt road to the tarred Gembu road, along which, after a stop at Ngoroje, we returned to Serti.

Large Mammals

Numerous ungulates (including buffalo, giant forest hogs, herds of hartebeest) and other wildlife (including colobus monkeys) are regularly observed on the trail from Kwano to Mayo Sabere. A large black monkey, sighted by VS in trees beside the river Gamgam and Mayo Sabere, fit the description of a grey-cheeked mangabey. These were previously unknown to occur in Taraba, but were subsequently also seen in Kwano forest (K. Arnold *et al.* pers. comm.).

Chimpanzee Distribution

Chimpanzee nests were seen near to the road between the river Gamgam and Mayo Sabere, and a small adult, perhaps female, chimpanzee was observed in the forest by AF. Vocalisations were heard twice from forests below Dutsin Dodo. The chimpanzees responded with pant-hoots to imitated vocalisations produced by AF. The steepness of the slopes made attempts to find the apes futile. The population between Kwano and the forests around Dutsin Dodo seems to be contiguous.

Conservation Prospects

Throughout grasslands leading from Mayo Sabere to Jauro Hammasale and Gangirwal, numerous Fulani pastoralist dwellings were encountered, consisting of lightly constructed grass and stick huts and rudimentary corrals. We arrested 2 boys grazing cattle clearly within the park boundary outside Sabere. Grazing is widespread and the impact very severe. The situation around Njawai indicates of how serious this process can become, with evidence of extreme erosion, including large gulleys and landslips. Across the northern Mambilla Plateau, the landscape is utterly barren. The abundance of chimpanzees throughout the forest stretching from Gamgam to Sabere and on to Chappal Wade is encouraging, but outlooks for the forest are pessimistic. The next two decades will need to see increased efforts to preserve this substantial population of *P. t. vellerosus*.

Gamgam Headwaters

Survey Rationale, Date and Team

This part of the survey aimed at the forests below the Chappal Wade escarpment, particularly regarding the distribution of chimpanzees, known to be present from the previously described hike. Rather than climbing above the area as previously, we decided to follow water-courses up from the Gamgam, hoping to be able to ascend the Chappal Wade escarpment from its western flank. We could, despite extensive discussions, not identify any park ranger or local who had previously entered these forests. There were intriguing rumours of the presence of a "large ape", reportedly not chimpanzee, in the foothills of Chappal Wade in the 1970s (Barnwell 1993b). The next known population of gorillas (*Gorilla gorilla diehli*) is about 200 km towards the south in the Bamenda highlands, but their historic range was certainly very close to the Mambilla Plateau. Thus, there is no pressing reason why these apes should not be found further north. The first (unsuccessful) attempt to climb Chappal Wade through the Gamgam headwaters took place from 10 – 14 Apr 05. It included AF, VS, senior GPP field assistant H. Guruza, guide S. Waziri, researchers J. Higham, D. Bennett, volunteer Aylin McNamara, park range officer Pepeh Kamaya, 2 rangers and 6 porters including camp manager Anthony David. A second (successful) attempt took place from 09 - 16 Mar 06. It included GPP researchers AF, VS, U. Buba and volunteer A. Dasgupta, senior field assistant H. Guruza, park range officer Pepeh Kamaya plus one additional ranger and nine porters including camp manager Anthony David.

Survey Narrative

We followed the same route as on the previous expedition, a day of walking from Kwano through the deserted village of Yakuba bringing us once more to camp beside the river Gamgam, where we passed the night. The following day, we moved up the basin, following a footpath in part, and the dry-season bed of a broad river course with overhanging trees on both sides forming perfect arcades for much of the trek. We then reached a large and spectacular waterfall that descended over several cascades, where we set up camp. A deserted camp of poachers with buffalo skeletal remains was found near. We destroyed the meat-drying rack there. Well-worn tracks leading away from this site suggested long-term use. Due to the difficulty of carrying large loads over precarious terrain in front of us, we left the food required for the return journey and some of the porters at this camp site. Several Nigerian members of the expedition were clearly relieved to be allowed to stay behind, as they had expressed anxiety about having to enter uncharted territory potentially inhabited by evil spirits. The terrain became extremely rugged and difficult once we began trekking up towards a bifurcation in the rocky river bed, where we took a right turn. The water-course forced us often to venture away from the river beds that required extensive trail-cutting through heavy undergrowth and low-lying vegetation. We navigated a deep gorge with beautiful dry-season ponds. The landscape was certainly "epic" in scale, surpassing that of any previously visited within the park. We only managed to penetrate one piece of forest with any success, and camped in a small river bed. After further slow progress following river beds, we were finally able to see Dutsin Dodo ahead of us – but then the first heavy rains of the wet season set in. Instantly, the rocks in the river beds became slippery and wet foot-wear caused severe problems to feet. We were thus forced to make our way back to the Sabere enclave. This took us about 9 h, much of it beneath a severe thunderstorm. We arrived late in the night, and were sheltered and fed by the enclave inhabitants. The next morning, we literally limped back towards the initial camping place at the river Gamgam. In the afternoon, we made our way towards Yakuba and then on to Kwano, beaten once again by rain. Due to fatigue and carelessness, one member of the party unfortunately became separated from the main group, necessitating an extensive search and recovery operation, which went throughout the night, involving the experienced guides Salamu Waziri and Hammaunde Guruza. We arrived at Kwano after a full day trekking, including several hours in the dark. Most of the western members of the expedition had suffered severe blisters and abrasions on their feet, which were treated in a native Nigerian way; with hot towels soaked in salt water. The vocalisations of the patients were remarkable throughout. A year later, we made another attempt to scale Chappal Wade from the west. This time, at the river bifurcation behind the Gamgam waterfall, we took a left turn. For a full day, we ventured up in the river bed, at times digressing towards its banks, until we were stopped by a sheer cliff, adorned by a high waterfall, right under the edge of Chappal Wade. We camped nearby. The next day, we made very little progress, with unsuccessful attempts to find a trail around the cliff that would lead to the peak. In the evening, we had to descend deep down into a narrow valley so as to find drinkable water. Morale was low, as again we could not be sure to scale the summit via this route. The next day, we summoned whatever energy remained. Around noon, with stamina and luck, we located a narrow passage leading high into the cliff. Curiously, a web site of US-creationists describes how meat-eating flying dinosaurs go on their hunts from here. The argument goes that the continued existence of these creatures (which scientists claim to be long extinct) proves theories about the evolution of life wrong; fortunately, we were unable to ascertain testimony to this and escaped from the place with our meat attached to the bones. After we had scrambled for many hours through undergrowth and heaps of boulders, we ultimately scaled the cliffs ledge. Soon and suddenly, the forest opened up, and we found ourselves on top of Chappal Wade. A strong sense of achievement overcame all members of the expedition. We camped on the escarpment as we had done on previous occasions. Over the next days, we headed back via Jauro Hammasale and Mayo Sabere to Kwano.

Large Mammals

Leopard tracks were seen in the sandy beds of rivers in the Gamgam headwaters on two separate occasions. No other large mammals were observed, perhaps due to the large size of the human expedition, but tracks and other signs of buffalo were frequently noticed. Baboons and putty-nosed monkeys were seen. Bird and small mammal foot prints abounded, particularly in the river beds. No evidence of gorillas was discovered, but much of montane forests on the western side of Chappal Wade seemed suitable for these apes. However, little of this potential habitat was surveyed, which raises the prospect of future expeditions.

Chimpanzee Distribution

Chimpanzee vocalisations were heard on several occasions between river Gamgam and the camp at the waterfall camp. At this spot, chimpanzees were seen near to the river by those who remained behind at the camp. During both expeditions, several recent nests were noted in the forests leading up to Chappal Wade, and vocalisations were heard. There can thus be little doubt that the population from Gashaka-Kwano towards Chappal Wade is contiguous.

Conservation prospects

The Gamgam headwater area beneath the Gangirwal peak harbour what must be the most inaccessible, remote and undisturbed forests of north-eastern Nigeria. This becomes increasingly evident, the further one moves away from the enclave of Mayo Sabere. On top of the ridges in the direction nearer to the enclave, no forest remains. There was severe encroachment into the forest by cattle from the direction the enclave, and forest had been clear-cut to establish farms at the outskirts of the Gamgam headwater forests. The vegetation leading up to the escarpment is dense. Further into the Dutsin Dodo basin and below the Chappal Wade cliffs, stands of primary forest remain and no signs of recent human activity were discovered. The steep slopes make access for hunters presumably difficult and prevent cattle on top of Chappal Wade to enter into the lower foothills. This relatively undisturbed area has an extent of about 20×10 km^2 and constitutes clearly a major asset of GGNP.

Eastern Enclaves from Chappal Delam to Filinga

Survey Rationale, Date and Team

The national park contains several enclaves with Fulani settlements. The most remote settlement at Chappal Delam on the western edge of the park right at the international border, is rarely visited by park authorities. A string of other enclave settlements – Filinga, Chappal Hendu, Selbe – leads through the highlands towards Gashaka. We intended to get an impression about the vegetation cover in these highland enclaves. The survey took place from 09 – 17 Mar 07. It included GPP researcher VS and volunteer N. Alberts; 2 representatives from GPP sponsor Chester Zoo (M. Pilgrim, R. Wilkinson), 1 guide (Saidu Isa Masabere), 2 rangers from the GGNP research unit (including head Joseph Bisong) and 8 porters including camp managers A. David and F. Vitalis.

Survey Narrative

The first 3 days of the hike followed the route through the Gamgam headwaters to the Chappal Wade escarpment established previously. We then proceeded over 2 days towards Chappal Delam, mostly upon deforested ridges that demarcate the international border with Cameroon. Vistas to the east into the neighbouring country went over a completely deforested landscape dominated by grassland and dotted with hamlets and some sizeable settlements. The views towards west into the national park revealed large chunks of forest in the distance, with groves remaining in steeper parts. Cattle treks lead deep into the park, and burning of grassland and forest was evident. The enclave of Chappal Delam seemed to be inhabited by just one extended family, which provided us with much genuine hospitality. This included provision of a horse plus handler for one of the non-Nigerians who found the trip particularly strenuous and for whom transport on horseback brought some relief. We trekked for another 3 days via Filinga and Selbe back to Kwano, over ridges and through valleys, where forest of any sort is only found along riverbeds, if at all. Trade relations with Cameroon are a major factor in this area. The vast majority of the landscape has been turned into pasture for cattle.

Large Mammals

The open landscape with its grasslands is utilised to graze cattle. We did not encounter many large wild mammals, except for some groups of baboon and isolated incidences of ungulates.

Chimpanzee Distribution

No signs of chimpanzees were recorded during the 6-day trek from Chappal Wade to Selbe.

Conservation Prospects

Huge chunks of landscape, which is original habitat for montane forest, have been converted into cattle pasture. Access to these vast grazing grounds seems to be monopolised by a few well-to-do Fulani clans. The demarcation of the enclaves is unclear in many parts, and thousands of heads of cattle stray freely into area designated as parkland. The cattle-grazing activities from within the Fulani settlements are not controlled by the park authorities. This would also seem a most improbable task, given the huge extend of the enclave areas. Grazing, combined with burning, turns the whole landscape slowly but steadily into grassland, thus depriving wildlife of its habitat.

Chapter 4
Monkeys and Apes as Animals and Humans: Ethno-Primatology in Nigeria's Taraba Region

Gilbert Nyanganji, Andrew Fowler, Aylin McNamara, and Volker Sommer

Abstract Nigeria's remote Taraba region harbours a wealth of wildlife. This diversity has been greatly reduced in many locales. We explored local attitudes towards monkeys and chimpanzees through a questionnaire survey, with the goal of identifying factors that may aid conservation measures. This so-called ethno-primatological approach ultimately aims to mitigate the cultural and perceptive isolation of non-local conservationists and primatologists.

Interviewees made clear distinctions between monkeys and chimpanzees. Monkeys reportedly raided crops more often, were more frequently hunted and eaten, and it was more commonly believed that chimpanzees were either protected or should be protected. Widespread folklore credits chimpanzees with human-like qualities. Eating of ape meat is therefore restricted by certain taboos, including an Islamic ban on eating primate meat. On the other hand, chimpanzees are killed precisely because of their human-like character, as their body parts are considered to be powerful ingredients for medicine. Because of their more human-like appearance, chimpanzee babies also make more desirable pets than monkeys. In any case, traditional taboos are increasingly breaking down because of an influx of immigrants from non-Muslim areas, and because of a commercialisation of the bush-meat trade. The sale of a single chimpanzee carcass provides hunters with roughly the equivalent of a month's salary for a government employee such as a national park ranger.

The results revealed at times striking differences between the Cameroon Border Area (more intense agriculture and plantations; largely influenced by Christian culture; cross-border trade activities; relatively remote from the national park) and the Park Support Zone (more pronounced pastoralist activities in the more extensive savannah-woodlands; largely influenced by Muslim culture; greater awareness concerning the existence and function of protected areas, in particular the neighbouring national park). Hunting, display of meat in markets and restaurants, display of primate pets and consumption of chimpanzees and monkeys is more frequently

V. Sommer (✉)
Department of Anthropology, University College London, London WC1E 6BT, UK
e-mail: v.sommer@ucl.ac.uk

V. Sommer and C. Ross (eds.), *Primates of Gashaka*, Developments in Primatology: Progress and Prospects 35, DOI 10.1007/978-1-4419-7403-7_4,

reported near the border, where protective laws are both less well known and less appreciated. Some of these differences may be due to a greater reluctance of interviewees living near the national park to admit to the existence of illegal primate killings and bush-meat consumption. However, most of the difference is probably due to cultural practices, more frequent law enforcement and better information about nature conservation in the vicinity of the park. As a result, the decrease in sightings of primates is more dramatic near the border than in the park vicinity. Also, near the border, informants could name an average of only 0.9 locations where primates occur, whereas those residing near the park named 1.3 locations.

Keywords Ethno-primatology • Folklore • Conservation • Primates

Introduction

Western views of "nature" are often based on a romantic perception of pristine ecosystems untouched by humans. This ideal might hold some truth for the depths of oceans and the vastness of Antarctica. However, for the great majority of eco-systems, this view is not correct (Markl 2007). For example, monkeys and apes exclusively populate regions, which are also populated and exploited by humans. Thus, there is geographic overlap with humans wherever non-human primates exist, i.e., these taxa are sympatric (Wolfe & Fuentes 2007). Of course, there are areas in the Amazon, the Himalayas or the Congo basin, into which relatively few humans venture, and then only occasionally. However, the vast majority of monkeys and apes are exposed to anthropogenic influences on a routine basis (Burton & Carroll 2005).

This sympatry can be either beneficial or detrimental for the survival of non-human primates. In some parts of the globe, human activities lead to a localised increase in primate populations as and when people feed free-ranging primates. A provisioned population of Barbary macaques thrives, for example, in Gibraltar. The feeding itself is largely a leisure activity of tourists, who entice and enjoy the human-like antics of the monkeys (Fa 1986). There is some mythical connotation attached to the Gibraltar monkeys, as their presence is said to guarantee the continued British governance of this territory. Provisioning of monkeys is particularly popular in Asia, for example in Japan (Japanese macaques; Kurita 1999) and Taiwan (Formosan macaques; Hsu & Lin 2001), where it is at least partly driven by religious sentiment. A very explicit religious motivation is behind the provisioning of monkeys in south Asia (Bishop *et al.* 1981), such as in India and Nepal (rhesus macaques, Hanuman langurs), Thailand (stump-tailed macaques, long-tailed macaques), and Indonesia (long-tailed macaques). Here, monkeys are cult-objects, and their sacred status connected with the *Ramayana* epic, in which the monkey general assists the God Rama to find and rescue his beloved wife from the hands of a demon, who had kidnapped her (Cadet 1971, Lal 1981, Sommer 1996). Human agricultural and horticultural activity can also make additional food resources available for non-human primates. This will then often lead to crop-raiding and thus a typical form of human-wildlife conflict (Paterson & Wallis 2005, Warren *et al.* this volume [Ch. 8]).

On the other hand, humans can be the party that exploits non-human primates. Monkeys and apes may, for example, be used as pets, as performing entertainers or in biomedical research (Fuentes & Wolfe 2002). This utilisation does not require that the non-human primate be immediately killed, as it is the case when their body parts are used as a source of meat, as ingredients for traditional medicine, or for fetishist purposes. Primates are relatively large mammals, and as such, are preferred targets for hunters (Fa *et al.* 2002). Interestingly, while there is growing evidence about how primate behaviour and morphology were shaped by long-lasting selection pressure caused by predation from mammalian carnivores, other non-human primates, reptiles and birds of prey (Miller 2002), there is little information on if, and how, continued exploitation by humans might have had similar effects (Wolfe & Fuentes 2007). The way humans prey on non-human primates is often mitigated by cultural factors. For example, peoples in the Manu region of Peru prefer to hunt large primates such as spider monkeys and woolly monkeys. Nevertheless, howler monkeys are sometimes spared, as they are considered to be less tasty and also somewhat lethargic – a characteristic that is feared to be passed on to children who consume their meat (Shephard 2002).

"Ethno-primatology" studies the complex interactions between humans and non-human primates (Wolfe & Fuentes 2007) that grow out of their often-sympatric existence as well as the tendency of humans to anthropomorphise these organisms, which resemble them most closely. We aim to contribute to this emerging discipline through a study of the interconnections of humans and non-human primates in north-eastern Nigeria. Here, humans have co-existed with several species of monkeys and apes for certainly thousands of years, if not much longer. However, the last 50 years or so are likely to mark a turning point in this relationship, as humans are increasingly threatening the very survival of non-human primates. The main causes are an increase in human population density combined with increases in agricultural and pastoralist activities, plus increased exploitation of non-humans as bush-meat through the use of shotguns (Peterson & Ammann 2003; Adanu *et al.* this volume [Ch. 3]).

We conducted a questionnaire survey of local communities in the wider Taraba area to explore the factual knowledge about the natural history of primates and contrasted them with records of traditional beliefs about these animals, which we collected at the same time. We also tried to ascertain the levels of hunting and bush-meat consumption, and if these differ with respect to species. Finally, we tried to assess local attitudes towards conservation. For this, we compared three areas with different levels of legal protection of wildlife and socio-economic background, i.e., the *Cameroon Border Area* (least protected, strong Christian influence), the *Park Support Zone* (medium protection, strong Muslim influence), and *Gashaka Gumti National Park* (GGNP; most protected; sparsely populated, almost exclusively by Muslims).

"Contextualizing myth and folklore combined with detailed assessments of the actual patterns of interactions" might be a helpful approach for primatologists who strive to encourage "people from places of sympatry" to "cooperate with government officials, conservation and management groups" (Wolfe & Fuentes 2007: 699). Our survey thus aims to contribute positively to efforts to secure the continued existence of non-human primates in north-eastern Nigeria (*Fig. 4.1*). Our particular concern is about the Nigerian chimpanzee (*P. t. vellerosus*), which has its remaining population stronghold in this area (Adanu *et al.* this volume [Ch. 3]). Chimpanzees are the closest living phylogenetic relatives of humans, and may thus merit special attention. Nevertheless, the particular concern for their survival is probably also embedded in a mental "scala naturae", which values organisms more if they resemble humans

Fig. 4.1 Primates of Taraba, north-eastern Nigeria: (**a**) black-and-white colobus, (**b**) mona monkey, (**c**) putty-nosed monkey (**d**) tantalus monkey, (**e**) olive baboon, (**f**) chimpanzee. Monkey photos taken in Gashaka-Kwano (David Bennett [**d, e**], Kate Arnold [**a, c**]; Aylin McNamara [**b**]); the chimpanzees live in the "Pandrillus" sanctuary, Afi Mountains, Cross-River state / Nigeria (photo: Yvonne Pohlner)

more closely (Janson 1952). Research about human attitudes towards chimpanzees is therefore in itself a topic that would merit an ethno-primatological analysis.

Materials and Methods

The questionnaire survey covered areas around GGNP where chimpanzees occurred at least in historic times, although local populations may have already gone extinct. Questions about chimpanzees were contrasted and supplemented with inquiries about sympatric monkeys, i.e., the more terrestrial baboons, patas monkeys and tantalus monkeys, as well as the more arboreal mona, putty-nosed and black-and-white colobus monkeys.

To test for reliability of the respondents, questions were also asked about gorillas, which are not known to occur in the area. The interview technique also aimed to determine whether and how chimpanzees occupy a special status amongst the local human populace. We therefore also collected narratives of folklore, tales and anecdotes (*Appendix*), as well as verbal expressions of attitudes towards the apes, e.g., whether people consider them to be fair game, if they are connected to hunting taboos, and what locals think about conservation.

Information was gathered from Feb 04 – Mar 05 in two major areas: 10 villages and towns in or near the support zone of the national park, and 8 villages and towns near the Cameroonian border (*Tab. 4.1*; *Fig. 4.2*). These areas were likely to reflect

Table 4.1 Location and dates of interviews

Area	Location	Date	Interviews (*n*)
Cameroon Border Area (south-west Taraba)			
	Gembu	10 Apr 05	2
	Buru	09 – 11 Mar 04	5
	Akwame	07 – 09 Mar 04	5
	Abong	05 – 07 Mar 04	8
	Baissa	04 – 06 Mar 04	7
	Bissaula	01 – 03 Mar 04	9
	Ussa	28 – 29 Feb 04	9
	Takum	26 – 27 Feb 04	9
Park Support Zone (west of Gashaka Gumti National Park)			
	Yelwa	30 Mar – 01 May 05	14
	Mai Samari	01 – 02 Apr 05	9
	Mayo Selbe	28 – 29 Apr 04	8
	Bodel	30 Apr 04	3
	Goje	27 – 28 Apr 04	5
	Beli	22 – 24 Mar 05	10
	Jamtari	23 – 26 Mar 05	11
	Jatau	22 – 24 Mar 05	9
	Gidin Dutse	25 Mar 05	1
	Kungana	23 – 25 Mar 05	10

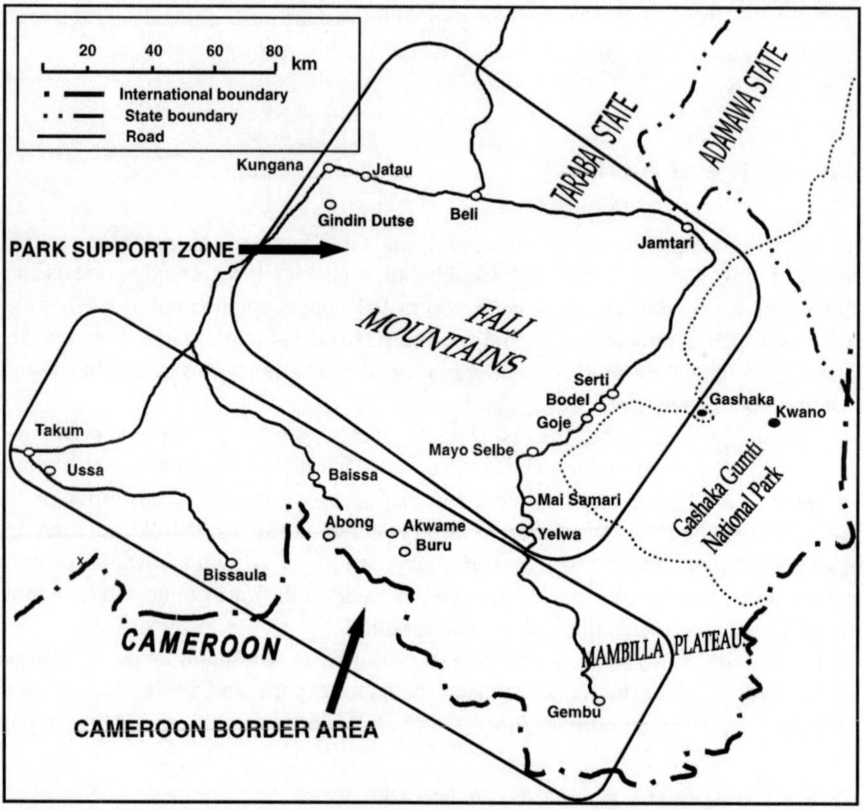

Fig. 4.2 Gashaka Gumti National Park and its geographical positioning in Taraba and Adamawa State in eastern Nigeria with locations of interviews in the Cameroon Border Area and Park Support Zone

different sets of attitudes towards wildlife and its protection. The dominating religion in the support zone is Islam, which instructs followers to observe certain food taboos (including not eating primates), and inhabitants are likely to be familiar with the existence of protection measures towards wildlife. The populace near Cameroon is less familiar with nature conservation measures and more influenced by Christianity, which places less emphasis on food taboos.

The survey focused on groups that might professionally come into contact with wildlife, such as hunters, farmers, forest guards or restaurant owners who sell bushmeat. An attempt was made to balance the composition of the pool of informants (*Tab. 4.2*) with respect to age, status, wildlife-oriented occupation, and religion. Ethnicity varies greatly from locale to locale and could thus not be balanced. Moreover, the societal structures in Taraba are male-centred, and men are more likely to occupy roles that come into contact with information about wildlife. We therefore did not attempt to balance the survey with respect to gender. Most informants were consequently male, except for some women such as restaurant owners who were affiliated with the wildlife trade.

Table 4.2 Demographic characteristics of interviewees

Category	Subcategory	Cameroon Border Area	Park Support Zone
Interviewees	*N*	54	80
Sex (%)			
	Male	96	100
	Female	4	
Age (%)			
	< 20 yrs		1
	20 – 29 yrs	11	9
	30 – 39 yrs	37	35
	40 – 49 yrs	13	24
	50 – 59 yrs	13	13
	60 – 69 yrs	17	13
	70+ yrs	9	6
Religion (%)			
	Christian	65	51
	Muslim	33	49
	African traditional	4	
Status, profession (%)			
	Hunter	38	21
	Farmer	30	45
	Ranger, forest guard	5	15
	Civil servant, teacher	11	11
	Restaurant operator, butcher, bush meat dealer	6	2
	Student, applicant	5	3
	Other (fisher, herb collector, carpenter, chief, business person)	6	3
Ethnicity			
	Tigun	33	1
	Nduro	11	10
	Fulani	4	14
	Mambilla	2	13
	Kaka		13
	Jibu		11
	Kuteb	11	1
	Jukun	9	5
	Tiv	2	5
	Chamba		5
	Ichen	7	
	Kambu	2	4
	15 other tribes	19	19

During the survey, laminated colour images (obtained from Kingdon 1997) of the following primates were shown to informants: baboons ("gogo", in Hausa-Fulani), patas monkeys ("jambiri"), tantalus monkeys ("kirikaa"), mona monkeys ("gim-chiki"), putty-nosed monkeys ("bakinbiri"), black-and-white colobus ("biri mai roro"), chimpanzees ("biri mai ganga"). While gorillas do not occur in the area, they were included as a control to determine the reliability of responses.

The questionnaires recorded date, location, basic information about the respondents (sex, tribe, age, religion, status / occupation), and whether or not the interviewer deemed the respondents reliable.

A first set of questions gathered semi-quantitative information on the occurrence of non-human primates in the area:

- Have you ever seen "with your own eyes" one of the following animals (represented by the laminated image)?
- Can you name the animals from the picture?
- Date of last sighting < 12 months or > 12 months?
- Give more accurate date if possible (day / month / year).
- Location of last sighting.
- How many animals have you seen?
- Is this your only sighting of this animal?
- If "no", where have you seen this animal before?
- Have your sightings of this animal increased or decreased since you have lived here?
- How far from the village (km, miles – provide scale) do these animals live?
- Which animals do raid crops? How much damage is done?

Other questions ascertained if respondents could discern monkey-specific behaviours from those of chimpanzees, such as nest-building, pant-hooting and drumming of buttress roots:

- Which of these animals is building a nest in a tree?
- Which animal cries out loud and drums?
- What do you know about chimpanzees? What do they do?
- If informant knows about drumming: How do chimps produce the sound?
- What do you know about the other animals?
- What do they do?

A final set of questions elicited information about hunting practices:

- Have you ever seen one of these animals dead?
- Have you ever seen adult animals or baby animals as pets?
- Have you ever seen baby animals for sale? Price?
- Which of these animals is hunted or killed?
- How is it hunted or killed (snare, trap, arrows, guns)?
- Have you ever hunted or killed one of these animals?
- Which of these animals is eaten?
- If "no": Why is the animal not eaten?
- Have you ever eaten any of these animals?
- Which meat is best?
- How expensive is the meat?
- Do these animals transmit diseases?

– Which of these animals do you believe is protected?
– Which of these animals should be protected?

The interviewer also recorded *ad libitum* other observations at the site, which seemed relevant to the topic (for example: Is bush-meat sold in the local market or served in restaurants? Are people carrying guns?).

The information was recorded in as near to the narrator's own words and style of telling as possible. The following guidelines concerning interviewing technique were adhered to:

– "During interviews, do not convey the impression that you are there to implement any conservation policy because people might be afraid of that."
– "Avoid asking leading questions, which would allow people to give you an answer they think you may want to hear."
– "Many answers may have nothing to do with the truth, and informants may well say what they think you want to hear. Just report what people say, but indicate whether or not you judge the information reliable."

The questionnaire survey was implemented by GN of NCF Gashaka, a native of the Mambilla Plateau, fluent in several local languages, who – as a completing undergraduate student in biology from Maiduguri University – had the educational background for such a task. The data were jointly analysed by the authors.

Results

General Knowledge About Monkeys and Apes

A large proportion of interviewees in both Cameroon Border Area and Park Support Zone reported having seen chimpanzees (85 %) or the 6 monkey species (73 %) about which questions were aimed (*Tab. 4.3: Q1*; *Fig. 4.3*). Nobody claimed to have seen gorillas – which do not occur in the area – indicating that the reports have a certain degree of reliability.

The good acquaintance with primates may be due to the fact that about 80 % of interviewees were hunters, farmers, rangers or forest guards who are likely to come into contact with wild primates (cf. *Tab. 4.2*). Wild apes are normally only seen by hunters or forest guards. Many sightings of chimpanzees therefore certainly refer to captive animals that are for sale or kept as pets.

The arboreal colobus, mona and putty-nosed monkeys were seen with lower frequencies than the more terrestrial tantalus monkeys and baboons. The much higher proportion of patas monkey sightings in the Park Support Zone is due to the fact that these terrestrial monkeys inhabit savannah-woodland, which is less prevalent near the Cameroon border.

Overall, sightings and general knowledge about primates were more pronounced near the park than the border. Inhabitants of the Park Support Zone had seen both

Table 4.3 Main interview questions: overall results for chimpanzees compared to average for 6 monkey species

Category	Question	Chimpanzees				Monkeys			
		C	P	Mean	test	C	P	Mean	Test
General knowledge									
	1. Have you ever seen this animal?	78	93	85	*	68	79	73	***
	2. Can you name the animal from the photo?	78	90	84	*	66	78	72	***
	3. Does this animal cry out loud and drums?	77	90	83	*	0	0	0	ns
	4. Does this animal build a nest in a tree?	76	89	83	ns	1	0	1	ns
Hunting									
	5. Have your sightings of this animal increased?	15	30	22	*	10	38	24	***
	6. Ever seen one of these animals dead?	67	46	56	*	94	69	82	***
	7. Is this animal hunted?	67	64	65	ns	94	83	88	***
	8. Have you ever hunted or killed this animal?	24	5	15	***	60	31	45	***
Consumption and trade									
	9. Is this animal eaten?	40	50	45	ns	96	87	91	**
	10. Have you ever eaten any of these animals?	7	12	10	ns	57	43	50	***
	11. Have you ever seen baby animals for sale?	67	3	35	***	37	17	27	***
	12. Ever seen adult or baby animals as pets?	69	45	57	**	55	28	42	***
Attitudes towards protection									
	13. Does this animal raid crops?	5	5	5	ns	74	60	67	**
	14. Do these animals transmit diseases?	2	11	7	*	0	0	0	ns
	15. Do you believe this animal is protected?	100	92	96	ns	26	90	58	***
	16. Should this animal be protected?	62	95	78	ns	14	91	52	***

C = Cameroon Border Area = 54 interviews; P = Park Support Zone = 80 interviews; all figures = % interviewees responding "yes" [or correctly = Q1–4]; values < 100 indicate "no answer" or "don't know"). test = 4-field test, df = 1; ns = non-significant, *$p < 0.05$, **$p < 0.01$, ***$p < 0.001$.

chimpanzees and monkeys significantly more often (*Tab. 4.3: Q1*), could correctly name specific primates significantly more often (*Tab. 4.3: Q2*). Here, significantly more respondents also knew that chimpanzees produce calls and drums (*Tab. 4.3: Q3*) and more knew that chimpanzees construct nests (*Tab. 4.3: Q4*).

Local knowledge about the natural history of primates and chimpanzees in particular is peppered with folkloristic elements (*Appendix*). The common Hausa name for the chimpanzee is "biri mai ganga" – "the monkey with the drum", which refers

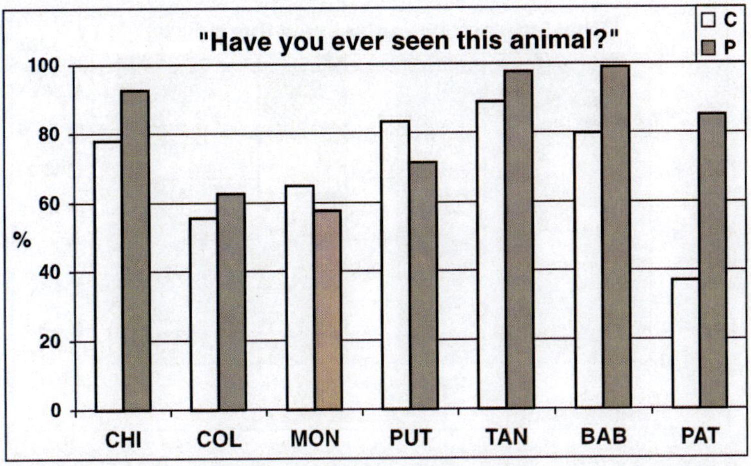

Fig. 4.3 Sightings of primate species. (**a**) Cameroon Border Area ("C", white bars, 54 interviews); (**b**) Park Support Zone ("P", shaded bars, 80 interviews). CHI = chimpanzee, COL = black-and-white colobus, MON = mona monkey, PUT = putty-nosed monkey, TAN = tantalus monkey, BAB = olive baboon, PAT = patas monkey

to the fact that chimpanzees strike the large buttressed roots of trees. According to folklore, such drumming was invented by a chimpanzee man so that his lost wife could find him (*App. [1]*). Drumming is also believed to represent a secret language (*App. [2]*) or is intended to signal that play is over and that it is time to go to sleep (*App. [3]*). There is a kernel of truth in these beliefs, as drumming is indeed probably used by chimpanzees to communicate their presence to each other. Another story maintains correctly that chimpanzees build a new nest every day, to then state that this aims to avoid being bitten by snakes who might settle in an old shelter (*App. [5]*).

More interviewees in the Park Support Zone had seen chimpanzees or monkeys within the last 3 months (*Fig. 4.4*). In addition, primates seem to live closer to human settlements near the park, as indicated by the much higher proportions found less than 10 km away from the village (*Fig. 4.5*). One reason for this could be the more extensive conversion of forests into farmland closer to Cameroon, which reduces the likelihood of spotting any of the more arboreal primates. Moreover, many villages within the Support Zone are < 3 km away from the actual national park where primates are potentially more abundant.

Respondents reported correctly that groups of tantalus monkeys and baboons had the most members. Quite realistic was also that 80 % of all chimpanzee sightings included < 10 animals (*Fig. 4.6*). For all 6 monkey species was true that large groups of monkeys above 25 members were seen more frequently in the Park Support Zone. Overall, these results seem to reflect a reliable knowledge of group sizes of local primates.

Only 10 – 38 % of respondents stated increased sightings of primates, and the higher figures were again significantly more common in the Park Support Zone (*Tab. 4.3*: Q5).

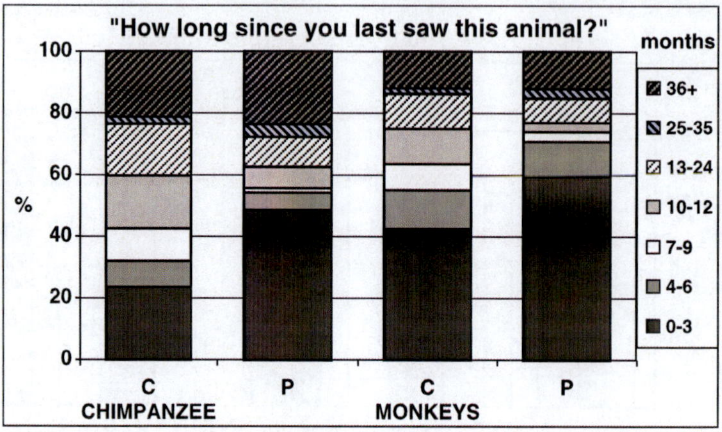

Fig. 4.4 Length of time since last sighting of different primate species (see *Fig. 4.3* for key)

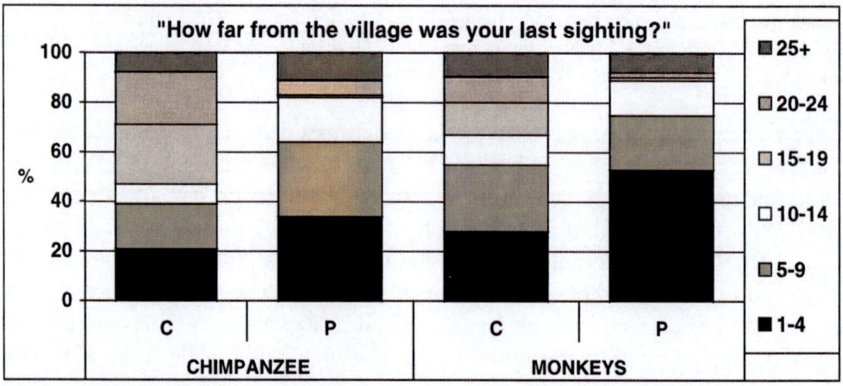

Fig. 4.5 Sightings of different primates: distances from the village of interviewee (see *Fig. 4.3* for key)

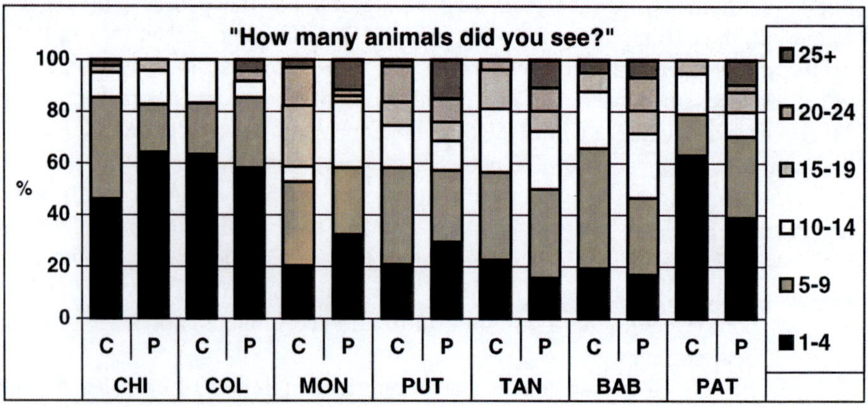

Fig. 4.6 Size of groups for different primate species observed at last sighting by interviewees (see *Fig. 4.3* for key)

Hunting of Primates

Hunting of all species was reported and the majority of respondents had seen dead animals. Hunting pressure is reportedly higher for monkeys than chimpanzees (*Fig. 4.7*). The latter might gain a certain protection from the ascription of human-like qualities to them (see below) as dead chimpanzees had been seen by only 56 % of respondents as opposed to 82 % who had seen dead monkeys (*Tab. 4.3: Q6*; *Fig. 4.8*). Similarly, hunting of chimpanzees was confirmed by 66 % of respondents whereas 89 % confirmed that monkeys were hunted (*Tab. 4.3: Q7*). Moreover, only 15 % admitted to having killed an ape, while the figure for monkeys was three times

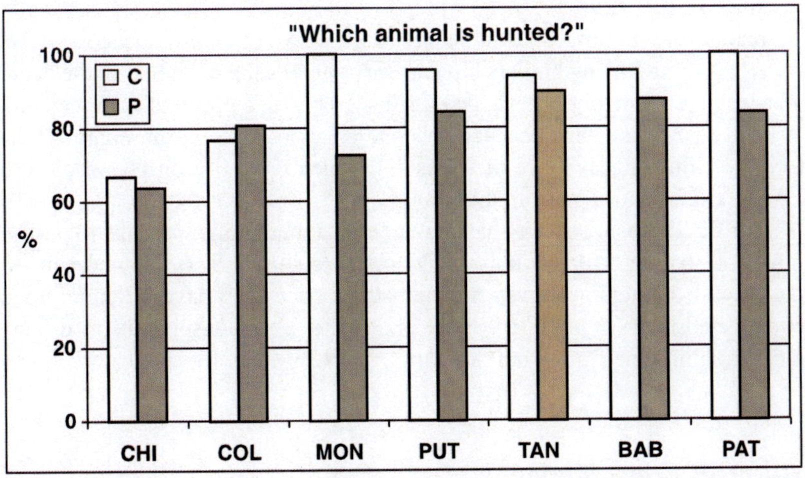

Fig. 4.7 Affirmative interview responses to whether different primate species are hunted (see *Fig. 4.3* for key)

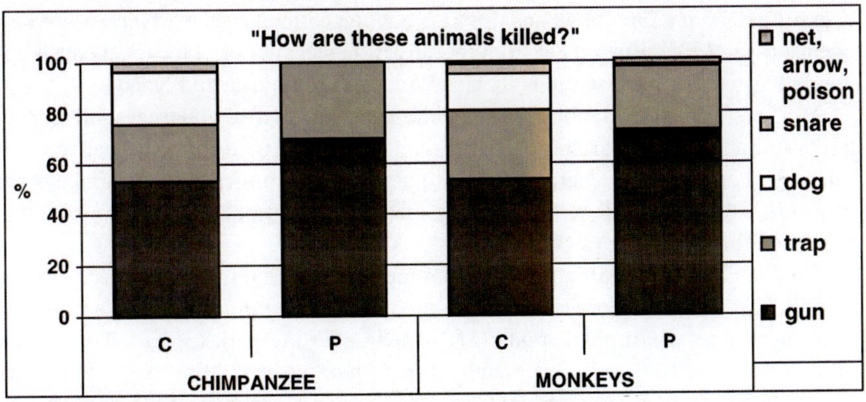

Fig. 4.8 Methods used to hunt different primate species (see *Fig. 4.3* for key)

as high (45 %; *Tab. 4.3: Q8*). Most of the above questions indicated that hunting was more common in the border area (*Tab. 4.3: Q5–10*).

In both areas, guns were the most commonly cited hunting weapon, followed by traps (*Fig. 4.8*). Hunting with dogs only occurred near the border, whereas arrows and poison, however rare, were only used in the park vicinity. Some interviewees said that hunters will stay for days near a water hole they have poisoned, watch which animals drink and then follow them to where they die.

Folklore states that chimpanzee hunting can be a tricky affair, as they dodge weapons thrown at them, to then turn around and use them against their hunters. To avoid this, hunters are advised to capitalise on the ape's tendency to imitate. Thus, a hunter should pretend to stab himself with a spear in the stomach. If the chimpanzee catches the spear, he will intend to launch it back – but not before fatally stabbing himself (*App. [26]*).

In reality, the hunting of chimpanzees follows a rather established pattern, as reported to AF and JA by hunters hired as survey guides in Bissaula: In the evening, from a vantage point, one listens, and if calls are heard, a group of men then quietly enters the bush and sleeps near the nest site or approaches in the night. At dawn, before the chimpanzees leave their nests, the men open fire on the whole group. They can kill an entire party in this way.

Killers of chimpanzees may have to undergo a quarantine, perhaps to purify the hunter from having killed a human-like creature (*App. [27]*). However, in some areas at least, hunters are given honorary titles once they have killed an ape and become celebrities (*App. [28]*), or are entitled to wear conspicuous bird feathers and drink palm wine from a buffalo horn (*App. [29]*).

Bush-meat Consumption

Primates are hunted because of their meat. Chimpanzees are also killed because body parts are used as medicine. For example, ape bones are believed to heal fractures (*App. [31]*), medicine made from chimpanzees can strengthen the body (*App. [32]*) or assist in labour (*App. [34]*), and drinking out of a chimpanzee skull increases intelligence (*App. [33]*). Bush-meat from the survey areas is not only for local consumption but also transported to cities such as Wukari, Jos, Onitsha and Kano.

A difference between monkeys and chimpanzees was also apparent with respect to consumption. Not even half of the respondents confirmed that chimpanzees are eaten by people (45 %) whereas twice as many (92 %) confirmed that monkeys are eaten (*Tab. 4.3: Q9*). Self-reported chimpanzee consumption (10 %) was also much rarer than that of monkeys (50 %; *Tab. 4.3: Q10*).

The likelihood of finding primate bush-meat in the local market is twice as high near the Cameroonian border (*Fig. 4.9*), and more people at the border also say that bush-meat is served in restaurants (51 %) than near the park (36 %). The locally used code name for chimpanzee and other primate meat is "fine boy". However, relatively few respondents were able to state their preference for specific primates. Still, the meat of putty-nosed, tantalus and patas monkeys was rated highest, whereas baboon meat was less preferred (*Fig. 4.10*). Mona and putty-nosed monkeys were considered to be the best meat near the border. As they raid the area-typical plantations (see below), they might be killed more often due to human-wildlife conflicts, and their meat might thus be in fashion. Colobus monkeys were the least desirable of the monkeys. Nevertheless, they are also hunted because their skins can be made into trophies, which were seen in several houses in the survey area, or for the manufacture of handbags. Chimpanzee was clearly the least preferred meat.

The prices for monkey meat in the two areas were very similar (*Tab. 4.4*) and, in general, followed the desirability of the meat. Thus, mona monkey was the most expensive, followed by putty-nosed monkey. These two species are often found in polyspecific association and so might also be hunted together. However, the price for chimpanzee meat was relatively high, particularly near the national park. This is surprising considering that ape meat is much less desired.

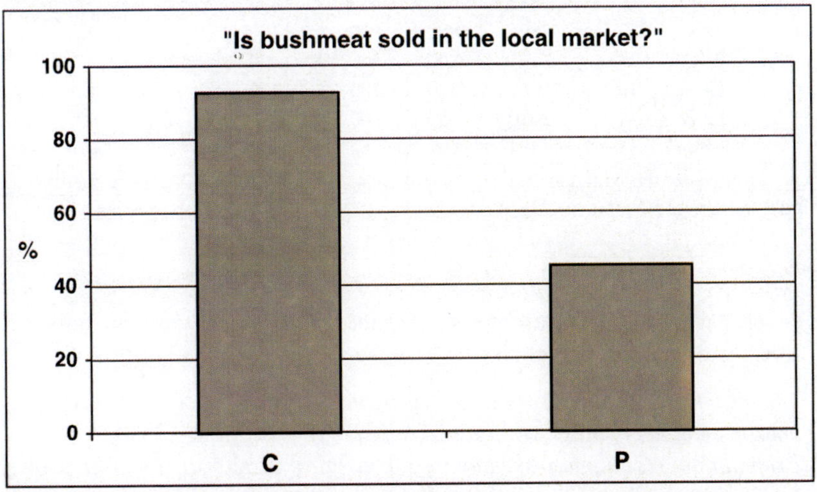

Fig. 4.9 "Is bush-meat sold in the local market?" Affirmative interview responses (see *Fig. 4.3* for key)

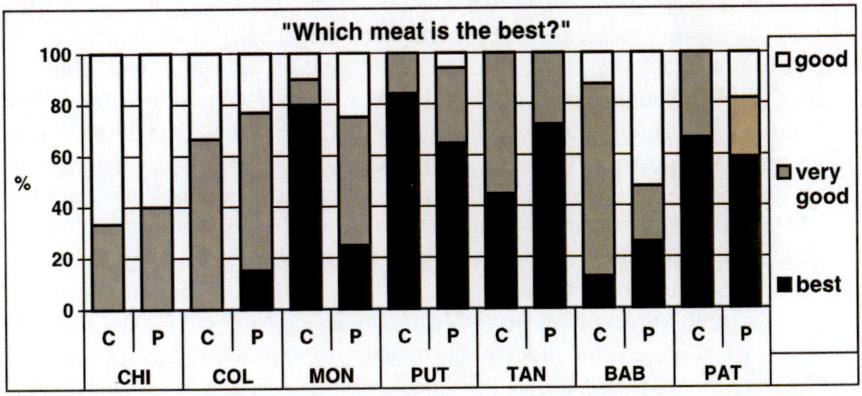

Fig. 4.10 Comparative rating of meat desirability between primate species (see *Fig. 4.3* for key)

Table 4.4 Price of meat for different primate species

Area	Species	Res-pondents (n)	Mean price (Naira) for whole carcass	Median price (Naira) for whole carcass	Minimum price (Naira)	Maximum price (Naira)	Pieces (n) in carcass	Mean price (Naira) per kg	Mean price (US-$) per kg
Cameroon	CHI	9	3578	3800	1200	5000	12	110	0.82
Border	COL	18	953	800	400	2500	2	58	0.43
Area	MON	32	859	800	100	1600	2	191	1.41
	PUT	36	867	800	200	2000	2	160	1.19
	TAN	38	814	800	20	1800	2	112	0.83
	BAB	31	2337	2500	300	6500	6	83	0.61
	PAT	15	964	900	60	1800	2	69	0.51
Park	CHI	17	5306	5000	2000	9000	12	163	1.21
Support	COL	26	883	800	300	1800	2	53	0.40
Zone	MON	20	865	800	500	1400	2	192	1.42
	PUT	28	791	800	500	1200	2	146	1.09
	TAN	43	810	800	400	1400	2	112	0.83
	BAB	44	2610	2650	900	3900	6	92	0.68
	PAT	35	864	800	450	2400	2	62	0.46

All prices in Naira, the Nigerian currency (approximate rates for 2004 – 2005; 1 £ Sterling = 250 Naira, 1 US-$ = 135 Naira; see *Fig. 4.3* for key).

People gave several reasons for *not* eating meat of primates, which are re-enforced or reflected by traditional beliefs.

- *The meat is not edible*. This reason ties in with popular beliefs such as that chimpanzee meat may cause an abortion (*App. [24]*).
- *Primates possess human-like qualities*. Human-like traits are not often explicitly credited to monkeys, but a particularly succinct story tells of a patas monkey that extended a helping tail to a lion caught in a pit. The ungrateful lion then attempted to eat his rescuer. Fortunately, a wise turtle taught the lion the ever pertinent lesson to "never threaten any one who has helped you before" (*App. [37]*). In contrast to monkeys, a general theme maintains that chimpanzees were originally humans who were cursed to live in the bush because of bad behaviour, such as abomination (*App. [15]*), fishing on Saturdays (*App. [16]*), quarrelling (*App. [17]*) or simply being stubborn (*App. [18]*). In tune with this topic, chimpanzees are reported to copy human behaviour, by, for example, carrying loads on their head (*App. [7]*), or that they are human-like because they cry (*App. [23]*); use tools to obtain honey or termites (*App. [10]*); share food such as fish (*App. [11]*); or have a refined technique of growing edible maggots on prey carcasses (*App. [12]*). Chimpanzees are also reported to be companions of humans, such as playmates (*App. [14]*) or allies in war (*App. [21]*). (The latter story finds a remarkable modern-day counterpart expressed in the magazine of Nigeria national parks. The author – perhaps tongue-in-cheek – is adamant that

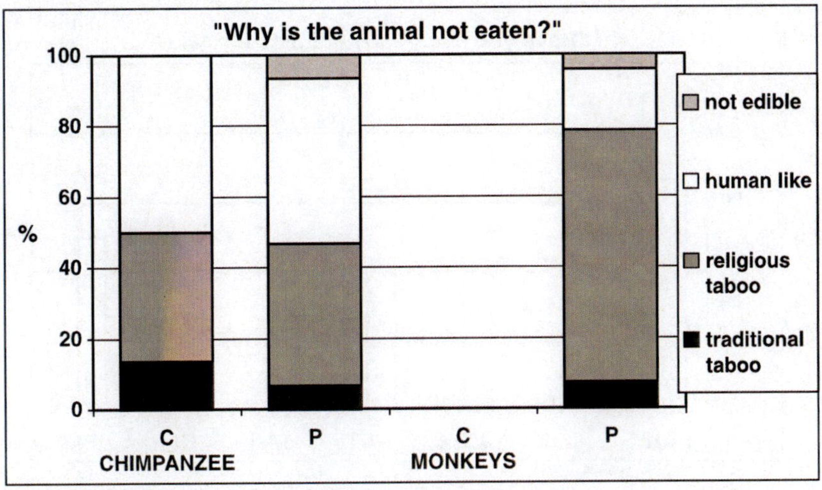

Fig. 4.11 Reasons given for the meat of primates to not be eaten (see *Fig. 4.3* for key)

the export of wild apes from Nigeria has to be stopped as they might be trained by other nations in warfare, which may catch Nigeria by surprise; Wari 2001).

- *Primates fall under Islamic religious taboos*. The Koran specifies that some animals are forbidden to be eaten ("haram"). Primates belong to this category, as they have human-like dentition, ears and eyes. In the Park Support Zone, 49 % of respondents were Muslim, compared to 33 % near the border.
- *Traditional taboos reflect specific tribal beliefs*. The human-ape similarity is stressed in many stories such as that humans have learned midwifery from them (*App. [22]*), that chimpanzees have shown mercy with a hunter (*App. [20]*), or that humans have realised how much chimpanzees are like themselves (*App. [19]*).

No restrictions against eating monkey meat were reported from the Cameroon border, whereas sentiments rooted in tradition or religion exist against eating chimpanzees (*Fig. 4.11*). In the Park Support Zone, these sentiments extend to both apes and monkeys, although human-like attributes are most often cited as the cause for not eating chimpanzees ("eating this animal is like eating a human being").

Primates as Pets

Baboons were the most popular monkeys offered for sale as pets, whereas colobus were the least popular (*Fig. 4.12*). Overall, chimpanzees were reported to be more often on sale as pets (35 %) than monkeys (27 %; *Tab. 4.3: Q11*) and more respondents had seen pet chimpanzees (57 %) compared to pet monkeys (42 %; *Tab. 4.3: Q12*). However, the pet trade of almost all species was significantly more pronounced near the border.

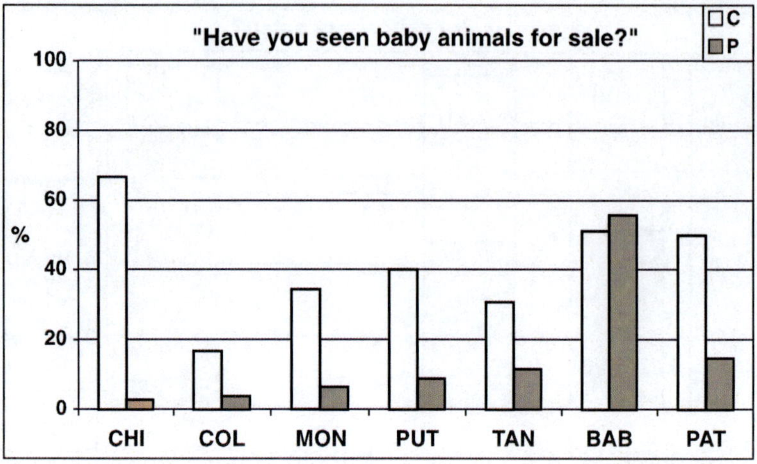

Fig. 4.12 Sightings of baby primates: affirmative interview responses (see *Fig. 4.3* for key)

Attitudes Towards Protection

Many primates species raid crops. Respondents in both survey areas identified tantalus monkeys, baboons and patas monkeys as the worst crop-raiders (*Fig. 4.13*). These species are the most terrestrial and eat crops such as maize, guinea corn and ground nut. Near Cameroon, a higher proportion of respondents perceived mona and putty-nosed monkeys to be crop-raiders. This is due to the different agricultural practices, as the border area has plantations of cocoa, orange, mango, palm nut and cola nut which can be exploited by these more arboreal primates. Folklore maintains that patas monkeys, which invaded a farm found a local brew from which they got drunk – upon which they could all be easily shot (*App. [35]*). It is also generally advised to eradicate crop-raiders such as putty-nosed and tantalus monkeys before starting a plantation (*App. [36]*). In contrast, very few respondents pointed to colobus monkeys and chimpanzees as being crop-raiders (*Fig. 4.13*).

Nobody in either survey area suggested that monkeys transmit diseases, although the majority of respondents said that they did not know. Only few believed that chimpanzees could transmit diseases such as AIDS while most denied this explicitly (*Tab. 4.3: Q14*).

Almost all respondents in both survey areas believed that chimpanzees were legally protected, whereas Park Support Zone respondents were significantly more likely to believe that monkeys were protected (*Tab. 4.3: Q15*). Only about a quarter of people near the border believed that monkeys were protected, with the surprising outlier that over 50 % of people thought that baboons were protected (when, in fact, they are not protected by Nigerian law). People dwelling near the national park were also more likely to state that both chimpanzees and monkeys *should* be protected by law (*Tab. 4.3: Q16*). Interestingly, respondents did not express less desire for the protection of species that raid crops most often, such as tantalus monkeys and baboons (*Fig. 4.14*).

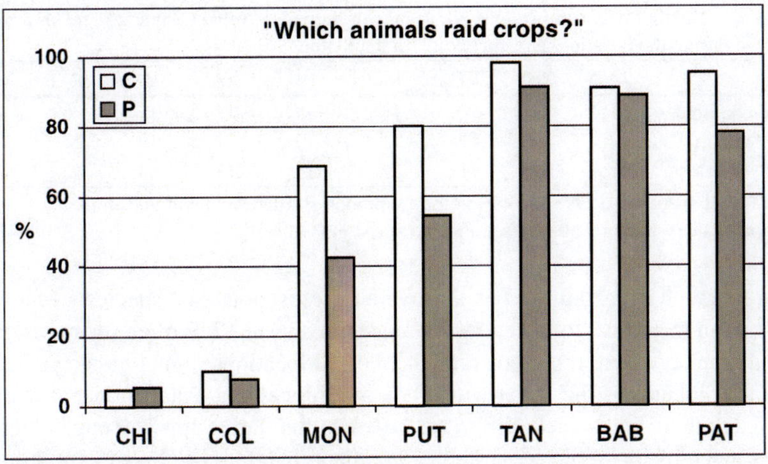

Fig. 4.13 Crop-raiding by different primate species: affirmative interview responses (see *Fig. 4.3* for key)

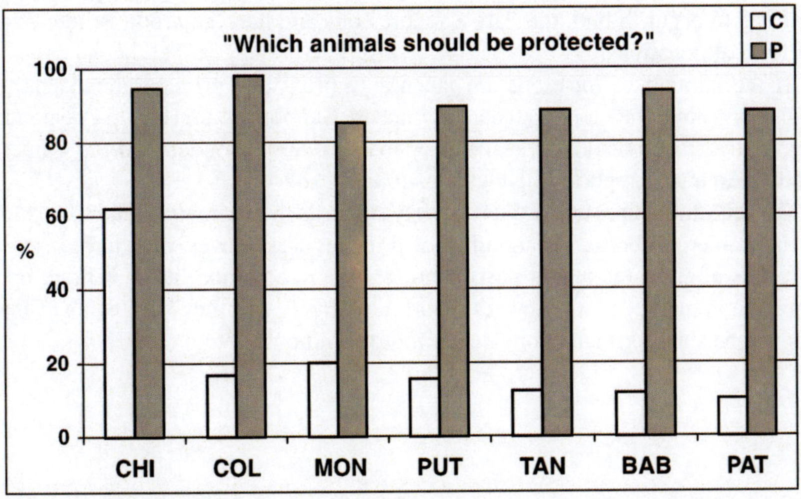

Fig. 4.14 "Which animals should be protected?" Affirmative interview responses (see *Fig. 4.3* for key)

Primate Populations in Taraba State

Respondents mentioned a total of 143 different locations at which primates occurred, 71 for the Cameroon Border Area and 72 for the Park Support Zone. For the 6 monkey species and the single ape species, this amounts to 1399 individual naming of locations. From this we calculated the percentage of locations in which the 7 different primate species reportedly occurred (*Tab. 4.5*), assuming that this provides a rough figure for their abundance.

Table 4.5 Number of locations with occurrence of individual monkey species and chimpanzees reported during the questionnaire survey in Taraba State

%	CHI	COL	MON	PUT	TAN	BAB	PAT	SUM
Cameroon Border Area	15	10	15	17	18	17	8	100
Park Support Zone	15	12	11	12	17	19	14	100
Total	15	11	12	14	17	18	12	100

Total for all species = 1399, i.e., 496 in the Cameroon Border Area and 903 in the Park Support Zone; all values are percentages (see *Fig. 4.3* for key).

In line with expectations, baboons were the most abundant species, occurring in 18 % of all locations, followed by tantalus monkeys in 17 % of all locations. Putty-nosed monkeys were reported from 14 % of all locations, mona monkeys from 12 %, and colobus was the rarest with 11 % of all locations. Patas monkeys averaged 12 %, but there was a clear difference between the Park Support Zone, which harbours much of its preferred savannah-woodland habitat (14 %), as opposed to the more moist Cameroon Border Area (8 %). A similar difference was observed for mona and putty-nosed monkeys, which were more commonly seen near the border, probably because of the greater number of plantations there. Chimpanzees were reported to occur in both the Park Support Zone and the Cameroon Border Area at 15 % of all locations.

If the number of locations are taken as a proxy for primate abundance, then primates seem to be more abundant in the Park Support Zone. Here, 54 respondents reported 72 locations (1.3 / head) as opposed to 80 respondents in the Cameroon Border Area who reported 71 locations (0.9 / head).

The questionnaire survey thus strongly suggests that primate abundance is lower at the Cameroon border as sightings of primates were fewer; they have been seen here less often in the recent past; hunting is more common; meat is more openly available in markets; interviewees report more freely that they have eaten primates; primate pets and primates for sale are more commonly found (see *Tab. 4.3*).

Discussion

Our study aimed to document local attitudes towards primates as well as the extent of hunting and bush-meat consumption in Nigeria's north-eastern Taraba region, not least because this region is the stronghold of the rarest chimpanzee subspecies, *P. t. vellerosus*.

Questionnaires revealed that wild primate populations face severe threats in the wider Taraba region. More than three quarters of interviewees, many of which come into close contact with these animals (hunters, wild-life wardens, farmers), stated that they saw primates more rarely now than in the past (cf. *Tab. 4.3: Q5*). This population decline can be related to a direct force of destruction, i.e., hunting (see below), and an indirect force, i.e., the alteration of habitat through fire damage of forests and cattle grazing (see Adanu *et al.* this volume [Ch. 3]).

Reduction of Forest Habitat

The regime of annual burning and extensive pastoralism have increased the proportion of savannah-woodland against lowland and riperian forest, thus creating vast areas of grassland, particularly at the expense of montane forests (Chapman *et al.* 2004). These dynamics might benefit more adaptable and terrestrial primate species, in particular olive baboons (Warren 2003) and tantalus monkeys. The latter avoid too densely forested areas, whereas baboons can thrive here as well, as long as the habitat is a mosaic of savannah-woodland and forest (see Ross *et al.* this volume [Ch. 9]). For patas monkeys, grassland is the exclusive biotope, and one can speculate if anthropogenic influence might not in fact have led to a wider distribution of this species in Taraba. Fire and grazing reduces, on the other hand, the habitat available to more arboreal primates such as black-and-white colobus, mona and putty-nosed monkeys (cf. *Tab. 4.5*). Chimpanzees also avoid open grassland. They do venture into savannah-woodland, but clearly seem to prefer forests as these provide more fruit and suitable nesting sites.

Hunting

Killing animals for bush-meat consumption has been a traditional source of protein for humans across tropical Africa. However, hunting is no longer a localised subsistence practice but is becoming increasingly commercialised (Wilkie 2001, Fa *et al.* 2003). Subsistence needs and monetary profit also drive hunting in Taraba. Guns were the preferred weapon (cf. *Fig. 4.8*). Although shots attract attention, they kill instantly, whereas snares and traps have to be laid and left. Returning to them brings the danger of being caught if they have been detected and wildlife guards are waiting. On the other hand, traps and snares are easy to conceal during transport and are less expensive than owning a gun and buying bullets. The relatively high frequency of traps and snares reported in the questionnaire survey may have been inflated due to recent tribal disputes. The Jukun and Kutep in particular have clashed near Takum and Ussa since 2002, and the Mambilla and Fulani were in often violent conflict on the Mambilla Plateau since 2003. Consequently, the government has restricted the ownership and use of guns and many have been confiscated, including hunting weapons that have indeed been involved in ethnic conflicts. This has possibly led people to using alternative means of hunting.

A more diverse mix of weapons is used near the border, including dogs. Not many dogs are owned in the Park Support Zone, perhaps because they are known to be used by hunters and so would attract attention from the park authorities. Instead, the use of poison and arrows was sometimes reported here, perhaps because they are discrete weapons, the arrow being silent and the poison difficult to detect.

Hunting was reportedly much more common in the border area (cf. *Tab. 03: Q6–10, Fig. 4.7*). However, the *level* of hunting cannot be ascertained from the results, as the question was binary, i.e., "yes" / "no". Thus, a few animals killed will give the same result as a high number. The responses therefore indicate only if hunting occurs or not, but they nevertheless clearly show a higher frequency at the border.

Similarly, bush-meat was more readily available in local markets and restaurants near the Cameroonian border (cf. *Fig. 4.9*). The finding might simply reflect that it is not illegal to sell bush-meat here whereas it contravenes the law near the national park. This would affect the level of meat being sold and the level of conspicuousness of meat for sale as well as the number of people who would admit that meat is being sold.

Few interviewees were able to state a preference for specific primate bush-meat, which reflects the difficulty of recognising species. As the meats are all smoked and dried together, few distinguishing features such as skin, hand, tail, head and feet are left on the sections of carcass that are sold. Only sometimes is the tail or head kept to verify the species. Moreover, hunters may present meat as that of another animal so that it can be sold at a higher price. Therefore, people may not know what type they have really eaten, nor indeed what is available. These circumstances go some way toward explaining why chimpanzee meat is more expensive than its low ratings for desirability suggest (cf. *Tab. 4.4*), as chimpanzee meat is often passed off as buffalo, which fetches a higher price. Therefore, chimpanzee meat is rare and the few people who specifically want to buy it – perhaps for medicinal or fetishist reasons – will find its price inflated.

Chimpanzee males weigh on average 43 kg (range 33 – 57 kg) and females 36 kg (range 28 – 49 kg; review in Sommer & Ammann 1998), yielding an average of 40 kg per ape. Killed animals are typically simply hacked into pieces, which means that lumps of bones, intestines and meat are sold together. A kg of ape meat sells for the equivalent of 0.8 – 1.2 US $ (cf. *Tab. 4.4*). A whole carcass is thus worth about 40 US $ or 5400 Naira, which, in a rural setting, is equivalent to a modest monthly salary. Killing a chimpanzee is therefore certainly worth a hunter's while, and much alternative incentive would be needed to stop a hunter from practising his trade.

Pet Trade

In many range countries, primates are traded as pets (Cowlishaw & Dunbar 2000). About half of our survey respondents had seen infant primates for sale, and about a third had seen primate pets, with chimpanzees being more popular than monkeys (cf. *Tab. 4.3: Q11–12, Fig. 3.12*). It has to be emphasised that the trade in chimpanzee babies is a by-product of the trade in chimpanzee meat (Teleki 1980, Peterson & Ammann 2003, Hughes *et al.* this volume [Ch. 14]). Infant chimpanzees sleep in nests with their mothers, or are transported clinging to her stomach. It thus sometimes happens that when the mother is killed, either in the nest or fleeing, the infant may not be directly hit. If it survives the fall to the ground, it may be collected alive for sale as a pet. Most probably all of the infant chimpanzees offered for sale in markets are obtained in this way, as it is difficult to selectively shoot mothers just to retrieve the infant. The vastly greater likelihood of seeing baby chimpanzees for sale near the Cameroonian border must therefore correlate with a vastly higher level of chimpanzee hunting.

Hunters like to draw attention to their exploits by bringing back baby chimpanzees to a town or village, where many inhabitants will go to see it. The preference for chimpanzees as pets and for sale is probably related to the fact that apes are more attractive to people than monkeys because of their greater resemblance to humans. Another factor is the likelihood of survival in captivity. This might explain why colobus monkeys are rarely seen as pets as it is difficult to provide the appropriate folivorous diet (Cowlishaw & Dunbar 2000). Baboons are the most popular pet monkey. Because they are omnivorous (e.g., Warren 2003 for Gashaka baboons), they are more hardy and tend to survive in captivity for a long time.

Smuggling of primate pets is widespread in Nigeria. The animals are sedated with drugs and infants can then be moved by public transport on the smuggler's stomach, who will simply wear a big coat over them to avoid arousing suspicion (cf. *App. [30]*). They are usually brought from places such as Baissa and Bissaula to Jos, Kano and other urban areas from where they often find their way to Europe or Asia (Sommer & Ammann 1998, Peterson & Ammann 2003).

Food Taboos

Many interviewees stated that primates and in particular chimpanzees fall under Islamic religious taboos and should not be eaten (cf. *Fig. 4.11*). Indeed, in those locations of Taraba with a greater Muslim influence, self-reported killing or consumption of chimpanzees was lower than in areas with predominantly Christian (or traditional African) influence (cf. *Tab. 4.3: Q8–10*).

From this one might conclude that religious or spiritual traditions can aid modern conservation goals. Surely, the preponderance of monkeys such as langurs and macaques in many Indian cities, towns and dwellings is a simple function of the population being overwhelmingly Hindu, who consider monkeys to be sacred (Sommer 1996). However, such religious conviction is rarely ecologically informed or indeed cares about the well-being of animals. Instead, believers are driven by their own selfish desires not to be punished for acts conceived to be against divine law. The prevalence of Hinduism on the Indian subcontinent has therefore not stopped the widespread destruction of natural habitat of wild animals (Sommer 2001). Similarly, while local Islamic practice discourages the *eating* of primates, it does not impose much of a restriction on the *killing* of primates for further sale of the meat, a logic that also holds true for the hunting of wild pigs. Thus, hunters and poachers of primates who are Muslim can be found in Taraba.

In any case, sentiments against eating primates seem to be weakening in the Taraba region. Several respondents noted that primates, especially chimpanzees, were not consumed in the survey area until fairly recently. Some interviewees explained how this practice has been introduced with the increase of population and the inward migration of new tribes. A chief hunter near the Cameroon border stated that "people never used to eat chimpanzee before. Only the Ibos and the Jukun ate them, but now many other tribes are eating". It is likewise possible that respondents will say one thing but do another as fewer admitted to having eaten primates compared to those that said primates were eaten by other people (cf. *Tab. 4.3: Q9–10*). This discrepancy might reflect the religious stigma attached to admitting to the consumption of primates.

There seems to be a similar stigma attached to hunting chimpanzees, given their human-like characteristics, as successful hunters may sometimes have to undergo a period of quarantine before they become celebrities for having killed an ape (cf. *App. [27]*). Such practices resemble purification rituals of warriors who kill other humans in battle. In this respect, it is important to note that, whereas some do not kill chimpanzees because they are human-like, others kill them for precisely that reason, since their human likeness either brings special status to the hunter, or the products of the dead chimpanzee, such as the hands or skull, may accord special powers to humans. People in Sierra Leone who were interviewed about attitudes towards conservation expressed concern that protection of chimpanzees was a scheme to cover up "chimpanzee business". According to this belief, power-seekers murder people to obtain their body parts to make "bad medicine", which confers political or economic powers. They do this through shape-shifting, magic powers or by disguising themselves as chimpanzees and using special knives that simulate bites done by the actual animals (Richards 2000). Animals such as chimpanzees can thus become stigmatised as human groups use them to define group boundaries (Douglas 1966).

It is therefore questionable if religious or mythological convictions alone (NCF / WWF 1995) can be an effective tool in conservation strategies. Still, there can be little doubt that the tendency of humans to anthropomorphise monkeys and apes can create not only certain inhibitions against killing them, but also a desire to see them living. This is certainly the case with the average viewer of nature documentaries. The insight that "television programmes can be a powerful means of raising awareness and triggering action to solve conservation problems" has led to ideas such as "The Great Ape Films Initiative", which screen films about non-human primates in habitat countries so that these can be seen "by the people who live and work in the places where the animals are and by decision-makers who control the fate of those habitats" (www.nutshellproductions.co.uk / gafi; accessed 12 Sep 08). It seems as if the persuasive power of such presentations (Westwood & Redmond 2008) rests less in references to mythology than in rather spontaneous empathy for human-like creatures. One would think that such "instinctive" empathy would have to be paired with an enlightened attitude about the dangers of eco-system destruction to trigger sustainable conservation initiatives. Such a state of mind seems to motivate many primatologists who have made conservation their personal and professional agenda.

Protected Areas and Law Enforcement

The questionnaire survey showed that people in the vicinity of the park had a greater awareness than inhabitants of the area near the Cameroonian border that Nigeria has laws protecting wildlife (cf. *Tab. 4.3: Q15)*. However, this finding might be due to the fact that people near the park may want to please the interviewer by displaying a more conservation-oriented attitude. Moreover, Park Support Zone inhabitants might deny that primates are illegally killed or that chimpanzee babies are available for sale because there is a greater risk of punishment for such activities than in the border area. Compared to the border zone where hunts are openly carried out and discussed, hunting near the park (and in the park itself) is a secret activity.

Nevertheless, other factors make it likely that at least some of the differences are "real", reflecting a greater reluctance of Park Support Zone inhabitants to be involved in the illegal trade in primates, and in particular chimpanzees, be it out of fear of punishment or because of moral or religious conviction. Even though hunting is not illegal in the Park Support Zone itself, licenses for hunting within the vicinity of the park have been revoked by the government. The consequences of being caught with meat or hunting equipment are as serious as if they were found in the park – the assumption being that any activity outside the park is strongly suggestive of similar activity within the park. Thus, hunting comes with a higher risk that people may not be prepared to take. In contrast, in the Cameroon Border Area, hunting permits can be bought from the state government Wildlife Department. However, even though it is a requirement, most hunters do not buy licenses as the risk of being caught and the fines involved are low.

In sum, it is important to realise for even the most idealistic conservationist that "nature" is "cultured" by humans (Markl 2007), and has been so for a long time, at least in places such as north-eastern Nigeria. Thus, like it or not, natural processes alone do not any more ensure suitable conditions for the desired biodiversity in reservoirs such as GGNP. Such places will thus be in an increasing need of "management" (Ausden 2007). The ethno-primatological approach can help "to overcome the cultural and perceptive isolation" of non-local conservationists and primatologists (Wolfe & Fuentes 2007: 700) and in this way assist conservation efforts by incorporating the challenges and opportunities embedded in local traditions and attitudes towards wildlife.

Acknowledgements Thanks are due to WWF-UK, who initiated the survey by part-sponsoring it.

Literature

Ausden, M. (2007). *Habitat Management for Conservation. A Handbook of Techniques*. Oxford: Oxford University Press.

Bishop, N., Hrdy, S. B., Teas, J. & Moore, J. (1981). Measures of human influence in habitats of South Asian monkeys. *International Journal of Primatology* 2: 153–167.

Burton, F. & Carroll, A. (2005). By-product mutualism: Conservation implications amongst monkeys, figs, humans, and their domesticants in Honduras. In Paterson, J. (ed.), *Commensalism and Conflict: The Human-Primate Interface*. Norman, OK: American Society of Primatology, pp. 24–39.

Cadet, J. M. (1971). *The Ramakien*. Tokyo, New York, San Francisco: Kodansha International.

Chapman, H., Olson, S. M. & Trumm, D. (2004). An assessment of changes in the montane forests of Taraba State, Nigeria, over the past 30 years. *Oryx* 38: 282–290.

Cowlishaw, G. & Dunbar, R. (2000). *Primate Conservation Biology*. Chicago: The University of Chicago Press.

Douglas, M. (1966). *Purity and Danger: An Analysis of the Concepts of Pollution and Taboo*. London: Routledge & Kegan Paul.

Fa, J. E. (1986). *Use of Time and Resources by Provisioned Troops of Monkeys: Social Behaviour, Time and Energy in the Barbary Macaque* (*Macaca sylvanus L.*) *at Gibraltar*. Contributions to Primatology: n° 23. Basel: Karger.

Fa, J. E., Currie, D. & Meeuwig, J. (2003). Bushmeat and food security in the Congo Basin: linkages between wildlife and people's future. *Environmental Conservation* 30: 71–78.

Fa, J. E., Peres, C. A. & Meeuwig, J. (2002). Bushmeat exploitation in tropical forests: an intercontinental comparison. *Conservation Biology* 16: 232–237.

Fuentes, A. & Wolfe, L. D. (2002). *Primates Face to Face*. Cambridge: Cambridge University Press.

Hsu, M. J. & Lin, J. (2001). Troop size and structure in free-ranging Formosan macaques (*Macaca cyclopis*) at Mt. Longevity, Taiwan. *Zoological Studies* 40: 49–60.

Janson, H. W. (1952). *Apes and Ape Lore in the Middle Ages and the Renaissance*. London: The Warburg Institute, University of London.

Kingdon, J. (1997). *The Kingdon Field Guide to African Mammals*. San Diego: Academic Press.

Kurita, H. (1999). Hourly changes in body weight in provisioned free-ranging Japanese macaques (*Macaca fuscata*). *Folia Primatologica* 70: 286–290.

Lal, P. (1981). *The Ramayana of Valmiki*. Delhi: Vikas Publishing House.

Markl, H. (2007). Kultur der Nachhaltigkeit. *Biologen heute* 1: 2–11.

Miller, L. (ed.) (2002). *Eat or be Eaten: Predator Sensitive Foraging Among Primates*. Cambridge University Press, Cambridge.

NCF & WWF (1995). *Islam and Ecology*. Lagos: NCF.

Paterson, J. & Wallis, J. (2005). *Commensalism and Conflict: The Human-Primate Interface*. Norman, Oklahoma: American Society of Primatologists.

Peterson, D. & Ammann, K. (2003). *Eating Apes*. California: University of California Press.

Richards, P. (2000). Chimpanzees as political animals in Sierra Leone. In Knight, J. (ed.), *People-wildlife conflicts in Anthropological Perspective*. London: Routledge, pp. 78–103.

Shephard, G. H. (2002). Primates in Matsigenka: subsistence and world view. In Fuentes, A & Wolfe, L. D. (eds.), *Primates Face to Face*. Cambridge: Cambridge University Press, pp. 101–136.

Sommer, V. (1996). *Heilige Egoisten. Die Soziobiologie indischer Tempelaffen*. Munich: Beck, C. H.

Sommer, V. (2001). Schlange, Adler, Affe & Co. Tierkult und Tierschutz im Hinduismus. In ZDF nachtstudio (ed.), *Mensch und Tier. Geschichte einer heiklen Beziehung*. Frankfurt / M.: Suhrkamp, pp. 177–193.

Sommer, V. & Ammann, K. (1998). *Die Großen Menschenaffen: Orang-utan, Gorilla, Schimpanse, Bonobo*. Munich: BLV.

Teleki, G. (1980). *Hunting and Trapping Wildlife in Sierra Leone: Aspects of Exploitation and Exportation*. Washington: WWF.

Wari, M. (2001). *Nigeria's Apes*. Nigeria Parks, the Magazine of the Nigeria National Parks.

Warren, Y. (2003). *Olive Baboons (Papio cynocephalus anubis): Behaviour, Ecology and Human Conflict in Gashaka Gumti National Park, Nigeria*. PhD thesis, Roehampton University.

Westwood, M. & Redmond, I. (2008). How can conservation films be a force for change and generate positive action? *Primate Eye* 96: 10.

Wilkie, D. S. (2001). Bushmeat trade in the Congo Basin. In Beck, B. B., Stoinski, T. S., Hutchins, M., Maple, T. L., Norton, B., Rowan, A., Stevens, E. F. & Arluke, A. (eds.), *Great Apes and Humans: the Ethics of Co-Existence*. Washington: Smithsonian Institution Press, pp. 86–109.

Wolfe, L. D. & Fuentes, A. (2007). Ethnoprimatology. Contextualizing human and nonhuman primate interactions. In Campbell, C. J., Fuentes, A., MacKinnon, K., Panger, M. & Bearder, S. K. (eds.), *Primates in Perspective*. New York, Oxford: Oxford University Press, pp. 691–702.

Appendix. Folklore About Chimpanzees and Monkeys in the Taraba Region, Nigeria

Based on interviews with local people, 2004 – 2005, by GN (*Tab. 4.1*). Narrated by GN to AF, transcribed by AF, and edited by VS (Apr 05). It was intended to preserve some of the traditional mode of storytelling.

Interpreting Naturalistic Behaviour of Chimpanzees

The Meaning of Drumming

[1] *A chimpanzee invents drumming to find his wife*. Told in Baissa. A chimpanzee was living with his family, and there was a shortage of food. So, he told his family that he would go and look for food. He said he would leave marks on the road, which they could follow in case he didn't come back. On his travels, he discovered a place with plenty of food and thought it would be good to transfer his family to that area. When he did not return, his wife decided to follow him as he had said. So, she followed but could not find him. And he was also looking around for her, because she had left the home. So, he decided to invent drumming. When she heard this, she knew it was him, and she went to find him. The drumming became a sign of contact and is also a sign of a good place to sleep. It is usually the male chimpanzee that drums and the females do the crying.

[2] *Drumming is a secret*. The drumming of the chimpanzees is a secret of the chimpanzees. That is why it is very difficult to discover.

[3] *Drumming tells everybody to go to sleep*. Told in Mayo Selbe. Drumming is a sign to show that playing is over, so everybody should go to sleep, and we will meet again tomorrow.

[4] *Drumming foretells death in the village*. Told in Abong. In the olden days, it was believed that when chimpanzees come very close to the village and cry and beat buttresses, this is a big sign that an adult from the village will die, and this will have to happen whether the person was sick or not. But nowadays, because of hunting pressure, the chimpanzees have all moved from close to their home and people no longer hear their cries.

Why Chimpanzees Build Nests

[5] *Chimpanzees avoid snake-bites*. Told in Bissaula. In those days, the chimpanzees normally spent many nights in each nest. They were suffering to make a new nest every night. So, one day, the chimpanzees went back to stay in a previous nest because they didn't go away too far. That afternoon a snake had invaded one of the nests. So, when the leader of the group went up to sleep, he was

bitten. From that day, they all learned their lesson: that it is not good to use a nest more than once, because a snake might be there. And they passed the message around, and that has continued until this generation.

The Human Nature of Chimpanzees

Human-like Qualities of Chimpanzees

[6] *Chimpanzees are afraid of guns.* Told near Akwame. If you point a gun at a chimpanzee, it will cry and shout, so chimpanzees behave like men.

[7] *Chimpanzees copy human behaviour.* Told near Akwame. Normally, in the old days, when a man was going to cut palm nuts, he left them near the trees for a few days for the water to reduce. When he went to collect them, he went with his wife. The woman would carry the bunch on her head. Usually the man would be holding at least two spears on his shoulders. In many instances in this village, female chimpanzees were seen carrying a bunch of nuts on their heads, and the male would break two sticks and carry them on his shoulders. This shows that the chimpanzees copy what human beings do exactly.

[8] *Chimpanzees do not raid crops.* Told in Bodel. According to the Jibawa people, chimpanzees are not known for crop-raiding or any damage on the farm. In the event of such on anybody's farm, it means the farmer has a problem, which he has to settle, and if he fails to do so, calamity will follow.

[9] *Chimpanzees forage like humans.* Told in Yelwa. The chimpanzees behave like human beings, using hands to fetch water and leaves sometimes. They use their hands to make the nest.

[10] *Chimpanzees use tools and share food.* Told in Bissaula. The chimpanzees also do group work like human beings. So, when they get food, they call out and wait for others to come around. It is normally honey or termites – which they hunt, using sticks. They will first give to the younger ones, then they will share it.

[11] *Chimpanzees fish and share the exploits.* Told in Gashaka. Chimpanzees fish from rivers. They toss all the fish to a leader chimpanzee who collects them and shares them out afterwards.

[12] *Chimpanzees grow maggots for food on hunted carcasses.* Told in Baissa. When chimpanzees kill an animal, antelope or something, they don't eat it. They cover it somewhere with leaves. A few days later, they will go on their normal hunting. At exactly the time that the maggots start forming in the dead animal they will come back to the same place and feast on the maggots. So, the chimpanzees do not eat the flesh. If human hunters find a pile of grass under which chimpanzees have hidden a killed antelope, they will take the prey but remove the intestines and leave them there to form the maggots for the chimpanzees. Also, when humans hunt, they will also leave intestines of their prey animal on a stone for the maggots to form for the chimpanzees.

[13] *Chimpanzee mothers hide insubordinate sons from the father.* Told in Baissa. It is in the culture of chimpanzees that when the female gives birth to a male chimpanzee, she will have to separate from the father to raise the male child because the young male chimpanzees are always insubordinate to their fathers. So the mother would always carry the male chimpanzee on her stomach and if it is a female, on her back. The reason for carrying the male on the stomach is so that the adult male will not see the private parts of the young chimpanzee. The male is kept by the mother until he grows, then he separates from her.

[14] *Chimpanzees as friends of humans.* Told in Bodel. In the past the Jibawa people used to make friends with the chimpanzees. They would go to the forest to play with the chimpanzees. The chimpanzees used to know them.

How Humans Came to Live in the Bush as Chimpanzees

[15] *Chimpanzees are humans cursed for abomination.* Told in Yelwa. Once upon a time, there was a newly married couple. The woman was left at home to cook while her in-laws went to the farm. So her mother-in-law gave her ingredients to make soup, including salt. But she misplaced the salt, so instead of waiting for the salt she replaced it with honey and sugar cane in the soup. She tasted it and it was sweet, so she knew that her mother-in-law would not like it. So she went to her mother-in-law's calabash that was placed on the top shelf in the kitchen. Inside the calabash she found some salt and added this to the sauce. She then took the food to the farm. The calabash, from which she took the salt, was not supposed to be used by anyone except the mother-in-law. So the calabash followed her to the farm. The woman knew that if she went to the farm with the calabash she would be queried, so she stopped and broke it with a stone. And then she threw the pieces in the river. She continued to the farm. By the time she reached the farm, the calabash fragments had reunited and it went down the river to the farm, arriving at the same time as the woman. So the people in the farm, including the mother-in-law, saw her and the calabash and worked out what had happened. Now, it is an abomination for a daughter-in-law to use the calabash of her mother-in-law without permission. So, as a result of this, everybody on the farm was transformed to chimpanzees. And they never came back home.

[16] *Chimpanzees are humans cursed for fishing on Saturdays.* Told in Takum. Monkeys, especially chimpanzees were formally human beings. So, when they were still people, there was a rule from God that no one goes to fish on Saturday. These chimpanzees, the first people to turn to chimpanzees, had a shortage of soup in the house to cook for the family. So, they went to fish on Saturday. Then, God found out and punished them to remain in the bush as chimpanzees. So they are shy because they cannot come back to be with the people.

[17] *Chimpanzees are humans cursed for quarreling amongst themselves.* Told in Mayo Selbe. Chimpanzees were formerly people before they were separated from normal human beings. Before that, there was a big misunderstanding and they could not resolve this problem. So, some of the group got so annoyed

with the other group and decided to stay in the forest away from other people. The ones who were annoyed went to the forest, carrying along their ugly faces with them. This is why the chimpanzees look as they do. After the separation, they also found a way to cut communication with the other people. So that is why the chimpanzees understand only themselves and cannot speak like normal humans. Using human language would be like bringing back the problem.

[18] *Chimpanzees are stubborn humans living in the bush.* Told in Baissa. The Ndoro people respect chimpanzees because chimpanzees were formally their early fathers or ancestors. It is their stubbornness that caused them to be separated from the new people now. That is why they remain in the bush.

Taboos and Customs Surrounding Hunting of Chimpanzees

Why Chimpanzees Are Not Hunted or Eaten

[19] *Humans realise that chimpanzees are like themselves.* Told in Baissa. Chief says his forefather used to say that chimpanzees are like humans, they will do anything people do. In 1950 he went to the farm with his father and heard the chimpanzees crying and drumming, and his father told him that these were also human beings but they live in the forest. So they went to see what the chimpanzees were doing. When they got there, the people sat down, the chimpanzees sat down, the people stood up and so did the chimpanzees. So his father told him, chimpanzees are human beings and should not be hunted.

[20] *A chimpanzee shows mercy with his hunter.* Told near Akwame. People used to hunt primates including chimpanzees in a group. They cut down trees in the forest to create a gap between the trees so the chimpanzees would not jump from one tree to the other. One day when they went hunting, one of the stubborn chimpanzees climbed up a big tall tree. All the styles to bring him down were abortive. So one of the young men climbed the tree so that he could disturb the chimpanzee and make him jump down. When the man went up, and got close to the chimpanzee, the chimpanzee quickly grabbed him and held him out trying to throw him down. The people on the ground started shouting and begging so the chimpanzee did not throw down the man. And the man was allowed to come down from the tree without being hurt. From this experience, they decided not to hunt chimpanzees again.

[21] *Chimpanzees can be allies in war.* Told in Ussa. People here believe the chimpanzees are strong and will help them during war when their enemies attack them. So they shouldn't kill or eat them.

[22] *Chimpanzees practice midwifery.* Told near Akwame. Tradition and culture of the Ndola people do not permit the eating of chimpanzees because in the early days, women were having problems in child bearing. So one day, a hunter spotted a chimpanzee giving birth, and the chimpanzees cut a leaf, which the mother

chimpanzee slept on. That knowledge was brought to their community, and so they do the same. If they do this, the woman will have a successful birth.

[23] *Chimpanzees cry like humans*. Told in Abong. A man went hunting in the seventies along the Nigeria – Cameroon border. He saw a chimpanzee and pointed a gun to shoot it. The chimpanzee did not run away but started to cry and shout. This behaviour is just like a human being, so he was discouraged and told his boys who also saw the action. From that day, they do not kill chimpanzees.

[24] *Chimpanzee meat causes abortion*. Told in Mayo Selbe. It is believed back in Ibo land that when a pregnant woman eats chimpanzee meat, it will cause an abortion.

[25] *Killing chimpanzees brings bad luck*. Told in Takum. If you kill a chimpanzee, you are likely to fall sick and if proper care is not taken, you may possibly die. Killing a chimpanzee will also get a hunter bad luck. So you won't easily kill any other animals.

Hunting Customs

[26] *Killing chimpanzees, capitalising on their imitations of human behaviour*. Told in Bissaula and Buru. The chimpanzees are like humans, but they are limited in mental ability and very difficult to hunt. There have been cases of chimpanzees in the bush using the same bow and arrow to shoot back at the hunters, because the chimpanzees normally dodge arrows and spears. So, to shoot one either by bow and arrow or by spear, you will first pretend to stab yourself in the stomach with the arrow, then shoot the arrow at the chimpanzee. He, being very strong, will dodge the arrow or catch it. When the chimpanzee gets the spear, he will do the same as the man did and stab himself in the stomach as he doesn't have the same control as the man. A similar story is told in Buru. When hunting chimpanzees with spears, you must first try to break the spear using your two hands and the head. This is because the chimpanzees will shoot back at you if you miss the target. But now the chimpanzee will do the same thing they saw you doing, and the spear will break.

[27] *A killer of chimpanzees has to undergo quarantine*. Told near Akwame. His forefathers always thought chimpanzees to be human, because they behave like humans, so if anybody happens to kill a chimpanzee the village will not be happy with that person. Some of the elders will go to the forest to collect leaves, called "Nsu", and these will be placed for the man to sleep on in isolation for some days. The same treatment is given to any man who kills a fellow human being by accident. The person will also sleep on those leaves for several days. Any man in that condition is fed by unhygienic food, which is the punishment for committing murder. This cleans up that person to allow him to associate with other people again. This is always done by people in the village Atta.

[28] *Killing a chimpanzee makes hunters great*. Told in Takum by an Ibo man. In the land of the Ibos, the greatest profession is hunting. These hunters are divided into two categories; the great and the less great. If you are a great hunter, you have killed animals like elephants, lions and chimpanzees, buffalo.

The other hunters are those who kill small animals like antelopes and so on. So, when a man kills any of these big animals, there is a big celebration in the village and his Royal Highness will be invited. The hunter is given a title, "Dinta", and the meat will be shared only among the hunters of that calibre. The skull of that particular animal will be given to the hunter. He will keep it in his house and show it to the generations to come. The remainder of the meat is then shared among other hunters. The hunter who killed that animal is not allowed to sleep in his house for 3 days. Although the celebration for the animal takes place in his compound, he is absent. This time he spends with the other great hunters and herbalists. Some traditional "stuffs" will be done and this will help to prevent any future attacks from the spirits of the forest. This also applies to the Yorubas, where the informant was in attendance at a "Dinta" celebration over the killing of a chimpanzee in 1999, in a village called Omok, Rivers State. Here, farmers do not go to their farms, everybody must stay and be a part of the celebration. If the hunter kills another chimpanzee the celebration takes place at the chief's palace.

[29] *Honours after killing a chimpanzee.* Told at Kob Vobye forest near Buru (recorded by J. D. Chapman in 1978). A hunter who killed a chimpanzee is "entitled to wear the red feather of Ngock, the violet plantain eater, and drink palm wine from a buffalo horn".

[30] *Chimpanzee smuggling in Nigeria.* Told in the Cameroon Border Area. Baby chimpanzees and other primates are usually transported from Baissa, Bissaula and other rural areas to Jos, Kano and other urban areas. The method of transportation involves the use of sedating drugs. The animals are injected or given a tablet, which makes them sleep all the way, without causing problems on the road or game guards or other people finding out. After the drugs are given, some people will carry it on their stomach. And wear a big coat over them and the primates will hold on to them while asleep. So this will not arouse any suspicions, because they will travel like any other passengers.

Medicinal Uses of Chimpanzee Body Parts

[31] *Chimpanzee bones heal fractures.* Told in Yelwa. Chimpanzee bone is used to treat compound fractures. The bone is tied to the area affected and reduces healing time.

[32] *Medicine from chimpanzee bones.* Told in Baissa. The informant's family does not eat chimpanzees. Anybody who does will develop mental illness. When a new child is born in that family, they will soak the bones of chimpanzees in water and the child will start drinking this when he is young to develop the bones. When the child grows up, his bones will hardly break.

[33] *Drinking out of chimpanzee skull confers intelligence.* Told in Mayo Selbe. The skull of the chimpanzee is used as a recipe for brilliance. Before a child is sent to school, a mixture is given to him every morning; the skull and the

hands of the chimpanzee in water. The child will drink from the skull, and when he goes to school, he will be very intelligent.

[34] *Medicine from chimpanzees assists in labour and gives strength.* Told in Mayo Selbe. The skin of a chimpanzee can be used to help a woman who is having difficulty in child bearing. The skin is spread for the woman to sit on top of, before the child-bearing process, and it will thereby be very simple for her. The bones are ground and mixed with some herbs, so when a person takes this mixture he will be very strong and can wrestle. All the heroes in the past were said to have been using these bones before they fight.

Stories About Monkeys

[35] *Hunting drunken patas monkeys.* Told in Ussa. When the informant was young, in the early 1970s, his grandmother had a farm and invited people to come and do community work. To entertain the people who work, a local drink called BKT is prepared from maize. She made this especially strong, so that people would work harder. But, the grandmother had two farms, and her son took the helpers to a different farm. She was preparing the drink at home and then took it to the farm where she wanted the work done. Normally in that farm, they had serious problems of crop-raiding by patas monkey. When she got there and did not see anybody, it dawned on her that she had gone to the wrong farm. So, she left everything there, went to the other farm, a small distance away, and told the workers "No I don't want you to work here, we must go to the other farm". But when she had left the drink there, the patas monkeys were watching her. They entered the farm, drank all and immediately became intoxicated. When the people finally came to work, they saw the farm full of monkeys who were not running away, as they were drunk and falling up and down. So, the people instead of working on the farm, began to kill the monkeys. There was so much meat, some people carried the monkeys away alive. Work could not continue that day because they changed parade. So, they called the woman the "blessed woman" until she died, because only on her farm had this happened.

[36] *Eradicating crop-raiders.* Told near Akwame. The first thing farmers do to farm cocoa is to locate any primates, especially tantalus and putty-nosed monkeys. They will go out at night and destroy the whole group before starting the plantation. Sometimes they get a lot of putty-nose and tantalus in one night.

[37] *Teaching the spiteful lion a lesson.* Told in Ussa. Once upon a time, in the jungle, a lion fell into a pit and he was looking for a way to come out. No one could help him. Many monkeys bypassed him without helping. The monkey with the long tail came by after many animals had refused to help the lion. The lion had been in the pit for some days. So the monkey with the long tail helped him out. But then the lion wanted to eat the monkey. They were still struggling when the tortoise came by and enquired what the problem was all about. And the monkey complained, "I helped this guy when no one would help him. And

now he threatens to take my life". And the tortoise asked how he had helped him. The monkey replied, I got him out of the pit. Then the lion said it was true. The tortoise said he didn't understand, can the lion get back in the pit so he could see how the monkey helped him. The lion got back into the pit. And the tortoise said to the monkey, "Next time you should know those to help". To the lion he said "Never threaten any one who has helped you before".

Chapter 5
The Bush as Pharmacy and Supermarket: Mechanisms and Functions of Plant Use by Human and Non-human Primates at Gashaka

Yianna Koutsioni and Volker Sommer

Abstract Both human and non-human primates exploit the plant resources of the woodland-savannah in Nigeria's Gashaka area. We generated a database on the usage of more than 300 plant species that serve as food (including beverages and seasoning), medicine, or implements. We supplemented previously unpublished reports with original survey data and data accumulated by primate researchers. We compared four consumer groups, i.e., humans, domestic animals, baboons, and chimpanzees. Two case studies refer to discernible medicinal effects of plant use by non-human primates. One case concerns baboons, which consume African black plum; this has a contraceptive effect and potentially reduces mortality during the rainy season. A second case concerns chimpanzees, which swallow intact leaves of a coarse herbaceous plant, a practice that expels parasitic worms. We also reflect on potential co-evolutionary processes that lead to a preference for certain plant families and plant parts. The Gashaka area is clearly still under researched, as many taxa with ascribed medicinal values are not yet included in a standard compendium of medicinal plants in Nigeria. Future work should also engage with traditional concepts of how to classify plants, and explore plant properties in more detail as this might affect their usage as nutrition, for treatments, or as equipment. One would also want to work towards a further merger of ethno-botany and zoo-botany with its emerging sub-discipline of animal self-medication.

Keywords Ethno-botany • Animal self-medication • Traditional medicine • Nigeria

V. Sommer (✉)
Department of Anthropology, University College London, Goldalming, London, WC1E 6BT, UK
e-mail: v.sommer@ucl.ac.uk

V. Sommer and C. Ross (eds.), *Primates of Gashaka*, Developments in Primatology: Progress and Prospects 35, DOI 10.1007/978-1-4419-7403-7_5,

Introduction

Food, Implements, Medicine: Plants as Resources

From vessels to cross oceans to objects of art, from arrow poisons that enable hunting to psychoactive compounds, from tea to soothe a cold to plum pudding, from wicker baskets to cotton shirts, from pounded yam to fillings for mattresses: Plant products have long provided nutrition, clothing, shelter, transportation, remedies, and paraphernalia to the peoples of the Earth. The targeted search for new species of plants that can be exploited (Balick & Cox 1997) represents perhaps the most ancient form of "applied research", dating back thousands, if not hundreds of thousands of years.

The systematic study of traditional plant use is associated with the term "ethnobotany", coined in 1895 by the American botanist John W. Harshberger (Cotton 1996, Balick & Cox 1997). The benefits of exploiting vegetation in the local surroundings can be grouped into three main categories: food, implements, and medicine.

- Wild foods constitute an essential component of many diets, as substitutes in times of scarcity, as snacks – especially for children, as flavourings, or to brew beverages including intoxicating drinks. Finally, wild plants are also used as food for domesticated animals.
- Plant products collected near settlements serve as fuelwood, as building material for e.g., livestock pens, huts, storage barns, furniture, and raw material for wood derivative products such as fibres, mats, baskets, tool handles, arms for hunting and defence, cooking equipment such as pestles and mortars, or ritualistic instruments to make music with or communicate with spirits.
- Finally, plant parts and ingredients are used as medicines. In traditional contexts, most people are familiar with plants in the vicinity of their village that can be used as first aid. Often, they are the only health care available in remote areas and traditional healers cater to this need. Again, plant-based medicine may also cure domestic animals.

Modern ethno-botany attempts to understand and document the multiple mechanisms and functions of plant utilisation, and has adopted an interdisciplinary approach that incorporates diverse methods such as those from mechanical engineering, molecular biology or medical anthropology.

However, not only humans use plants, other animals do too. Conceivable synergies could therefore emerge by combining and integrating ethno- and zoo-botanical knowledge. This logic already informs nutritional biology, as human and non-human animals possess similar needs, in particular if they are as closely related to us as monkeys and apes. But human and non-human primate interests do not only converge in the area of food. Apes in particular may also use plant parts as implements, for example, as tools for extractive foraging (Fowler *et al.* this volume [Ch. 13]) or to construct resting and sleeping platforms from bent branches (Fruth & Hohmann 1996). Recently, yet another partner-discipline of ethno-botany has emerged, which focuses on the third grand benefit that plants can provide for non-human animals: medicine. This field originally went by the name of *zoopharmacognosy* (Rodriguez & Wrangham 1993), alluding to the Greek words *zoon* (animal), *pharmacon* (medicine) and *gnosis* (knowledge). However, as there are numerous non-pharmacological means to improve health, it is now referred to as "animal self-medication" (Huffman 2007), and includes phenomena such as dietary selection and the ingestion or external application of substances, most of them plant-based (see also Krief *et al.* 2006).

The documentation of the rich repertoire of resources for people and animals provided by biomes such as "the grassland", "the savannah", "the bush", "the forest" or "the jungle" can be read as an inventory for what are, in effect, natural pharmacies and supermarkets – only shoppers do not pay in cash.

Unfortunately, ancestral vegetation covers disappear at alarming rates due to habitat conversion and deforestation, adding a momentum of urgency to the task of recording and preserving knowledge about varied plant uses (Iwu 1993). Our study aims to document the diverse utilisation of plants by humans and non-human primates in an area of high biodiversity, i.e., in and around Gashaka Gumti National Park in Nigeria (*Fig. 5.1*). We also intend to describe the mechanisms behind the selection of plants or plant parts and relate them to potential functions.

Fig. 5.1 Plant samples from the forests and savannah-woodland of Gashaka. *Cola gigantea* (red fruit in brown pod, top l.), *Taebermontana pachysiphon* (large brown and small green fruit, top r.), *Annona senegalensis* (speckled green fruit, middle l.), *Costus afer* (berries, middle centre), *Ficus* spp. (green fruit, bottom l.), *Afzelia africana* (white seed pod, bottom l.), *Irvingia* spp. (green and yellow fruit, middle l.), *Landolphia* spp. (orange fruit, middle centre), *Pseudospondias microcarpa* (single black fruit, bottom centre), *Pycnanthes angloensis* (single red fruit, bottom centre), *Leea guinensis* (orange flower and brown berries, bottom r.), *Strephonema mannii* (middle r.) (photo: VS)

Why Can Plants Be Medicinal? And Why Do Fruits Taste Good?

Our research is biased towards medicinal plants, as these provide most of the available information that also includes interesting cases with respect to monkeys and apes. Thus, we first briefly sketch important trajectories in the fields of traditional medicine, ethno-veterinary medicine and animal self-medication.

Tropical forests abound with plants that offer a selection of nutrients and toxins. Traditional foraging models emphasised the need to maximise energy intake, taking into account the effects of predation and competition with conspecifics and other foragers, but the importance of selection pressures generated by parasitic stress is increasingly recognised (Lozano 1998). This begs one important initial question: Why is the promotion of well-being specifically associated with plant consumption – and not, let's say, meat or fish (Jackson 1991)? Many animal taxa evolved based on their ability to feed on plants. This will often damage the latter and interfere with their reproduction, as animals destroy bark, leaves, flowers and seeds. The result is an evolutionary arms race, in which plants developed defence mechanisms, from nettles and spikes, stings and prickles to chemical agents, designed to discourage predators (Engel 2002). Moreover, plants produce so-called secondary compounds including alkaloids, saponins, phenolics, terpenoids, and non-protein amino acids. These allo-chemicals are metabolically expensive to synthesise and often interfere with the metabolism of plant-eaters, thus deterring them.

Such phyto-chemical propensities, however, also have an upside for plant-eaters, as they can likewise impede the metabolism of pathogens that infected an animal host. This explains, for example, the anti-microbial or anti-helminthic functions of certain plants. Moreover, some secondary compounds may be designed to attract animals – for instance the scent of flowers or the taste of ripe fruit, thus aiding pollination or seed dispersal. In fact, the primate preference for sweet taste with its associated nutritional benefits probably evolved as flowering plants developed fleshy fruits, rich in sugar, specifically aimed to attract them. It is thus probably no coincidence that diversification of flowering plants during the Eocene (55 – 40 mya) has a parallel in the radiation of primate taxa.

The influence of secondary compounds on consumers will depend on various factors, such as ontogenetic stage, nutritional status, dosage and presence of other compounds. These chemicals thus create a mosaic of beneficial and harmful traits that affect health, growth, and behaviour of plant eaters as well as pathogens they host (Johns 1990).

The tropics are particularly rich in flowering plants, predominantly in woody forms, and harbour about 90 % of all animal taxa, including the vast majority of primate species (Cronquist 1968, Engel 2002). Given the co-evolutionary processes just outlined, it comes as no surprise that the tropics are also home to a much greater diversity of medicinal plants than temperate zones.

Traditional Plant-based Medicine

Lists of medicinally valuable plants date back thousands of years, such as those assembled by the Chinese emperor Shen Nung before 2000 BC (Balick & Cox 1997, Sofowora 1993), or the Ebers papyrus records from Egypt from around 1500 BC (Iwu 1993, Sofowora 1993). Such compilations can certainly not be disregarded as backward or superstitious, because three quarters of pharmaceutical drugs are the result of chemical analyses of traditional medicinal plants (Johns 1990). Two well known examples are digitalis and quinine. Digitalis can treat congestive heart failure and was extracted from dried foxglove leaves by William Withering in the 18th century after he consulted a local healer in Shropshire. Quinine and its anti-malarial propensities became known to the West through the Indians of Peru. A compound in *Artemisia annua* had been a febrifuge in China for over 2000 years, before its compound artemisinin was analysed so that it can now be widely used in malaria endemic countries. Nevertheless, less than one percent of flowering plant species have been studied for chemical composition and medical potential (Balick & Cox 1997).

Plants, whether or not their health-promoting phyto-chemical qualities have been established, are used by the majority of the world population in one or the other way as a remedy. In fact, for more than 3.5 billion in the developing world, this is the primary mode to seek a cure, as plants are more readily available than pharmaceutical drugs – and often effective.

Thus, various countries recently decided to integrate traditional medicine into their health care systems (Balick & Cox 1997, Falconer 1992). Nigeria, the most populous nation in Africa, is one of them (Federal Ministry of Health 1988). Firstly, because it harbours a particularly rich culture of traditional practices (Sofowora 1986, Odugbemi 2006), secondly, because few people have appropriate access to Western medicine, and thirdly, because the successful development of a traditional remedy to an internationally marketable drug could be a source of revenue. Referring to the success story of *Artemisia*, Lateed Salako, Emeritus Professor of Pharmacology at the University of Ibadan, expresses his hope "that Nigeria could also have a plant with such potential waiting to be discovered and developed" (Odugbemi 2006: viii).

Ethno-veterinary Medicine

In many traditional contexts people rely on the fertility, health and productivity of domestic livestock, such as cattle or poultry. Ethno-veterinary medicine studies how people try to enhance the welfare and treat diseases of animals they raise or manage. This investigation covers material used, their preparation and administration, as well as related folk knowledge and magico-religious practices and beliefs.

Researchers aim in particular to document plant species of medicinal or nutritional importance for livestock, to then screen them for potential chemo-therapeutical values and toxicological implications (McCorkle1986).

Often, the same plants treat a disease that affects both humans and other animals. For example, African trypanosomiasis transmitted by tse-tse flies is easily fatal for domesticates such as cattle and donkeys, but, as sleeping sickness, affects also an estimated 250000 – 300000 humans. Drugs are either too toxic or too expensive – a fact that led to screening of plants that, according to indigenous knowledge, may treat humans as well as animals (Freiburghaus *et al.* 1996, Adewunmi *et al.* 2001, Atawodi *et al.* 2002).

Animal Self-medication

Folklore and mythology of many cultures report on animals that heal themselves or others. A prime example is told in the Indian epic *Ramayana* (Lal 1981) about Hanuman's journey to the Himalayas, where the monkey-General obtains medicinal herbs that salvage the mortally wounded brother of Rama, the epic's hero. Biologists have assembled evidence that animals including bears, cattle, birds, and primates may indeed be able to medicate (Rodriguez & Wrangham 1993, Huffman 2001, 2007). They don't seem to heal others, as the mythological monkey, but self-medicate, through ingestion or application of non-nutritional substances such as plant secondary compounds, bark or soil.

For African apes in particular, about 3 dozen plant species at 13 study sites have been recorded to be situation-specifically ingested (Huffman 1997, 2001). Certain behaviour patterns are strong indicators of likely medicinal purposes, in particular (a) infrequent intake of species not regularly part of the diet, (b) the habit of ingestion, such as folding and swallowing them unchewed, which minimises any nutritional benefit, (c) plant use is associated with periods of high-risk of parasite infection, (d) illness or infection at time of ingestion, (e) apparently healthy conspecifics show no interest in the plant, and (f) a positive change in condition after ingestion.

A well-documented example is leaf-swallowing, e.g., of plants of the genus *Aspilia* (*Compositae*; Wrangham & Nishida 1983). East African chimpanzees, for example, roll up *Aspilia* leaves, fold them with tongue and palate to ease swallowing, suck them for few seconds, to then swallow them whole without chewing (Wrangham & Goodall 1989). This happens typically in the morning, i.e., on a relatively empty stomach, and the peak of consumption occurs during the rainy season. Chimpanzees will sometimes make a special journey to get to the leaves but ignore them at other times even if they are nearby. They are excreted undigested, often together with parasitic worms (Glander 1994, Huffman & Wrangham 1994).

Apes also consume the bitter pith of *Vernonia amygdalina*, another *Compositae*, for assumed anti-helmintic parasite control. An obviously ill female chimpanzee was observed to recover fully within 24 h after she had carefully extracted the pith and chewed it (Huffman & Seifu 1989, Huffman *et al.* 1997). The pith contains toxic substances, but in much lower concentrations than bark or leaves. Interestingly, these plant parts were avoided by the sick chimpanzee.

Another interesting behaviour is the energetic application of foreign substances into the fur. Wild capuchin monkeys use material from different plants of the genus *Citrus*. According to ethnographic records, these plants treat mainly skin problems. Their phyto-pharmaceutical properties include anaesthetic, insecticidal, anti-septic, fungistatic, anti-allergic, and anti-inflammatory activities. Fur-rubbing was indeed frequent during the wet season when the rise in temperature and humidity increased the risk of bacterial and fungal infection (Baker 1996). Some South American primates rub their fur with millipedes, again mainly during the rainy season. The secretion of these arthropods contains benzoquinon, a repellent for insects such as mosquitos (Huffman 2007).

Research into animal self-medication attempts to unravel complex animal-plant-parasite interactions and ultimately aims to understand the evolution of self-medication in early hominins (Huffman 2001). Comparative evidence from various animal populations across their range of distribution is indispensable for such agenda.

Study Aims

We explore the varied uses of plants in an area of particularly high biodiversity, i.e., the Gashaka area of north-eastern Nigeria in and around the vast Gashaka Gumti National Park. Humans and non-human primates, because of shared evolutionary history and similar physiology, exploit similar sources in their environment. This should be particularly true for primates that share the human trait of considerable ecological flexibility – such as olive baboons and chimpanzees, two species that, since ancient times, lived in close proximity with humans of the Gashaka region. Habitats occupied by baboons and chimpanzees overlap those of early hominins. Understanding their modes of resource utilisation can therefore aid to construct models of hominin ecology and evolution (Johns 1990,Huffman 2001, Peters *et al.* 1981).

We also intend to describe mechanisms behind the selection of particular plants or plant parts for certain purposes, and relate them to potential functions. We therefore compare plant utilisation by local people – for their own benefit or that of domestic animals – with that by chimpanzees and baboons. This includes the use of particular trees by chimpanzees for nest-building purposes, as well as the (unintended or deliberate) ingestion of plants with medicinal effects by both baboons and chimpanzees.

For this, we compile and scrutinise scattered information accumulated from unpublished reports, original ethno-botanical field work as well as ongoing primate research in the area. Our database is designed to encourage future use by students, scientists, naturalists, traditional healers, conservationists and policy makers. We also hope to draw attention to the danger that rich traditions based on plant diversity may soon disappear if the destruction of natural vegetation covers cannot be halted.

Materials and Methods

Study Area

Data were collected in and around the southern sector of Gashaka Gumti National Park (GGNP) in north-eastern Nigeria, about half way up the international border with Cameroon. The area is renowned for its wildlife (Dunn 1999), which includes monkeys and apes (Adanu *et al.* this volume [Ch. 3]). Non-governmental conservation and research activities are coordinated by the Nigerian Conservation Foundation (NCF) and the *Gashaka Primate Project* (GPP). GPP maintains a research station near the village of Gashaka at the park boundary and, about 11 km away, a field station inside the national park at Kwano (07°19′ N – 11°35′ E).

The ethnic composition of the Gashaka sector (Adanu *et al.* this volume [Ch. 3], Nyanganji *et al.* this volume [Ch. 4]), is diverse, although the dominant groups are Fulani cattle herders who speak Fulfulde and farmers who speak mostly Hausa. The predominant religion is Islam, especially amongst the Fulani, although Christian denominations exist, particularly in the town of Serti. The area is largely inaccessible by road, except for the south-western perimeter of the park. People practise subsistence agriculture. Main crops include maize, guinea corn, cassava, yam, groundnut, sugar cane, palm oil, and millet. Within the park exist 6 Fulani-dominated enclaves, mainly in the highlands, where livestock grazing and cultivation are permitted (Bennett & Ross this volume [Ch. 6]). The central market town is Serti, which, at the time of the study, offered drug dispensaries but only a very basic and underresourced hospital. Lack of primary health care facilities and the remoteness of many settlements mean that Western drugs are rarely used, except for tablets available in some village shops (aspirin, paracetamol, ciproflaxin). However, these are often counterfeits or contaminated. Many people therefore prefer or resort to traditional medicine.

The climate of GGNP fluctuates between a dry and a wet season. Rains are often completely absent from mid November till mid March, when a dry and dusty wind, the Harmattan, will frequently blow down from the Sahara Desert. Heavy downpours occur between mid April till mid October, averaging 1935 mm (range 1683 – 2337 mm). The mean minimum temperature is 20.9 °C, the coolest day 12 °C, the mean maximum temperature 32.2 °C, and the hottest day 42 °C (data for 2000 – 2008 measured at the research stations of GPP at Gashaka and Kwano; see Sommer & Ross this volume [Ch. 1]). The terrain is undulating and rugged with altitudes of 300 – 2400 m. GGNP is an important water catchment area for the Benue River, as abundant rivers flow continuously, even throughout the marked dry season.

The area is located in the sub-Saharan Guinea zone and its vegetation cover represents a mosaic of montane grasslands, montane forests, savannah-woodlands, lowland and gallery forests. For overviews of floristic composition and taxonomy see Dunn (1999) and Chapman & Chapman (2001).

Montane forests, often misty, grow upwards of 1650 m. Trees reach a height of 15 – 10 m, with open canopy and abundant epiphytes such as orchids and ferns.

Common species include *Syzygium guineense* (rose apple), *Prunus africana* (of the Rosaceae family, and on the IUCN Red List) as well as the genera *Ficus* and *Albizia*. Montane grassland is at least partly human-made through grazing and annual burning from the onset of the dry season. The ground is then covered with a layer of ash until fresh grass appears. Common genera, rarely exceeding 60 cm, include *Chloris*, *Eragrostis* and *Sporobolus*.

Woodland is dominated by a ground cover of tall coarse grasses, 2 – 3 m high (genera *Andropogon*, *Hyparrhenia*). Trees are relatively small, dispersed and often fire-resistant, including *Lophira lanceolata* (ironwood), *Daniellia oliveri* (balsam tree), *Afzelia africana* (pod mahogany), *Crossopteryx febrifuga* (sandcrown berry), *Piliostigma thonningii*, *Annona senegalensis* (African custard apple), *Parkia biglobosa* (locust bean) and various species of the genus *Terminalia*.

The lowland rain forest is often stratified. Emergents such as *Khaya senegalensis* (mahogany) may grow up to 40 m and include trees with characteristic buttress roots such as *Ceiba pentandra* (white silk cotton tree) and the genus *Terminalia*. At 15 – 35 m, trees of genera such as *Pseudospondias* and *Afzelia* may form a closed canopy. Trees found in the next lower storey include *Erythrophleum suaveolens* (sasswood) and the genera *Albizia* and *Celtis*. Still below thrive small trees of 2 – 8 m height and shrubs such as *Cnestis ferruginea*, and *Monodora tenuifolia*. Lianas and epiphytes are abundant throughout the forest, and moister parts of the forest floor may be covered by a herbaceous layer as well as mosses, ferns and orchids.

Gallery forests along muddy stream banks include the unmistakable *Pandanus candelabrum* (candelabrum tree) and *Costus* spp., while those along rivers and streams contain *Brachystegia eurycoma*, *Erythrophleum suaveolens*, *Phoenix reclinata* as well as the genera *Berlinia* and *Blighia*.

Data Pool

The study compiles five major sources of information about plants and their exploitation by humans and non-human primates in and around GGNP:

(a) *Herbarium*. We consulted a reference collection at Gashaka maintained by NCF and GPP.
(b) *Forest phenology*. We accessed data on tree diversity in the major study area of GPP around the Kwano research station. Here, 2 straight-line transects of 4 km length had been established (Sommer *et al.* this volume [Ch. 12]). Along these transects, about 1000 trees with diameters of more than 10 cm were tagged and identified to the level of genus or species. The composition reflects tree cover in altitudes of about 400 – 700 m.
(c) *Ecology of chimpanzees and baboons*. We used data collected by GPP primate researchers on plants consumed or utilised by baboons and chimpanzees. The chimpanzees (*Pan troglodytes vellerosus*) of Gashaka-Kwano have been studied since 2000. The apes were frequently encountered directly, although continuous follows were not possible at the time of this investigation. Foraging party size

averaged 4 animals (range 1 – 17) and nest-groups 6 (range 1 – 23; Sommer *et al.* 2004). Many tree species in which night-nests were constructed were known (Adanu 1997). Feeding remains, direct observations, and faecal samples found under night-nests were utilised to identify chimpanzee food species – including potential medicinal plants (Fowler 2006 pers. comm., Fowler *et al.* 2007, Hohmann *et al.* 2006). Two groups of olive baboons (*Papio anubis*) were already fully habituated to human observers at the time of this study. A wild-feeding troop with an average size of 28 members ranged near Kwano, and a troop with an average of 19 members near Gashaka village, where the monkeys also raided crops. The majority of information on feeding behaviour came from direct observations (Warren 2003 pers. comm., Warren *et al.* this volume [Ch. 8], supplemented by data from Higham 2006, Higham *et al.* 2007 on hormone metabolites extracted from faeces).

(d) *Existing ethno-botanical records.* We mined various unpublished data sets on vegetation cover, non-timber forest products, ethno-botanical and ethno-veterinary uses of plants of the GGNP area collated between 1996 – 1998. These were available as reports to NCF by Nigerian botanists (Akinsoji 1996, Ayanbamiji 1996, NCF n.d.) as well as dissertations by masters students from University College London (Martin 1996, Pellaumail 1998).

(e) *Ethno-botanical field work.* YK collected original ethno-botanical data during field work from Jan – Apr 03 in 9 locations in and around GGNP. These included the town of Serti and 5 villages straddling the south-eastern border of the national park (Karamti, Bodel, Mayo Selbe, Gashaka, Do Mayo). The remaining locations were situated in highland enclaves at altitudes of 1650 – 1850 m. The first was Selbe, a market village in the Tale enclave on top of the Chappal Tale mountain; the second was Chappal Hendu in the neighbouring highland enclave of Hendu; the third location was Filinga, a market village in the Filinga enclave situated in a low-lying plateau near the border with Cameroon. Field surveys relied on staff from GPP and NCF, native to the area, who enabled logistics in what was often difficult terrain, helped recruit informants and acted as translators. One assistant was male and Muslim, the other female and of a Christian background. Rapid Rural Appraisal techniques were employed, aimed at gaining information and insight from rural people about rural conditions through participant observations combined with informal as well as formal conversations and interviews (see Benett & Ross this volume [Ch. 6]). Some previous reports used in this study had relied on similar techniques. Informants were asked to identify herbarium voucher specimen or plants on photographs. Most plants were chosen based on their inclusion in unpublished reports about ethno-botanical usage. Informants were also asked about any other plants they used and for which neither voucher specimens nor pictures were available. For each plant, information was recorded on growth habit, habitat / usages, plant parts utilised, season of collection, preparation and applications, as well as how they had acquired knowledge of the plants' properties. In total, 52 woman and 57 men between the ages of 25 and 70 were interviewed. Of these, 19 were active traditional healers, 2 of them women and 17 men.

Data Processing

All information was compiled and transferred into a master table (see *Appendix)*. For each plant entry information is given as to data sources; botanical family, genus and species names; English name; growth habit and habitat of the plant; parts used; usage category; and method of administration. Botanical and English names were verified and / or determined with the help of floristic compendia specialising on West Africa (Dalziel 1963, Dalziel & Hutchinson 1963, Burkill 1985, 1994, 1995, 1997, 2000, Keay 1989, Chapman & Chapman 2001), voucher specimens present in the Gashaka herbarium, and with the expertise of a leading botanist on West African flora (Emmanuel Obot of NCF). The information was edited for obvious mistakes in spelling and nomenclature, and gaps in sources were filled as far as possible.

Plant parts used are listed for the 4 consumer groups *humans* (H), *baboons* (B), *chimpanzees* (C), and domestic *animals* (A; referring to cattle and sometimes chicken). Usage is described in terms of the general categories *food* (F), *medicine* (M), and *implements* (I), with various subcategories.

Some entries are ambiguous or problematic. For example, magico-religious usage of plants is included under "implements", although it could be argued that the effects are linked to health issues and should therefore fall under "medicine". Moreover, magico-religious entries may refer to the cause of a disease, such as when wild custard apple (*Annona senegalensis*) is used against the "evil spirit", whereas the symptom could be anything from fever to headache to skin conditions. Further mix-ups presented themselves with disease diagnoses, and with whether or not to lump or split them into certain categories. For instance, "stomach ache" or "stomach problems" could refer to various gastro-intestinal symptoms such as constipation, diarrhoea and stomach ache. Similarly, "cholera" was mentioned sometimes, but the informant specifically referred to diarrhoea, a symptom of cholera. Such confusion arises easily as lay people and traditional healers do not always know the pathology of diseases – or they may have a clear concept but focus instead on the symptoms (Sofowora 1993, Wall 1988). Moreover, some of the indigenous concepts used may be entirely different from those of Western medicine.

Finally, it is not uncommon for different plants to be known by the same local name. For example, the Hausa name "yibal" is attributed to various species of the genus *Ficus*, the fig tree. In such cases, we used the definition "spp.". Conversely, the same plant may be known by different names, especially in languages as rich as Hausa and Fulfulde. Occasionally, informants would also use a name from a dialect used in their place of origin.

The data are biased in the sense that there is far more information available on plant use by humans than by domestic animals, baboons or chimpanzees. Data on non-human primates in particular accumulate only slowly and depend much on the degree of habituation and length of study (Sommer *et al*. 2004).

Of course, traditional uses of forests encompass more than plant exploitation – to which the term Non-Timber-Forest-Products tries to do justice (Falconer 1992).

NTFPs are biological forest products collected for subsistence, commercial or cultural purposes (excluding, as the name suggests, timber, which is often traded by non-local companies). Our study also omits faunal matter, e.g., animal products such as honey or edible larvae, hides, trophies or bush-meat (see Nyanganji *et al.* this volume [Ch. 4]). We also do not attempt to quantify how much NTFPs contribute to income generation when sold directly to households or on markets (Balick & Cox 1997).

Instead, our study intends to document any usage of plants and plant parts, irrespective of frequency. Therefore, as quantification is not an issue, there are few problems with biases typically caused by an uneven distribution of human informants with respect to age, sex, religion, occupation, and so on – as well as with respect to age, sex, social status or reproductive condition of non-human animals. In any case, the compilation provided here is intended as a database on which future work can be built.

Results

Natural Occurrence of Exploited Plants

A total of 309 species were recorded as being exploited in various ways for food, medicine, or implements by one or more of the four consumer groups (*App.*). Information was available about the *habitat* of 119 species. Woodland-savannah is occupied by 59.7 % of these species, lowland forest by 28.6 %, and montane forest by 6.7 %; 5.0 % refer to domestic plants or crops. Information on *growth habit* covers 235 species. More than half – (58.3 %) – are trees, 16.1 % shrubs, 12.3 % herbs, 6.8 % lianas, 3.4 % crops, and the remaining 3.0 % are grasses.

Plant Families

Of 309 documented species, 62 could not be scientifically identified, as only local names – in Hausa and / or Fulfulde – were available. The 247 known species belong to 81 families of which 11 (13.6 %) make up more than half (51.0 %) of all species. These dominant families include 7 – 20 species whereas the vast majority of the remaining 70 families covers just 1 or 2 species. Families with at least 5 % of all species are Leguminosae-Caesalpinioidea (8.1 %), Euphorbiaceae (5.7 %), Leguminosae-Mimosoidea (4.9 %), Leguminosae-Papilionoideae (4.9 %) and Combretaceae (4.5 %).

A total of 517 different usages are recorded for the 4 consumer groups (*Tab. 5.1*). Again, only 11 families make up more than half (50.7 %) of all entries – more or less the taxa with the most species. Leguminosae-Caesalpinioidea top the list with 10.6 % of all usages, followed by Leguminosae-Mimosoidea (5.6 %), Euphorbiaceae (5.0 %) and Moraceae (4.8 %).

Table 5.1 Number of plant species per family with recorded usages as food, medicine, and implements. H = humans, A = domestic animals, B = baboons, C = chimpanzees

	Species	Food					Medicine					Implements				Total species		
Family	*n*	H	A	B	C	Sum	H	A	B	C	Sum	H	A	C	Sum	*n*	%	Cumulative %
Leguminosae – Caesalpinioideae	20	6	2	10	4	22	16	3			19	11	1	2	14	55	10.6	10.6
Leguminosae – Mimosoideae	12	4		5	2	11	9	2			11	7			7	29	5.6	16.2
Euphorbiaceae	14	4	1	7	2	14	6				6	6			6	26	5.0	21.3
Moraceae	10	4	2	8	1	15	5	2			7	2	1		3	25	4.8	26.1
Anacardiaceae	7	4		4	2	10	6				6	5			5	21	4.1	30.2
Palmae	7	5		3	2	10	4				4	6			6	20	3.9	34.0
Combretaceae	11			3		3	8	2			10	5		1	6	19	3.7	37.7
Sterculiaceae	8	3	1	1	3	8	3				3	2		5	7	18	3.5	41.2
Leguminosae – Papilionoideae	12			6		6	6			1	7	3		1	4	17	3.3	44.5
Rubiaceae	9	1		4	1	6	6	2			8	3			3	17	3.3	47.8
Gramineae	9	3	1	6		10	1				1	4			4	15	2.9	50.7
Burseraceae	4	2	2	1	1	6	4				4	4			4	14	2.7	53.4
Apocynaceae	7	3		5	3	11	1				1	2			2	14	2.7	56.1
Compositeae	6	2	1			3	6	1			7	3			3	13	2.5	58.6
Myrtaceae	4	2		2	1	5	2	1			3	1		3	4	12	2.3	60.9
Annonaceae	4	2		2	2	6	2				2	2		2	4	12	2.3	63.2
Zingiberaceae	5	2		5	2	9	2				2					11	2.1	65.4
Meliaceae	4		1			1	3	2			5	2		2	4	10	1.9	67.3
Verbenaceae	2	1	1	1	1	4	2	1	1		4	1		1	2	10	1.9	69.2
Rutaceae	5	1				1	4				4	1		1	2	7	1.4	70.6
Sapotaceae	2	1	1	2		4	1	1			2			1	1	7	1.4	72.0
Bignoniaceae	3	1				1	3	1			4	2			2	7	1.4	73.3
Bombacaceae	3	1		1		2	2				2	3			3	7	1.4	74.7

(continued)

Table 5.1 (continued)

	Species	Food					Medicine					Implements				Total species		
Family	*n*	H	A	B	C	Sum	H	A	B	C	Sum	H	A	C	Sum	*n*	%	Cumulative %
Ebenacae	3	1		1	1	3	1				1			2	2	6	1.2	75.8
Loganiaceae	3	2		1	1	4	2				2					6	1.2	77.0
Tiliaceae	2	2		1		3	1				1	1			1	5	1.0	77.9
Olacaceae	2	1		1	1	3	1				1			1	1	5	1.0	78.9
Malvaceae	3	1				1	1				1	3			3	5	1.0	79.9
Lecythidaceae	2			2		2	1				1	1			1	4	0.8	80.7
Ochnaceae	1			1		1	1	1			2	1			1	4	0.8	81.4
Rhamnaceae	1	1				1	1	1			2	1			1	4	0.8	82.2
Sapindaceae	2	1				1	1				1	1		1	2	4	0.8	83.0
Solanaceae	2						2	1			3	1			1	4	0.8	83.8
Amaranthaceae	2	2		1		3										3	0.6	84.3
Araceae	2	1				1	2				2					3	0.6	84.9
Caricaceae	1	1		1		2	1				1					3	0.6	85.5
Costaceae	2			2		2	1				1					3	0.6	86.1
Labiatae	2	2				2	1				1					3	0.6	86.7
Moringaceae	1	1				1	1				1	1			1	3	0.6	87.2
Musaceae	2	1		1		2	1				1					3	0.6	87.8
Araliaceae	2			1		1	1				1	1			1	3	0.6	88.4
Asclepiadaceae	1	1				1	1				1	1			1	3	0.6	89.0
Balanitaceae	1	1				1	1				1	1			1	3	0.6	89.6
Balanophoraceae	1	1				1	1				1					2	0.4	89.9
Boraginaceae	1	1				1	1				1					2	0.4	90.3
Chrysobalanaceae	2	1		1		2										2	0.4	90.7
Cochlospermaceae	1						1				1	1			1	2	0.4	91.1
Cucurbitaceae	1	1		1		2										2	0.4	91.5
Cyperaceae	2	1				1	1				1					2	0.4	91.9
Guttiferae	2						1				1			1	1	2	0.4	92.3

Hypoxidaceae	1		1		1	1		1				2	0.4	92.6
Icacinaceae	1		1	1	2							2	0.4	93.0
Iridaceae	1	1			1	1		1				2	0.4	93.4
Leeaceae	1		1	1	2							2	0.4	93.8
Lythraceae	1					1		1	1		1	2	0.4	94.2
Pandanaceae	1		1		1				1		1	2	0.4	94.6
Polygalaceae	1					1		1	1		1	2	0.4	95.0
Taccaceae	1	1			1				1		1	2	0.4	95.4
Nyctaginaceae	2					2		2				2	0.4	95.7
Amaryllidaceae	1					1		1				1	0.2	95.9
Ampelidaceae	1					1		1				1	0.2	96.1
Cannabaceae	1						1	1				1	0.2	96.3
Capparidaceae	1	1			1							1	0.2	96.5
Cecropiaceae	1					1		1				1	0.2	96.7
Commelinaceae	1								1		1	1	0.2	96.9
Convolvulaceae	1		1		1							1	0.2	97.1
Dioscoreaceae	1	1			1							1	0.2	97.3
Dipterocarpaceae	1								1		1	1	0.2	97.5
Dracaenaceae	1								1		1	1	0.2	97.7
Flacourtiaceae	1								1		1	1	0.2	97.9
Liliaceae	1					1		1				1	0.2	98.1
Marantaceae	1		1		1							1	0.2	98.3
Menispermaceae	1					1		1				1	0.2	98.5
Myristicaceae	1		1		1							1	0.2	98.6
Nympheaceae	1	1			1							1	0.2	98.8
Pedaliaceae	1	1			1							1	0.2	99.0
Rhizophoraceae	1									1	1	1	0.2	99.2
Salvadoraceae	1					1		1				1	0.2	99.4
Simaroubaceae	1									1	1	1	0.2	99.6

(continued)

Table 5.1 (continued)

	Species	Food					Medicine					Implements				Total species		
Family	*n*	H	A	B	C	Sum	H	A	B	C	Sum	H	A	C	Sum	*n*	%	Cumulative %
Smilacaceae	1			1		1										1	0.2	99.8
Ulmaceae	1													1	1	1	0.2	100.0
Subtotal	247	82	13	99	32	226	141	22	1	1	165	97	2	27	126	517	100.0	
Unknown	62																	
Total	309																	

Of 226 recorded usages for *food*, Leguminosae-Caesalpinioideae score 9.7 %. They are followed by Moraceae (6.6 %) and Euphorbiaceae (6.2 %), mainly due to their popularity with baboons. Leguminosae-Mimosoideae (4.9 %) and Apocynaceae (4.9 %) are also well represented. Humans represent 36.3 % of all food usages. Interestingly, Combretaceae and Leguminosae-Papilionoideae who are amongst the most popular families overall, do not feature as human food; their popularity is more related to medical uses. Baboons have the highest proportion of food entries (43.8 %), given their well-studied diet. Chimpanzee food represents just 14.1 %, and only 5.7 % of all entries relate to food provisioned to domestic animals.

With respect to 165 entries for *medicine*, Leguminosae-Caesalpinioideae score 11.5 %, followed by Leguminosae-Mimosoideae (6.7 %) and Combretaceae (6.1 %). The majority of medicinal uses refer to humans (85.4 %), and 13.3 % to domestic animals. Only two records (1.2 %) consider potential medicinal effect for non-human primates.

Out of 126 entries for *implements*, Leguminosae-Caesalpinioideae score 11.1 %, followed by Leguminosae-Mimosoideae (5.6 %) and Sterculiaceae (5.6 %). More than three quarters (77.0 %) of all entries refer to humans, and 1.6 % to magico-religious implements applied to domestic animals. The remaining 21.4 % relate to construction of night nests by chimpanzees.

Selective Use: The Example of Trees

The floristic composition of the 8-km straight line transect through savannah-woodland and forests near the Kwano research station can be taken as a random representation of tree cover in the Gashaka area. The transect encompasses 985 trees of which 925 are scientifically identified, representing 158 different species. Of these, 52 (32.9 %) are also used as food, medicine or implements by one of the four consumer groups. These species represent 46.6 % of all individual transect specimen. This ratio of 1.4 points towards a preferential use of certain tree species while others are neglected. In fact, just 17 species (10.8 %) represent more than half (51.3 %) of all specimens on the transect. Of these dominant species, only 9 (52.9 %) are used by the consumer groups. This indicates again a rather pronounced discrimination, which, moreover, is *not* based on tree frequency. Unused species of dominant trees are, for example, *Trichilia martineani* (6.0 % of all transect specimen), *Strephenoma manii* (3.5 %), and *Sterculia oblonga* (3.2 %). Dominant trees, which *are* found in the records for various usages include *Anogeissus leiocarpus* (6.6 % of all transect specimen), *Crossopteryx febrifuga* (5.3 %) and *Cola gigantea* (3.3 %).

Plant Parts Used

We were able to ascertain 917 records for the use of specific plant parts (*Tab. 5.2*). Almost half of them pertain to three major categories, fruits (19.2 %), leaves (17.1 %) and bark (13.0 %). However, the picture for the general usage categories is more differentiated.

Table 5.2 Number of plant species from which different parts were used as food, medicine, and implements. H = humans, A = domestic animals, B = baboons, C = chimpanzees

	Food				Medicine				Implements			Total		
Parts used	H	A	B	C	H	A	B	C	H	A	C	n	%	Cumulative %
FR: Fruit	51	2	76	31	14	1	1					176	19.2	19.2
LF: Leaf	24	7			76	5		1	16		28	157	17.1	36.3
BA: Bark	2	2	2		76	12			24	1		119	13.0	49.3
RO: Root	4		10		80	3			7			104	11.3	60.6
SE: Seed, bean	21		26	2	15	2			5	1		72	7.9	68.5
WO: Wood					4				43			47	5.1	73.6
ST: Stem	3	1	15		9				14			42	4.6	78.2
WP: Whole plant (a)	7	2			7	1			17	1		35	3.8	82.0
TR: Bole (trunk)									6		28	34	3.7	85.7
TW: Twig					4				1		28	33	3.6	89.3
FL: Flower	1		11		1	1			2			16	1.7	91.1
FP: Fruit pulp	7		2	3	2							14	1.5	92.6
LY: Leaf, young			9	1	4							14	1.5	94.1
LA: Latex / Sap	4				6				2			12	1.3	95.4
TU: Tuber	3				3				1			7	0.8	96.2
FJ: Fruit juice					5							5	0.5	96.7
PI: Pith					5							5	0.5	97.3
FO: Frond									4			4	0.4	97.7
FI: Fibre					1				2			3	0.3	98.0
FS: Floss (mostly in fruit)					1				2			3	0.3	98.4
GU: Gum			2		1							3	0.3	98.7
NU: Nut			3									3	0.3	99.0
PO: Pod					2				1			3	0.3	99.3
SH: Shoot					2							2	0.2	99.6
BU: Bulb, Corm					1							1	0.1	99.7

FD: Fruit rind					1							1	0.1	99.8
NE: Nectar			1									1	0.1	99.9
TH: Thorn									1			1	0.1	100.0
Total	127	14	157	37	320	25	1	1	148	3	84	917	100.0	

(a) Including grass

With respect to *food*, fruit is indeed most popular (17.4 % of all species), followed by seeds (5.3 %) and leaves (3.4 %). Latex, tuber, nut, gum, and nectar are the least frequent food items. Domestic animals are mainly fed leaves, which is not surprising, given that cattle are ruminants.

Parts most commonly exploited as *medicine* are, with similar proportions, bark (9.6 %), roots (9.1 %), and leaves (8.9 %). The bark in particular is most often applied to treat domestic animals. The fruit of one species and the leaves of another constitute the few entries for non-human primates.

Parts mostly employed as *implements* are the leaf (4.8 %), followed by wood (4.7 %) and the bole (3.7 %). Parts listed as implements for domestic animals (bark, seed, whole plant) refer to magico-religious purposes. Chimpanzees use branches and – as they nest above ground – the bole of 28 tree species (3.1 %) to construct sleeping platforms.

Multiple Usages of Same Plant Species

Of 81 recorded families, 55 (67.9 %) are drawn on in multiple ways, i.e., are utilised in more than one of general usage category. Of these, 10 families (18.2 %) make up more than half of the overlap entries (51.5 %). Leguminosae-Caesalpinioideae are top (12.2 %), followed by Mimosoideae (5.9 %), Anacardiaceae (5.4 %), Moraceae (5.1 %), Euphorbiaceae 4.6 %) and Palmae (4.6 %).

Of 309 recorded species, 136 (44.0 %) are utilised in more than one of general usage category. Of these, 122 species have been scientifically identified (*Tab. 5.3*).

Humans use 82 species as food, 141 as medicine, and 97 for implements. Use as food *and* medicine is recorded for 57 species. This means that 69.5 % of all food plants are also used as medicine, but only 40.4 % of all medicine plants are also used as food. This indicates a greater exclusivity of medicinal plants, perhaps because many of their active substances would be dangerous if consumed in larger quantities, as might be the case if eaten. Use as food and implements is recorded for 43 species, and overlap between medicine and implements for 64 species. This means, that 44.3 % of all implement species are also used as food plants, whereas 66.0 % are also used as medicinal plants. Implements often make use of woody plant parts; these are less suitable for food, but a relatively greater proportion can still be utilised to brew medical concoctions.

The different consumer groups exploit several of the same taxa. This is particularly true for *food* species. The overlap in 8 species between humans and domestic animals means that humans eat from all but one of the 9 species they feed to livestock. Overlap in 42 species between humans and baboons means that 42.4 % of all 99 baboon food species are also eaten by humans. Overlap between humans and chimpanzees in 24 cases means that 75.0 % of all known chimpanzee food species are also consumed by humans. This might reflect a greater physiological similarity between humans and chimpanzees compared to chimpanzees and baboons. The overlap for chimpanzees and baboons in 31 cases means that all but one of the 32 known chimpanzee food items are also known to be eaten by baboons.

Table 5.3 Plant species with multiple utilisation as food, medicine and implements. X indicates at least one entry per category of usage. H = humans, A = domestic animals, B = baboons, C = chimpanzees

	Food				Medicine				Implements			Total species		
Species	H	A	B	C	H	A	B	C	H	A	C	*n*	%	Cumulative %
Vitex doniana	x	x	x	x	x	x	x		x		x	9	2.3	2.3
Afzelia africana	x	x	x	x	x	x			x			7	1.8	4.1
Syzygium guineense	x		x	x	x	x			x		x	7	1.8	5.9
Ficus spp.	x		x	x	x	x			x			6	1.5	7.5
Ficus sur	x	x	x		x	x				x		6	1.5	9.0
Parkia biglobosa	x		x	x	x	x			x			6	1.5	10.6
Vitellaria paradoxa	x	x	x		x	x					x	6	1.5	12.1
Annona senegalensis	x		x	x	x				x			5	1.3	13.4
Borassus aethiopum	x		x	x	x				x			5	1.3	14.7
Bridelia ferruginea	x		x	x	x				x			5	1.3	16.0
Canarium schweinfurthii	x		x	x	x				x			5	1.3	17.3
Cola millenii	x		x	x					x		x	5	1.3	18.6
Daniellia oliveri			x	x	x	x			x			5	1.3	19.8
Khaya senegalensis		x			x	x			x		x	5	1.3	21.1
Phoenix reclinata	x		x	x	x				x			5	1.3	22.4
Prosopis africana	x		x	x	x				x			5	1.3	23.7
Xylopia spp.	x		x	x	x				x			5	1.3	25.0
Adansonia digitata	x		x		x				x			4	1.0	26.0
Aframomum angustifolium	x		x	x	x							4	1.0	27.1
Aframomum spp.	x		x	x	x							4	1.0	28.1
Anogeissus leiocarpus			x		x	x			x			4	1.0	29.1
Brachystegia eurycoma	x		x		x				x			4	1.0	30.2
Cola gigantea	x		x	x							x	4	1.0	31.2
Detarium microcarpum	x		x		x				x			4	1.0	32.2
Diospyros mespiliformis	x		x	x	x							4	1.0	33.2

(continued)

Table 5.3 (continued)

Species	Food				Medicine				Implements			Total species		
	H	A	B	C	H	A	B	C	H	A	C	*n*	%	Cumulative %
Elaeis guineensis	x		x		x				x			4	1.0	34.3
Entada africana	x				x	x			x			4	1.0	35.3
Erythrophleum suaveolens			x		x				x		x	4	1.0	36.3
Ficus thonningii	x	x			x				x			4	1.0	37.4
Grewia mollis	x		x		x				x			4	1.0	38.4
Hymenocardia acida	x	x			x				x			4	1.0	39.4
Landolphia landolphioides	x		x	x					x			4	1.0	40.5
Lophira lanceolata			x		x	x			x			4	1.0	41.5
Magnifera indica	x		x		x				x			4	1.0	42.5
Panicum maximum	x	x	x						x			4	1.0	43.6
Piliostigma reticulatum			x		x	x			x			4	1.0	44.6
Pseudospondias microcarpa	x		x	x					x			4	1.0	45.6
Tamarindus indica	x		x		x				x			4	1.0	46.6
Terminalia laxiflora			x		x	x			x			4	1.0	47.7
Vernonia amygdalina	x				x	x			x			4	1.0	48.7
Ximenia americana	x		x	x	x							4	1.0	49.7
Ziziphus spina-christi	x				x	x			x			4	1.0	50.8
Acacia sieberiana			x		x				x			3	0.8	51.5
Alchornea cordifolia			x		x				x			3	0.8	52.3
Anacardium occidentale	x				x				x			3	0.8	53.1
Aspilia africana		x			x				x			3	0.8	53.9
Balanites aegyptiaca	x				x				x			3	0.8	54.6
Boswellia dalzielii	x				x				x			3	0.8	55.4
Calotropis procera	x				x				x			3	0.8	56.2
Carica papaya	x	x			x							3	0.8	57.0
Commiphora kerstingii		x			x				x			3	0.8	57.7
Commiphora spp.		x			x				x			3	0.8	58.5

Detarium senegalense	x	x	x					3	0.8	59.3
Dialium spp. (Strephonema manii ?)		x	x			x		3	0.8	60.1
Dichrostachys cinerea	x			x		x		3	0.8	60.8
Erythrina senegalensis		x		x		x		3	0.8	61.6
Haematostaphis barteri	x			x		x		3	0.8	62.4
Hibiscus spp.	x			x		x		3	0.8	63.1
Lannea barteri		x	x	x				3	0.8	63.9
Lasianthus batangensis	x	x	x					3	0.8	64.7
Moringa oleifera	x			x		x		3	0.8	65.5
Napoleona vogelii		x		x		x		3	0.8	66.2
Newbouldia laevis	x			x		x		3	0.8	67.0
Paullinia pinnata	x			x		x		3	0.8	67.8
Piliostigma thonningii		x		x		x		3	0.8	68.6
Pseudocedrella kotshyi				x	x	x		3	0.8	69.3
Psidium guajava	x	x		x				3	0.8	70.1
Senna obtusifolia	x				x		x	3	0.8	70.9
Solanum aculeastrum				x	x	x		3	0.8	71.6
Strychnos spinosa	x	x	x					3	0.8	72.4
Uapaca spp.	x	x	x					3	0.8	73.2
Uapaca togoensis	x	x				x		3	0.8	74.0
Vernonia ambigua	x			x		x		3	0.8	74.7
Acacia nilotica		x		x				2	0.5	75.3
Alstonia boonei				x		x		2	0.5	75.8
Amorphophallus abyssinicus	x			x				2	0.5	76.3
Apodytes dimitiata ?		x	x					2	0.5	76.8
Bauhinia rufescens				x		x		2	0.5	77.3
Ceiba pentandria				x		x		2	0.5	77.8
Celosia trigyna	x	x						2	0.5	78.4

(continued)

Table 5.3 (continued)

Species	Food				Medicine				Implements			Total species		
	H	A	B	C	H	A	B	C	H	A	C	*n*	%	Cumulative %
Chamaecrisda mimosoides		x							x			2	0.5	78.9
Cochlospermum planchonii					x				x			2	0.5	79.4
Cola ?				x	x							2	0.5	79.9
Cola verticillata					x				x			2	0.5	80.4
Cordia africana	x				x							2	0.5	80.9
Costus afer			x		x							2	0.5	81.4
Crossopteryx febrifuga					x				x			2	0.5	82.0
Cucurbita pepo	x		x									2	0.5	82.5
Circuligo pilosa			x		x							2	0.5	83.0
Cussonia arborea			x		x							2	0.5	83.5
Desmodium gangeticum					x			x				2	0.5	84.0
Ensete gilletii	x				x							2	0.5	84.5
Ficus ovata	x		x									2	0.5	85.1
Ficus sycomorus			x		x							2	0.5	85.6
Gardenia erubescens					x				x			2	0.5	86.1
Gardenia nitida			x		x							2	0.5	86.6
Gardenia triacantha					x	x						2	0.5	87.1
Gladiolus primulinus	x				x							2	0.5	87.6
Imperata cylindrica					x				x			2	0.5	88.1
Landolphia macrantha ?			x	x								2	0.5	88.7
Landolphia owariensis	x		x									2	0.5	89.2
Landolphia togolana ?			x	x								2	0.5	89.7
Lannea nigritana			x		x							2	0.5	90.2
Lannea schimperi					x				x			2	0.5	90.7
Lawsonia inermis					x				x			2	0.5	91.2
Leea guineensis			x	x								2	0.5	91.8

Mucuna pruriens			x		x							2	0.5	92.3
Nauclea latifolia			x		x							2	0.5	92.8
Ocimum gratissimum	x				x							2	0.5	93.3
Pandanus candelabrum			x						x			2	0.5	93.8
Pterocarpus erinaceus			x		x							2	0.5	94.3
Raphia hookeri	x								x			2	0.5	94.8
Raphia sudanica	x								x			2	0.5	95.4
Securidaca longepedunculata					x				x			2	0.5	95.9
Sterculia setigera	x				x							2	0.5	96.4
Stereospermum kunthianum					x	x						2	0.5	96.9
Strychnos innocua	x				x							2	0.5	97.4
Tacca involucrata	x								x			2	0.5	97.9
Terminalia spp.			x		x							2	0.5	98.5
Thonningia sanguinea	x				x							2	0.5	99.0
Zanthoxylum leprieurii					x				x			2	0.5	99.5
Zanthoxylum zanthoxyloides	x				x							2	0.5	100.0
Total	70	13	69	32	97	21	1	1	75	2	7	388	100.0	

Overlap in terms of *medicine* is likewise interesting. Thus, humans treat themselves with all but 2 of the 22 species, which they use as a remedy for their livestock. The two species consumed by non-human primates for potential medicinal effects are both also used for medicinal purposes by humans.

Overlap in terms of *implements* is more restricted. The 2 species employed for magico-religious purposes related to domestic animals are also used for human purposes. Interestingly, the overlap between humans and chimpanzees is restricted to 5 species. Thus, humans exploit only 18.5 % of the chimpanzee nesting tree species – indicating a high selectivity on part of the apes.

Categories of Usage: Food, Medicine, Implements

Records for subcategories of utilisation by all four consumer groups total 1050 (*Tab.* 5.4, *Tab* 4.5). The most common single usages are nutrition (22.9 %), followed by gastro-intestinal problems (12.8 %), fevers (6.2 %), skin treatment (5.4 %), implemental use as timber and furniture (5.2 %) and crafts and skill (4.6 %).

The vast majority of the 259 entries for *food* refers to the edibility of plants or plant parts (92.7 %). This purpose dwarfs the next most common usages: seasoning (2.3 %) and brewing beverages (1.9 %).

There are 591 entries under *medicine*. The subcategory stomach malaise is most common with 10.7 % (made up by combining 33 entries for stomach ache, 22 for stomach problems and 8 for stomach upset). Loose bowel motions constitute 6.9 % (by combining 24 entries for dysentery and 17 for diarrhoea). Worms (internal parasites) accrue 6.8 % of entries. This is the only health issue that applies to all four consumer groups. Skin conditions follow with 4.7 %. Finally, one might not want to fail mentioning that 1.4 % of entries come with the promise to "strengthen the penis". Although most medicinal uses refer to humans, the following case studies illustrate how baboons and chimpanzees eat plants that affect their health status.

Of 193 entries under *implements*, chimpanzee nesting species top the list with 14.5 %, followed by human usages such as cooking tools (7.7 %), firewood (7.7 %), and materials for construction (6.2 %) and fences (5.7 %).

Baboon Case Study: Contraceptive Effects of African Black Plum

Studies of olive baboons at Gashaka (Higham *et al.* 2007) established a strong link between changes in sexual behaviour and consumption of African black plum (*Vitex doniana*). Around the time of ovulation, female baboons develop ano-genital swellings, conspicuously pink, which attract male attention and lead to increased copulation frequencies. Onset and maintenance of swellings is oestrogen-dependent, and they disappear during the luteal phase of the menstrual cycle with rising levels of progesterone (review in Higham 2006).

Table 5.4 Number of records for sub-categories in general usage areas food, medicine, implements. Combined figures derived from entries in Table 5.5

Category	Usages (*n*)	%	Cumulative %
Edible	240	22.9	22.9
Gastro-intestinal	134	12.8	35.6
Fevers	65	6.2	41.8
Skin	57	5.4	47.2
Timber, furniture	55	5.2	52.5
Crafts, skills	48	4.6	57.0
Parasites	44	4.2	61.2
Fertility, genitalia	41	3.9	65.1
Endemic, epidemic	39	3.7	68.9
Oral health	36	3.4	72.3
Female reproduction	35	3.3	75.6
Analgesic	35	3.3	79.0
Health promotion	34	3.2	82.2
Nest	28	2.7	84.9
Internal organs, respiration	23	2.2	87.0
Magico-religious	21	2.0	89.0
Blood	20	1.9	91.0
Firewood	16	1.5	92.5
Hunting, fishing	13	1.2	93.7
Seasoning, processing	11	1.0	94.8
Antidotes, repellents	10	1.0	95.7
Hygiene, cosmetics	10	1.0	96.7
Beverages	8	0.8	97.4
Bones	6	0.6	98.0
Psychiatric diseases	3	0.3	98.3
Soil conditioning	2	0.2	98.5
Unknown	16	1.5	100.0
Total	1050		

Females of two baboon study troops exhibited major seasonal increases in levels of faecal progesterone metabolites, and consequently, swellings and copulations ceased. Behavioural observations between Apr 04 – Apr 05 were backed up by collection of 733 faecal samples from all 13 mature females (8 from the Kwano troop, 5 from the Gamgam troop). These were analysed with Enzyme-Immuno-Assays for oestrogen and progesterone metabolites. Major increases of progesterone began in August, lasted for up to 3 months, with a second, smaller peak in the Kwano troop around Jan – Mar. Increases were higher than maximal levels measured in pregnancies, and occurred also during lactation (*Fig. 5.2*, *Fig. 5.3*).

A comparison with feeding data from 1003 h of observations of both troops between May 01 – Apr 02 revealed that a single species was consumed by both troops during the seasonal increase of progesterone and not during any other time of the year: the African black plum. During the peak consumption period from Aug – Oct, these fruits constituted up to 4.1 % of all feeding time in the Kwano troop, and up to 2.9 % in the Gamgam troop. The second peak in the Kwano troop around

Table 5.5 Number of plant species recorded for specific usage categories (food, medicine, implements). H = humans, A = domestic animals, B = baboons, C = chimpanzees

	Species (*n*)					Species (*n*)					Species (n)			
	H	A	B	C		H	A	B	C		H	A	B	C
Food					Medicine (ctd.)					Implements				
Beverages					Female reproduction					Crafts, skills				
beverage	5				abortion	1				broom	3			
palm wine	3				breast feeding	2				carving	2			
Edible					dysmenorrhoea	1				drums	1			
Food	79	15	112	34	labour	9				dye	5			
Seasoning, processing					lactation	2				mat	5			
drink flavouring	1				mastitis		1			musical instrument	1			
food processing	3				menstrual problems	5				needles	1			
salt substitute	1				milk production		1			roofing	7			
seasoning	6				post-partum	10				rope	14			
					productivity		3			rubber ball	1			
Medicine					Gastrointestinal					slates	4			
Analgesic					bloat		1			sponge	1			
body pain	7				constipation	6				starch for clothes	2			
dizziness	1				diarrhoea	16	1			walking sticks	1			
earache	3				dysentery	22	2			Firewood				
eye problem	2				flatulence	3				charcoal	1			
headache	1				haemorrhoids	6				firewood	15			
headache w. dizziness	1				heartburn	2				Hunting, fishing				
headache w. nose bleeding	1				intestinal infections	1	1			arrow poison	1			
nose, head problems		1			purgative	5				bows, arrows	1			
Pain	4				stomach ache	33				fish poison	8			
side pain	3				stomach problems	22				fishing spear	1			
swollen glands	1				stomach upset	8				gun powder	1			
swollen parts	4				stool greenish, watery	1				spears	1			

Antidotes, repellents			vomiting	4		Hygiene, cosmetics		
insecticide	2		Health promotion			baby bath	3	
scorpion bite	1		antibiotic	1		bathe children	1	
snake bite	6		disease prevention	1		cosmetic	3	
Blood			fatness	1		female hygiene	1	
Bleeding	2		general health	10		soap	1	
blood clot	4		hengre		4	tattoo	1	
blood formation	1		lick		2	Magico-religious		
blood tonic	6		prophylaxis		2	evil spirit	8	
haemoglobinurea	1		stimulant	3		good luck	5	
haemorragic septicaemia		1	strength	5		juju	1	
hypertension	3		weakness	4	1	leadership		1
increase blood	1		Internal organs, Respiration			orientation	2	
internal injury	1		appendicitis	1		spirits		1
Bones			bile		1	witchcraft	3	
bone weakness	2		chest pain	4		Soil conditioning		
dislocation	2		heart attack	1		soil conditioner	1	
rheumatism	1		heart problems	2	1	soil fertility	1	
sprain, fracture	1		jaundice	9		Timber, Furniture		
Endemic, epidemic			liver		1	boat carving	2	
chickenpox	2		lung problems	1		canoe carving	3	
cholera	6		pneumonia	1		construction	12	
leprosy	1		tuberculosis	2		fence	11	
measles	3		Oral health			furniture	5	
mumps	1		chewing stick	5		implements	15	
small pox	2		dental care	1		shade tree	4	
whooping cough	1		mouth sores	1		stuffing	3	

(continued)

Table 5.5 (continued)

	Species (*n*)					Species (*n*)					Species (n)			
	H	A	B	C		H	A	B	C		H	A	B	C
Food					Medicine (ctd.)					Implements				
yellow fever	23				mouth ulcers	1				Nest tree (chimpanzee)				
Fertility, genitalia					mouthwash	1				nest tree				28
aphrodisiac	2	1			teeth problem	2								
fertility	2	3			toothache	25								
gonorrhea	19				Parasites									
impotence	1				bilharzia	1								
painful to urinate	1				worms	34	4	1	1					
problems w. urination	1				Psychiatric diseases									
reproduction		1			convulsions	1								
strengthen penis	5				pacify		1							
syphilis	1				seizures	1								
testicular problems	4				Skin									
Fevers					boils	6								
catarrh	2				burns	3								
colds	16	1			cuts	2								
cough	12	1			filariasis	1								
fever	25				mange	1								
malaria	6				rashes	1								
malaria fever	2				scabies	2								
					skin conditions	28								
					sores	1								
					wounds	11	1							

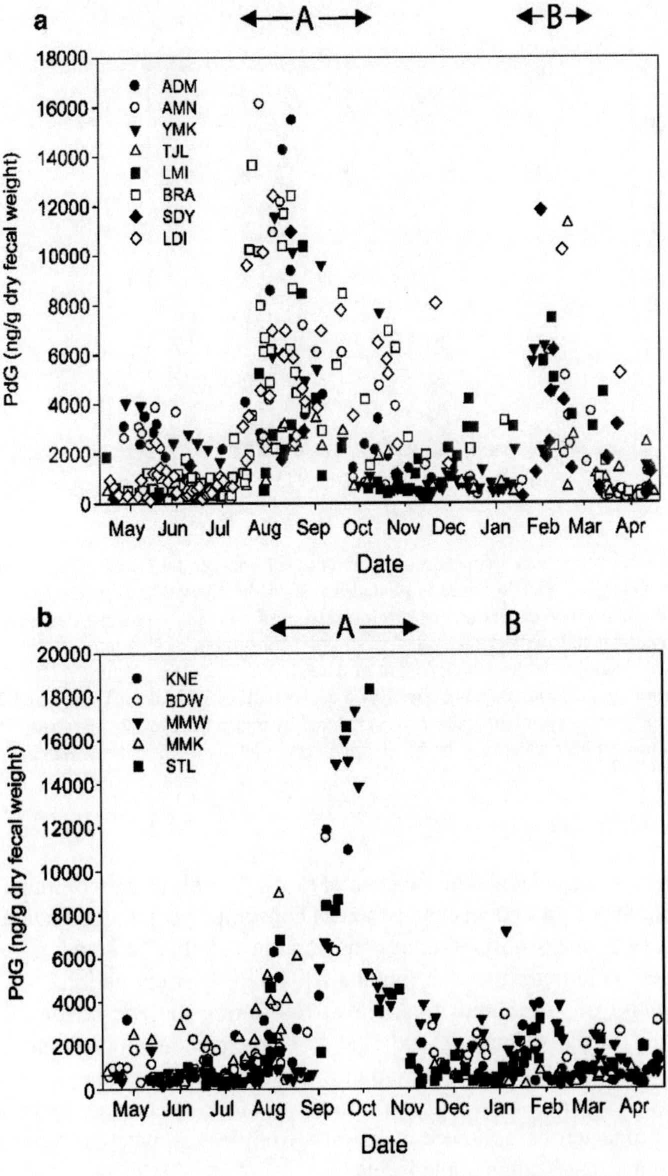

Fig. 5.2 Faecal progestogen concentration (as measured using a PdG assay) in wild olive baboons. Clear seasonal peaks in PdG levels correlate with the consumption of *African plum* fruit (**A**) and immature leaf consumption (**B**). Combined plots for 8 females of Kwano troop (**a**) and 5 females of Gamgam troop (**b**). Different females represented by symbols and 3-letter identification code (from Higham *et. al.* 2007)

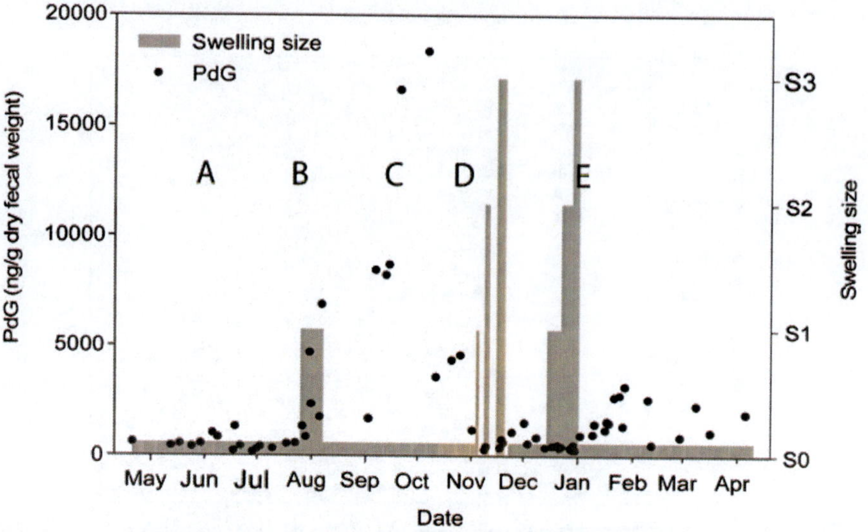

Fig. 5.3 Faecal progestogen concentrations and size of ano-genital swellings for a baboon female from troop Gamgam. Swelling size is plotted as 0.1 to distinguish "no swelling" from "no data". Lactational amenorrhea occurs (**A**) before resumption of swelling (**B**). An extended inter-swelling period (**C**) corresponds with the period of *black plum* consumption. This inter-swelling interval lasts 101 days, compared to a mean cycle length of 41 ± 2.5 days. Swelling returned (**D**) at the end of *V. doniana* fruiting, and the female underwent two cycles with ovulations on 18 Nov and 29 Dec), conceiving in the second cycle. Pregnancy (**E**) is marked by raised PdG levels; these are, however, much lower than levels observed during the consumption of *V. doniana* fruits (from Higham *et al.* 2007)

Feb – Mar was associated with substantial feeding levels (> 1%) of immature leaves of *V. doniana* for up to 6 weeks – whereas consumption in the Gamgam troop was restricted to 2 weeks only. Similar consumption patterns, although not quantified, were observed in both troops during the 2004 – 2005 study period.

V. doniana is a deciduous tree; leaves fall during the end of the rainy season, reappearing in early January, and ripe fruit are present only from Aug – Oct. Samples of fruit and leaves of black plum and some other feeding plants were tested with assays. Extracts of African plum fruits were found to contain by far the highest levels of progesterone, followed by extracts from leaves, whereas those from other plants did not reach comparable levels.

Consumption of African plum fruit and leaves are therefore highly likely to act on cycling females as both a hormonal contraceptive (in that pregnancy is simulated) and social contraceptive (in that swellings are suppressed, which renders the females unattractive for males). As an important corollary of these findings, none of the 30 births recorded over a 5-year period in either troop occurred between 20 Mar – 03 May. Pregnancies last about 6 months, which suggests a period of infertility from mid September till the end of the next month.

Chimpanzee Case Study: Leaf-swallowing as Self-medication

Inspection of faecal samples from individuals of the Kwano-Gashaka chimpanzee study community strongly suggests that they were self-medicating against parasitic worms (Fowler *et al.* 2007).

Every evening, chimpanzees build a new night nest in the branches of trees; early next morning, they defecate over the rim of the sleeping platform. On 7 occasions, undigested leaves were discovered in fresh faeces collected from beneath night nests. Of 12 individual samples, 4 contained folded herbaceous leaves whereas 8 contained both folded herbaceous leaves (up to 25) and grassy leaves or clumps of grass (*Fig. 5.4*). The varying weight of the bolus (mean 101 g, range 29 – 193 g) suggests that leaf-swallowing was not restricted to a certain age-sex class. At least one sample was diarrhetic. The frequency of leaf-swallowing was calculated by comparing how often they were found in 299 faecal samples inspected between Apr 02 – Apr 03. Accordingly, unchewed herbaceous leaves were present in 3.7 % of samples (11 / 299), and grass in 2.3 % (7 / 299). All leaf-swallowing was recorded during the rainy season, about 1 ½ months after the onset of the rains (*Fig. 5.5*).

The leaves belonged to *Desmodium gangeticum* var. *maculatum* (*Leguminosae-Papilionoideae*). Three sets of leaves were measured (set A of 25 leaves and set B of 17 recovered from a faecal sample; set C of 11 leaves from an undisturbed *Desmodium* bush), indicating little variation in dimension (average length x width [cm] and ratio average length / average width: A = 8.6 × 4.3, ratio 2.0; B = 8.8 × 5.1, ratio 1.7; C = 8.8 × 5.1, ratio 1.7). Tissue had disintegrated from most of set A but was intact in half of those from set B. Grass recovered from faecal samples was not identified. There were very few grass stems, and many sharp-edged grassy leaves. Obvious signs of chewing could not be detected in either the herbaceous or grassy leaves.

Two small worms of 0.6 – 0.8 mm width in midbody and 20 – 23 mm length found on the outer surface of leaves that had passed through the chimpanzee guts (*Fig. 5.4c*) were identified (by Hideo Hasegawa, Oita, Japan) as *Oesophagostomum stephanostomum*. The same or similar worms were present in at least one more sample.

Discussion

Plants are exploited widely by humans and other animals. The accelerating demise of natural vegetation covers in vast areas of the globe calls for urgent measures to document this diversity, its mechanisms and functions. Our research in the Gashaka area of north-eastern Nigeria contributes to these aims by providing a compilation of what is known till date about plant use by four consumer groups, i.e., humans, domestic animals, baboons and chimpanzees.

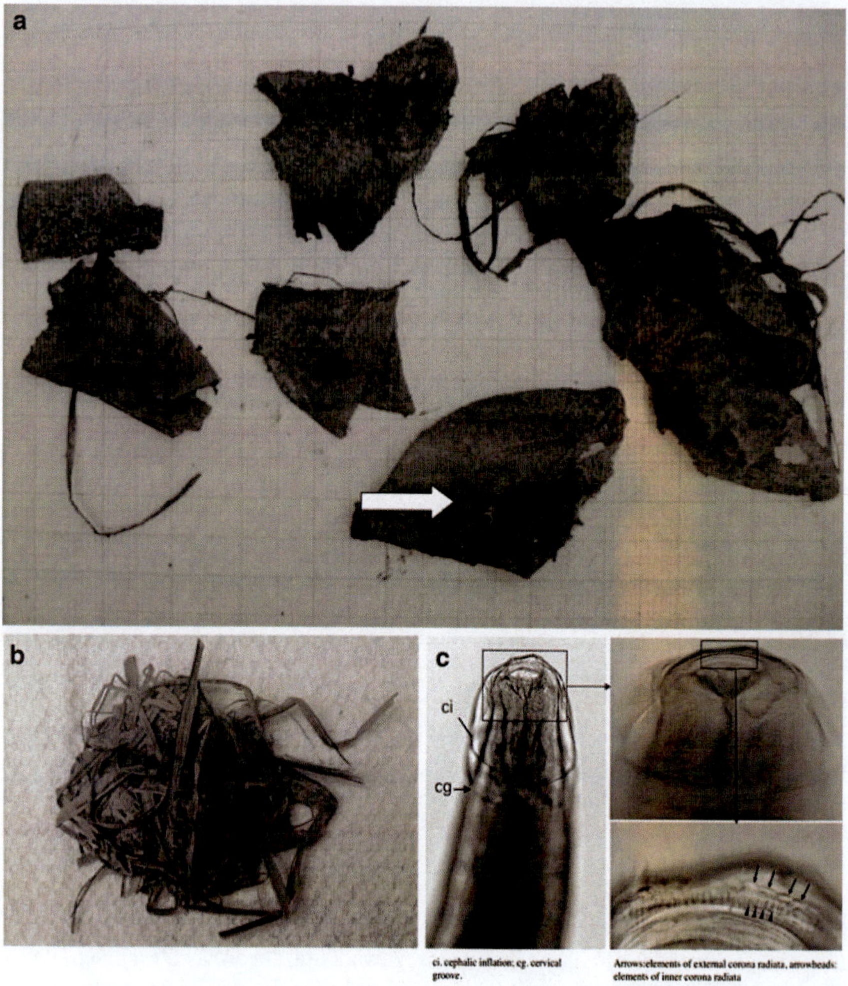

Fig. 5.4 Leaves, swallowed undigested, and recovered from faeces of chimpanzees of the Gashaka-Kwano community. (**a**) Folded leaves of *Desmodium* plus several unfolded unchewed grassy leaves (arrow = *Oesophagostomum* worm on surface of lower leaf). (**b**) Grassy leaves. (**c**) Microscopic image of *Oesophagostomum* parasite extracted from faeces of Gashaka-Kwano chimpanzees (photos: VS [**a, b**], Hideo Hasegawa [**c**])

We were able to identify 309 species and 1050 different types of usage in the broad areas of food, medicine, or implements. This is a minimum estimate, particularly for non-human primates. Records about baboon food plants are probably reasonably complete (Warren 2003), but those for chimpanzees are certainly not (Sommer *et al.* this volume [Ch. 12]). Moreover, we could document just 2 cases of probable self-medicating behaviours for non-human primates (Fowler *et al.* 2007, Higham *et al.* 2007); these are likely to increase as behavioural studies progress. Baboons do not use detached tools, nor construct night nests, and entries for implements of non-human

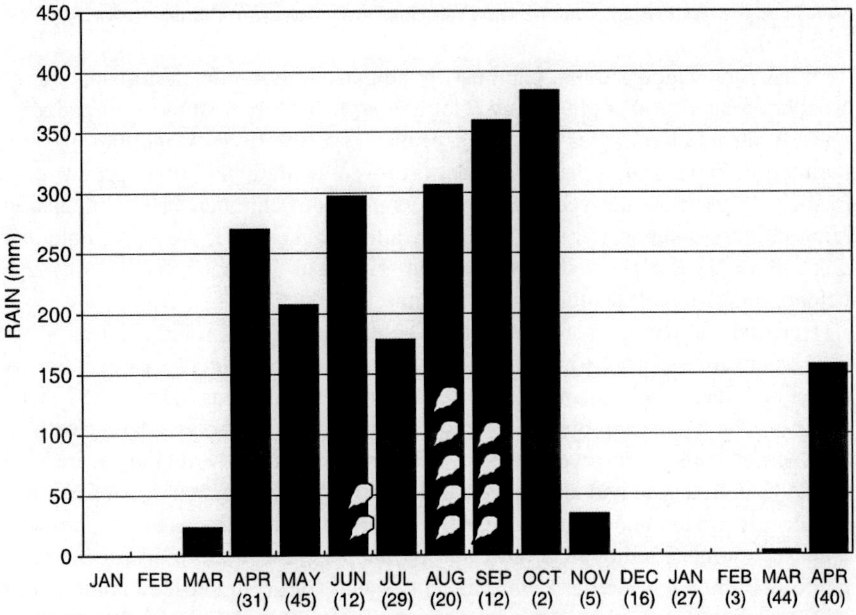

Fig. 5.5 Incidences of leaf-swallowing in Gashaka chimpanzees. Rainfall at the study site (Nov 01 – Apr 03) and (in brackets) total number of recovered faecal samples for each month. Symbols on monthly bars indicate number of samples with unchewed leaves

primates are therefore restricted to chimpanzees utilising certain trees to construct sleeping platforms. Chimpanzees at Gashaka also employ a varied tool-kit for extractive harvesting of army ants and honey produced by bees (Fowler *et al.* this volume [Ch. 13]); however, plants from which the wooden tools are manufactured have not yet been identified, and are therefore missing from our records.

In the following, we will discuss emerging trends of plant use, case studies of important plants, potential mechanisms of how the knowledge about plant utilisation is acquired in human and non-human primate societies, and, finally, directions for future research.

Emerging Trends

Taxonomic Pattern

Flowering plants consist of two subgroups: the primitive dicotyledons and the derived monocotyledons. Dicotyledons are generally woody and monocotyledons predominantly herbaceous (with Palmae as a notable exception). The most primitive subclass of dicotyledons are the Magnoliidae from which all other flowering taxa evolved. The derived clades include the Hammamelidae, Caryophyllidae, and Dilleniidae. A fourth derived clade are the Rosidae, which gave rise to the youngest subclass, the Asteridae. Part of the Asteridae are the Compositae, order Asterales (Cronquist 1968).

It has been suggested that plant use by human and non-human primates follows a pattern related to this phylogeny (Gottlieb *et al.* 1996). A study of three indigenous Amazonian societies and three primate species (pygmy marmoset, spider monkey, muriqui) found that food plant taxa were evolutionary more primitive and medicinal taxa more derived. The classic examples of chimpanzee self-medication reported above support this trend, as swallowed *Aspilia* leaves (Wrangham & Goodall 1989) and chewed *Vernonia* pith (Huffman & Wrangham 1994) indeed belong to the advanced subclass Compositae.

However, our data do not show such dichotomy as major families are used as both food *and* medicine by all consumer groups. Particularly prominent are Leguminosae-Caesalpinioideae, most primitive of the Leguminosae, which are part of the subclass Rosidae. The Moraceae, belonging to the more primitive subclass Hammamelidae, are another example for overlap. Taxa with relatively little overlap are represented by the Combretaceae and Leguminosae-Papilionoideae, both again members of the Rosidae. However, none of the 11 species of Combretaceae are eaten by humans or domestic animals, although 3 are eaten by baboons and 8 are used for medicinal purposes by humans. Similarly, none of the 12 species of Leguminosae-Papilionoideae are eaten by humans, although 7 are eaten by baboons, and 6 are used as medicine by humans and 1 by chimpanzees.

Our study is biased towards medicinal plants, mainly because these were the focus of earlier reports integrated into our data base. We would therefore have expected to find more evidence for the use of Compositae – although we do not know if this taxon is perhaps underrepresented in the study area. In any case, most medicinal plants belong to more primitive taxa. Our data do thus not support the aforementioned pattern. It would be interesting to know if the pattern holds true for biomes outside Amazonia, or if perhaps an overall different floristic composition predisposes consumers to develop a divergent pattern.

Plant Parts

Our research confirms earlier studies (Moerman 1996, Peters *et al.* 1981), which found that plant parts used for food are preferentially fruits, seeds and leaves, whereas bark, roots and leaves are mostly used for medicine (cf. *Tab.* 5.2). This seems plausible to the extent that fruit contains carbohydrates, which provide energy and young, tender leaves contain protein. Bark, on the other hand, has high concentrations of toxic chemicals, and processing is required to allow humans to use its medicinal constituents (Johns 1990).

Secondary compounds are often concentrated in plant parts more vulnerable to predation. For example, high concentrations of toxins are found in the peel of cassava roots and potatoes, and in the seeds of apples and pears. Young leaves (e.g., cassava) and seedlings (e.g., grain sorghum) can also contain high concentrations of toxins, in which case they decrease as the plant matures (Jackson 1991). Secondary compound concentrations can therefore vary throughout the year, the age of the plant, or even its sex; it also seems that female plants are not grazed as heavily as male plants, implying that the former contain more deterrents (Engel 2002).

Nevertheless, humans seem to possess a certain tolerance for tannins and toxic substances. The introduction of considerable amounts of meat protein and concentrated carbohydrates from tubers and seeds into the diet selected for humans who were able to tolerate associated toxins. For example, although tannins impede the absorption of protein, iron and other minerals, humans consume tannin-rich sources such as tea, coffee, red wine and broad beans. Tannins are known to stimulate in mammalian herbivores the production of saliva saturated with proline-rich proteins, which detoxify tannins. This points to an evolutionary past of tannin consumption (Gottlieb 1996, Jackson 1991, Johns 1990).

Food versus Medicine

Our data demonstrate a considerable overlap in species exploited for nutritional as well as medicinal purposes for all four consumer groups. The most frequently recorded plant families are used as both food and medicine (cf. *Tab. 5.3*). Nevertheless, different parts of a given plant will often serve as food or medicine, such as fruit or bark (Moerman 1996). Dosage and context can also lead to a differentiation (Etkin 1997). A dichotomy between food and medicine is deeply engrained in modern Western societies whereas traditional African medicine distinguishes far less there between foods, beverages, condiments, medicines, stimulants, psychoactive agents and toxins. Healthy food and a medicine are hard to distinguish as many food plants can become medicine and vice versa (Iwu 1993, Johns 1990).

Interactions with plant substances are, at least from a chemical-ecological perspective, more of a continuum, with high-energy foods at one end and bioactive substances at the other. Foods and medicines are therefore in many ways undistinguishable, as exemplified by a glass of orange juice rich in vitamin C. In any case, good nutrition can have considerable health benefits in terms of coping with pathogens. This is illustrated by the well-fed crop-raiding olive baboons of the Gamgam troop near Gashaka. While confronted with a higher parasite richness than their wild-feeding counterparts near Kwano, they exhibit less mortality and higher birth-rates (Warren *et al.* this volume [Ch. 8]).

What counts as "medicine" is thus largely a social construct – in that a substance is considered a medicine when its consumption is determined by the intention to cure, treat or prevent a disease.

Case Studies of Plants with Multiple Usages

Several plants exploited in the Gashaka area can serve as showcases for multiple usage potential and alleged medicative functions for both humans and animals.

Desmodium gangeticum

Chimpanzees of the Gashaka-Kwano community were observed to swallow and excrete undigested leaves, particularly of the herbaceous plant *Desmodium gangeticum*. Our field study adds *P. t. vellerosus* to the list of leaf-swallowing apes, which so far included the three other chimpanzee subspecies, as well as bonobos and Eastern low-land gorillas (Huffman 1997, 2001). The main benefit seems to lie with the expulsion of parasitic worms, which are known to cause loss of weight, anaemia, abdominal pain, bleeding cysts and blocked colons.

Medicinal benefits of leaf-swallowing were originally associated with pharmacological anti-parasitic properties, such that phytochemicals decrease the ability of worms to attach to the mucosal lining of the intestines, and / or that chemotaxis attracts worms into the folds of leaves (review in Huffman 1997). However, many different plant taxa are now known to be swallowed, rendering it unlikely that they all have similar chemical properties. Instead, these plants appear to have a consistent physical property, i.e., the rough surface of the leaves sports many hooked and spiky hairs. This is also true for the leaves of *D. gangeticum* swallowed by the Gashaka chimpanzees.

Indeed, the plant is known in Hausa and Fulfulde as "takamahi" ("the one that sticks on you"), because its leaves and seeds due to their coarse surface attach to people's clothes as they walk by. Physical properties such as those had therefore led to the hypothesis that worms become attached to the bristly surface as if to "velcro", and that the coarse structure may also function as a rasping plug, dislodging parasites from the intestines (review in Huffman 1997).

However, the current view of the parasite expulsion mechanism of leaf swallowing does not focus on the physical trapping of worms. Instead, material that is not digested well may trigger a purging response of the gastro-intestinal tract by rapidly inducing diarrhoea. This decreases gut transit time, causes worm expulsion and interrupts the life-cycle of parasites (Huffman & Caton 2001). A similar mechanism might explain occurrences of undigested grass in the faeces of such diverse taxa as birds, carnivores and primates; sedge grass (Cyperacea) in particular can be high in fibre, sharp-edged and coarse (review in Huffman 1997). This is perhaps also an explanation for why dogs and cats eat grass – just to expel it shortly after through vomiting.

The grass eaten by the chimpanzees could not be identified, but it possesses sharp-edged leaves similar to sedge grass swallowed in at least two other populations of wild chimpanzees. With *D. gangeticum*, we have a first record that leaves from the family *Leguminosae-Papilionoideae* are swallowed by apes. There is no indication that chimpanzees select leaves of a particular size, as swallowed leaves and those from an intact bush of *Desmodium* had similar dimensions.

Various patterns – even in the absence of direct behavioural observations – suggest that chimpanzees at Gashaka do ingest the leaves for medicinal purposes (cf. Huffman 1997, Huffman & Caton 2001, Dupain *et al.* 2002): (a) The leaves are either sharp-edged (grass) or have a coarse surface (herb), indicating suitability for dislodging intestinal parasites and / or triggering a purging response of the gastro-intestinal tract to rapidly expel material that does not digest well; (b) the leaves are not chewed, and excreted whole, suggesting that they are not ingested for nutritional purposes; (c) nematode worms were found together with excreted leaves; (d) leaf-swallowing was restricted to the rainy season, during which parasite re-infections are more common.

Interestingly, several informants including healers reported "takamahi" as having medicinal qualities. Thus, juice from squeezed "takamahi" leaves is drunk to remove threadworm, leaves are pounded for use as an enema to remove parasites, and mixed with Vaseline as ointment for haemorrhoids. Leaves pounded with beans are also poured into nostrils to stop bleeding. A tea from boiled leaves is drunk to cleanse blood after labour, to reduce a heavy menstrual flow, and to purify infected mother's milk. The root is also boiled and drunk for skin rashes, dysentery and to help children become strong.

Unlike with self-medication in chimpanzees, these practices seem to reflect pharmacological properties. Their exact nature is unknown, but the plant might have varied phytochemical potentials. Gangetin, a substance isolated from the roots of *D. gangeticum* suppresses fertility in male rats, which, however, return to normal after the end of the treatment. Extract of the leaves of the closely related *D. adscendens* seems to possess hypothermic and analgesic actions. The species is used in African traditional medicine to treat asthma (observed to be effective in clinical trials), pain and epilepsy (N'gouemo *et al.* 1996). Leaves of *D. ramosissimum* are also soaked in water, which is then drunk for 2 – 3 days to induce abortion (Noumi & Tchakonang 2001). This species is used in Ayurvedic medicine as a tonic, stimulant, digestive and anti-pyretic (Latha *et al.* 1997).

These processes have little to do with the mechanism that seems to benefit chimpanzees, i.e., a purging response of the gastro-intestinal tract induced by coarse, folded leaves. It remains to be seen if and how mechanical and pharmacological effects are interrelated.

Vitex doniana

The "black plum" or "African olive" is a common tree in Nigeria found in savannah, forest and montane forest. Several of its parts are used by all four animal consumer groups of our study: humans eat the fruit, leaves, and seeds; cattle are provisioned with leaves; chimpanzees eat the fruit; and baboons eat the leaves and the fruit. The fruit is nutritious, as it has a high lipid content (Glew *et al.* 1997), and the pulp contains vitamin C. The seed is a source of oil (Ajiwe *et al.* 1998), and vitamin A and proteins are also present in the plant (Iwu 1986). Humans ascribe varied medicinal functions to different plant parts. Thus, the bark is given to cattle to treat diarrhoea, dysentery and liver problems. The bark is boiled and the water drunk by locals to treat yellow fever, stomach ache, and skin conditions, to "promote health" and to remove parasites from children. Women eat the leaves after labour to "cleanse the blood", and both leaves and fruit treat dysentery. Tea made from the stem is given to children for stomach problems. Interestingly, one female informant maintained that baboons eat the fruit "to remove worms" – a statement that has not yet been corroborated. The tree also provides for implements, as humans use the wood as fuel and to make drums. Finally, chimpanzees construct night nests in its branches.

The consumption of fruits and leaves may have effects other than nutritional – as illustrated by the described contraceptive consequences for baboons (cf. *Fig. 5.2.*, *Fig. 5.3.*). The plant became a particular object of research, once it was noted that the sexual behaviour of female olive baboons of two study groups at Gashaka was strongly influenced by the consumption of *Vitex* fruit (Higham *et al.* 2007). Much less sexual activity took place while the baboons fed extensively on the black plum. The females did not develop ano-genital swellings, and consequently, males showed little interest in them. Analyses of faecal samples demonstrated that plum consumption is associated with high levels of phyto-progestogens, which have a similar effect to the synthetic progesterone used in contraceptive pills.

Vitex fruits have been linked since ancient times to contraception. The species *V. agnus castus* is sold in health food stores as a remedy against menopausal symptoms and pre-menstrual stress. Nigerian traditional medicine also associates *V. doniana* with female reproduction, as the plant is used in the Gashaka area to treat post-partum problems ("to cleanse blood after birth"). A comprehensive guide to medicinal plants in Nigeria, however, does not mention any connection to reproductive functions (Odugbemi 2006).

At first glance, black plum consumption seems to be maladaptive, as it impedes reproduction in baboons. However, the main period of progesterone excretion and plum consumption is also the peak of the rainy season, with monthly downpours averaging 299 mm. This period is associated with high mortality of infants and adult females (Higham *et al.* 2009). Plants of the genus *Vitex* are renowned for their medicinal properties that combat bacteria, viruses, funghi, insects, snake-poison, cancer, as well as being analgesic (sources in Higham *et al.* 2007). It has therefore been suggested that "any medicinal properties of *V. doniana* are likely to be particularly valuable at this time […] despite its seemingly negative effects on reproduction in the short-term" (Higham *et al.* 2007: 389). This explanation interprets the contraceptive effect as a by-product of the medicinal benefits of plum consumption. However, there is also the possibility that the contraceptive effect in itself is directly advantageous. During a season with high probability of declining health, there may be an increased risk of becoming infected by sexually transmitted diseases, and it is therefore beneficial (or at least, of low cost) to abstain from sexual behaviour during such times (Nunn & Altizer 2006). Moreover, any pregnancy would probably also carry a higher risk of abortions than conceptions at other times. It is thus conceivable that contraception is a positively selected effect of plum consumption, instead of just a by-product of medicinal properties. On the other hand, baboons also consume the new leaves of this plant, with concomitant effects on hormones in Jan – Feb, which may indicate that the contraceptive effect is not selectively advantageous. Several authors have previously suggested an effect of phyto-oestrogens on the reproductive physiology in primate females, usually by increasing the likelihood of conception (Glander 1994). The connection between plum consumption and contraception is thus the reverse of how plant secondary compounds influence, for example, Brazil's muriquis, large New World monkeys (Strier 1993). Faeces were monitored for levels of oestrogen and progesterone to detect links between the diet and birth season. At the beginning of the rainy season – which is also the onset of the mating season – muriquis expend considerable energy to come within reach of "monkey ear" (*Enterolobium contortisiliquum*), which grows at the periphery of the forest. Despite these efforts, they eat only few of the available fruits – perhaps because of stigmasterol, a steroid in the fruits that can assist in the synthesis of progesterone, and thus aid conception.

Clearly, many causal connections between plant consumption and primate reproduction still await discovery.

Erythrophleum suaveolens

The bark of the "sasswood" or "poison wood tree" contains saponins, steroids, tannins and alkaloids, two of which are erythrophleine and cassaine (van Wyk *et al.* 2000). Both are highly toxic and have cardiotonic activities similar to digitalis. The bark is known to possess anti-inflammatory and analgesic properties, which support its use in traditional medicine (Dongmo *et al.* 2001).

At Gashaka, humans pound and drink the bark as analgesic, to treat stomach ache and for skin conditions. Also, a very small amount of ground bark is reported to treat seizures. The trunk is used to make canoes. Traditionally, people used to wash with the water that got into the canoe as a remedy for skin conditions. Nowadays the use of ground bark as fish poison is also very popular. Moreover, people boil the bark and use the water to counteract the effects of witchcraft. This reflects the traditional African use of the poison wood tree as a source for an ordeal brew to establish whether a person is guilty of witchcraft (De Smet 1998, Dongmo *et al.* 2001). (The procedure ends with the likely death of the candidate: her innocence is proven if she dies from the highly toxic brew; if she survives, she is a witch, who will be persecuted.)

Gashaka baboons eat the seeds. Thus, either they are not toxic but have some nutritional value, in which case one could assume that humans would have found a way to take advantage of that plant part too. Or, baboons have a way to detoxicate the seeds. Chimpanzees, in any case, play it safe: they only use the branches of the poison wood tree to construct nests in.

Erythrina senegalensis

Similarly toxic is the "coral tree" or "yellow fever tree", medium sized, 3 – 4 m high, with spectacular bright red flowers. Again, baboons eat the seeds. *E. senegalensis* is widely used by people for its anti-bacterial, analgesic and anti-inflammatory properties. People in the Gashaka area report boiling a tea from the leaves or bark to treat fever and jaundice. A large number of alkaloids known as Erythrina alkaloids have been isolated from the plant, with properties similar to those of curare as muscle relaxants (Iwu 1993, van Wyk *et al.* 2000). Moreover, the bark has weak anti-malarial activity and strong analgesic and anti-inflammatory effects (Saidu *et al.* 2000).

Vitellaria paradoxa

The name of the "shea butter tree" derives from the oil extracted from its seeds. Shea butter is a common ingredient of meals throughout much of Africa and also in the Gashaka area. The seed extracts are mixed into creams – popular with costumers patronising Western health stores. They are applied to cure rheumatism, sprains, fractures and also to smoothen the skin. A sud from the bark used as a mouthwash soothes sores, toothache, and an oil extracted from the bark helps against skin infections. Throughout much of Africa, implements are manufactured from the tree's hard wood that have to withstand pounding – such as pestles, mortars or drum bodies. The fruit and leaves make fodder for cattle, and baboons help themselves to the fruits and flowers.

Cola spp

Kola nuts contain about 2.5 % caffeine and small quantities of theobromine along with phenolic substances and tannins (Iwu 1993). The substances act as a muscle relaxant, which reduces the sense of fatigue, although high doses can cause depression. Kola nuts have always played an important part in African social life, particularly as a stimulant that is chewed. They are also given as a gift by a man to his future father-in-law, they are offered as a treat to valued guests and they have been used extensively in trade and ritual (De Smet 1998, Iwu 1993, Wall 1988).

Six *Cola* varieties have been reported in our study areas. The fruits of various species (*C. verticillata, C. hispida*, and "*ataras*" or Cameroon cola [perhaps *C. anomala*, the Bamenda kola; De Smet 1998]) are chewed as a stimulant and to help digestion. The bark of an unidentified plant (possibly of the genus *Cola)* known as "dan Cameroon" ("of Cameroon") is also chewed as a stimulant. Female informants did not recognise the name or actual bark specimen. On the contrary, it is very sought after by men as it is believed to enhance virility and to help men carry on their farm work. It is common to see men chewing a piece of "dan Cameroon" bark together with some kola nut but they avoid chewing the bark in the evening as it brings insomnia.

However, varied other uses are attached to *Cola* species. *C. hispida* is burned as an incense to drive away evil spirit. The fruits of *C. gigantea* and *C. millenii* are edible, although the latter is rather sour. The pod of *C. verticillata*, known as *goro* in Hausa and Fulfulde, is soaked in water, which is drunk by women during labour to facilitate delivery and also after birth to restore their system. The exceptionally big pods of "ataras" are used against menstrual problems.

Non-human primates make various uses of *Cola* varieties. Apparently, mountain gorillas have a taste for kola nuts as they perhaps provide cardiac stimulants necessary for a life in montane region with less atmospheric oxygen (Cousins 2002). The Gashaka baboons eat *C. millenii* and "dan Cameroon", and the Gashaka chimpanzees fruits of *C. gigantea, C. millenii* – mostly during the rainy season –, "ataras" and also the fruits of "dan Cameroon". Moreover, various *Cola* trees are popular with chimpanzees to construct night nests in.

Knowledge Acquisition: Which Plants Are Beneficial?

The question of how properties of plants are discovered and how this knowledge is passed on is relevant for human as well as non-human animals. The exact mechanisms of primate cognition in general and social transmission in particular are heavily debated, but both individualistic and social learning certainly plays a role (for reviews, see Byrne 1995, Tomasello & Call 1997, Wynne 2001, Hurley & Nudds 2006).

The initial "discovery" of a beneficial quality associated with nutritional or medicinal properties – and, vice versa, the aversion of toxic substances – can be individualistic and is probably linked to discernible effects of the chemical compounds in the plant. These generate physiological effects that are largely genetically determined. This is not surprising, as monkeys and apes rely on a mainly plant-based nutrition, and are thus chronically exposed to specific plant chemicals with resulting consequences for their genomic diversity (Jackson 1991). The gustatory system of primates will thus generate immediate responses to food composition (Hladik *et al.* 2002). Taste perception is a major guide here, as an adaptation that allows the detection of beneficial and toxic substances in potential foods (Jackson 1991).

Two main categories generate very specific reactions of humans and non-human primates, those elicited by sugars and those by the bitter substances quinine and tannins. The latter are generally toxic secondary compounds of plants. The primate preference for a sweet taste with its associated nutritional benefits probably co-evolved with flowering plants that developed fleshy fruits, rich in sugar, which could attract primates who would then function as seed dispersers (Chapman 1995). When it comes to bitter substances, things are not so clear-cut. Compounds that to humans taste especially unpleasant (quinine) or astringent (tannins), are often part of non-human primates' environments and diets. As not all bitter substances are toxic, recognition thresholds among primates vary, presumably to allow for a more thorough exploitation of feeding niches (Sommer *et al.* this volume [Ch. 12]). This may explain why certain fruits regularly eaten by wild chimpanzees look perfectly delectable to the human eye – just to taste disgusting.

Nevertheless, humans and non-human primates are vulnerable to many of the same parasites and infectious diseases (Johns 1986, 1990). As explained above, because plants with a strong, bitter and slightly unpleasant taste may contain health promoting agents that can kill or disable a pathogen, sick primates, including humans, might benefit if they seek out *mildly* toxic substances, or ingest toxic substances in moderate dosages.

More or less accidentally ingested plants that lead to a sense of well-being – due to reduction of nausea, pain, fever, anxiety or because of a stimulating effect – can create a conditioned response in the nervous system. The relief a specific plant provides will then be associated with a taste preference. Such non-mentalistic causations would also suffice to explain the consumption of *Vitex*-fruit by baboons as well as the ingestion of *Desmodium*-leaves by chimpanzees. Nevertheless non-human primates could well make a causal connection between the consumption of a particular plant and its effects, at least, if the feedback is rather immediate, such as parasite control.

Conditioned preferences – whether or not connected with mentalistic reflections – do not require knowledge about the biological basis of an ailment nor about the chemical constituents of a plant medicine; all that is needed is removal of the symptoms. After all, people started isolating the active principals of plant medicines less than two hundred years ago (Johns 1990).

Individualistic experiences with plant exploitation, pleasant and unpleasant, thus predate the development of communal "cultures" of plant utilisation. Social transmission of experiences is a more effective means of building up knowledge. It can replace more solitary processes such as simple conditioning, trial-and-error, and "insight" into cause and effect in favour of processes mitigated by group mates, such as *stimulus enhancement* and *social facilitation*. These mechanisms are at work, if an individual pays attention to a location or stimulus simply because it is near to another individual engaged with them – a major way young primates learn how to locate, select, and process foodstuffs and other resources. This may take the form of *emulation* learning, which focuses on the changes that take place in the environment and not on the instrumental behaviour that brought them about, thus reproducing the observed end result without copying the behavioural method. For example, a baboon may learn that the nut of an oil palm can be opened but may not necessarily follow the same biting technique as a troop mate.

More advanced modes of transmission require the ability to understand somebody else's intentions and knowledge deficiencies (theory of mind). Learning through *imitation* involves comprehending both the method and the end result, but also, to an extent, the intentional relation between the two (Tomasello & Call 1997). It is not clear to which degree monkeys and apes employ "true" imitation. A famous text-book example for imitation is the practice of Japanese macaques to collect sweet potatoes strewn onto a beach by researchers, carry them to the edge of the water and wash the sand off them. Nevertheless, the spread of this behaviour throughout the population is more likely based on individual learning, as the rate of diffusion was extremely slow (Whiten 2000). Full-blown *teaching* is probably extremely rare in primates other than humans. It would be more a case of facilitation if, for instance, a chimpanzee mother leaves a tool near a nut for her infant to use. Active teaching has only been described for the nut-cracking chimpanzees of the Ivory Coast, where a mother seems to clearly instruct her infant how to, for example, position a nut on an anvil to then hit it with a hammer (Boesch & Tomasello 1998). Certainly, mothers play a pivotal role in shaping an infant's feeding behaviour. A telling case is a healthy chimpanzee mother who stopped her infant from chewing on the pith of bitter leaf, something the latter had apparently observed on a previous occasion (Huffman 1997).

An interesting interplay between individualistic and social learning is illustrated by the mechanisms of feeding selectivity in mantled howler monkeys. These primates concentrated on only 15 of the 62 available tree species. Chemical analyses revealed that they prefer plant parts high in protein while avoiding high concentrations of secondary compounds. Observations of feeding implied social learning, as infants would only eat leaves other group members ate. However, when the feeding bout involved fruits, the infant would watch females in only half the cases, and proceeded instead with eating any fruit. The two separate mechanisms seem to reflect two evolutionary forces. Leaves require a more conservative pattern of consumption, as they are chemically protected against predators. Fruit, on the other hand, are safer to ingest, as plants "aim" to induce consumption by primates to facilitate seed dispersal (Whitehead 1986).

In any case, social transmission facilitates the successful non-random selection of beneficial plants. To cite a human example, the Yucatec Maya identify medicinal plants because of symbolic signatures such as that its morphology resembles an element of the disease or the diseased part of the body – similar to medieval herbalists who treated liver problems with herbs displaying liver-shaped flowers. The Maya also judge a plant based on its taste and smell. In fact, more often than not will medicinal plants have a strong odour, usually pleasant, whereas non-medicinal plants will typically not smell. Interestingly, the Maya do not use bitter taste as a guide towards identifying medicinal traits – whereas many ethnic groups do. Thus, a plant's perception is a combination of smell, taste, colour, form, and texture (Ankli *et al.* 1999).

Still, there is certainly no "fool proof" method to identify medicinally functional plants, and while tests established chemical effects for 60 % of plants with alleged medicinal value, there are 40 %, for which such phytochemical effects could not be demonstrated. Nevertheless, while a plant may not relieve yellow fever because it has yellow flowers, it may have an effect, because those who prescribe and utilise the plant *believe* in it. Traditional medicine could have a weak link to physiology, but would still embody healing powers because psychosocial needs are satisfied through quasi-magical principles (Fabrega & Silver 1973). This link would reflect the psychosomatic powers of placebo-effects (Frank 1961, Krebs *et al.* 1988).

Future Research

Our compilation, despite its extensive coverage, is still incomplete but gaps will hopefully be filled in as long-term research proceeds in the Gashaka area.

Our research also does not analyse the cognitive framework of human plant use in the Gashaka region, particularly not with respect to ethno-medicine – a point linked to psychosomatic aspects just discussed. For example, traditional medical systems will often encompass *naturalistic* as well as *personalistic* frameworks. The naturalistic approach is similar to Western science in that it is based on causes, responses and treatments that are condition-specific and symptom-based and rely on actual pharmaceutical potentials of the plants. The personalistic approach encompasses magico-religious aspects related to extranatural phenomena and sensate beings such as deities, demons, spirits, ancestors, sorcerers, etc., and often involve healing rituals and prayers (Foster & Anderson 1978, Berlin *et al.* 1996).

In this vein, one would like to further systematise the pharmacopeia of the Gashaka region (as well as the nutritional and implemental portfolio). For this, one would need to expand the existing database. A three-pronged approach would be useful (see Berlin *et al.* 1996). Data on *medical ethno-botany* would encompass collection of medicinal plants, and identification of the most frequently used by knowledgeable informants. Data on *ethno-medicine* would be gathered by taking a "travelling herbarium" with these plants through the region, and use it in individual interviews and group-based sessions. With this, one would gather plant-specific information relating to signs, symptoms, prognosis, seasonality, treatment schedule and outcomes. This should be complemented by retrospective ethno-epidemiological surveys of a certain number of households, to determine illnesses that have occurred and how they were treated. A third step would be *ethno-pharmacology* focused on pharmacological analyses of the most important plants.

We would also like to know more about potential self-medication of baboons – even if practices such as the contraceptive consumption of black plum fruits might not reflect mentalistic abilities, contrary to the probably "thoughtful" processing of *Desmodium* leaves by chimpanzees. Local people in Ethiopia believe that consumption of the desert date (*Balanites aegyptiaca*) kills the bilharzia-transmitting parasite. Baboons in the same area are also exposed to schistosomiasis, and likewise eat *Balanites* fruits (Phillips-Conroy 1986). People and baboons in Gashaka are as well infected with schistosomiasis (Weyher *et al.* 2006, Warren *et al.* this volume [Ch. 8]). However, baboons here have not yet been observed to consume desert date fruits. People eat them, but do not ascribe a bilharzia-related function to the dates – although other plant parts are used as medicine. It would be interesting to explore the connection between bilharzia – a serious problem in the Gashaka region – and potentially self-medicating behaviour of people and non-human primates further. Similarly, chacma baboons in Namibia and Botswana consume plants with euphoric and hallucinogenic properties, perhaps for therapeutic purposes, as hallucinogens are often toxic to gastro-intestinal parasites (Hamilton *et al.* 1978, Johns 1990). There are various candidate plants with hallucinogenic qualities in the Gashaka region that would merit further investigation, from *E. suaveolens* and *E. senegalensis* (eaten by baboons) to *Cannabis sativa*. (Note: Marijuana is allegedly not provided to humans, but only to the happy chickens of Gashaka, to "promote their health" – a rather clear case of bias in our records!)

With increased habituation and more direct observations, we will also learn more about potential medicinal plant use of Nigerian chimpanzees. "Candidate plants" can be derived from our ethno-medical and ethno-veterinarian inventory. For example, Gashaka chimpanzees are not known to use plants other than *Desmodium* with a similar scabrous surface – although they *are* swallowed by apes in other populations (Huffman 1997), and are also consumed at Gashaka by humans, cattle and baboons. For example, *Aspilia africana* is fed to cattle, while olive baboons consume young leaves of the "sandpaper tree", *Ficus exasperata* (Warren 2003). We do not know if Gashaka chimpanzees eat these plants. The same holds true for *V. amygdalina* of which apes elsewhere consume the bitter pith for assumed pharmacological purposes related to parasite control (Huffman & Seifu 1989, Huffman *et al.* 1997). *Vernonia* is frequently consumed by locals at Gashaka, going by the name "bitter leaf" – and they are bitter, indeed! As has been mentioned, bark and leaves contain toxic substances in much higher dosage than the pith. Therefore, bitter leaf is washed 6 – 7 times, creating a distinct greenish-white foam, before the leaves are boiled. The sauce is very popular and accompanies staples such as yams, maize or rice.

Moreover, it will be interesting to compare the source species of the tool kit, which chimpanzees use to extract resources from nests of social insects with criteria humans employ when they manufacture implements. For example, certain tools such as digging sticks have to be sturdy, whereas others, such as dipping sticks to obtain honey, have to be pliable and will often be frayed to increase the surface (Fowler *et al.* this volume [Ch. 13]). Such criteria should also concern humans, as the type of material needed for, lets say, needles, drums and spears, is different from that required for mats, stuffing and brooms. Similar considerations apply to trees preferred by chimpanzees to construct their nests in, as one wonders if their branches are particularly pliable, if the leaves are particularly soft, or, vice versa, if tree species that are avoided have less suitable properties for the construction of a platform (Fruth 1995). Completely unexplored is also the possibility that nest tree vegetation has potent insect-repellent properties (Largo *et al.* 2009).

A further issue is seasonality, as quality and quantity of active constituents of plants vary throughout the year (Sofowora 1993). Parasite pressures are much greater during rainy seasons, which should be reflected in the selection of foodstuffs as well as potentially medicinal plants.

As records become more complete, we might also be able to discern plants or plant parts that are *not* used by certain consumer groups. It would be interesting to learn, if non-use reflects a lack of intrinsic qualities of a given plant, or the need to avoid them, because they are not suitable or are dangerous to consume.

The comprehensive guide-book *Outlines and Pictures of Medicinal Plants From Nigeria* (Odugbemi 2006) lists 621 species. We have records for approximately 309 species. Of these, about 186 are used in traditional medicine. Of the 140 species known to us by their scientific name, 75 (53.6 %) are also listed in the guide-book, and the guide lists a further 29 species (20.7 %) of at least the same genus. Therefore, about three quarters (74.3 %) of medicinal plants recorded at Gashaka are somehow listed in the guide. Another 37 species / 18 genera listed in the guide do occur at Gashaka, but are only used for food or implements; we might reasonably assume underreporting here, i.e., that a fair number of these taxa might have some medical application that has not been reported to us. This is at least likely for prominent species such as *Vitellaria paradoxa* and *Piliostigma thonningii*.

Vice versa, only about a quarter of the guide-book entries (159 / 621; 25.6 %) are also recorded at Gashaka. There are several explanations for this discrepancy, such as that numerous medicinal species at Gashaka are only known in Hausa and / or Fulfulde ($n = 46$) or that species may be listed under alternative names. More importantly, Nigeria is a vast country with numerous different biomes, and the Gashaka region represents not all of them – a fact that probably explains the majority of missing entries. The other way around, at least 36 species reported as medicinal at Gashaka (25.7 % of our scientific entries) are *not* listed in the guide. Amongst these are taxa whose bark treats many ailments and which are also economically important, such as the shea butter tree, *Vitellaria paradoxa,* and the nut-producing *Detarium microcarpum*. This indicates that the Nigerian highland area is not comprehensively covered by the guide.

Clearly, Nigeria's ethno-botany is in need of further development. Our study will hopefully close some of the gaps in knowledge. Studies such as ours can also be of applied value. For example, many medically important bacteria and fungi have developed drug resistances, and one would hope that traditional medicinal plants could be recognised as potent sources for new drugs (Aibinu 2006). Similar considerations can be extended to food plants and the need to develop new staple sources to feed humans worldwide. On the other hand, one has to recognise the important contribution of localised, non-commercialised exploitation of wild plants to the well-being and day-to-day subsistence of humans – and the wildlife that shares their range. We thus hope that the display of the fantastic diversity of plant use in our study area will contribute towards efforts to preserve as much of its natural vegetation cover as possible.

Acknowledgements The Nigeria National Park Service granted a research permit to the *Gashaka Primate Project*. The Chester Zoo Nigeria Biodiversity Programme generously supported the field work. – For identifications, we thank Emmanuel Obot of NCF (herbs) and Hideo Hasegawa, Department of Infectious Diseases (Biology), Faculty of Medicine, Oita University, Japan (parasites). Michael Huffman, Primate Research Institute, Kyoto University, made helpful comments. Thanks are due to Andrew Fowler, co-author of the original study on chimpanzee self-medication. Some unpublished data on non-human primate plant utilisation were kindly provided by Ymke Warren and Andrew Fowler. – YK expresses her gratitude to Nigerian Conservation Foundation (NCF) / WWF-UK project managers Abdullahi Zabudum and Aaron Nicholas; to all staff of the NCF Gashaka Gumti Project; to the national park staff at Gashaka barracks and the people of Gashaka; for hospitality, to the Ardo of Chappal Hendu and his family and the Mai'ungwa of Selbe and of Filinga; to Doris Nyanganji, Maigari Ahmadu, and Amina Mohammed for assistance and guidance during field trips; to Tukur Idi, Chairman of the Healer's Association in Serti, for trust and support; to GPP staff Hammaunde Guruza, Sam Yusufu, Bobbo Buba, Felix Vitalis, and Ali Tappare. This study was partially funded by a masters award from the Department of Anthropology, to YK.

Literature

Adanu, J. (1997). *Determination of Life-span of Chimpanzee Nests (Nest-aging) in Gashaka-Gumti National Park*. Report to NCF & WWF-UK.

Adewunmi, C. O., Agbedahunsi, J. M., Adebajo, A. C., Aladesanmi, A. J., Murphy, N. & Wando, J. (2001). Ethno-veterinary medicine: screening of Nigerian medicinal plants for trypanocidal properties. *Journal of Ethnopharmacology* 77: 19–24.

Aibinu, I. (2006). Medicinal plants as antimicrobials. In Odugbemi, T. (ed.), *Outlines and Pictures of Medicinal Plants from Nigeria*. Lagos: University of Lagos Press, pp. 53–64.

Ajiwe, V. I. E., Okeke, C. A., Ogbuagu, J. O., Ojukwu, U. & Onekama, V. I. (1998). Characterization and applications of oils extracted from *Canerium schweinfurtii, Vitex doniana* and *Xylopia aethiopica* fruits / seeds. *Bioresource Technology* 64: 249–252.

Akinsoji, A. (1996). *Vegetation Types and Ethnobotanical Studies of Gashaka Gumti National Park, Nigeria*. Report in conjunction with Department of Biological Sciences, University of Lagos, Nigeria, to NCF & WWF-UK.

Ankli, A., Sticher, O. & Heinrich, M. (1999). Yucatec Maya medicinal plants versus nonmedicinal plants: Indigenous characterization and selection. *Human Ecology* 27: 557–580.

Atawodi, S. E., Ameh, D. A., Ibrahim, S., Andrew, J. N., Nzelibe, H. C., Onyike, E. O., Anigo, K. M., Abu, E. A., James, D. B., Njoku, G. C. & Sallau, A. B. (2002). Indigenous knowledge system for treatment of trypanosomiasis in Kaduna state of Nigeria. *Journal of Ethnopharmacology* 79: 279–182.

Ayanbamiji, T. A. (1996). *Economic Plants and Vegetation Cover of Gashaka-Gumti National Park*. Report in conjunction with Department of Biological Sciences, University of Lagos / Nigeria, to NCF & WWF-UK.

Baker, M. (1996). Fur rubbing: use of medicinal plants by capuchin monkeys (*Cebus capucinus*). *American Journal of Primatology* 38: 263–270.

Balick, M. J. & Cox, P. A. (1997). *Plants, People, and Culture: the Science of Ethno Botany*. New York: Scientific American Library.

Berlin, E. A., Berlin, B., Lozoya, X., Meckes, M., Tortoriello, J., & Villarreal, M. L. (1996). The scientific basis of gastrointestinal herbal medicine among the highland Maya of Chiapas, Mexico. In Nader, L. (ed.), *Naked Science. Anthropological Inquiry into Boundaries, Power, and Knowledge*. New York, London: Routledge, pp. 43–68.

Boesch, C. & Tomasello, M. (1998). Chimpanzee and human cultures. *Current Anthropology* 39: 591–614.

Burkill, H. M. (1985). *The Useful Plants of West Tropical Africa, Vol. 1*. Kew: Royal Botanical Gardens.

Burkill, H. M. (1994). *The Useful Plants of West Tropical Africa, Vol. 2*. Kew: Royal Botanical Gardens.

Burkill, H. M. (1995). *The Useful Plants of West Tropical Africa, Vol. 3*. Kew: Royal Botanical Gardens.

Burkill, H. M. (1997). *The Useful Plants of West Tropical Africa, Vol. 4*. Kew: Royal Botanical Gardens.

Burkill, H. M. (2000). *The Useful Plants of West Tropical Africa, Vol. 5*. Kew: Royal Botanical Gardens.

Byrne, R. W. (1995). *The Thinking Ape. Evolutionary Origins of Intelligence*. Oxford: Oxford University Press.

Chapman, C. A. (1995). Primate seed dispersal: coevolution and conservation implications. *Evolutionary Anthropology* 20: 74–82.

Chapman, J. D. & Chapman, H. M. (2001). The Forest Flora of Taraba and Adamawa States, Nigeria. An Ecological Account and Plant Species Checklist. New Zealand: University of Canterbury.

Cotton, C. M. (1996). *Ethnobotany: Principles and Applications*. Chichester, New York: John Wiley & Sons.

Cousins, D. (2002). African apes and ethno medicine. *Gorilla Journal* 25: 10–12.

Cronquist, A. (1968). *The Evolution and Classification of Flowering Plants*. London: Thomas Nelson Ltd.

Dalziel, J. M. (1963) *A Hausa Botanical Vocabulary*. London: T. Fisher Unwin Ltd.

Dalziel, J. M. & Hutchinson, J. (1963). *Flora of West Tropical Africa*. London: Crown Agents.

De Smet, P. A. G. M. (1998). Traditional pharmacology and medicine in Africa: ethnopharmacological themes in sub-Saharan art objects and utensils. *Journal of Ethnopharmacology* 63: 1–179.

Dongmo, A. B., Kamanyi, A., Anchang, M. S., Nkeh, B. C.-A., Njamen, D., Nguelefack, T. B., Nole, T. & Wagner, H. (2001). Anti-inflammatory and analgesic properties of the stem bark extracts of *Erythrophleum suaveolens* (Caesalpiniaceae), Guillemin & Perrottet.*Journal of Ethnopharmacology* 77: 137–141.

Dunn, A. (1999). *Gashaka Gumti National Park: a Guidebook*. Lagos: Gashaka Gumti National Park, NCF / WWF-UK.

Dupain, J., van Elsaker, L., Nell, C., Garcia, P., Ponce, F., Huffman, M. A. (2002) New evidence for leaf swallowing and *Oesophagostomum* infections in bonobos (*Pan paniscus*). *International Journal of Primatology* 23: 1053–1062.

Engel, C. (2002). *Wild Health: How Animals Keep Themselves Well and What We Can Learn from Them*. Boston, MA: Houghton Mifflin Company.

Etkin, N. L. (1997). Antimalarial plants used by Hausa in northern Nigeria. *Tropical Doctor* 27 (supplement 1): 12–16.

Fabrega, H. J. & Silver, D. B. (1973). *Illness and Shamanistic Curing in Zinacantán*. Stanford, CA: Stanford University Press.

Falconer, J. (1992). *Non-Timber Forest Products in Southern Ghana: Main Report*. Kent: Natural Resources Institute.

Federal Ministry of Health (1988). *The National Health Policy and Strategy to Achieve Health for All Nigerians*. Lagos: Federal Ministry of Health.

Foster, G. M. & Anderson, B. G. (1978). *Medical Anthropology*. New York: John Wiley & Sons.

Fowler, A & Sommer, V. (2007). Subsistence technology in Nigerian chimpanzees. *International Journal of Primatology* 28: 997–1023.

Fowler, A., Koutsioni, Y. & Sommer, V. (2007). Leaf-swallowing in Nigerian chimpanzees. Assumed evidence for self-medication. *Primates* 48: 73–76.

Frank, D. J. (1961). *Persuasion and Healing: A Comparative Study of Psychotherapy*. Baltimore: Johns Hopkins University Press.

Freiburghaus, F., Kaminsky, R., Nkunya, M. H. H. & Brun, R. (1996). Evaluation of African medicinal plants for their in vitro trypanocidal activity. *Journal of Ethnopharmacology* 55: 1–11.

Fruth, B. (1995). *Nests and Nest Groups in Wild Bonobos (Pan paniscus): Ecological and Behavioural Correlates*. Aachen: Shaker.

Fruth, B. & Hohmann, G. (1996). Nest building behaviour in the great apes: the great leap forward? In McGrew, W. C., Marchant, L. F. & Nishida, T. (eds.), *Great Ape Societies*. New York: Cambridge University Press, pp. 225–240.

Glander, K. E. (1994). Nonhuman primate self-medication with wild plant foods. In Etkin, N. L. (ed.), *Eating on the Wild Side: The Pharmacologic, Ecologic, and Social Implications of using Noncultigens*. Tucson: The University of Arizona Press, pp. 227–239.

Glew, R. H., VanderJagt, D. J., Lockett, C., Grivetti, L. E., Smith, G. C., Pastuszyn, A. & Millson, M. (1997). Amino acid, fatty acid, and mineral composition of 24 indigenous plants of Burkina Faso. *Journal of Food Composition and Analysis* 10: 205–217.

Gottlieb, O. R., Borin, M. R. de M. B., Bosisio, B. M. (1996). Trends of plant use by humans and nonhuman primates in Amazonia. *American Journal of Primatology* 40: 189–195.

Hamilton, W. J. III, Buskirk, R. E. & Buskirk, W. H. (1978). Omnivory and utilization of food resources by chacma baboons, *Papio ursinus*. *The American Naturalist* 112: 911–924.

Higham, J. P. (2006) *The Reproductive Ecology of Female Olive Baboons (Papio hamadryas anubis) at Gashaka-Gumti National Park, Nigeria*. PhD thesis, Roehampton University.

Higham, J. P., Ross, C., Warren, Y., Heistermann, M. & MacLarnon, A. M. (2007). Reduced reproductive function in wild olive baboons (*Papio hamadryas anubis*) related to natural consumption of the African black plum (*Vitex doniana*). *Hormones and Behavior* 52: 384–390.

Higham, J. P., Warren, Y., Adanu, J., Umaru, B., MacLarnon, A., Sommer, V. & Ross C. (2009). Life on the edge: Life-history of olive baboons at Gashaka-Gumti National Park, Nigeria. *American Journal of Primatology* 71: 293–304.

Hladik, C. M., Pasquet, P. & Simmen, B. (2002). New perspectives on taste and primate evolution: the dichotomy in gustatory coding for perception of beneficent versus noxious substances as supported by correlations among human thresholds. *American Journal of Physical Anthropology* 117: 342–348.

Hohmann, G., Fowler, A., Sommer, V. & Ortmann, S. (2006). Frugivory and gregariousness of Salonga bonobos and Gashaka chimpanzees: the abundance and nutritional quality of fruit. In Hohmann, G. (ed.), *Feeding Ecology of Apes and Other Primates*. Cambridge: Cambridge University Press, pp. 123–159.

Huffman, M. A. (1997) Current evidence for self-medication in primates: a multidisciplinary perspective. *Yearbook of Physical Anthropology* 40: 171–200.

Huffman, M. A. (2001). Self-medicative behavior in the African Great Apes: An evolutionary perspective into the origins of human traditional medicine. *BioScience* 51: 651–661.

Huffman, M. A. (2007). Primate self-medication. In Campbell, C. J., Fuentes, A., MacKinnon, K., Panger, M. & Bearder, S. K. (eds.), *Primates in Perspective*. New York, Oxford: Oxford University Press, pp. 677–690.

Huffman, M. A. & Caton, J. M. (2001). Self-induced increase of gut motility and the control of parasistic infections in wild chimpanzees. *International Journal of Primatology* 22: 329–346.

Huffman, M. A. & Seifu, M. (1989) Observations on the illness and consumption of a possibly medicinal plant *Vernonia amygdalina* (Del), by a wild chimpanzee in the Mahale Mountains National Park, Tanzania. *Primates* 30: 51–63.

Huffman, M. A. & Wrangham, R. W. (1994). Diversity of medicinal plant use by chimpanzees in the wild. In Wrangham, R. W., McGrew, W. C., de Waal, F. B. M. & Heltne, P. G. (eds.), *Chimpanzee Cultures*. Cambridge, MA: Harvard University Press, pp. 129–148.

Huffman, M. A., Gotoh, S., Turner, L. A., Hamai, M. & Yoshida, K. (1997) Seasonal trends in intestinal nematode infection and medicinal plant use among chimpanzees in the Mahale mountains, Tanzania. *Primates* 38: 111–125.

Hurley, S. & Nudds, M. (2006) (eds.). *Rational Animals*? Oxford: Oxford University Press.

Iwu, M. M. (1986). Empirical investigations of dietary plants used in Igbo ethno medicine. In Etkin, N. L. (ed.), *Plants in Indigenous Medicine and Diet: Biobehavioral Approaches*. New York: Bedford Hills, pp. 131–150.

Iwu, M. M. (1993). *Handbook of African Medicinal Plants*. Boca Raton: CRC Press.

Jackson, F. L. C. (1991). Secondary compounds in plants (allelochemicals) as promoters of human biological diversity. *Annual Review of Anthropology* 20: 505–546.

Johns, T. A. (1986). Chemical selection in Andean domesticated tubers as a model for the acquisition of empirical plant knowledge. In Etkin, N. L. (ed.), *Plants in Indigenous Medicine and Diet: Biobehavioral Approaches*. Bedford Hills, New York, pp. 266–288.

Johns, T. A. (1990). *With Bitter Herbs They Shall Eat it: The Origins of Human Diet and Medicine*. Tucson: University of Arizona Press.

Keay, R. W. (1989). *Trees of Nigeria*. Oxford: Clarendon Press.

Krebs, D., Denton, K. & Higgins, N. C. (1988). On the evolution of self-knowledge and self-deception. In Mac Donald, K. B. (Hg.) (ed.), *Sociobiological Perspectives on Human Development*. New York, Berlin, Heidelberg: Springer, pp. 103–139.

Krief, S., Wrangham, R. W. & Lestel, D. (2006). Diversity of items of low nutritional value ingested by chimpanzees from Kanyawara, Kibale National Park: an example of the etho-ethnology of chimpanzees. *Social Science Information* 45: 227–263.

Lal, P. (1981). *The Ramayana of Valmiki*. Delhi: Vikas Publishing House.

Largo, C. J., Florez, J. D. & van Schaik, C. P. (2009). Nest site selection by Bornean orangutans? Abstracts, 11. Meeting, *Gesellschaft für Primatologie*, 24 – 26 Feb 2009, Hannover.

Latha, P., Govindasamy, S. & Balakrishna, K. (1997). Effect of gangetin on fertility of male rats. *Phytotherapy Research* 11: 372–375.

Lozano, G. A. (1998). Parasitic stress and self-medication in wild animals. *Advances in the Study of Behavior* 27: 291–317.

Martin, M. (1996). *The Use of Rural Appraisal Methods in the Study of Ethnoveterinary Medicine*. MSc thesis, University of Edinburgh.

McCorkle, C. M. (1986). An introduction to ethnoveterinary research and development. *Journal of Ethnobiology* 6: 129–149.

Moerman, D. E. (1996). An analysis of the food plants and drug plants of native North America. *Journal of Ethnopharmacology* 52: 1–22.

N'gouemo, P., Baldy-Moulinier, M. & Nguemby-Bina, C. (1996). Effects of an ethanolic extract of *Desmodium adscendens* on central nervous system in rodents. *Journal of Ethnopharmacology* 52: 77–83.

NCF (no date; ca. 1998). *Plants of Medicinal Value in Gashaka Gumti National Park*. Unpublished compilation for the Nigerian Conservation Foundation & WWF-UK.

Noumi, E. & Tchakonang, N. Y. C. (2001). Plants used as abortifacient in the Sangmelima region of Southern Cameroon. *Journal of Ethnopharmacology* 76: 263–268.

Nunn, C. L. & Altizer, S. (2006). *Infectious Diseases in Primates: Behavior, Ecology and Evolution* (Oxford Series in Ecology & Evolution). Oxford: Oxford University Press.

Odugbemi, T. (ed.) (2006). *Outlines and Pictures of Medicinal Plants from Nigeria*. Lagos: University of Lagos Press.

Pellaumail, K. (1998). *Women and Non-Timber Forest Products in the Gashaka Gumti National Park's area, Nigeria*. MRes Environmental Sciences thesis, University College London.

Peters, C. R., O'Brien, E. M., Boaz, N. T., Conroy, G. C., Godfrey, L. R., Kawanaka, K., Kortlandt, A., Nishida, T., Poirier. F. E. & Smith, E. O. (1981). The early hominid plant-food niche: insights from an analysis of plant exploitation by *Homo*, *Pan*, and *Papio* in eastern and southern Africa. *Current Anthropology* 22: 127–140.

Phillips-Conroy, J. E. (1986). Baboons, diet, and disease: food plant selection and schistosomiasis. In Taub, D. & King, F. (eds.), *Current Perspectives in Primate Social Dynamics*. New York: Van Nostrand Reinhold, pp. 287–304.

Rodriquez, E., Wrangham, R. W. (1993) Zoopharmacognosy: The use of medicinal plants by animals. In Downum, K. R., Romeo, J. T., Stafford, H. (eds.), *Recent advances in phytochemistry, vol 27: Phytochemical potential of tropic plants*. New York: Plenum, pp 89–105.

Saidu, K., Onah, J., Orisadipe, A., Olusola, A., Wambebe, C. & Gamaniel, K. (2000). Antiplasmodial, analgesic, and anti-inflammatory activities of the aqueous extract of the stem bark of *Erythrina senegalensis*. *Journal of Ethnopharmacology* 71: 275–280.

Sofowora, A. (1986). *The State of Medicinal Plants Research in Nigeria (Proceedings of a Workshop)*. Nigeria: Nigerian Society of Pharmacognosy.

Sofowora, A. (1993). *Medicinal Plants and Traditional Medicine in Africa*. Ibadan / Nigeria: Spectrum Books.

Sommer, V., Adanu, J., Faucher, I. & Fowler, A. (2004). The Nigerian Chimpanzee (*Pan troglodytes vellerosus*) at Gashaka: Two years of habituation efforts. *Folia Primatologica* 75: 295–316.

Strier, K. B. (1993). Menu for a monkey. *Natural History* 102: 34–43.

Tomasello, M. & Call, J. (1997). *Primate Cognition*. Oxford: Oxford University Press.

van Wyk, B. E., Bosch van, O. & Nigel, G. (2000). *Medicinal Plants of South Africa*. Pretoria: Briza.

Wall, L. L. (1988). *Hausa Medicine: Illness and Well-being in a West African Culture*. Durham: Duke University Press.

Warren, Y. (2003). *Olive Baboons (Papio cynocephalus anubis): Behaviour, Ecology And Human Conflict In Gashaka Gumti National Park, Nigeria*. PhD thesis, Roehampton University.

Weyher, A. H., Ross, C. & Semple, S. (2006). Gastrointestinal parasites in crop raiding and wild foraging Papio anubis in Nigeria. *International Journal of Primatology* 27: 1519–1534.

Whitehead, J. M. (1986). Development of feeding selectivity in mantled howling monkeys, *Alouatta palliata*. In Else, J. G. & Phyllis, C. L. (eds.), *Primate Ontogeny, Cognition and Social Behaviour*. Cambridge: Cambridge University Press, pp. 105–117.

Whiten, A. (2000). Primate culture and social learning. *Cognitive Science* 24: 477–508.

Wrangham, R. W. & Goodall, J. (1989). Chimpanzee use of medicinal leaves. In Heltne, P. G. & Marquardt, L. A. (eds.), *Understanding Chimpanzees*. Cambridge, MA: Harvard University Press, pp. 22–37.

Wrangham, R. W. & Nishida, T. (1983) *Aspilia* spp. leaves: A puzzle in the feeding behavior of wild chimpanzees. *Primates* 24: 276–282.

Wynne, C. D. L. (2001). *Animal Cognition. The Mental Lives of Animals*. New York: Palgrave.

Appendix. Use of Plants as Food, Medicine, and Implements by Humans, Baboons, and Chimpanzees in the Gashaka Area: A Compilation of Current Knowledge

Family, Species (main references)

Chapman, J. D. & Chapman, H. M. (2001). *The Forest Flora of Taraba and Adamawa States, Nigeria. An Ecological Account and Plant Species Checklist.* Univ. of Canterbury / NZ: WWF / DFID

Dalziel, J. M. & Hutchinson, J. (1963). *Flora of West Tropical Africa.* London: Crown Agents

Keay, R. W. (1989). *Trees of Nigeria.* Oxford: Clarendon Press

Obot, Emmanuel; pers. comm. of Director of Technical Programs of NCF, Lekki / Lagos (leading expert on Nigerian Flora; visited Gashaka Jan – Feb 03)

English, Hausa, Fulfulde (Fulani)

Vernacular names in English and predominant local languages

Habit

T: Tree
L: Liana (woody climber)
S: Shrub
G: Grass
H: Herb
F: Fern
A: Animal
C: Crop
M: Mushroom

Habitat

MG: Montane grassland
MF: Montane forest

OS: Open savannah
WS: Woodland savannah
RF: Riperian forest
LF: Lowland forest
F: Forest
S: Savannah
FE: Forest edge
DC: Domesticated / Crop

Food, Medicine, Implements: Consumers

H: Humans
A: Animal (domestic, mostly cattle)
B: Baboon
C: Chimpanzee

Food, Medicine, Implements, Usage, Description: Plant Parts Used

? part unknown
BA: Bark
BU: Bulb, Corm
FD: Fruit rind
FI: Fibre
FJ: Fruit juice
FL: Flower
FM: Fruit, ripe (mature)
FO: Frond
FP: Fruit pulp
FR: Fruit
FS: Floss (mostly in fruit)
FU: Fruit, unripe
GR: Grass
GU: Gum
LA: Latex / Sap
LF: Leaf
LO: Leaf, old
LY: Leaf, young
NE: Nectar

NU: Nut
PI: Pith
PO: Pod
RO: Root
SE: Seed, bean
SH: Shoot
ST: Stem
TH: Thorn
TR: Bole (trunk)
TU: Tuber
TW: Twig
WO: Wood
WP: Whole plant
Example. H: BA (= bark, used by humans)

Usage: Description, Category

F: Food
M: Medicine
I: Implement
H: Humans
A: Animal (domestic, mostly cattle)
B: Baboon
C: Chimpanzee
Examples. HM: (= human medicine); FB: (= baboon food)

Miscellaneous

&: and
*: to be clarified
?: uncertain (note, that this is not the same as *)
hf.: handful
O: Other
w.: with
w/o: without
GGH specimen: sample present in collection of Gashaka Gumti Herbarium

Sources

Unpublished reports (UR) to NCF (Nigerian Conservation Foundation, Gashaka Gumti Project) and personal communications

Adanu, Jeremiah (1998). *Determination of Life-span of Chimpanzee Nests (Nest-aging) in Gashaka Gumti National Park.* UR in conjunction with Dept. of Forestry and Wildlife Management, Univ. of Technology, Yola, to: NCF& WWF-UK

Akinsoji, Aderopo (1996). *Vegetation Types and Ethnobotanical Studies of Gashaka Gumti National Park, Nigeria.* UR in conjunction with Dept. of Biological Sciences (Botany Unit), Univ. of Lagos, to: NCF& WWF-UK

Ayanbamiji, Taiwo Ayanniyi (1996). *Economic Plants and Vegetation Cover of Gashaka-Gumti National Park.* UR in conjunction with Dept. of Biological Sciences (Botany Unit), Univ. of Lagos, to: NCF& WWF-UK

Fowler, Andrew (p.c.). Unpubl. data collected for Ph.D. dissertation at Dept. of Anthropology, UCL, on chimpanzee socioecology at Gashaka-Kwano, Jan 01 - Mar 03. Field assistant: Hammounde Guruza

Martin, Marina (1996). *The Use of Rural Appraisal Methods in the Study of Ethnoveterinary Medicine.* Unpubl. dissertation, MSc in Tropical Animal Production and Health, Centre for Tropical Veterinary Medicine, Univ. of Edinburgh

NCF (n.d. = no date; ca. 1998). *Plants of Medicinal Value in Gashaka Gumti National Park.* Unpubl. Compilation for NCF & WWF-UK

Pellaumail, Karine (1998). *Women and Non-Timber Forest Products in the Gashaka Gumti National Park's area, Nigeria.* Unpubl. dissertation, MSc in the Anthropology of Development, Dept. of Anthropology, UCL, London

Warren, Ymke (p.c.). Unpubl. data collected for Ph.D. dissertation at Univ. of Surrey-Roehampton on baboon socioecology at Gashaka-Kwano, Jan 01 - Dec 02. Field assistant: Bobbo Buba

Local informants

Interviews. Conducted by YK during field visit to Gashaka area, Jan - Mar 03; interpreters: Doris Nyanganji, NCF women officer; Maigari Ahmadu, GPP assistant)

Locations. BD: Bodel; CH: Chappal Hendu; DM: Do Mayo; FI: Filinga; GK: Gashaka; KM: Karamti; MS: Mayo Selbe; SB: Selbe; ST: Serti

Individuals. F: Female; M: Male; H: Healer; A: assistant to Gashaka Primate Project (Hammaunde Guruza, Bobbo Buba, Maigari Ahmadu)

Examples. LI (ST-FH): Local Informant, Serti, Female Healer; LI (GK-MA): Local Informant, Gashaka, Male Assistant to Gashaka Primate Project

Appendix

Family	Species	English	Hausa	Fulfulde	Habit	Habitat	Food	Medicine	Implement	Usage Description	Usage Category	Description in Original Source	Source
Leguminosae - Mimosoideae	*Acacia spp.*			Gerofu*	T								
								H: RO		MH: Breast feeding	MH: Female reproduction	Purify women's breast: RO w. potash	Ayanbamiji 1996
Leguminosae - Mimosoideae	*Acacia gourmaensis*		Kama muraba		ST	S F							
								H: BA		MH: Body pain	MH: Analgesic	Body pain: boil BA w. potash & take decoction	Akinsoji 1996, NCF n.d.
Leguminosae - Mimosoideae	*Acacia nilotica*	Egyptian mimosa	Gabaruwa, Bagaruwa	Gabde, Gaudi	T	S							
							B: SE			FB: Edible	FB: Edible	Food	Warren p.c.
								H: BA		MH: Constipation	MH: Gastrointestinal	Constipation: boil or soak BA into water, mix w. corn flour & eat 3 times a day	LI (GK-F)
								H: BA LF		MH: Diarrhoea	MH: Gastrointestinal	Diarrhoea: boil LF & bathe or drink: also enema, 4-5 days	Akinsoji 1996, NCF n.d., LI (GK-F), LI (GK-MA)
								H: LF		MH: Diarrhoea	MH: Gastrointestinal	Diarrhoea: boil LF & bathe or drink; also enema, 4-5 days	LI (GK-F)
								H: LF		MH: Chest pain	MH: Internal organs & Respiration	Chest pain: grind fresh LF & rub on chest or in the body	Akinsoji 1996, NCF n.d.
								H: SE		MH: Toothache	MH: Oral health	Toothache: drink like tea	LI (KV-F), LI (KM-M)
								H: LF FR		MH: Worms	MH: Parasites	Worms: pound LF mix w. rice & eat or soak FR in water & drink	LI (GK-F)
Leguminosae - Mimosoideae	*Acacia seyal*	White whistling thorn	Dushe	Bulki	ST	F S							
								H: BA		MH: Aphrodisiac	MH: Fertility & Genitalia	Aphrodisiac: boil BA w. potash & take decoction	Akinsoji 1996, NCF n.d.
								H: RO		MH: Stomach problems	MH: Gastrointestinal	Stomach ache: mix RO w. flour to make porridge	LI (S3-F)
								H: BA		MH: Toothache	MH: Oral health	Toothache, tooth problems in kids: boil BA w. potash & drink decoction or wash mouth	Akinsoji 1996, NCF n.d.
Leguminosae - Mimosoideae	*Acacia sieberiana*	White thorn	Farar kaya, Bauii	Gie daneji	T	S							
							B: SE			FB: Edible	FB: Edible	Food	Warren p.c.
								H: RO		MH: Bone weakness	MH: Bones	Bone weakness: drink RO infusion	Akinsoji 1996, NCF n.d.
									H: TW	IH: Needles	IH: Crafts & Skills	Needles	LI (GK-F)
Compositae (Asteraceae)	*Acanthospermum hispidum*	Starbur	Kashiyawo	Yakamji									
								H: LF		MH: Haemorrhoids	MH: Gastrointestinal	Haemorrhoids: drink LF decoction	Akinsoji 1996
Bombacaceae	*Adansonia digitata*	Baobab	Kuka	Boko	T	S							
							B: FR			FB: Edible	FB: Edible	Food	Warren p.c.
							H: LF FR			FH: Edible	FH: Edible	Soup: dry & pound LF & FR & make soup	Akinsoji 1996, LI (GK-MA), LI (GK-F), LI (SB-M)
							H: LF FR			FH: Edible	FH: Edible	Food: boil LF	LI (S3-M)
							H: FP			FH: Drink flavouring	FH: Seasoning & Processing	High vitamin C content: use FP as drink flavouring	Akinsoji 1996
								H: RA		MH: Increase blood	MH: Blood	Increase blood: boil RA	LI (MS-F)

								H: ?		MH: Yellow fever	MH: Endemic & Epidemic Diseases		Martin 1996
								H: ?		MH: Malaria	MH: Fevers	Malaria: make infusion & bathe child	Martin 1996
								H: BA		MH: Vomiting	MH: Gastrointestinal	Stop vomiting: BA	Ayanbamiji 1996
								H: RO		MH: Stomach upset	MH: Gastrointestinal	Stomach upset: RO	NCF n.d.
								H: BA		MH: Fatness	MH: Health promotion	To fatten: use its parasite	Ayanbamiji 1996
								H: BA		MH: Strength	MH: Health promotion	Increase weight on unhealthy babies: boil BA & drink 3 cups per day	LI (GK-F)
								H: BA		MH: Toothache	MH: Oral health	Toothache: boil BA w. potash	Akinsoji 1996, NCF n.d.
									H: BA	IH: Ropo	IH: Crafts & Skills	Make rope & mat: use inner BA	Akinsoji 1996
Leguminosae - Papilionoideae	*Adenodolichos paniculatus*		Kwiwa	Kilikainawa	S	S							
								H: RO		MH: Blood tonic	MH: Blood	Blood tonic in pregnant women: boil RO w. potash & take decoction	Akinsoji 1996, NCF n.d.
Zingiberaceae	*Aframomum* spp.	Forest ginger	Jabbasuka, Borkonun daji, Jabba leinde	Jabbasuka, Yamasuka	H	MF							
							B: FR ST RO			FB: Edible	FB: Edible	Food	Warren p.c.
							C: FR LY			FC: Edible	FC: Edible	Food	Fowler p.c.
							H: FR			FH: Edible	FH: Edible	Soup: dry LF & mix w. pepper	Fowler p.c.
								H: LF		MH: Fever	MH: Fevers	Fever	Pellaumail 1998
Zingiberaceae	*Aframomum angustifolium*	Madagascar cardamon	ditto	ditto	H								
							B: FR			FB: Edible	FB: Edible	Food	LI (GK-MA)
							C: FR			FC: Edible	FC: Edible	Food	LI (GK-MA)
							H: FR			FH: Beverage	FH: Beverages	Quench thirst	Akinsoji 1996
							H: FR			FH: Edible	FH: Edible	Food: particularly dry season	LI (S3-F), LI (GK-F), LI (KM-F), LI (KM-M)
							H: LF			FH: Eedible	FH: Edible	Eat raw	LI (S3-M)
								H: RO		MH: Stomach problems	MH: Gastrointestinal	Stomach problems: boil RO & drink	LI (GK-MA)
								H: SE		MH: Haemorrhoids	MH: Gastrointestinal	Haemorrhoids: grind SE & add to porridge	Akinsoji 1996
								H: PI		MH: General health	MH: Health promotion	Unknown benefit: chew pith	LI (GK-F)
Zingiberaceae	*Aframomum latifolium*	Savannah ginger	Borkonun daji	Jabbasuka, Yamasuka	H								
							B: FR ST RO			FB: Edible	FB: Edible	Food	Warren p.c.
Zingiberaceae	*Aframomum sulcatum ?*	Forest ginger	ditto	ditto	H								
							B: FR ST RO			FB: Edible	FB: Edible	Food	Warren p.c.
Leguminosae - Caesalpinoideae	*Afzelia africana*	African mahogany, Mahogany bean	Kawo	Petoje, Petoji, Gayohi	T	S						GGH specimen	
							A: LF			FA: Edible	FA: Edible	Fodder	LI (GK-MA), LI (SB-M)
							B: SE			FB: Edible	FB: Edible	Food	Warren p.c., LI (GK-MA)

(continued)

Appendix (continued)

Family	Species	English	Hausa	Fulfulde	Habit	Habitat	Food	Medicine	Implement	Usage Description	Usage Category	Description in Original Source	Source
							C: FR			FC: Edible	FC: Edible	Food	Fowler p.c.
							C: SE			FC: Edible	FC: Edible	Food	Salamu
							H: SE			FH: Edible	FH: Edible	Food: grind SE & burn in fire	LI (S3-M), LI (GK-MA)
							H: SF			FH: Edible	FH: Edible	Oil: make oil from SF	LI (GK-MA)
								A: LF		MA: Milk production	MA: Female reproduction	Increase milk production in cows: LF	Akinsoji 1996
								H: SE		MH: Side pain	MH: Analgesic	Side pain: grind 4 SE	Pellaumail 1998
								H: SE		MH: Swollen parts	MH: Fertility & Genitalia	Swollen parts: make porridge & eat	LI (FI-M)
								H: BA		MH: Colds	MH: Fevers	Cold: take 1 hf	Pellaumail 1998
								H: SE		MH: Colds	MH: Fevers	Cold: dry & pound SE, soak in hot water & drink	LI (GK-FH)
								H: BA		MH: Stomach ache	MH: Gastrointestinal	Stomach ache: take 2 hf	Pellaumail 1998
								H: SE		MH: Boils	MH: Skin	Boils: collect in dry season. Grind, pound & roast SE. Take 2x daily for 10 days	LI (GK-F)
								H: SE		MH: Skin conditions	MH: Skin	Skin problems: dry & pound SE & add water to make soup	LI (GK-F), LI (GK-FH)
									H: WO	IH: Implements	IH: Timbers & Carving	Utensils & agricultural implements: WO	Akinsoji 1996
Leguminosae - Mimosoideae	*Albizia coriaria*		Dorowar Mahalbi	Gawjo	T								
									H: BA	IH: Fish poison	IH: Hunting & Fishing	Fish poison: BA	Akinsoji 1996
Leguminosae - Mimosoideae	*Albizia zygia*		Masdubiya rafi	Banuhi	T	F							
									H: WP	IH: Shade tree	IH: Timbers & Carving	A shade tree	Akinsoji 1996
Euphorbiaceae	*Alchornea spp.*												
								H: LF		MH: Small pox	MH: Endemic & Epidemic diseases	Small pox: boil LF & bathe children	Ayanbamiji 1996
Euphorbiaceae	*Alchornea cordifolia*	Christmas bush	Bambami	Bori lay*, Karkandam	LS								
							B: FR			FB: Edible	FB: Edible	Food	Warren p.c.
								H: BA		MH: Yellow fever	MH: Endemic & Epidemic Diseases	Yellow fever: boil & drink BA	LI (CH-F)
								H: LF		MH: Fever	MH: Fevers	Fever: boil LF & dinya LF or jambajue & kolijol & bathe	(LI (GK-F)
								H: LF ST		MH: Colds	MH: Fevers	Skin swelling caused by cold: tie LF & ST 7 times, boil, bathe. Cover with towel & inhale	(LI (GK-F)
								H: LF		MH: Dysentery	MH: Gastrointestinal	Dysentery: mix w. galbije, boil LF for enema	(LI (GK-F)
								H: LF		MH: Stomach ache	MH: Gastrointestinal	Stomach: boil LF & dinya LF & bathe	(LI (GK-F)
									H: ?	IH: Rope	IH: Crafts & Skills	Make rope	(LI (GK-F)
Apocynaceae	*Alstonia boonei*	Stool wood, Pattern wood			T								
								H: BA		MH: Fever	MH: Fevers	Fever: boil BA w. potash & drink	Akinsoji 1996
								H: BA		MH: Toothache	MH: Oral health	Toothache: boil BA w. potash & drink	Akinsoji 1996
									H: WO	IH: Carving	IH: Timbers & Carving	Carving: WO	Akinsoji 1996
Amaranthaceae	*Amaranthus spinosus*	Spiny amaranth	Namijin gasaya	Anyaka, Ha-kon diyam, Camgam	H								
							B: ST			FB: Edible	FB: Edible	Food	Warren p.c.
							H: WP			FH: Edible	FH: Edible	FH: Pot herb	Akinsoji 1996
Araceae	*Amorphophallus abyssinicus*	Barter's arum											

							H: TU			FH: Edible	FH: Edible	Food: soak TU in water to leach out the toxic substance	Akinsoji 1996
								H: TU		MH: Snake bite	MH: Antidotes & Repellents	Snake bite: TU	Akinsoji 1996
Anacardiaceae	*Anacardium occidentale*	Cashew nut	Kanju		T								
							H: FR SF			FH: Edible	FH: Edible	Edible: FR; SF	Akinsoji 1996
								H: LF		MH: Fever	MH: Fevers	Fever: LF	Akinsoji 1996
									H: SE	IH: Cosmetic	IH: Hygiene & Cosmetics	Tattoo: unripe exudate of SE	Akinsoji 1996
Araceae	*Anchomanes difformis*												
								H: TU		MH: Smallpox	MH: Endemic & Epidemic diseases	Smallbox: make decoction of TU & bathe	Akinsoji 1996
Annonaceae	*Annona senegalensis*	Wild custard apple	Gwandar daji	Dukkuje lad-de, Dukuhi, Dukual	T	S MF						GGH specimen	
							B: FR FL			FB: Edible	FB: Edible	Food	Warren p.c., LI (GK-MA)
							C: FR			FC: Edible	FC: Edible	Food, especially rainy season	LI (GK-MA
							H: FR			FH: Edible	FH: Edible	Food	Akinsoji 1996, LI (GK-MA), LI (SB-M), LI (SB-F)
								H: LY		MH: Snake bite	MH: Antidotes & Repellents	Snake bite: swallow 7 LY & 7 LY of Piliostigma	NCF n.d.
								H: RO		MH: Snake antidote	MH: Antidotes & Repellents	Snake antidote: boil RO	LI (SB-M)
								H: BA		MH: Pregnancy	MH: Female reproduction	Ease pregnancy pains: give after 7th month	LI (CH-F)
								H: RO		MH: Testicular problem	MH: Fertility & Genitalia	Testicular problem: take 1 mouthful (2 RO)	Pellaumail 1998
								H: BA RO		MH: Gonorrhoea	MH: Fertility & Genitalia	Gonorrhoea: boil & drink BA (inner layer) & RO	Akinsoji 1996, NCF n.d.
								H: BA SH		MH: Fever	MH: Fevers	Fever: grind BA & new shoot & BA & LF of Piliostigma & administer to children	NCF n.d.
								H: BA		MH: Stomach problems	MH: Gastrointestinal	Stomach: use if sugar is bad for stomach	LI (CH-F)
								H: BA RO		MH: Dysentery	MH: Gastrointestinal	Dysentery: boil BA w. potash & take decoction; alternatively: RO decoction	Akinsoji 1996
								H: FI BA		MH: Dysentery	MH: Gastrointestinal	Dysentery	LI (MS-F)
								H: LF		MH: Dysentery	MH: Gastrointestinal	Dysentery: boil LF & drink 2 cups per day for about 3 days	LI (GK-F)
								H: PI		MH: Diarrhoea	MH: Gastrointestinal	Runny stomach: eat raw pith, or mix pith w. corn flour & fill the ST & roast in fire	LI (GK-MA)
								H: RO		MH: Appendicitis	MH: Internal organs & Respiration	Appendicitis: make RO decoction	Akinsoji 1996, Ayanbamiji 1996
								H: RO BA		MH: Diarrhoea	MH: Gastrointestinal	Diarrhoea: take 1 hf	Pellaumail 1998
								H: RO BA		MH: Stomach ache	MH: Gastrointestinal	Stomach ache: take 1 hf	Pellaumail 1998
								H: FI BA		MH: Worms	MH: Parasites	Worms & dysentery: boil BA or chew FI	LI (MS-F)
								H: BA		MH: Wound	MH: Skin	Wound healing: tie BA around wound	Ayanbamiji 1996
									H: BA	IH: Good luck	IH: Magico-religious	Good luck: boil BA & drink 3 times & bathe	NCF n.d.
									H: RO	IH: Evil spirits	IH: Magico-religious	Evil spirit: boil & drink RO & eastern RO of Securidaca	NCF n.d.

(continued)

Appendix (continued)

Family	Species	English	Hausa	Fulfulde	Habit	Habitat	Food	Medicine	Implement	Usage Description	Usage Category	Description in Original Source	Source
									H: RO	IH: Orientation	IH: Magico-religious	Know way in the bush: RO & eastern RO of Securidaca	NCF n.d.
Combretaceae	*Anogeissus leiocarpus*	Axlewood	Marike	Kojoli, Kujoli	T	S F							
							B: FR			FB: Edible	FB: Edible	Food	Warren p.c.
								A: BA		MA: ?	MA: ?	For animals: pound BA, boil & add to limestone	Martin 1996
								A: BA		MA: Nose & head problems	MA: Analgesic	Animal nose & head problems: boil BA, mix w. potash & administer through nose	Martin 1996
								H: BA		MH: Headache	MH: Analgesic	Headache: pound BA & make tea	Martin 1996
								H: LF		MH: Pain	MH: Analgesic	Relieve pain: burn LF	Ayanbamiji 1996
								H: BA		MH: Yellow fever	MH: Endemic & Epidemic Diseases	Yellow fever: take 1-3 hf	Pellaumail 1998
								H: LF		MH: Yellow fever	MH: Endemic & Epidemic Diseases	Yellow fever: burn LF	Ayanbamiji 1996
								H: BA		MH: Gonorrhoea	MH: Fertility & Genitalia	Gonorrhoea: boil BA w. potash & drink decoction	Akinsoji 1996
								H: BA		MH: Cough	MH: Fevers	Cough: boil BA w. potash & take decoction	Akinsoji 1996, NCF n.d.
								H: BA		MH: Purgative	MH: Gastrointestinal	Purgative: boil BA w. potash & take decoction	Akinsoji 1996
								H: BA		MH: Stomach ache	MH: Gastrointestinal	Stomach ache (infants esp.): take 1-3 hf	Pellaumail 1998
								H: LF		MH: Weakness	MH: Health promotion	Weakness: burn LF	Ayanbamiji 1996
								H: RO		MH: Chewing stick	MH: Oral health	Chewing stick: RO	Ayanbamiji 1996
								H: BA		MH: Worms	MH: Parasites	Anthelmintic: boil BA w. potash & take decoction	Akinsoji 1996
								H: BA		MH: Worms	MH: Parasites	Worms: take 1-3 hf	Pellaumail 1998
								H: BA		MH: Worms	MH: Parasites	Worms: boil BA & bathe or drink	LI (GK-F)
								H: BA		MH: Boils	MH: Skin	Boils: boil BA & drink 2 cups per day for about 7 days	LI (GK-F)
								H: BA		MH: Scabies	MH: Skin	Scabies: boil BA & drink 2 cups per day for about 7 days	LI (GK-F)
								H: BA		MH: Skin disease	MH: Skin	Skin disease: take 3 hf per week	Pellaumail 1998
								H: BA		MH: Skin conditions	MH: Skin	Skin problems: boil BA & bathe children or drink	LI (GK-F)
								H: LF		MH: Skin disease	MH: Skin	Skin problems: burn LF	Ayanbamiji 1996
									H: WO	IH: Implements	IH: Timbers & Carving	Implements e.g. hoe handles: WO	Akinsoji 1996
Leguminosae - Caesalpinoideae	*Anthonotha macrophylla*				T								
									C: TW LF TR	IC: Nest-tree	IC: Nest	10% of all nest-trees	Adanu 1998
Icacinaceae	*Apodytes dimidiata* ?	Korn tree*	Korn, Namiji, (Black) Korn*		T								
							B: FL FR			FB: Edible	FB: Edible	Food	Warren p.c.
							C: FR			FC: Edible	FC: Edible	Food	Fowler p.c.
Leguminosae - Papilionoideae	*Arachis hypogaea*	Ground nut		Biriyu	C								
							B: NU			FB: Edible	FB: Edible	Food	Warren p.c.
Liliaceae	*Asparagus flagellaris*		Kayar kadangare	Nyaranyaro									
								H: RO		MH: Breastfeeding	MH: Female reproduction	Contaminated milk: boil RO w. potash & give decoction to babies that took contaminated milk from mother's breast	Akinsoji 1996

Compositae (Asteraceae)	*Aspilia africana*	Wild sun-flower, Haemorrhage plant	Jamajina	Nanyarki, Nyarki	T								
					T		A: LF			FA: Edible	FA: Edible	Fodder: LF	Ayanbamiji 1996
					T			H: RO		MH: Swollen parts	MH: Analgesic	Swollen parts: drink the RO	LI (CH-F)
					T			H: LF		MH: Wound	MH: Skin	Wounds: LF	Akinsoji 1996
					T				H: LF	IH: Construction	IH: Timbers & Carving	Smooth wood: use LF as sandpaper	LI (S3-M)
					T				H: WO	IH: Construction	IH: Timbers & Carving	Construction	LI (CH-F)
Meliaceae	*Azadirachta indica*	Neem	Dogon yaro, Maggoi	Dalbejiya	T	S							
								H: LF		MH: Fever	MH: Fevers	Fever: boil LF boil	NCF n.d.
								H: LF		MH: Jaundice	MH: Internal organs & Respiration	Jaundice: pound LF, soak in water & drink	NCF n.d.
Balanitaceae	*Balanites aegyptiaca*	Soapberry tree, Desert date	Aduwa	Tanne, Tanni	T	S							
							H: LF FR SE			FH: Edible	FH: Edible	Soup: LF, edible: FR, oil extraction: stony SE	Akinsoji 1996, Ayanbamiji 1996, LI (CH-F), LI (GK-MA)
								H: RO		MH: Gonorrhoea	MH: Fertility & Genitalia	Gonorrhoa: drink RO decoction	Akinsoji 1996, NCF n.d.
								H: FP		MH: Purgative	MH: Gastrointestinal	Purgative: FP	Akinsoji 1996
									H: WO	IH: Slates	IH: Crafts & Skills	Slates for Arabic school: WO	Akinsoji 1996
									H: BA	IH: Fish poison	IH: Hunting & Fishing	Fish poison: BA	Akinsoji 1996
									H: TH	IH: Cosmetic	IH: Hygiene & Cosmetics	Eye-pencil: TH	Akinsoji 1996
									H: WO	IH: Boat carving	IH: Timbers & Carving	Boats: WO	Akinsoji 1996, Ayanbamiji 1996
									H: WO	IH: Furniture	IH: Timbers & Carving	Light furniture: WO	Akinsoji 1996
Leguminosae - Caesalpinoideae	*Bauhinia rufescens*		Jirga	Namarehi	T								
								H: LF		MH: Dysentery	MH: Gastrointestinal	Dysentery: drink LF decoction	Akinsoji 1996
									H: BA	IH: Rope	IH: Crafts & Skills	Rope: BA	Akinsoji 1996
Nyctaginaceae	*Boerhavia spp.*		Babba jibji	Burgu yolde, Nyakkabre, Dakalde	H								
								H: RO		MH: Toothache	MH: Oral health	Toothache: RO w. potash	Ayanbamiji 1996
Nyctaginaceae	*Boerhavia diffusa*	Hogweed	ditto	ditto	H								
								H: RO		MH: Toothache	MH: Oral health	Toothache: RO decoction	Akinsoji 1996
Bombacaceae	*Bombax costatum*	Kapok	Kurya	Kuruhi, Djohi*									
									H: FS	IH: Stuffing	IH: Timbers & Carving	Stuffing of mattresses & pillows: FS in FR	Akinsoji 1996
Palmae	*Borassus aethiopum*	Palm	Giginya	Dubbi, Dubbal	T	S							
							B: FR			FB: Edible	FB: Edible	Food	Warren p.c., LI (GK-M)
							C: FR			FC: Edible	FC: Edible	Food	LI (GK-MA)
							H: FR			FH: Edible	FH: Edible	Food, especially dry season	Akinsoji 1996, LI (SB-M), LI (KM-F), LI (KM-M), LI (GK-F)

(continued)

Appendix (continued)

Family	Species	English	Hausa	Fulfulde	Habit	Habitat	Food	Medicine	Implement	Usage Description	Usage Category	Description in Original Source	Source
							H: RO			FH: Edible	FH: Edible	Food: RO of the seedling (radicle) is edible	Ayanbamiji 1996
								H: LF		MH: Bleeding	MH: Blood	Recover blood: LF	Ayanbamiji 1996
								H: LF		MH: Internal injury	MH: Blood	Internal injury: boil LF stalk & drink	NCF n.d.
								H: RO		MH: Blood clot	MH: Blood	Dissolves blood clot in injuries: RO decoction w. potash	Akinsoji 1996, Ayanbamiji 1996
								H: RO		MH: Testicular problem	MH: Fertility & Genitalia	Swollen testicles: RO decoction w. potash	Akinsoji 1996, Ayanbamiji 1996
								H: RO		MH: Aphrodisiac	MH: Fertility & Impotence	Aphrodisiac: soak RO & drink w. Gardenia RO	Akinsoji 1996, NCF n.d.
								H: FS		MH: Dental care	MH: Oral health	Clean teeth: FS in FR	Akinsoji 1996
									H: I F ST	IH: Roofing	IH: Crafts & Skills	House roofing	Ayanbamiji 1996
									H: FO	IH: Construction	IH: Timbers & Carving	Construction	LI (KV-F), LI (KM-M)
									H: ST	IH: Boat carving	IH: Timbers & Carving	Carving boats	Ayanbamiji 1996
Burseraceae	*Boswellia dalzielii*	Frankincense tree	Hano, Arrarabi	Antakehi, Adakihu, Janauhi		S							
							H: LF			FH: Edible	FH: Edible	Food	LI (BD-F), LI (BD-M)
								H: BA		MH: Body pain	MH: Analgesic	Body pain: boil BA & potash & drink	Akinsoji 1996, NCF n.d., LI (BD-F), LI (BD-M)
								H: BA		MH: Blood tonic	MH: Blood	Lack of blood: boil BA	LI (CK-MA)
								H: BA		MH: Cholera	MH: Endemic & Epidemic diseases	Cholera: pound BA & drink	LI (CH-MH)
								H: BA		MH: Yellow fever	MH: Endemic & Epidemic Diseases	Yellow fever: pound BA & drink	LI (CH-MH)
								H: BA		MH: Fever	MH: Fevers	Feverish conditions: drink BA decoction w. potash	Akinsoji 1996
								H: BA		MH: Malaria	MH: Fevers	Malaria: pound BA & drink	LI (CH-MH)
								H: BA		MH: General health	MH: Health promotion	General health: pound BA & drink	LI (CH-MH)
								H: BA		MH: Boils	MH: Skin	Boils: grind BA (second layer) grind & put in tea	Akinsoji 1996, NCF n.d.
								H: BA		MH: Jedi-jedi	MH: Unknown	Jedi-jedi: boil BA & bathe patient	NCF n.d.
									H: WO	IH: Fence	IH: Timbers & Carving	Live fence in villages	Akinsoji 1996
Leguminosae - Caesalpinoideae	*Brachystegia eurycoma*		Wambo	Wambo	T	F LF							
							H: SE			FH: Seasoning	FA: Seasoning & Processing	Soup: SE processed into seasoning	Akinsoji 1996
							B: LY FL SE			FB: Edible	FB: Edible	Food	Warren p.c.
							H: SE			FH: Edible	FH: Edible	Food: ground SE important source of income: 300 - 500 Naira per modo	LI (MS-F)
								H: BA		MH: Cough	MH: Fevers	Cough: boil BA & drink	LI (MS-M)
								H: BA		MH: Dysentery	MH: Gastrointestinal	Dysentery: boil or soak BA, put water in basin & sit in to be absorbed via rectum	LI (MS-F)
								H: BA		MH: Worms	MH: Parasites	Worms: take 1 hf	Pellaumail 1998
								H: BA		MH: Skin conditions	MH: Skin	Skin problems: boil or soak BA & bathe baby	LI (MS-F)
									H: RO	IH: Rope	IH: Crafts & Skills	Rope: BA	Akinsoji 1996
Euphorbiaceae	*Bridelia spp.*		Kirni, Kizni, Kizmi, Maakuba	Mburum mburum, Marchi	TS								
							B: FR			FB: Edible	FB: Edible	Food	Warren p.c.
Euphorbiaceae	*Bridelia ferruginea*		ditto	ditto	TS							GGH specimen	

								H: BA		?	?	People use BA	LI (SB-F)
							B: FR			FB: Edible	FB: Edible	Food	Warren p.c. LI (GK-MA)
							C: FR			FC: Edible	FC: Edible	Food	LI (GK-MA)
							H: FR			FH: Edible	FH: Edible	Food	LI (SB-F), LI (GK-MA), LI (ST-MH), LI (SB-M)
								H: BA		MH: Eye problem	MH: Analgesic	Eye problem: pound BA, soak in water & wash face	LI (ST-MH)
								H: LF		MH: Earache	MH: Analgesic	Earache: mix LF w. dukkuje ladde LF, soak in water & use drops	LI (GK-F)
								H: BA		MH: Fever	MH: Fevers	Fever: BA infusion	Akinsoji 1996
								H: BA		MH: Constipation	MH: Gastrointestinal	Constipation in children: boil BA & use as enema	LI (GK-F)
								H: BA LF		MH: Diarrhoea	MH: Gastrointestinal	Diarrhoea: boil BA or LF & drink	LI (GK-MA), LI (GK-F)
								H: BA		MH: Dysentery	MH: Gastrointestinal	Dysentery: BA decoction w. potash	Akinsoji 1996
								H: BA		MH: Toothache	MH: Oral health	Mouthwash: boil BA for toothache	LI (GK-F), LI (SB-M)
								H: LF		MH: Mouthwash	MH: Oral health	Mouthwash: prepare liquid w. LF	Akinsoji 1996
									H: WP	IH: Dye	IH: Crafts & Skills	Decorating: process into paint	Akinsoji 1996
									H: BA	IH: Fish poison	IH: Hunting & Fishing	Make fish poison	LI (ST-MH)
Leguminosae - Caesalpinoideae	*Burkea africana*	Wild seriga	Bakin makarfo	Kokobi	T								
								H: BA		MH: Leprosy	MH: Endemic & Epidemic Diseases	Leprosy: boil BA w. potash	Akinsoji 1996
								H: BA		MH: Heartburn	MH: Gastrointestinal	Heartburn: BA decoction	Akinsoji 1996
								H: BA RO		MH: Heart attack	MH: Internal organs & Respiration	Heart attack: BA & RO	Ayanbamiji 1996
Palmae	*Calamus deeratus*	Rattan	Kwagiri	Kwagiri	T								
									H: ST	IH: Furniture	IH: Timbers & Carving	Furniture: use ST	Akinsoji 1996
Asclepiadaceae	*Calotropis procera*	Giant milk weed	Tumfafia, Tumfafiya	Papahi. Tumfafia, Babambel	S	S							
							H: LA			FH: Food processing	FA: Seasoning & Processing	Cheese making: add LA to cow milk to thicken	Akinsoji 1996
								H: LF		MH: Snake poison	MH: Antidotes & Repellents	Snake poison: use LF as antidote	Akinsoji 1996
								H: FR		MH: Mumps	MH: Endemic & Epidemic Diseases	Mumps: break FR at the affected part (mumse or hanoum [ill])	NCF n.d.
								H: RO		MH: Stomach problems	MH: Gastrointestinal	Stomach: cut & boil RO or soak in hot water. SOFT LF	LI (CH-M)
								H: LA		MH: Skin disease	MH: Skin	Skin disease: LA	Akinsoji 1996
									H: BA	IH: Rope	IH: Crafts & Skills	Rope: BA	Akinsoji 1996
									H: FS	IH: Stuffing	IH: Timbers & Carving	Stuffing for pillows: FS in FR	Akinsoji 1996
Burseraceae	*Canarium schweinfurthii*	African elemi	Atele, Atile	Atele, Atile	T	F							
							B: RO			FB: Edible	FB: Edible	Baboons dig out RO	LI (GK-MA)
							C: FR			FC: Edible	FC: Edible	Food	LI (GK-MA)
							H: ST			FH: Beverage	FH: Beverages	Make simi (local drink)	LI (GK MA)
							H: FR			FH: Edible	FH: Edible	Food: soak FR in water and eat	LI (GK-MA, LI (SB-M), LI (KM-F), LI (KM-M), LI (GK-F), LI (ST-MH)
								H: BA		MH: Dysentery	MH: Gastrointestinal	Dysentery: BA decoction	Akinsoji 1996
								H: BA		MH: Toothache	MH: Oral health	Toothache: boil BA w. potash & wash mouth	Akinsoji 1996, NCF n.d.

(continued)

Appendix (continued)

Family	Species	English	Hausa	Fulfulde	Habit	Habitat	Food	Medicine	Implement	Usage Description	Usage Category	Description in Original Source	Source
									H: SE	IH: Musical instrument	IH: Crafts & Skills	Shaker, Chakara: make net w. SE and fasten around calabash	LI (ST-MH)
									H: BA	IH: Juju	IH: Magico-religious	Drive away funny dreams: pound BA, dry & burn, mix w. other trees. Inhale smoke covered with a towel before betime	LI (GK-MA)
									H: LF	IH: Evil spirit	IH: Magico religious	Evil spirit: burn LF & inhale smoke	LI (ST-MH)
Cannabinaceae	*Cannabis sativa*	Marijuana			S								
								A: LF		MA: Prophylaxy	MA: Health promotion	Prophylaxy: grind LF, add water, give to chicken	Martin 1996
Caricaceae	*Carica papaya*	Pawpaw	Gwandan dukuje, Gwanda	Dukkuje	T	S							
							B: FR			FB: Edible	FB: Edible	Food	Warren p.c.
							H: FR			FH: Edible	FH: Edible	Edible: ripe FR	Akinsoji 1996
								H: LF		MH: Yellow fever	MH: Endemic & Epidemic Diseases	Yellow fever: boil LF w. other plants & drink	Akinsoji 1996, LI (SP-M)
								H: LF		MH: Labour	MH: Female reproduction	Prolonged labour: pound LF & mix w. water & drink	Akinsoji 1996, NCF n.d.
								H: FR		MH: Antibiotic	MH: Health promotion	Antibiotic: soak unripe FR in water	Akinsoji 1996
								H: LF		MH: Jaundice	MH: Internal organs & Respiration	Jaundice: boil LF & drink	NCF n.d.
Apocynaceae	*Carissa edulis*	Carisse	Bagozaki, Lemun tsuntsu										
							H: RO			FH: Seasoning	FH: Seasoning & Processing	Make honey potent & confer medicinal properties: pound RO & add to honey beer	Akinsoji 1996
Leguminosae - Caesalpinoideae	*Cassia mannii*			Bochimaibe, Sabimaibe	T	S LF							
								H: ?		MH: Stomach ache	MH: Gastrointestinal	Stomach ache	Martin 1996
								H: RO BA		MH: Stomach ache	MH: Gastrointestinal	Stomach ache: take 1-3 hf	Pollaumail 1998
Leguminosae - Caesalpinoideae	*Cassia sieberiana*	African laburnum	Gama fada	Malgahi	T								
								H: FP		MH: Purgative	MH: Gastrointestinal	Purgative: FP	Akinsoji 1996
								H: RO		MH: Dysentery	MH: Gastrointestinal	Dysentery: RO	Akinsoji 1996
								H: RO		MH: Haemorrhoids	MH: Gastrointestinal	Haemorroids: RO	Akinsoji 1996
Rhizophoraceae	*Cassipouree ugandensis*				T								
									C: TW LF TR	IC: Nest-tree	IC: Nest	<1% of all nest-trees	Adanu 1998
Bombacaceae	*Ceiba pentandra*	Silk cotton tree, Kapok	Rimi	Rantahi	T	S F							
								H: BA		MH: Toothache	MH: Oral health	Toothache: boil BA w. potash	Akinsoji 1996, NCF n.d.
								H: WO		MH: Skin conditions	MH: Skin	Red & black skin disease: people used to bathe in water that got in the canoe	LI (GK-MA)
									H: FL	IH: Stuffing	IH: Timbers & Carving	Stuffing for pillows & mattresses: FL in FR	Akinsoji 1996
									H: TR	IH: Canoe carving	IH: Timbers & Carving	Make canoes	LI (GK-MA)
Amaranthaceae	*Celosia trigyna*	Celosia											
							H: WP			FH: Edible	FH: Edible	FH: Pot herb	Akinsoji 1996
Ulmaceae	*Celtis philippensis**				T	F							
									C: TW LF TR	IC: Nest-tree	IC: Nest	1% of all nest-trees	Adanu 1998

Leguminosae - Caesalpinoideae	*Chamaecrista mimosoides*	Tea senna	Bagaruwar kasa	Gawri lesdi	H		A: WP			FA: Edible	FA: Edible	Fodder for goats	Ayanbamiji 1996
									H: BA	IH: Dye	IH: Crafts & Skills	Leather dye: BA	Ayanbamiji 1996
Menispermaceae	*Cissampelos owariensis*	Velvet leaf	Damar gwaraji, Fivaka	Magirahi	L							GGH specimen	
								H: LF		MH: Diarrhoea	MH: Gastrointestinal	Diarrhoea: LF	Ayanbamiji 1996
Ampelidaceae	*Cissus quadrangularis*		Gadali	Sembal									
								H: ST		MH: Cuts	MH: Skin	Cuts: mix ST w. kerozene & ash & apply	LI (GK-F)
								H: ST		MH: Wounds	MH: Skin	Wounds: burn the stem over fire & squeeze the juice on wounds to heal & stop bleeding	LI (GK-MA)
Rutaceae	*Citrus spp.*	Orange	Lemu		T	S							
								H: FJ		MH: Jaundice	MH: Internal organs & Respiration	Jaundice: drink	NCF n.d.
Rutaceae	*Citrus aurantifolia*	Lime	ditto		T	S							
								H: FD		MH: Insecticide	MH: Antidotes & Repellents	Insecticide: FD	Akinsoji 1996
								H: FJ		MH: Cough	MH: Fevers	Component of cough mixture: FJ	Akinsoji 1996
Annonaceae	*Cleistopholis patens*	Salt and oil tree			T								
									C: TW LF TR	IC: Nest-tree	IC: Nest	20% of all nest-trees	Adanu 1998
Capparidaceae	*Cleome viscosa*	Wild mustard		Kinasti	C		H: WP			FH: Edible	FH: Edible	FH: Pot herb	Akinsoji 1996
Cochlosperma-ceae	*Cochlospermum planchonii*	Cotton plant	Sunzuma, Balge, Rawaya	Ambulolowol	S	S							
								H: RO		MH: Yellow fever	MH: Endemic & Epidemic Diseases	Yellow fever: grind RO into powder & add to porridge or mixed w. lemon juice	Akinsoji 1996, NCF n.d.
									H: BA	IH: Rope	IH: Crafts & Skills	Rope: BA	Akinsoji 1996
Sterculiaceae	*Cola ?*		Ataras	Ataras	T								
							C: FR			FC: Edible	FC: Edible	Food	LI (GK-MA)
								H: PO		MH: Menstrual problems	MH: Female reproduction	Menstrual problems	LI (GK-F)
								H: FR		MH: Stimulant	MH: Health promotion	Stimulant: eat FR	LI (SB-M), LI (GK-MA)
Sterculiaceae	*Cola caricifolia*	Fig-leaved kola, Monkey kola			ST								
									C: TW LF TR	IC: Nest-tree	IC: Nest	2% of all nest-trees	Adanu 1998
Sterculiaceae	*Cola gigantea*		Bokoko		ST								
							A: FR			FA: Edible	FA: Edible	Food: FR	Ayanbamiji 1996
							C: SE			FC: Edible	FC: Edible	Food: SE in PU. Rainy season	LI (GK-MA)
							H: FR			FH: Edible	FH: Edible	Food: FR	Ayanbamiji 1996
									C: TW LF TR	IC: Nest-tree	IC: Nest	Occasional nest-tree	LI (GK-MA)
Sterculiaceae	*Cola hispida*		Kwalatam, Bunsuru		T	F S							
									H: LF BA	IH: Evil spirits	IH: Magico-religious	Drive witchcraft & evil spirit away: grind BA & LF & smoke the room	Akinsoji 1996, NCF n.d.

(continued)

Appendix (continued)

Family	Species	English	Hausa	Fulfulde	Habit	Habitat	Food	Medicine	Implement	Usage Description	Usage Category	Description in Original Source	Source
Sterculiaceae	*Cola millenii*	Monkey kola		Bo.ohi	ST								
							B: FR			FB: Edible	FB: Edible	Food	Warren p.c., LI (GK-MA)
							C: FR			FC: Edible	FC: Edible	Food	LI (GK-MA)
							C: PU			FC: Edible	FC: Edible	Food, rainy season	LI (GK-MA)
							I: FR			FI: Edible	FI: Edible	Food: sour FR	LI (GK-MA)
									C: TW I F TR	IC: Nest-tree	IC: Nest	1% of all nest-trees	Adanu 1998
									H: ST	IH: Rope	IH: Crafts & Skills	Rope: sucklings	LI (GK-MA)
Sterculiaceae	*Cola verticillata*	Kola nut	Goro, Hanunruwa	Goro	T								
								H: PO		MH: Post-partum	MH: Female reproduction	Cleansing: soak PO in cold water & drink after birth to remove blood	LI (GK-F)
								H: PO		MH: Labour	MH: Female reproduction	Faciliate labour: soak PO in cold water & give mother the day she's expecting	LI (GK-FH)
								H: FR		MH: Antiemetic	MH: Gastrointestinal	To help keep food in stomach: 1 nut only.	LI (S3-F)
								H: FR		MH: Stimulant	MH: Health promotion	Stimulant: chew	LI (S3-M)
									C: TW LF TR	IC: Nest-tree	IC: Nest	5% of all nest-trees	Adanu 1998
Combretaceae	*Combretum ghasalense*		Bakin taramniya	Buski, Bodehi									
									H: WO	IH: Firewood	IH: Firewood	Firewood	Akinsoji 1996
Combretaceae	*Combretum molle*		Wuyan damo	Damoruhi		S							
								H: RO		MH: Body pain	MH: Analgesic	Body pain: soak RO & drink	NCF n.d.
								I: I F		MI: Intestinal infections	MI: Antidotes & Repellents	Intestinal infections: I F decoction	Akinsoji 1996
								H: RO		MH: Weakness	MH: Health promotion	Weakness: soak RO & drink	NCF n.d.
Combretaceae	*Combretum nigricans*		Chiriri, Ciriri	Buski									
									H: WO	IH: Firewood	IH: Firewood	Firewood	Akinsoji 1996
									I: WO	II: Construction	II: Timbers & Carving	Construction	Akinsoji 1996
Commelinaceae	*Commelina erecta*		Balasar-dawaki, Balasa										
									I: WP	II: Soil fertility	II: Soil conditioning	Soil fertility: green manure	Ayanbamiji 1996
Burseraceae	*Commiphora spp.*		Bozana*, Hana gobara, Barumagada	Kaoiwal, Kaobije	T								
							A: LF			FA: Edible	FA: Edible	Fodder: LF	Ayanbamiji 1996
								H: LF BA		MH: Toothache	MH: Oral health	Toothache: LF, BA boiled	Ayanbamiji 1996
									H: ST	IH: Fence	IH: Timbers & Carving	Fencing: ST	Ayanbamiji 1996
Burseraceae	*Commiphora kerstingii*		ditto	ditto	T								
							A: WP			FA: Edible	FA: Edible	Fodder for goats: LF	Akinsoji 1996
								H: I F		MH: Fever	MH: Fevers	Fever: I F decoction	Akinsoji 1996
									I: WO	II: Fence	II: Timbers & Carving	Fence: trees	Akinsoji 1996
Tiliaceae	*Corchorus olitorius*	Bush okra	Lalo	Lalo, Botoko	C								
							H: LF			FH: Edible	FH: Edible	Food	LI (KM-F), LI (KM-M), LI (GK-MA)

							H: WP			FH: Edible	FH: Edible	FH: Pot hero	Akinsoji 1996
Boraginaceae	*Cordia africana*		Aliliba	Lilbani	T	S							
							H: FP			FH: Edible	FH: Edible	Food: add FP to porridge	Akinsoji 1996
								H: LF		MH: Whooping cough	MH: Endemic & Epidemic Diseases	Whooping cough: mix LF & local tobacco & smoke	Akinsoji 1996, NCF n.d.
Zingiberaceae	*Costus afer*	Ginger lily, Bush cane		Lamokoe, Kakizuwa	H							GGH specimen	
							B: FL ST			FB: Edible	FB: Edible	Food	Warren p.c.
								H: ST		MH: Cough	MH: Fevers	Cough: boil ST w. potash & drink decoction	Akinsoji 1996
Zingiberaceae	*Costus spectabilis*				H								
							B: RO			FB: Edible	FB: Edible	Food	Warren p.c.
Leguminosae - Papilionoideae	*Craibia atlantica*				T								
									C: TW LF TR	IC: Nest-tree	IC: Nest	15% of all nest-trees	Adanu 1998
Amaryllidaceae	*Crinum zeylanicum*	Beautiful crinum	Albasar kwandi	Tiniere, Burru*, Albace dawadi									
								H: FR		MH: Skin infections	MH: Skin	Skin infections: pound FR w. red potash and apply mixture to skin	Akinsoji 1996
Rubiaceae	*Crossopteryx febrifuga*	African bark	Kasfiya, Kashin awaki	Rima jogohi, Rimajoga	T								
								A: BA		MA: Fertility	MA: Fertility & Genitalia	Increase fecundity: crush BA & mix w. corn	Ayanbamiji 1996
									H: WO	IH: Slates	IH: Crafts & Skills	Wooden slates for Koranic schools	Akinsoji 1996
									H: ST	IH: Implements	IH: Timbers & Carving	ST for mortar and pestle	Ayanbamiji 1996
Euphorbiaceae	*Croton macrostachyus*				T	S							
									H: WO	IH: Construction	IH: Timbers & Carving	Used in construction	Akinsoji 1996
Cucurbitaceae	*Cucurbita pepo*	Pumpkin	Kabewa	Vaigore									
							B: FR			FB: Edible	FB: Edible	Food	LI (GK-MA)
							H: LF FR SE			FH: Edible	FH: Edible	Food: LF & FR edible, SE for soup	Ayanbamiji 1996, LI (GK-MA), LI (CH-M)
Hypoxidaceae	*Curculigo pilosa*	African crocus		Bir ji jire	H								
							B: ST			FB: Edible	FH: Edible	Food	Warren p.c.
								H: TU		MH: Purgative	MH: Gastrointestinal	Purgative: TU	Akinsoji 1996
Araliaceae	*Cussonia arborea*		Takanda giwa, Gwabsa	Burnarlahi	T								
							B: FR			FB: Edible	FB: Edible	Food	Warren p.c.
								H: RO		MH: Colds	MH: Fevers	Colds: drink RO decoction	Akinsoji 1996
								H: LF		MH: Purgative	MH: Gastrointestinal	Purgative: drink LF decoction	Akinsoji 1996
								H: BA		MH: Skin conditions	MH: Skin	Black skin disease (e.g., itchy skin): dry BA & drink or make soup or bathe	LI (FI-M)
Cyperaceae	*Cyperus denudatus**		Gajiji*	Goyemanjo*									
								H: ST		MH: Headache	MH: Analgesic	Headache: boil rhizome & drink	Akinsoji 1996

(continued)

Appendix (continued)

Family	Species	English	Hausa	Fulfulde	Habit	Habitat	Food	Medicine	Implement	Usage Description	Usage Category	Description in Original Source	Source
Cyperaceae	*Cyperus esculentus*	Tiger nut	Arigiza	Ayaare									
							H: TU			FH: Edible	FH: Edible	Food: TU (tiger nuts)	Akinsoji 1996
Leguminosae - Caesalpinoideae	*Daniellia oliveri*	West African copal	Maje	Kaharlahi, Karlahi	T	S							
							B: LY FL SE NE			FB: Edible	FB: Edible	Food	Warren p.c.
							C: FR			FC: Edible	FC: Edible	Food	LI (GK-MA)
								A: BA		MA: Haemorrhagic septicaemia	MA: Blood	Haemorrhagic septicaemia: concoct BA w. sumsum, sonve & bumehi	Martin 1996
								A: ?		MA: Henore	MA: Health promotion	Healthy calves. henore	Martin 1996
								A: BA		MA: Lick	MA: Health promotion	Lick: pound BA & mix w. limestone or salt or dilute in water & drink	Martin 1996
								A: ?		MA: Heart problems	MA: Internal organs & Respiration	Heart problems	Martin 1996
								A, H: ?		MA, MH: Worms	MA, MH: Parasites	Worms	Martin 1996
								H: BA		MH: Colds	MH: Fevers	Colds: boil BA & drink	LI (GK-MA)
								H: ?		MH: Stomach problems	MH: Gastrointestinal	Stomach	Martin 1996
								H: ?		MH: Heart problems	MH: Internal organs & Respiration	Heart problems	Martin 1996
								H: BA		MH: Skin disease	MH: Skin	Skin diseases (infants esp.): 1 bowl	Pellaumail 1998
									H: LA	IH: Starch for clothes	IH: Crafts & Skills	Starch: use LA for clothes	Ahmadu
									H: WO	IH: Firewood	IH: Firewood	Fuel: WO	Akinsoji 1996
									H: BA	IH: Evil spirits	IH: Magico religious	Evil spirits: boil BA boil & drink	Akinsoji 1996, NCF n.d.
									H: BA	IH: Implements	IH: Timbers & Carving	Make mortar & internal construction: WO	Akinsoji 1996
Leguminosae - Papilionoideae	*Desmodium gangeticum (velutinum ?)*		Takamahi, Damgere, Danka dafi	Takamahi									
								C: LF		MC: Worms	MC: Parasites	Worms	LI (GK-MA)
								H: RO		MH: ?	MH: ?	People dig the RO but we do not know what for	LI (GK-MA)
								H: LF		MH: Headache w. nosebleed	MH: Analgesic	Nosebleed: pound LF w. beans. pour into nostrils. Bleeding stops after 20 minutes	LI (ST-MH)
								H: LF		MH: Post-partum	MH: Female reproduction	Cleanse blood after labour: boil LF & drink	LI (ST-MH)
								H: LF		MH: Lactation	MH: Female reproduction	Infected mother's milk: mix LF w. mburum mburum, boil & give mother to drink	LI (GK-F)
								H: LF		MH: Menstrual problems	MH: Female reproduction	Heavy menstrual flow: boil LF & drink	LI (ST-MH)
								H: LF		MH: Haemorrhoids	MH: Gastrointestinal	Haemorrhoids: pound LF, mix w. vaseline or oil & rub on anus for 3 days	LI (ST-MH)
								H: RO		MH: Dysentery	MH: Gastrointestinal	Dysentery: boil RO w. limestone. Drink once & no more than 3 times	LI (ST-MH)
								H: RO		MH: Strength	MH: Health promotion	Strength: boil RO & give to children to become strong	LI (CH-F)
								H: LF		MH: Worms	MH: Parasites	Threadworm: squeeze LF & drink	LI (ST-MH)
								H: LF		MH: Worms	MH: Parasites	Worms: pound LF add water & use as enema	LI (ST-MH)
								H: RO		MH: Rashes	MH: Skin	Rashes: boil RO & drink	LI (ST-MH)
Leguminosae - Caesalpinoideae	*Detarium microcarpum*	Tallow tree	Taura	Konkehi	T	S							
							B: FR			FB: Edible	FB: Edible	Food	LI (GK-MA)

					B: SE			FB: Edible	FB: Edible	Food	Warren p.c.
					H: FR			FH: Edible	FH: Edible	Food	Akinsoji 1996, Ayanbamiji 1996, LI (SB-M), LI (GK-M)
						H: BA		MH: ?	MH: ?	People use BA	LI (GK-MA)
						H: RO		MH: ?	MH: ?	People use RO	LI (GK-MA)
						H: BA RO		MH: Painful to urinate	MH: Fertility & Genitalia	Urination: boil BA & RO & drink when it is painful to urinate	LI (ST-MH)
						H: BA		MH: Fever	MH: Fevers	Fever: mix BA w. BA of marike & kiriya. boil & bathe. Can also drink	LI (GK-F)
						H: ?		MH: Dysentery	MH: Gastrointestinal	Dysentery	NCF n.d.
						H: BA		MH: Diarrhoea	MH: Gastrointestinal	Dysentery: mix BA w. BA of marike & kiriya. boil & bathe for fever. Can also drink	LI (GK-F)
						H: BA		MH: Diarrhoea	MH: Gastrointestinal	Diarrhoea: mix w. galbije, boil BA & bathe child or use as enema for kurka. Mother can drink as well.	LI (DM-M)
						H: BA		MH: Dysentery	MH: Gastrointestinal	Dysentery: BA is boiled w. potash & decoction drunk	Akinsoji 1996
						H: BA		MH: Haemorrhoids	MH: Gastrointestinal	Pile in children: boil BA w. potash & drink	Akinsoji 1996, NCF n.d.
						H: BA		MH: Heartburn	MH: Gastrointestinal	Heartburn: BA	Ayanbamiji 1996
						H: BA		MH: Stomach problems	MH: Gastrointestinal	Stomach: mix BA w. kojoli BA & bathe children or give to drink	LI (GK-F)
						H: BA		MH: Stomach ache	MH: Gastrointestinal	Stomach ache: 1-2 hf	Pellaumail 1998
						H: BA		MH: Stomach ache	MH: Gastrointestinal	Stomach ache: boil BA & give to children	LI (GK-F)
						H: LF		MH: Dysentery	MH: Gastrointestinal	Dysentery: mix w. LF karkandam, galbije & barkehi, boil & use as enema	LI (GK-F)
						H: SE		MH: Stomach ache	MH: Gastrointestinal	Stomach ache: SE	Ayanbamiji 1996
						H: BA TW		MH: Disease prevention	MH: Health promotion	Prevent disease: boil BA & TW & bathe children or give as a drink	LI (GK-M)
						H: ?		MH: Skin conditions	MH: Skin	Red skin disease (e.g. cut finger, from cattle, spider)	LI (FI-M)
						H: BA		MH: Skin disease	MH: Skin	Skin disease: 1-3 hf - 1/2 basin (combined w. kohal, for bathing infants)	Pellaumail 1998
						H: BA		MH: Skin disease	MH: Skin	Skin disease: boil BA & drink	LI (S3-M)
						H: BA		MH: Abonakeke	MH: Unknown	Abonakeke = pain in joints?: boil BA & BA of kiriya & LF of jan yaro & use as enema or drink; 3 hf for boy, 4 hf for girl, 4 hf for mother who breastfeeds for 7 days	LI (GK-F), LI (GK-M)
							H: WO	IH: Firewood	IH: Firewood	Firewood	Ayanbamiji 1996
Leguminosae - Caesalpinoideae	*Detarium senegalense*	Abondo		T							
					B: FR			FB: Edible	FB: Edible	Food	LI (GK-MA)
					B: SE			FB: Edible	FB: Edible	Food	Warren p.c.
					C: FR			FC: Edible	FC: Edible	Food	LI (GK-MA)
					H: SE			FH: Edible	FH: Edible	Soup: SE	LI (GK-MA)
Leguminosae - Caesalpinoideae	*Dialium spp. (Strephonema manii ?)*	Kom	Kom	T							
					B: FR			FB: Edible	FB: Edible	Food	Warren p.c.
					C: FR			FC: Edible	FC: Edible	Food	LI (GK-MA)
							H: TR	IH: Construction	IH: Timbers & Carving	Construction	LI (S3-M)

(continued)

Appendix (continued)

Family	Species	English	Hausa	Fulfulde	Habit	Habitat	Food	Medicine	Implement	Usage Description	Usage Category	Description in Original Source	Source
Leguminosae - Mimosoideae	*Dichrostachys cinerea*	Sickle bush, Marabou thorn	Dundu	Burli, Burle	T S	S							
							H: FR			FH: Edible	FH: Edible	Food	LI (BD-F)
								H: LF		MH: Dysentery	MH: Gastrointestinal	Dysentery: pound LF & add to any beverage	Akinsoji 1996, NCF n.d.
									H: SE	IH: Implements	IH: Timbers & Carving	Knives: SE	NCF n.d.
									H: WO	IH: Construction	IH: Timbers & Carving	Internal construction: WO	Akinsoji 1996
Dioscoreaceae	*Dioscorea spp.*	Bush yam	Doyar daji	Bulumje ladde	L								
							H: TU			FH: Edible	FH: Edible	Alternative to crop	LI (GK-F)
Ebenacae	*Diospyros canaliculata*	Flint bark	Mechowu		T								
									C: TW LF TR	IC: Nest-tree	IC: Nest	2% of all nest-trees	Adanu 1998
Ebenaceae	*Diospyros mespiliformis*	West African ebony	Kanya, Kaiwa	Nelbi	T								
							B: FL FR			FB: Edible	FB: Edible	Food	Warren p.c.
							C: FR			FC: Edible	FC: Edible	Food	Fowler p.c.
							H: FP			FH: Beverage	FH: Beverages	Make fermented beverage: FP	Akinsoji 1996
							H: FR			FH: Edible	FH: Edible	Food	LI (S3-M), LI (ST-MH)
								H: RO		MH: Strengthen penis	MH: Fertility & Impotence	Virility: cook RO w. wood ash to strengthen penis. Like da Kamerun, you cannot sleep at night	LI (FI-M)
								H: BA		MH: Worms	MH: Parasites	Worms: boil BA & drink 2 days, 3 times daily	LI (ST-MH)
Ebenacae	*Diospyros piscatoria*				T	LF							
									C: TW LF TR	IC: Nest-tree	IC: Nest	1% of all nest-trees	Adanu 1998
Dracaenaceae	*Dracaena arborea*					F							
									H: WO	IH: Fence	IH: Timbers & Carving	Fencing	Akinsoji 1996
Palmae	*Elaeis guineensis*	Oil palm	Kwa kwa	Darle	T	S VF							
							B: FR NU			FB: Edible	FB: Edible	Food	Warren p.c.
							H: LA			FH: Palm wire	FH: Beverages	Palm wine: tree	Akinsoji 1996
							H: FR			FH: Edible	FH: Edible	Palm oil: FR	Akinsoji 1996
								H: RO		MH: Stomach ache	MH: Gastrointestinal	Stomach ache: 4-5 RO	Pellaurrail 1998
								II: SE		MI: General health	MI: Health promotion	SE palm kernel oil known to have medicinal properties	Akinsoji 1996
								H: WO		MH: Chest pain	MH: Internal organs & Respiration	Chest pain mostly smokers: take ash of the burnt WO through soap & wash chest	NCF n.d.
									H: LF	IH: Broom	IH: Crafts & Skills	Broom: leaflet	Akinsoji 1996
									H: FO	IH: Sheds	IH: Timbers & Carving	Sheds: FO	Akinsoji 1996
Gramineae	*Eleusine indica*	Indian goose grass	Ciyawar tuji	Sargade, Tuji	G								
									H: WP	IH: Mat	IH: Crafts & Skills	Mat for table (palpala)	Ayanbamiji 1996
Musaceae	*Ensete gilletii*	Bush banana	Ayaban daji		H								
							H: FR			FH: Edible	FH: Edible	FR Edible	Akinsoji 1996
								H: RO		MH: Jaundice	MH: Internal organs & Respiration	Jaundice: boil RO w. potash	NCF n.d.

Leguminosae - Mimosoideae	*Entada africana*		Tawatsa, Tawarsa	Pade wan-duhi, Fadu-wanduhi, Peluwahi	T	S							
							H: LF			FH: Food processing	FA: Seasoning & Processing	Haste ripening: mix LF w. unripe banana in air-tight bag	Akinsoji 1996
								A: BA		MA: Cough	MA: Fevers	Animal coughs: administer BA filtrate through the nose	Martin 1996
								H: BA		MH: Dysmenorrhoea	MH: Female reproduction	Dysmennorhoea: pound BA & mix w. yoghurt	Martin 1996
								H: RO		MH: Gonorrhoea	MH: Fertility & Genitalia	Gonorrhoea: RO decoction	Akinsoji 1996
								H: ?		MH: Diarrhoea	MH: Gastrointestinal	Diarrhoea	Martin 1996
								H: RO		MH: Diarrhoea	MH: Gastrointestinal	Diarrhoea: boil RO & drink	NCF n.d.
									H: BA	IH: Fish poison	IH: Hunting & Fishing	Fish poison: BA	Akinsoji 1996
Sapindaceae	*Eriocoelum kerstingii*				T								
									C: TW LF TR	IC: Nest-tree	IC: Nest	5% of all nest-trees	Adanu 1998
Leguminosae - Papilionoideae	*Erythrine senegalensis*	Yellow fever tree, Coral tree	Majiriya, Minjiriya	Wibonolong*	T								
							B: SE			FB: Edible	FB: Edible	Food	Warren p.c.
								H: LF		MH: Fever	MH: Fevers	Fever: boil LF	LI (KM-F), LI (KM-M)
								H: BA		MH: Jaundice	MH: Internal organs & Respiration	Jaundice: boil BA w. potash & drink	Akinsoji 1996
								H: LF		MH: Jaundice	MH: Internal organs & Respiration	Jaundice: boil LF	LI (KM-F), LI (KM-M)
									H: WO	IH: Fence	IH: Timbers & Carving	Live fence: yard demarcation	Akinsoji 1996
Leguminosae - Papilionoideae	*Erythrina sigmoidea*			Borodong, Burdenahi	T								
								H: BA		MH: Yellow fever	MH: Endemic & Epidemic Diseases	Yellow fever: boil BA w. potash	Akinsoji 1996
Leguminosae - Caesalpinoideae	*Erythrophleum suaveolens*	Sasswood, Poison wood tree	Gwaska	Mber	T	F							
							B: SE			FB: Edible	FB: Edible	Food	Warren p.c.
								H: BA		MH: Body pain	MH: Analgesic	Body pain: soak BA & use as enema or drink	LI (GK-FH)
								H: BA		MH: Stomach ache	MH: Gastrointestinal	Pound BA, add water & drink for stomach ache	LI (DM-M)
								H: BA		MH: Chest pain	MH: Internal organs & Respiration	Chest pain: pound BA, add water & drink	LI (DM-M)
								H: BA		MH: Seizures	MH: Psychiatric Diseases	Seizures: grind BA & take VERY little	LI (GK-MA)
								H: WO		MH: Skin conditions	MH: Skin	Red & black skin disease: people used to bathe in water that got in the canoe	LI (GK-MA)
									C: TW LF TR	IC: Nest-tree	IC: Nest	1% of all nest-trees	Adanu 1998
									H: BA	IH: Fish poison	IH: Hunting & Fishing	Fish poison: grind BA & put in water	Akinsoji 1996, NCF n.d.
									H: BA	IH: Witchcraft	IH: Magico-religious	Witchcraft (maye): boil BA, add cold water & bathe person	LI (GK-F)
									H: TR	IH: Canoe carving	IH: Timbers & Carving	Carve canoes	LI (GK-MA)

(continued)

Appendix (continued)

Family	Species	English	Hausa	Fulfulde	Habit	Habitat	Food	Medicine	Implement	Usage Description	Usage Category	Description in Original Source	Source
Euphorbiaceae	*Euphorbia hirta*	Asthma herb	Nonon karkiya, Nonnon guriya	Kossam poli, Eudamyol	H	S						GGH specimen	
								H: LA		MH: Scorpion bite	MH: Antidotes & Repellents	Scorpion bite antidote: LA	Akinsoji 1996
								H: LF		MH: Scorpion bite	MH: Antidotes & Repellents	Scorpion bite: LF crushed & rubbed on bite	Ayarbamij 1996
								H: WP		MH: Scorpion bite	MH: Antidotes & Repellents	Scorpion sting: rub plant on scorpion sting or pound WP & rub sting	NCF n.d., LI (GK-MA)
								H: WP		MH: Dislocation	MH: Bones	Dislocation: tie plant w. palm oil	Ayarbamij 1996
								LI: LF		MLI: Skin disease	MLI: Skin	Skin diseases: squeeze LF & rub juice on skin	Akinsoji 1996
Euphorbiaceae	*Euphorbia kamerunika*	Fence cactus	Kerana	Kuldehi	T								
									H: WO	IH: Fence	IH: Timbers & Carving	Fencing	Akinsoji 1996
Moraceae	*Ficus spp.*		Baura, Cediya, Chediya	Litahi, Yibal, Yibe, Ibal, Ib	T								
							B: FR			FB: Edible	FB: Edible	Food	Warren p.c.
							C: FR			FC: Edible	FC: Edible	Food	Fowler p.c., LI (GK-MA)
							H: FR			FH: Edible	FH: Edible	Food: FR	Ayarbamij 1996
							LI: FR LF			FLI: Edible	FLI: Edible	Vegetable soup w. LF, also w. fish	LI (GK-MA)
								A: ?		MA: Productivity	MA: Female reproduction	Increase productivity	Martin 1996
								H: LF		MH: Post-partum	MH: Female reproduction	Labour: bathe woman w. LF	Ayarbamij 1996
								H: RO		MH: Cough	MH: Fevers	Cough & difficulty in breathing: cut RO & drink the juice that comes out	LI (GK-MA)
								H: BA		MH: Diarrhoea	MH: Gastrointestinal	Diarrhoea: boil BA & drink	LI (GK-MA)
								H: RO		MH: Tuberculosis	MH: Internal organs & Respiration	Tuberculosis: cut RO & use to collect water that drips	Ayarbamij 1996
									H: BA	IH: Rope	IH: Crafts & Skills	Rope: BA	Ayarbamij 1996
Moraceae	*Ficus exasperata*	Sandpaper tree		Nyamfargo	T								
							B: LY			FB: Edible	FB: Edible	Food	Warren p.c.
Moraceae	*Ficus iteophylla*		Shiriya	Sekehi, Boodehi	T	S I F							
								H: BA LF		MH: Pain	MH: Analgesic	Pain: 1/2 basin (for bathing)	Pellaumail 1998
								H: BA LF		MH: Stomach ache	MH: Gastrointestinal	Stomache ache: 1/4 modo	Pellaumail 1998
								LI: LF		MLI: Stool greenish & watery	MLI: Gastrointestinal	Greenish & watery stool in children: boil LF, drink & bathe	Akinsoji 1996, NCF n.d.
								H: BA LF		MH: Toothache	MH: Oral health	Toothache: 1/4 modo	Pellaumail 1998
								H: BA LF		MH: Worms	MH: Parasites	Worms: 1/2 basin (for bathing)	Pellaumail 1998
Moraceae	*Ficus natalensis*	Bark-cloth tree	Cediya, Chediya	Biskehi, Biskoje	T								
							B: FR			FB: Edible	FB: Edible	Food	Warren p.c.
Moraceae	*Ficus ovata*	Fig tree	Camji	Dundehi	T								
							B: FR			FB: Edible	FB: Edible	Food	Warren p.c.
							LI: FR			FLI: Edible	FLI: Edible	Food	LI (FI-F)

Moraceae	*Ficus polita ?*		Durumi	Durmihi	T								
							B: FR			FB: Edible	FB: Edible	Food	Warren p.c.
Moraceae	*Ficus sur*	Bush fig	Farin baushe, Uwar yara, Dulu	Rima bichehi	T	F S							
							A: FR			FA: Edible	FA: Edible	Food	LI (CH-M), LI (SB-M)
							B: FR			FB: Edible	FB: Edible	Food	Warren p.c.
							H: FR			FH: Edible	FH: Edible	Food	LI (SB-M)
								A: RO		MA: Worms	MA: Parasites	Worms: boil RO w. potash & drink	NCF n.d.
								H: PI		MH: Colds	MH: Fevers	Colds: give children to drink juice of SI	LI (CH-M)
								H: RO		MH: Worms	MH: Parasites	Worms: boil RO w. potash & drink decoction	Akinsoji 1996, NCF n.d.
									A: BA	IA: Leadership	IA: Magico-religious	Increase popularity: cut BA, infuse in cold water	Martin 1996
									A: BA SE	IA: Leadership	IA: Magico-religious	Leadership: pound BA & SE, infuse & give to bull, will lead the herd	Martin 1996
Moraceae	*Ficus sycomorus*	Sycamore fig		Yibal, Yibe, Ibal, Ibi	T							GGH specimen	
							B: FR			FB: Edible	FB: Edible	Food	Warren p.c.
								H: LA		MH: Cough	MH: Fevers	Cough: LA obtained from RO	Akinsoji 1996
Moraceae	*Ficus thonningii*	Wild fig	Cediya, Chediva	Biskehi, Biskele	T								
							A: LF			FA: Edible	FA: Edible	Fodder for goats: LF	Akinsoji 1996
							H: FR			FH: Edible	FH: Edible	Edible: FR	Akinsoji 1996
								H: LF		MH: Yellow fever	MH: Endemic & Epidemic Diseases	Yellow fever: drink 1 cup of LF, also soak LF & bathe	LI (SB-F), LI (GK-F)
								H: BA		MH: Pregnancy	MH: Female reproduction	Tummy ache: boil BA & give after 6th month of pregnancy	LI (GK-F)
								H: LF		MH: Colds	MH: Fevers	Colds: pound & boil LF and bathe children	LI (CH-F)
								H: RO		MH: Dysentery	MH: Gastrointestinal	Dysentery	LI (MS-F)
								H: RO		MH: Stomach ache	MH: Gastrointestinal	Stomach ache	LI (MS-F)
									H: WP	IH: Shade tree	IH: Timbers & Carving	Shade tree	Akinsoji 1996
Moraceae	*Ficus vallis-choudae*	Fig tree	Dulu	Yibe danejum	T								
							B: FR			FB: Edible	FB: Edible	Food	Warren p.c.
Flacourtiaceae	*Flacourtia flavescens*	Niger plum		Sofare*	ST								
									H: WO	IH: Implements	IH: Timbers & Carving	Implements	Akinsoji 1996
Guttiferae	*Garcinia afzelii*	Bitter kola			T								
									C: TW LF TR	IC: Nest-tree	IC: Nest	3% of all nest-trees	Adanu 1998
Rubiaceae	*Gardenia erubescens*		Gaude	Dingale	ST								
								H: RO		MH: Gonorrhoea	MH: Fertility & Genitalia	Gonorrhoea: boil RO w. potash & drink decoction	Akinsoji 1996, NCF n.d.
									H: FR	IH: Cosmetic	IH: Hygiene & Cosmetics	Cosmetic: FR used by women	Akinsoji 1996
									H: WO	IH: Implements	IH: Timbers & Carving	Farm instruments: WO	Akinsoji 1996
Rubiaceae	*Gardenia nitida*		Gagara giwa	Dingale	ST								
							B: FR NU			FB: Edible	FB: Edible	Food	Warren p.c.
								H: RO		MH: Strengthen penis	MH: Fertility & Impotence	Strengthen penis	LI (BD-F)
								H: PI		MH: Strength	MH: Health promotion	Strength: boil PI & give to babies	LI (GK-FH)

(continued)

Appendix (continued)

Family	Species	English	Hausa	Fulfulde	Habit	Habitat	Food	Medicine	Implement	Usage Description	Usage Category	Description in Original Source	Source
Rubiaceae	*Gardenia triacantha*		Gauden kuura	Dinyali gori		S							
								A: ?		MA: Aphrodisiac	MA: Fertility & Genitalia	Aphrodisiac: mix w. kahi & cook	Martin 1996
								H: RO BA		MH: Yellow fever	MH: Endemic & Epidemic Diseases	Yellow fever: 2 hf	Pellaumail 1998
								H: RO BA		MH: Testicular problem	MH: Fertility & Genitalia	Testicular problem: 2 hf	Pellaumail 1998
								H: BA		MH: Stomach ache	MH: Gastrointestinal	Stomach ache: 1-2 hf	Pellaumail 1998
Iridaceae	*Gladiolus primulinus**				H								
							H: ?			FH: Edible	FH: Edible	Food, corm	Akinsoji 1996
								H: BU		MH: General health	MH: Health promotion	Medical properties: corm	Akinsoji 1996
Tiliaceae	*Grewia mollis*		Dargaza	Bori	TS								
							B: FR			FB: Edible	FB: Edible	Food	Warren p.c.
							H: BA			FH: Edible	FH: Edible	Edible: soak inner BA in water & cook into (jibawa) soup. Also, add juice to kose (bean) cakes	Akinsoji 1996, LI (GK-F), LI (SB-M)
							H: FR			FH: Edible	FH: Edible	Food	LI (BD-F), LI (BD-M)
							H: SE			FH: Edible	FH: Edible	Edible: SE to bake cake & to make soup	LI (KM-F), LI (KM-M)
							H: SE BA			FH: Edible	FH: Edible	SE edible, BA for draw soup	Ayanbamiji 1996
								H: BA		MH: Dysentery	MH: Gastrointestinal	Dysentery: BA	Ayanbamiji 1996
								H: BA		MH: Stomach ache	MH: Gastrointestinal	Stomach pain: BA	Ayanbamiji 1996
								H: FR		MH: Stomach problems	MH: Gastrointestinal	Stomach ache	LI (BD-F), LI (BD-M)
									H: ?	IH: Rope	IH: Crafts & Skills	Rope	LI (BD-F), LI (BD-M)
									H: WO	IH: Firewood	IH: Firewood	Fuelwood	Akinsoji 1996
									H: SE	IH: Bows, arrows	IH: Hunting & Fishing	Bows & arrows: SE	Akinsoji 1996
Combretaceae	*Guiera senegalensis*		Shabara	Geloki		S							
								H: LF		MH: Cough	MH: Fevers	Cough: take LF decoction	Akinsoji 1996
								H: RO		MH: Haemorrhoids	MH: Gastrointestinal	Pile: make RO decoction w. potash & drink	Akinsoji 1996
								H: RO		MH: Jedi-jedi	MH: Unknown	Jedi-jedi: boil RO w. potash, drink & bathe	NCF n.d.
Anacardiaceae	*Haematostaphis barteri*	Blood plum	Jan danya, Jinin kafir	Tursuhi, Iyamhi		S							
							H: FR			FH: Edible	FH: Edible	Edible: FR	Akinsoji 1996, Ayanbamiji 1996
							H: SE			FH: Edible	FH: Edible	Oil extraction: SE	Akinsoji 1996
								H: BA		MH: Bleeding	MH: Blood	Bleeding: 1 hf	Pellaumail 1998
								H: LF BA		MH: Blood tonic	MH: Blood	Blood tonic: LF & BA decoctions	Akinsoji 1996, Ayanbamiji 1996
								H: BA		MH: Chest pain	MH: Internal organs & Respiration	Chest pain: 1 hf	Pellaumail 1998
									H: ST	IH: Gun powder	IH: Hunting & Fishing	Gun powder: burn ST	Ayanbamiji 1996
Simaroubaceae	*Hannoa klaineana*		Takandar giwa, Namijin gwabsa	Bummere badi	T	F							
									C: TW LF TR	IC: Nest-tree	IC: Nest	10% of all nest-trees	Adanu 1998
Malvaceae	*Hibiscus spp.*	Rama fibre	Ram, Rama	Gebo, Gabal	S	S							

							H: LF			FH: Edible	FH: Edible	Soup: LF	Ayanbamiji 1996, LI (GK-MA), LI (BD-F)
								H: SE		MH: Gonorrhoea	MH: Fertility & Genitalia	Gonorrhoea: boil cover of SE boil & potash & drink	NCF n.d.
									H: FI	IH: Rope	IH: Crafts & Skills	Rope: FI	Ayanbamiji 1996
									H: WP	IH: Rope	IH: Crafts & Skills	To tie up beds	LI (GK-MA)
Euphorbiaceae	*Hymenocardia acida*		Jan yaro, Jan itache	Sa:natahi, Yawa soloje	T	S							
							A: LF			FA: Edible	FA: Edible	Fodder	Ayanbamiji 1996
							H: RO			FH: Beverage	FH: Beverages	Beverage: pound RO & add to simi (local beverage made w. honey)	Akinsoji 1996, Ayanbamiji 1996
								H: RO		MH: Impotence	MH: Fertility & Impotence	Strengthen male organ: RO w. other plant	Ayanbamiji 1996
								H: RO		MH: Strength	MH: Health promotion	Recover strength: boil RO w. potash	Ayanbamiji 1996
								H: BA		MH: Wound	MH: Skin	Heal wound: fresh BA	Akinsoji 1996, Ayanbamiji 1996
								H: SE BA		MH: Wound	MH: Skin	Treat wounds: 15 SE	Pellaurrail 1998
									H: ST	IH: Roofing	IH: Crafts & Skills	Roofing: ST	Ayanbamiji 1996
									H: WO	IH: Firewood	IH: Firewood	Firewood	Ayanbamiji 1996
									H: WO	IH: Construction & firewood	IH: Timbers & Carving	Local construction & fuelwood	Akinsoji 1996
Labiatae	*Hyptis spp.*				H								
							H: LF			FH: Seasoning	FA: Seasoning & Processing	Seasoning: LF	Akinsoji 1996
Gramineae	*Imperata cylindrica*	Spear grass	Tohuwa, Tofa	Gasa kigere	G	S							
								H: RO		MH: Toothache	MH: Oral health	Toothache: boil RO w. potash & drink	NCF n.d.
									H: WP	IH: Construction	IH: Timbers & Carving	Building houses & fences	Ayanbamiji 1996
Leguminosae - Papilionoideae	*Indigofera spp.*	Indigo			I								
							B: SE			FB: Edible	FB: Edible	Food	Warren p.c.
Convolvulaceae	*Ipomoea batatas*	Sweet potato	Dan kai	Dan kai	C								
							B: RO			FB: Edible	FB: Edible	Food	Warren p.c.
Leguminosae - Caesalpinoideae	*Isoberlinia tomentosa*		Farar doka		T	S							
								H: BA		MH: Diarrhoea	MH: Gastrointestinal	Runny stomach in children: boil BA (second layer) & drink	NCF n.d.
Gramineae	*Jardinea congoensis*		Diwa. Shada		G								
									H: WP	IH: Mat	IH: Crafts & Skills	Mats	Akinsoji 1996
									H: WP	IH: Fence	IH: Timbers & Carving	Fencing	Akinsoji 1996
Euphorbiaceae	*Jatropha curcas*	Purging nut	Ba ni da zugu	Ko akolaji, Kwolkwclaio	T S	S							
								H: LA		MH: Blood clot	MH: Blood	Clot blood: LA	NCF n.d.
								H: BA RO		MH: Gonorrhoea	MH: Fertility & Genitalia	Gonorrhoea: boil RO &/or BA w. potash & drink decoction	Akinsoji 1996, Ayanbamiji 1996. NCF n.d.
								H: BA RO		MH: Stomach ache	MH: Gastrointestinal	Stomach pain: BA or RO w. potash	Ayanbamiji 1996
								H: LF		MH: Stomach problems	MH: Gastrointestinal	Stomach problems: boil LF	LI (BD-M), LI (BD-F)
								H: FJ LA		MH: Mouth sores	MH: Oral health	Mouth sores: squeeze LA out of LF & use juice	Akinsoji 1996

(continued)

Appendix (continued)

Family	Species	English	Hausa	Fulfulde	Habit	Habitat	Food	Medicine	Implement	Usage Description	Usage Category	Description in Original Source	Source
								H: FJ LA		MH: Wound	MH: Skin	Fresh wound: squeze LA out of LF & use juice	Akinsoji 1996
								H: LA		MH: Burns	MH: Skin	Burns: LA	Ayanbamiji 1996
								H: LA		MH: Sores	MH: Skin	Sores: LA	Ayanbamiji 1996
								H: LF		MH: Wound	MH: Skin	Wound: crush LF to massage a wound & stop bleeding	Ayanbamiji 1996
Meliaceae	*Khaya anthotheca*	White mahogany			T								
									C: TW LF TR	IC: Nest-tree	IC: Nest	3% of all nest-trees	Adanu 1998
Meliaceae	*Khaya senegalensis*	African mahogany	Madaci, Madachi	Daleh, Kahi, Daloh	T	S							
								H: SE LF		MH: ?	?	Medicine: use SE juice, LF oil as drops	Akinsoji 1996
							A: BA			FA: Edible	FA: Edible	Food: mix BA w. salt	LI (CH-M
								A: ?		MA: Fertility	MA: Fertility & Genitalia	Fertility: mix w. dugugi, rimajoga & sembal, pound & cook	Martin 1996
								A: BA		MA: Hengre	MA: Health promotion	Hengre: concoct BA, take w. sumsum, sonye & bumchi	Martin 1996
								A: BA		MA: Bile	MA: Internal organs & Respiration	Bile disease in cattle: pound BA	NCF n.d.
								A: BA		MA: Worms	MA: Parasites	Worms: pound BA, add salt	Martin 1996
								H: LA		MH: Earache	MH: Analgesic	Ear pain: apply LF LA	Ayanbamiji 1996
								H: SE		MH: Chicken pox	MH: Endemic & Epidemic diseases	Chicken pox: ingest SE oil	Ayanbamiji 1996
								H: BA		MH: Stomach ache	MH: Gastrointestinal	Stomach pain: boil BA or soak in water & drink	Ayanbamiji 1996, LI (BD-F), Martin 1996
								H: BA		MH: Stomach upset	MH: Gastrointestinal	Stomach upset: pound BA	NCF n.d.
								H: RO BA		MH: Stomach ache	MH: Gastrointestinal	Stomach ache: 1 hf	Pellaumail 1998
								H: RO TW		MH: Chewing stick	MH: Oral health	Chewing stick: rootlets & branchlets	Akinsoji 1996
								H: BA		MH: Worms	MH: Parasites	Worms: soak BA in cool water for 4-6 hours & drink	LI (FI-F), LI (GK-F), LI (SB-F)
								H: BA		MH: Skin disease	MH: Skin	Skin disease: 1 hf for skin disease	Pellaumail 1998
								H: BA		MH: Skin conditions	MH: Skin	Skin conditions: mix w. cow's urine & bathe	LI (FI-F)
								H: WO		MH: Skin conditions	MH: Skin	Red & black skin disease: people used to bathe in water that got in the canoe	LI (GK-MA)
									C: TW LF TR	IC: Nest-tree	IC: Nest	1% of all nest-trees	Adanu 1998
									H: TR	IH: Canoe carving	IH: Timbers & Carving	Carve canoes	LI (GK-MA)
Bignoniaceae	*Kigelia africana*	Sausage tree	Rawuya	Jirlahi	T								
								H: RO BA		MH: Stomach ache	MH: Gastrointestinal	Stomach pain: take RO & BA decoction	Akinsoji 1996
									H: WO	IH: Implements	IH: Timbers & Carving	Handles for implements: WO	Akinsoji 1996
Compositae	*Lactuca capensis*	Lettuce	Nonokwarai	Kaade-kaade	H								
								H: WP		MH: Worms	MH: Parasites	Worms: boil & drink	LI (SB-F)
Apocynaceae	*Landolphia landolphioides*	Rubber vine	Tiboko	Bel bel, Kogi	I								
							B: FR			FB: Edible	FB: Edible	Food	Warren p.c.
							C: FP			FC: Edible	FC: Edible	Staple food: early rainy season (March), chimps eat seeds as well, whereas monkeys spit them out. Chimps eat it unripe as well	LI (GK-MA)
							H: FR			FH: Edible	FH: Edible	Food	LI (CH-M), LI (GK-MA)
							H: FP			FH: Edible	FH: Edible	Food	LI (GK-MA)

								IH: LA	IH: Rubber ball	IH: Timbers & Carving	Make rubber ball to play from LA	LI (GK-MA)
Apocynaceae	*Landolphia macrantha ?*		Tiboko	Kogi, Bel bel	L							
						B: FR			FB: Edible	FB: Edible	Food	Warren p.c.
						C: FR			FC: Edible	FC: Edible	Food	Fowler p.c.
Apocynaceae	*Landolphia owariensis*	White rubber vine	Ciwo	Citta lainde, Bom bom	L							
						B: FR			FB: Edible	FB: Edible	Food	Warren p.c., LI (GK-MA)
						FI: FR			FI: Edible	FI: Edible	Edible: mix FR w. pepper (tamunji). Difficult to eat by itself, has to be mixed w. something. Make soup or mix w. corn flour	LI (SB-F)
Apocynaceae	*Landolphia togolana ?*		Tiboko	Be bel	L							
						B: FR RO			FB: Edible	FB: Edible	Food	Warren p.c.
						C: FR			FC: Edible	FC: Edible	Food	Fowler p.c.
Anacardiaceae	*Lannea barteri*		Faru, Faruhi	Murututa	T							
						B: FI FR			FB: Edible	FB: Edible	Food	Warren p.c.
						C: FR			FC: Edible	FC: Edible	Food	LI (GK-MA)
							H: RO		MH: Constipation	MH: Gastrointestinal	Constipation: boil RO & drink	LI (GK-MA)
							H: RO		MH: Diarrhoea	MH: Gastrointestinal	Diarrhoea: dry & pound RO, add to porridge	Akinsoji 1996
							H: LF		MH: Toothache	MH: Oral health	Toothache: mix LF w. limestone & place on tooth	LI (ST-MH)
Anacardiaceae	*Lannea nigritana*	Wodier wood	Faru, Faruhi	Sonyi	T							
						B: FR GU			FB: Edible	FB: Edible	Food	Warren p.c., LI (GK-MA)
							H: BA		MH: Cholera	MH: Endemic & Epidemic diseases	Cholera: boil & drink BA	LI (CH-M)
							H: BA		MH: Malaria	MH: Fevers	Malaria: boil & drink BA	LI (CH-M)
							H: BA		MH: Worms	MH: Parasites	Worms: drink or give as enema to children	LI (CH-F)
Anacardiaceae	*Lannea schimperi*		Farin doya*	Tsorohi	T							
							H: SF		MH: Worms	MH: Parasites	Worm expeller: SF oil	Akinsoji 1996
								H: BA	IH: Rope	IH: Crafts & Skills	Rope: BA	Akinsoji 1996
Rubiaceae	*Lasianthus batangensis*		Hako dalam		H							
						B: FR			FB: Edible	FB: Edible	Food	Warren p.c.
						C: FR			FC: Edible	FC: Edible	Food	Fowler p.c.
						FI: LF			FI: Edible	FI: Edible	Soup	LI (SB-F)
						H: LF			FH: Salt substitude	FH: Seasoning & Processing	Salt substitute: burn LF, put the ash in pot & filter water through	LI (GK-MA)
Lythraceae	*Lawsonia inermis*	Henna	Lalle	Lalle	TS							
							H: RO		MH: Abortion	MH: Female reproduction	Procure abortion: boil RO w. potash	Akinsoji 1996
							H: RO		MH: Boils	MH: Skin	Swollen parts or boils: grind fresh RO to break swollen part of the body or boils	NCF n.d.
								H: LF	IH: Cosmetic	IH: Hygiene & Cosmetics	Nail varnish: LF	Akinsoji 1996
Leeaceae	*Leea guineensis*				ST							
						B: FR ST			FB: Edible	FB: Edible	Food	Warren p.c.
						C: FR			FC: Edible	FC: Edible	Food	Fowler p.c.

(continued)

Appendix (continued)

Family	Species	English	Hausa	Fulfulde	Habit	Habitat	Food	Medicine	Implement	Usage Description	Usage Category	Description in Original Source	Source
Ochnaceae	*Lophira lanceolata*	Red iron wood	Namijin kadanya	Karehi gori, Saktohi	T	S							
							B: LY			FB: Edible	FB: Edible	Food	Warren p.c.
								A: BA		MA: Weakness	MA: Health promotion	Weak animal (bunchude): grind BA, mix w. water & leave overnight	Martin 1996
								H: BA		MH: Yellow fever	MH: Endemic & Epidemic Diseases	Yellow fever: boil BA	NCF n.d.
								H: RO		MH: Stomach ache	MH: Gastrointestinal	Stomach ache: RO	Akinsoji 1996
								H: SE		MH: Worms	MH: Parasites	Worm expeller: oil from SE	Ayanbamiji 1996
									H: SE	IH: Soap	IH: Hygiene & Cosmetics	Make soap: SE oil	Akinsoji 1996
Anacardiaceae	*Mangifera indica*	Mango	Mangoro	Mangoro	T	S							
							B: FR SE			FB: Edible	FB: Edible	Food	Warren p.c.
							H: FR			FH: Edible	FH: Edible	Edible: FR	Akinsoji 1996
								H: BA LF		MH: Colds	MH: Fevers	Colds: boil BA & LF & dala & bathe. Not drinkable	LI (DM-F), LI (DM-M)
								H: LF BA		MH: Fever	MH: Fevers	Fever: boil LF & BA	Akinsoji 1996
								H: BA		MH: Diarrhoea	MH: Gastrointestinal	Diarrhoea: BA infusion	Akinsoji 1996
								H: BA		MH: Dysentery	MH: Gastrointestinal	Dysentery: BA infusion	Akinsoji 1996
								H: FJ		MH: Jaundice	MH: Internal organs & Respiration	Jaundice: drink FJ	NCF n.d.
								H: BA		MH: Skin conditions	MH: Skin	Scabies (koro koro): boil BA, mix w. dala (toka) & bathe children	LI (GK-F)
									H: WP	IH: Shade tree	IH: Timbers & Carving	Shade tree	Akinsoji 1996
Euphorbiaceae	*Manihot esculenta*	Cassava	Rogo	Mbaji	C								
							B: RO			FB: Edible	FB: Edible	Food	Warren p.c.
Sapotaceae	*Manilkara obovata*	African pearwood	Kadanya rafi	Kareje, Kareji	T								
							B: FR			FB: Edible	FB: Edible	Food	Warren p.c.
Leguminosae - Papilionoideae	*Milletia thonningii*		Turburku		T							GGH specimen	
									H: WO	IH: Implements	IH: Timbers & Carving	Implements	Akinsoji 1996
Annonaceae	*Monodora tenuifolia*				T								
									C: TW LF TR	IC: Nest-tree	IC: Nest	2% of all nest-trees	Adanu 1998
Diptero-carpaceae	*Monotes kerstingii*		Hantso, Wasami*	Jaingi	T	WS							
									H: WO	IH: Firewood	IH: Firewood	Firewood	Akinsoji 1996
Moringaceae	*Moringa oleifera*	Oil of Ben tree, Horse-raddish tree	Zogalla gandi, Zogale	Kabije, Koma, Giligandja*	T	S							
							H: LF			FH: Edible	FH: Edible	Edible: cook LF as vegetable	Akinsoji 1996, Ayanbamiji 1996
								H: LF		MH: Eye problem	MH: Analgesic	Eye problem: squeeze the LF & apply the water to the eye	Akinsoji 1996, NCF n.d.
								H: RO		MH: Body pain	MH: Analgesic	Pain: crush RO w. kerosene & salt & apply on body	Ayanbamiji 1996
								H: RO		MH: Colds	MH: Fevers	Cold: crush RO w. kerosene & salt & apply on body	Ayanbamiji 1996

							H: LF	IH: Witchcraft	IH: Magico-religious	Detecting if one is bewitched: pound LF & apply on the body	NCF n.d.
							H: WO	IH: Fence	IH: Timbers & Carving	Live fence	Akinsoji 1996
Leguminosae - Papilionoideae	*Mucuna pruriens*	Velvet bean, Cowage	Karara	Nyanya rigaduru, Nyanya lay*	L						
							B: LY SE ST	FB: Edible	FB: Edible	Food	Warren p.c.
							H: LF	MH: Headache w. dizziness	MH: Analgesic	Headache w. dizziness: pound LF w. limestone, shave head & rub on mixture. Leave some time to dry	LI (ST-MH)
							H: LF RO	MH: Side pain	MH: Analgesic	Sharp pain on side: pound LF, mix w. limestone, cut area w. razor & rub mixture on soot	LI (ST-MH)
							H: SE	MH: Snake bite	MH: Antidotes & Repellents	Snake bite antidote: swallow 1 SE per year	Martin 1996, LI (ST-MH)
							H: LF	MH: Post-partum	MH: Female reproduction	Cleanse blood after labour: pound LF w. limestone & drink first thing in the morning	LI (ST-MH)
							H: LF RO	MH: Cough	MH: Fevers	Cough: pound & boil LF & RO, drink 2 daily, morning-evening	LI (ST-MH)
							H: LF	MH: Stomach upset	MH: Gastrointestinal	Stomach upset: chew LF	LI (ST-MH)
							H: LF	MH: Toothache	MH: Oral health	Toothache: pound LF & chew where pain is to kill worm causing toothache	LI (ST-MH)
							H: LF	MH: Worms	MH: Parasites	Worms: chew or boil LF; pound LF, mix w. tamarind or lemon, soak overnight, drink 1-2 times	LI (ST-MH)
							H: LF	MH: Skin conditions	MH: Skin	Itchy skin: pound LF & rub on body	LI (ST-MH)
Musaceae	*Musa sapientum*	Banana	Ayaba	Ayaba	C						
							B: FR	FB: Edible	FB: Edible	Food	Warren p.c.
Cecropiaceae	*Myrianthus serratus*		Farin ganye			S					
							H: GU	MH: Toothache	MH: Oral health	Toothache: GU	NCF n.d.
							H: LF ST	MH: Abonakeke	MH: Unknown	Abonakeke: boil LF & ST & bathe or drink. As taura	LI (GK-F)
Lecythidaceae	*Napoleona imperialis*	Napoleona	Mabungi		TS	F					
							B: FR	FB: Edible	FB: Edible	Food	Warren p.c.
Lecythidaceae	*Napoleona vogelii*	Napoleona			ST						
							B: FR	FB: Edible	FB: Edible	Food	Warren p.c.
							H: RO	MH: Fever	MH: Fevers	Fever: RO decoction	Akinsoji 1996
							H: BA	IH: Arrow poison	IH: Hunting & Fishing	Arrow poison: BA	Akinsoji 1996
Rubiaceae	*Nauclea latifolia*	African peach	Tafashiya	Bakurehi	ST	SMF				GGH specimen	
							B: FR	FB: Edible	FB: Edible	Food	Warren p.c.
							H: RO	MH: Yellow fever	MH: Endemic & Epidemic Diseases	Yellow fever	LI (FI-F)
							H: RO BA	MH: Yellow fever	MH: Endemic & Epidemic Diseases	Yellow fever: take 1 hf (2-5 RO) in decoction	Pellaumail 1998
							H: ST	MH: Cholera	MH: Endemic & Epidemic diseases	Cholera: boil ST & drink	LI (S3-F)
							H: PI	MH: Labour	MH: Female reproduction	Labour: boil PI, give women when labour is near	LI (GK-F)

(continued)

Appendix (continued)

Family	Species	English	Hausa	Fulfulde	Habit	Habitat	Food	Medicine	Implement	Usage Description	Usage Category	Description in Original Source	Source
								H: RO		MH: Gonorrhoea	MH: Fertility & Genitalia	Gonorrhoea: boil RO w. potash	Akinsoji 1996, NCF n.d.
								H: LF		MH: Fever	MH: Fevers	Fever: soak LF & bathe	LI (GK-F)
								H: BA		MH: Diarrhoea	MH: Gastrointestinal	Diarrhoea: mix w. kojoli & galbije & kohal & use as enema or drink for 1-2 month old babies w. diarrhoea	LI (GK-F)
								H: BA RO		MH: Stomach ache	MH: Gastrointestinal	Stomach ache: boil BA or RO & drink	Martin 1996, LI (GK-F)
								H: LF		MH: Flatulence	MH: Gastrointestinal	Induce defecation: boil LF & drink for bloated stomach & to induce defecation	LI (GK-F)
								H: PI		MH: Stomach ache	MH: Gastrointestinal	Stomach ache: soak PI in cold water or boil & drink	LI (GK-F)
								H: RO		MH: Stomach problems	MH: Gastrointestinal	Stomach problems: boil RO & drink	LI (GK-MA)
								H: RO		MH: Stomach ache	MH: Gastrointestinal	Stomach ache: boil RO & drink	LI (GK-MA), NCF n.d.
								H: RO		MH: Stomach upset	MH: Gastrointestinal	Stomach upset: boil RO w. potash	Akinsoji 1996, NCF n.d.
								H: RO BA		MH: Diarrhoea	MH: Gastrointestinal	Diarrhoea: take 1 hf (2-5 RO) in decoction	Pellaumail 1998
								H: RO BA		MH: Stomach ache	MH: Gastrointestinal	Stomach ache: take 1 hf (2-5 RO) in decoction	Pellaumail 1998
								H: TW		MH: Chewing stick	MH: Oral health	Chewing stick	Ayanbamiji 1996
								H: PI		MH: Worms	MH: Parasites	Worms: mix PI w. kanwa (potash)	LI (GK-FH)
								H: RO		MH: Worms	MH: Parasites	Worms: boil or soak for 3 hours & drink or chew RO	LI (FI-F), LI (FI-M)
								H: RO BA		MH: Worms	MH: Parasites	Worms: take 1 hf (2-5 RO) in decoction	Pellaumail 1998
Bignoniaceae	*Newbouldia laevis*	Tree of life	Aduruku		T	F							
							H: LF SE			FH: Edible	FH: Edible	Food	Ayanbamiji 1996
							H: LF			FH: Food processing	FH: Seasoning & Processing	Make drink more potent: use LF to wash containers in which "burukutu" (local brew from cereals) is fermented.	Akinsoji 1996
								H: LF		MH: Fever	MH: Fevers	Fever: boil LF	LI (KM-F), LI (KM-M)
								H: LF		MH: Wound	MH: Skin	Wounds: LF	Ayanbamiji 1996
									H: WO	IH: Firewood	IH: Firewood	Firewood	Ayanbamiji 1996
									H: LF	IH: Baby bath	IH: Hygiene & Cosmetics	Bathe babies: LF decoction	Akinsoji 1996
									H: WO	IH: Fence	IH: Timbers & Carving	Live fence	Akinsoji 1996
									H: ST	IH: Construction	IH: Timbers & Carving	Graving (burial), fencing & roofing	Ayanbamiji 1996
									H: TR	IH: Fence	IH: Timbers & Carving	Live fence	LI (KM-F), LI (KM-M)
Nympheaceae	*Nymphaea lotus*	Water-lily	Bado	Tabbere									
							H: FR			FH: Edible	FH: Edible	Food	Akinsoji 1996
Labiatae	*Ocimum gratissimum*	Fever bush	Dai doya gida		S								
							H: LF			FH: Edible	FH: Edible	Soup: LF	Ayanbamiji 1996
								H: LF		MH: ?	MH: Health promotion	Medicine: LF	Ayanbamiji 1997
Octoknemaceae	*Octoknema borealis*				T								
									C: TW LF TR	IC: Nest-tree	IC: Nest	1% of all nest-trees	Adanu 1998
Gramineae	*Oryza sativa*	Rice	Cinkafa	Malori	GC								
							B: ST			FB: Edible	FB: Edible	Food	Warren p.c.
Pandanaceae	*Pandanus candelabrum*	Screw pine	Abutu	Butoltingong	T								
							B: FR SE RO			FB: Edible	FB: Edible	Food	Warren p.c.
									H: LF	IH: Mat	IH: Crafts & Skills	Mats: LF	Akinsoji 1996
									H: RO	IH: Sponge	IH: Crafts & Skills	Sponge: pound stilt RO	Akinsoji 1996

Graminae	*Panicum maximum*	Guinea grass	Tolore	Tolore, Tulore	G								
							A: LF			FA: Edible	FA: Edible	Fodder	LI (CH-F)
							B: ST			FB: Edible	FB: Edible	Food	Warren p.c.
							H: WP			FH: Edible	FH: Edible	Food: burn WP & add the ash to cook meat. Mix w. anyaka to have dala soup. Also mix w. cassava LF to make soup	LI (CH-F)
									H: WP	IH: Furniture	IH: Timbers & Carving	Items of furniture such as beds: culm*	Akinsoji 1996
Chrysobalana-ceae	*Parinari curatellifolia*	Mobola plum	Rura	Nawarre badi	T	S							
							H: FR SE			FH: Edible	FH: Edible	Food	Akinsoji 1996
Chrysobalana-ceae	*Parinari excelsa*	Grey plum, Guinea plum	Tuwonbiri, Rura naude	Nowde	T	F							
							B: FR SE			FB: Edible	FB: Edible	Food	Warren p.c.
Leguminosae-Mimosoideae	*Parkia biglobosa, Syn. clappertoniana*	Locust bean tree	Dorowa	Nareji, Narehi	T	S							
							B: FL SE FP			FB: Edible	FB: Edible	Food	Warren p.c., LI (GK-MA)
							C: FP			FC: Edible	FC: Edible	Food, before & during rainy season	LI (GK-MA)
							H: FR SE FL			FH: Edible	FH: Edible	Food: FR edible, use SE for soup, FL pod for soup	Ayanbamiji 1996
							H: LF			FH: Edible	FH: Edible	Substitute for potash: burn LF & use the ash	Akinsoji 1996
							H: SE			FH: Edible	FH: Edible	Local maggi cubes (dadawa): ferment SE	LI (GK-F), LI (GK-MA), LI (KM-F), LI (KM-M)
							H: SE PU			FH: Edible	FH: Edible	Food: PU of FR raw but ripe	LI (GK-MA), LI (SB-F)
								A: LF		MA: Bloat	MA: Gastrointestinal	Bloat: pound LF, add to water, filter & administer through the nose	Martin 1996
								H: BA		MH: Stomach upset	MH: Gastrointestinal	Stomach upset: boil BA	NCF n.c.
								H: LF		MH: Weakness	MH: Health promotion	Weakness: dry LF & smoke	NCF n.c.
								H: BA		MH: Toothache	MH: Oral health	Toothache: boil BA & use as mouthwash	Ayanbamiji 1996, LI (GK-MA)
								H: BA		MH: Worms	MH: Parasites	Worms: mix BA w. barks from other trees (galbijo & kolohi), boil & give to babies or use as enema	LI (GK-F), LI (GK-MA)
									H: BA	IH: Dye	IH: Crafts & Skills	Source of dye to paint & decorate walls	Akinsoji 1996
									H: WO	IH: Firewood	IH: Firewood	Firewood	Ayanbamiji 1996
									H: BA	IH: Fish poison	IH: Hunting & Fishing	Fish poison	LI (GK-MA)
									H: FL	IH: Fish poison	IH: Hunting & Fishing	Fish poison	Ayanbamiji 1996
									H: PO	IH: Fish poison	IH: Hunting & Fishing	Fishpoison: pound PO w. other leaves	LI (GK-MA)
									H: FL	IH: Construction	IH: Timbers & Carving	Use pod as cement	Ayanbamiji 1996
Sapindaceae	*Paullinia pinneta*	Five fingers, water cola	Yasarbiyar	Kolijoi, Gorolamido	L	MF						GGH specimen	
							H: FR			FH: Edible	FH: Edible	Food	LI (SB-M)
								H: LF		MH: Pain	MH: Analgesic	Pain: take 1 hr (only collected on Sundays)	Pellaumail 1998
								H: LF		MH: Yellow fever	MH: Endemic & Epidemic Diseases	Yellow fever: bathe	LI (FI-F)
								H: LF		MH: Colds	MH: Fevers	Colds: boil LF & give to children to drink or bathe. Can mix w. mango BA	LI (CH-F), LI (SB-F)
								H: LF		MH: Malaria	MH: Fevers	Malaria: bathe	LI (FI-F)
								H: LF		MH: Worms	MH: Parasites	Worms: bathe	LI (FI-F)

(continued)

Appendix (continued)

Family	Species	English	Hausa	Fulfulde	Habit	Habitat	Food	Medicine	Implement	Usage Description	Usage Category	Description in Original Source	Source
								H: LF		MH: Skin conditions	MH: Skin	Bathe to remove white substance from babies' skin	LI (FI-F), LI (GK-MA)
								H: LF		MH: Skin conditions	MH: Skin	Boil LF & bathe children for skin problems	LI (GK-MA)
									H: LF	IH: Female hygiene	IH: Hygiene & Cosmetics	Bathe women and children	LI (GK-MA), LI (S3-M)
Palmae	*Phoenix mulleriinus**		M'nf'rya*		T	S							
								H: BA		MH: Malaria fever	MH: Fevers	Malaria	LI (GK-MA)
								H: BA		MH: Jaundice	MH: Internal organs & Respiration	Jaundice: BA & potash & make porridge	NCF n.d.
Palmae	*Phoenix reclinata*	Wild date palm	Kajinji, Kajinjiri	Bali, Bali darle	T	S							
							B: FR			FB: Edible	FB: Edible	Food	LI (GK-MA)
							B: SE			FB: Edible	FB: Edible	Food	Warren p.c.
							C: FR			FC: Edible	FC: Edible	Food	LI (GK-MA)
							H: FR ST			FH: Edible	FH: Edible	Edible: FR & apical bud of ST	Akinsoji 1996, LI (GK-MA)
								H: LY		MH: Yellow fever	MH: Endemic & Epidemic Diseases	Yellow fever: boil & drink LY	LI (GK-MA)
								H: RO		MH: Swollen parts	MH: Analgesic	Swollen parts: boil RO & drink	LI (FI-M)
								H: RO		MH: Toothache	MH: Oral health	Toothache: soak RO in hot water & use as mouthwash	LI (FI-M)
									H: WP	IH: Broom	IH: Crafts & Skills	Broom: use inflorescence	Akinsoji 1996
									H: FO	IH: Mat	IH: Crafts & Skills	Mats: juvenile FO	Akinsoji 1996
Leguminosae - Caesalpinoideae	*Piliostigma reticulatum*	Camel's foot	Kargo	Barkehi	T	S MF							
							B: FR			FB: Edible	FB: Edible	Food	LI (GK-MA)
								A: FL		MA: Prophylaxy	MA: Health promotion	Prophylactic for: pound FL, mix w. salt	Martin 1996
								H: RO		MH: Menstrual problems	MH: Female reproduction	Menstrual pain: grind & boil RO	LI (FI-M)
								H: RO		MH: Fertility	MH: Fertility & Impotence	Cleanse reproductive system: mix RO w. rimajoga & bauri, boil & give to barren women	LI (ST-FH)
								H: BA		MH: Cough	MH: Fevers	Cough: boil BA & drink	LI (GK-F)
								H: LF		MH: Colds	MH: Fevers	Colds: boil LF & drink for runny nose & sore throat. Also, boil LF in big pot w. kargo & other leaves & breathe covered w. towel	LI (S3-F)
								H: LF		MH: Cough	MH: Fevers	Cough: boil LF & drink for 4-5 days	Martin 1996, LI (GK-F)
								H: LF RO		MH: Colds	MH: Fevers	Colds: boil RO & LF & drink	LI (CH-F)
								H: LY		MH: Catarrh	MH: Fevers	Catarrh: boil LY & give to children	LI (GK-F)
								H: LY		MH: Fever	MH: Fevers	Fever w. cough: boil LY & bathe children or give as a drink	LI (GK-F)
								H: BA		MH: Stomach problems	MH: Gastrointestinal	Stomach pain	LI (GK-MA)
								H: LF		MH: Mouth ulcers	MH: Oral health	Mouth ulcers: rub JU of LF on lips & tongue	LI (GK-F)
								H: BA		MH: Skin conditions	MH: Skin	Skin problems: boil BA & drink	LI (GK-F)
								H: LA		MH: Cuts	MH: Skin	Cuts: apply LA onto cuts	LI (GK-F)
								H: LF		MH: Cuts	MH: Skin	Cuts: boil LF & wash cuts	LI (GK-F)
								H: LF		MH: Skin conditions	MH: Skin	Skin problems: boil LF & drink	LI (GK-F)
									H: WP	IH: Good luck	IH: Magico religious	Good luck to find food	LI (FI-M)

Leguminosae - Caesalpinoideae	*Piliostigma thonningii*	Camel's foot	Kargo	Barkehi	T	S MF							
							B: SE FP			FB: Edible	FB: Edible	Food	Warren p.c.
								H: LF SH		MH: Snake bite	MH: Antidotes & Repellents	Snake bite: swallow LF, now SH & LY of Annona	NCF n.d.
								H: RO		MH: Yellow fever	MH: Endemic & Epidemic Diseases	Yellow fever: drink RO infusion	Akinsoji 1996, NCF n.d.
								H: LY		MH: Cough	MH: Fevers	Cough: drink LY decoction	Akinsoji 1996, NCF n.d.
								H: SE LF		MH: Catarrh	MH: Fevers	Catarrh: take 2 LF	Pellaumail 1998
								H: RO		MH: Chest pain	MH: Internal organs & Respiration	Chest pain: boil RO w. BA of Khaya senegalensis	Akinsoji 1996, NCF n.d.
								H: LY		MH: Toothache	MH: Oral health	Toothache: use decoction of LY	Akinsoji 1996
									H: ST RO	IH: Dye	IH: Crafts & Skills	Dye to decorate pottery: ST & RO	Akinsoji 1996
									H: ST BA	IH: Rope	IH: Crafts & Skills	Rope: ST & BA	Akinsoji 1996
									H: WO	IH: Implements	IH: Timbers & Carving	Implements e.g. hoe handles: WO	Akinsoji 1996
Araliaceae	*Polyscias fulva*				T	MF							
									H: WO	IH: Fence	IH: Timbers & Carving	Fence	Akinsoji 1996
Leguminosae - Mimosoideae	*Prosopis africana*	Ironwood	Kiriya	Kohi, Kohal	T	S LF							
							B: FR			FB: Edible	FB: Edible	Food	Warren p.c., LI (GK-MA)
							B: SE			FB: Edible	FB: Edible	Food	Warren p.c.
							C: FR			FC: Edible	FC: Edible	Food	LI (GK-MA)
							H: FR			FH: Edible	FH: Edible	Food	Akinsoji 1996, LI (KM-F), LI (KM-M)
							H: SE			FH: Seasoning	FH: Seasoning & Processing	Seasoning: use to make "dadawa"	Akinsoji 1996
								H: BA		MH: Body pain	MH: Analgesic	Body pain: 2-3 hf	Pellaumail 1998
								H: BA		MH: Strengthen penis	MH: Fertility & Impotence	Strengthen penis	LI (GK-MA)
								H: BA		MH: Fever	MH: Fevers	Fever: give children for fever	LI (GK-F), LI (KM-F), LI (KM-M)
								H: BA		MH: Diarrhoea	MH: Gastrointestinal	Diarrhoea: boil BA, take 2-3 hf	Pellaumail 1998, LI (GK-F)
								H: BA		MH: Dysentery	MH: Gastrointestinal	Dysentery: soak the BA in water & drink	NCF n.d.
								H: BA		MH: Dysentery	MH: Gastrointestinal	Dysentery: 2-3 hf	Pellaumail 1998
								H: BA		MH: Stomach ache	MH: Gastrointestinal	Stomach ache (infants esp.): 2-3 hf	Pellaumail 1998
								H: BA		MH: Stomach ache	MH: Gastrointestinal	Stomach ache: 3 hf (combined w. kojokohi & kohal)	Pellaumail 1998
								H: BA		MH: Stomach ache	MH: Gastrointestinal	Stomach ache: boil BA & drink or bathe	LI (GK-F)
								H: BA		MH: General health	MH: Health promotion	Medicine for children	LI (GK-MA)
								H: ?		MH: Toothache	MH: Oral health	Toothache: use as mouthwash	LI (CH-M)
								H: BA		MH: Toothache	MH: Oral health	Toothache: boil BA w. potash & wash mouth	Ayanbamiji 1996, LI (GK-MA), NCF n.d.
								H: BA		MH: Worms	MH: Parasites	Worms in children: pound and dry BA & use as enema	LI (SB-F)
								H: BA		MH: Skin disease	MH: Skin	Skin disease: 1 hf-1/2 basin (combined w. konkehi, for bathing infants)	Pellaumail 1998
									H: RO WO	IH: Charcoal	IH: Firewood	Make charcoal: blacksmiths burn the RO and make charcoal for long burning	Akinsoji 1996, LI (GK-MA)
									H: BA	IH: Baby bath	IH: Hygiene & Cosmetics	Bathe babies: boil BA	Akinsoji 1996

(continued)

Appendix (continued)

Family	Species	English	Hausa	Fulfulde	Habit	Habitat	Food	Medicine	Implement	Usage Description	Usage Category	Description in Original Source	Source
Meliaceae	*Pseudocedrela kotschyi*	Dry zone cedar	Tuna	Bodi, Wada-wurohi, Boderlahi	T	SF							
								A: BA		MA: ?	?	Boil BA	LI (GK-MA)
								A: WP BA		MA: Wound	MA: Skin	Cuts & wounds: grind WP & apply to cuts & wounds. Alternatively pound & dry BA & apply to wounds	Martin 1996
								H: ?		MH: Measles	MH: Endemic & Epidemic Diseases	Measles: bathe w. it & drink	NCF n.d.
								H: BA		MH: Cholera	MH: Endemic & Epidemic diseases	Cholera: boil BA & calcium & drink. Calcium facilitates drinking.	LI (FI-M)
								H: ST		MH: Measles	MH: Endemic & Epidemic Diseases	Measles: make ST infusion & bathe	Akinsoji 1996
								H: BA		MH: Malaria fever	MH: Fevers	Malaria fever: boil BA & drink	LI (CH-M)
								H: LF		MH: Fever	MH: Fevers	Fever: boil LF	LI (KM-F), LI (KM-M)
								H: RO		MH: Fever	MH: Fevers	Fever: boil or chew RO	Ayanbamiji 1996
								H: BA		MH: Stomach problems	MH: Gastrointestinal	Stomach ache: boil BA or soak & drink	LI (GK-MA)
								H: RO		MH: Constipation	MH: Gastrointestinal	Constipation: chew RO or boil	Ayanbamiji 1996
								H: ST		MH: Stomach ache	MH: Gastrointestinal	Stomach ache: soak ST & drink infusion	Akinsoji 1996, NCF n.d.
								H: BA		MH: Worms	MH: Parasites	Boil BA & drink	LI (CH-M)
								H: WP BA		MH: Wound	MH: Skin	Cuts & wounds: grind WP, apply to cuts & wounds. Alternatively pound & dry BA & apply to wounds	Martin 1996
									H: WO	IH: Implements	IH: Timbers & Carving	Implements: WO	Akinsoji 1996
Anacardiaceae	*Pseudospondias microcarpa*		Mugum		T	F							
							B: FR			FB: Edible	FB: Edible	Food	Warren p.c.
							C: FR			FC: Edible	FC: Edible	Food	Fowler p.c., LI (GK-MA)
							H: FR			FH: Edible	FH: Edible	Food	LI (GK-MA)
									H: WP	IH: Firewood	IH: Firewood	Firewood	LI (GK-MA)
Myrtaceae	*Psidium guajava*	Guava	Gwoba lande, Gualba	Gwoba lande, Goyaban	T	S							
							B: FR			FB: Edible	FB: Edible	Food	Warren p.c., LI (GK-MA)
							II: FR			FII: Edible	FII: Edible	Food	LI (BD-F), LI (BD-M), LI (FI-F)
							H: LF			FH: Edible	FH: Edible	Food. (Domesticated form, not lainde-bush)	LI (FI-F)
								H: LF		MH: Yellow fever	MH: Endemic & Epidemic Diseases	Yellow fever: boil LF w. calcium & drink	LI (FI-F)
								H: LF		MH: Diarrhoea	MH: Gastrointestinal	Diarrhoea: eat raw / LF, never mix w. anything else	LI (FI-F)
								H: LF		MH: Dysentery	MH: Gastrointestinal	Dysentery: boil LF	LI (BD-F), LI (BD-M)
								II: FJ		MII: Jaundice	MII: Internal organs & Respiration	Jaundice: drink FJ	NCF n.d.
Leguminosae - Papilionoideae	*Pterocarpus erinaceus*	African rosewood	Madobiya, Dorowan kurmi	Banuhi, Yamhi*	T	S							
							B: FL			FB: Edible	FB: Edible	Food	Warren p.c.
								H: BA		MH: Post partum	MH: Female reproduction	Post partum problems: boil BA w. potash	Martin 1996

Sterculiaceae	*Pterygota macrocarpa*				T	LF		C: TW LF TR	IC: Nest-tree	IC: Nest	1% of all nest-trees	Adanu 1998
Myristicaceae	*Pycnanthus angolensis*	African nutmeg			T	LF	B: FR		FB: Edible	FB: Edible	Food	Warren p.c.
Palmae	*Raphia hookeri*	Raphia-palm		Duje, Patawal	T	F	H: LA		FH: Palm wine	FH: Beverages	Wine is tapped from tree LA	Akinsoji 1996
								H: WO	IH: Furniture	IH: Timbers & Carving	Furniture	Akinsoji 1996
Palmae	*Raphia sudanica*		Tukuruwa	Gongola	T	S	H: LA		FH: Palm wine	FH: Beverages	Palm wine from LA	LI (BD-F), LI (BD-M), LI (GK-MA)
							H: FR		FH: Edible	FH: Edible	Food	LI (BD-F), LI (BD-M)
								H: FI	IH: Mat	IH: Crafts & Skills	Mats: fibres	Akinsoji 1996
								H: LF	IH: Roofing	IH: Crafts & Skills	Thatching: LF	Akinsoji 1996
								H: ST	IH: Roofing	IH: Crafts & Skills	Beams in roofing: ST	Akinsoji 1996
								H: FO	IH: Construction	IH: Timbers & Carving	Light construction	LI (BD-F), LI (BD-M)
								H: LF	IH: Furniture	IH: Timbers & Carving	Furniture: pinnae	Akinsoji 1996
Euphorbiaceae	*Ricinus communis*	Castor oil plant	Zurma, Kula kula		S		H: RO		MH: Yellow fever	MH: Endemic & Epidemic Diseases	Yellow fever: take RO concoction	Akinsoji 1996
Rubiaceae	*Rothmannia whitfieldii*				ST	F					GGH specimen	
								H: ?	IH: Dye	IH: Crafts & Skills	Source of dye to decorate pottery	Akinsoji 1996
Gramineae	*Rottboellia exaltata* ?	Grass ("Buffalo grass")	Cewa, Ceyawa*	Fafe, Gene bana, Sanba	G							
							B: SE ST		FB: Edible	FB: Edible	Food	Warren p.c.
Salvadoraceae	*Salvadora persica*	Mustard tree, Toothbrush tree	Farin ashuwaki, Asawaki		ST	S						
							H: RO		MH: Gonorrhoea	MH: Fertility & Genitalia	Gonorrhoea: boil RO w. potash & drink	Akinsoji 1996, NCF n.d.
Solanaceae	*Schwenkia americana*		Dandana	Dupuhi	H		H: WP		MH: Worms	MH: Parasites	Worms: boil WP w. potash & give to children	Ayanbamiji 1996, LI (GK-MA)
Polygalaceae	*Securidaca longepedunculata*	Violet tree	Uwan magunguna, Sanwa	Alali	ST	S						
							H: RO		MH: Body pain	MH: Analgesic	Body pain: soak RO & RO of Annona & burogali (Hausa)	NCF n.d.
							H: RO		MH: Headache	MH: Analgesic	Headache: pound RO, dry. Add to water & drink or inhale smoke by placing over charcoal	Martin 1996
							H: LF		MH: Colds	MH: Fevers	Colds: boil LF & drink	LI (GK-MA)
							H: RO		MH: Colds	MH: Fevers	Colds: rub the body or drink like tea	LI (CH-F)
							H: RO BA		MH: Stomach problems	MH: Gastrointestinal	Stomach problems: use w. potash	LI (BD-F)

(continued)

Appendix (continued)

Family	Species	English	Hausa	Fulfulde	Habit	Habitat	Food	Medicine	Implement	Usage Description	Usage Category	Description in Original Source	Source
								H: RO		MH: Weakness	MH: Health promotion	Weakness: soak RO & RO of Annona & burogali (Hausa)	NCF n.d.
								H: BA		MH: Kumburan kayi	MH: Unknown	Kumburan kayi*: dry BA dry & sniff	NCF n.d.
									H: LF	IH: Evil spirit	IH: Magico-religious	Bad spirits: pound LF, boil & drink 2xday to send away bad spirit	LI (S3-F)
									H: LF	IH: Evil spirit	IH: Magico-religious	Worrying: boil LF, mix w. rimajoga & barkeji & inhale the vapour when something worries you	LI (GK-MA)
									H: RO	IH: Evil spirit	IH: Magico-religious	Drive away witches, wizards & evil spirits: RO infusion	Akinsoji 1996, NCF n.d.
									H: RO	IH: Orientation	IH: Magico-religious	Know way in the bush when missing: RO & RO of Annona	NCF n.d.
Leguminosae - Caesalpinoideae	*Senna alata*	Ringworm tree			S								
								H: LF		MH: Skin disease	MH: Skin	Skin diseases: rub LF on skin	Akinsoji 1996
Leguminosae - Caesalpinoideae	*Senna obtusifolia*	Foetid cassia	Tafasa, Raidauri	Yakamji, Jambajohi, Mamatasba, Tasba	S								
							H: LF			FH: Edible	FH: Edible	Make soup	LI (GK-MA), LI (FI-M)
								H: RO		MH: Blood	MH: Blood	Blood becomes white: boil RO & drink	LI (H-M)
								H: WP		MH: Swollen glands	MH: Analgesic	Swollen parts: grind WP, mix w. ash & rub onto skin for swollen glands on throat	LI (GK-M)
								H: LF		MH: Colds	MH: Fevers	Cough, colds: bathe, in the rainy season	LI (TI-F)
								H: LF		MH: Fever	MH: Fevers	Fever: mix w. jabbasuka, boil LF & bathe	LI (GK-F)
								H: BA		MH: Stomach ache	MH: Gastrointestinal	Ache near navel: boil BA & give to children	LI (CK-F)
								H: RO		MH: Vomiting	MH: Gastrointestinal	Stop vomiting: pound RO & mix w. nes-cafe & drink	LI (FI-M)
								H: RO		MH: General health	MH: Health promotion	Boil RO w. dandana & kimba, give to children	LI (GK-MA)
								H: RO		MH: Pneumonia	MH: Internal organs & Respiration	Pneumonia: mix w. other tree	LI (GK-MA)
								H: LF		MH: Worms	MH: Parasites	Ringworm: LF	Ayanbamiji 1996
								H: BA		MH: Boils	MH: Skin	Boils: rub BA over body	LI (KV-F), LI (KM-M)
								H: LF		MH: Skin infection	MH: Skin	Eczema: LF	Ayanbamiji 1996
									A: WP	IA: Spirits	IA: Magico-religious	Spirit problems: squeeze WP, rub over cow & drink sap to relieves spirit problems	Martin 1996
Leguminosae - Caesalpinoideae	*Senna occidentalis*	Nigerian senna	Mama tasba	ditto	S								
								H: RO		MH: Stomach ache	MH: Gastrointestinal	Stomach ache: boil RO w. potash	Akinsoji 1996
								H: LF		MH: Worms	MH: Parasites	Ring worm: rub LF on skin	Akinsoji 1996
Pedaliaceae*	*Sesamum indicum*	Sesame, Benniseed	Karikashi, Ridi	Malasiri	H								
							H: WP			FH: Edible	FH: Edible	Pot herb	Akinsoji 1996
Malvaceae	*Sida spp.*		Miyan tsanya	Sadori	L	F							
									H: WP	IH: Broom	IH: Crafts & Skills	Brooms	Akinsoji 1996
									H: WP	IH: Rope	IH: Crafts & Skills	Rope	Akinsoji 1996
Zingiberaceae	*Siphonochilus aethiopicus*	Wild ginger	Lanyar kwadi	Borihi	H								
							B: FR ST			FB: Edible	FB: Edible	Food	Warren p.c.
Smilacaceae	*Smilax anceps*	West African sarsparilla	Kayar kusu	Libano	T								

Solanaceae	*Solanum aculeastrum*	Poison apple	Ido shanu	Gite nai	H		B: FR			FB: Edible	FB: Edible	Food	Warren p.c.
								A: SE		MA: ?	MA: ?	For chickens	LI (BD-F)
								H: RO		MH: Toothache	MH: Oral health	Toothache: boil RO & use as mouthwash	LI (GK-MA)
								H: FR		MH: Skin disease	MH: Skin	Skin disease. Wounds: slash FR & rub on skin	Akinsoji 1996
									H: WP	IH: Fence	IH: Timbers & Carving	Fencing: cultivate around plots to protect against invasion of animals	Akinsoji 1996
Gramineae	*Sorghum bicolor*	Guinea corn	Farin janjaro, Karan dai		C								
							B: ST			FB: Edible	FB: Edible	Food	Warren p.c.
Rubiaceae	*Spermacoce spp.*		Gogamasa	Nyamyarugel	H	LFS							
								H: BA		MH: Yellow fever	MH: Endemic & Epidemic Diseases	Yellow fever: 1 cup	Pellaumail 1998
								H: WP		MH: Gonorrhoea	MH: Fertility & Genitalia	Gonorrhoea: boil WP & potash	NCF n.d.
								H: BA		MH: Skin diseases	MH: Skin	Skin diseases: 1 cup	Pellaumail 1998
Loganiaceae	*Spigelia anthelmia*	Worm grass			H								
								H: BA		MH: Toothache	MH: Oral health	Toothache: boil BA w. potash	Akinsoji 1996
Verbenaceae	*Stachytarpheta indica*	Brazilian tea	Tsarkiyar kusu	Duwerehi	H								
								H: LF		MH: Gonorrhoea	MH: Fertility & Genitalia	Gonorrhoea: boil LF w. potash	Akinsoji 1996
Sterculiaceae	*Sterculia setigera*	Karaya gum tree	Kukuka, Kukuki	Boboli, Bobori	T	S							
							H: SE			FH: Edible	FH: Edible	Edible: SE	Akinsoji 1996
								H: RO		MH: Swollen parts	MH: Analgesic	Swollen parts: boil RO & drink	LI (FI-M)
								H: BA		MH: Blood clot	MH: Blood	Stimulate blood flow after birth to avoid retention of blood clot in uterus: boil BA w. potash	Akinsoji 1996
								H: BA		MH: Menstrual problems	MH: Female reproduction	Menstrual problems: boil BA	LI (BD-F)
								H: BA		MH: Fertility	MH: Fertility & Impotence	Fertility: boil BA & give to barren women	LI (BD-F)
								H: RO		MH: Problems with urination	MH: Fertility & Genitalia	For problems with urination	LI (FI-M)
Bignoniaceae	*Stereospermum kunthianum*		Sansami	Gilohi, Golombi	T	S							
								A: BA		MA: Worms	MA: Parasites	Worms: boil BA mix w. limestone or potash, give through nose	Martin 1996
								H: BA		MH: Dizziness	MH: Analgesic	Dizziness: BA decoction	Akinsoji 1996, NCF n.d.
								H: BA		MH: Cholera	MH: Endemic & Epidemic diseases	Cholera: soak BA in water & drink	LI (CH-M)
								H: BA		MH: Dysentery	MH: Gastrointestinal	Dysentery: BA decoction w. (red) potash	Akinsoji 1996, NFC n.d.
								H: BA		MH: Wound	MH: Skin	Wound healing: grind BA into powder & administer on open wound	Akinsoji 1996, NCF n.d.
Loganiaceae	*Strychnos innocua*		Kokirmo, Kokiyar biri, Kokon biri	Djatibolohi	ST	S							
							H: FR			FH: Edible	FH: Edible	Food	Akinsoji 1996

(continued)

Appendix (continued)

Family	Species	English	Hausa	Fulfulde	Habit	Habitat	Food	Medicine	Implement	Usage Description	Usage Category	Description in Original Source	Source
								H: FR		MH: Testicular problem	MH: Fertility & Genitalia	Swollen testicles: take FR decoction	Akinsoji 1996
Loganiaceae	*Strychnos spinosa*	Green monkey apple, Monkey ball	Kokom biri, Kok ya	Toumande-badi, Kumbija	ST	WS							
							B: FR			FB: Edible	FB: Edible	Food	Warren p.c., LI (GK-MA)
							C: FR			FC: Edible	FC: Edible	Food	LI (GK-VA), LI (KM-F), LI (KM-M)
							H: FR			FH: Edible	FH: Edible	Food	LI (GK-VA)
Myrtaceae	*Syzygium guineense*	Waterberry, Rose apple	Cikakondo, Malmo, Jonga	Bunehi, Buneji, Sumsum, Asurahi*	T	LF MF						GGH specimen	
							B: FR			FB: Edible	FB: Edible	Food	LI (GK-VA)
							B: FR BA GU			FB: Edible	FB: Edible	Food	Warren p.c.
							C: FR			FC: Edible	FC: Edible	Food	LI (GK-VA)
							H: FR			FH: Edible	FH: Edible	Food	Akinsoji 1996, LI (BD-F), LI (GK-MA), LI (KM-F), LI (KM-M), LI (SB-F)
								A: ?		MA: Hengre	MA: Health promotion	Hengre	Martin 1996
								H: BA LF		MH: Fever	MH: Fevers	Fever: 3 pieces of BA +1L	Pellaumail 1998
								H: ?		MH: Dysentery	MH: Gastrointestinal	Dysentery: mix w. potash & add to breakfast porridge for dysentery	Martin 1996
								H: BA		MH: Flatulence	MH: Gastrointestinal	Prevent flatulence: boil BA & drink before meal	LI (FI M)
								H: BA LF		MH: Stomach ache	MH: Gastrointestinal	Stomach ache: 5 modos	Pellaumail 1998
								H: BA LF RO		MH: Stomach problems	MH: Gastrointestinal	Stomach ache: boil & drink 2 cups daily for 3 days for stomach ache. Also bathe baby	LI (KM-F), LI (KM-M)
								H: FR RO		MH: Stomach problems	MH: Gastrointestinal	Stomach ache	LI (BD-F)
								H: BA		MH: Skin disease	MH: Skin	Lumpy skin: add BA to palm oil & use as ointment	Martin 1996
								H: BA		MH: Wound	MH: Skin	Treat wound: BA decoction	Akinsoji 1996
									C: TW LF TR	IC: Nest-tree	IC: Nest	<1% of all nest-trees	Adanu 1998
									H: WO	IH: Roofing	IH: Crafts & Skills	Roofing & internal construction: WO	Akinsoji 1996
									H: BA	IH: Fish poison	IH: Hunting & Fishing	Fish poison: 4 hf-1 basin	Pellaumail 1998
Myrtaceae	*Syzygium owariense*		Malmoa, Malmon rafi		T								
									C: TW LF TR	IC: Nest-tree	IC: Nest	1% of all nest-trees	Adanu 1998
Myrtaceae	*Syzygium staudtii*				T	MF							
									C: TW LF TR	IC: Nest-tree	IC: Nest	1% of all nest-trees	Adanu 1998
Apocynaceae	*Tabernaemontana pachysiphon*	Cow-tree			T							GGH specimen	
							B: FR			FB: Edible	FB: Edible	Food	Warren p.c.
Taccaceae	*Tacca involucrata*	African arrowroot lily	Giginyar biri	Dubbi wandu									
							H: TU			FH: Edible	FH: Edible	Edible after proper preparation	Akinsoji 1996
									H: TU	IH: Starch for clothes	IH: Crafts & Skills	Starch for clothes: dry TU & grind into powder	Akinsoji 1996

Leguminosae - Caesalpinoideae	*Tamarindus indica*	Tamarind	Tsamiya	Jabbe, Jatami, Dafara	T	S DC							
							B: SE			FB: Edible	FB: Edible	Food	Warren p.c.
							H: FR LF FP			FH: Edible	FH: Edible	Edible: cook FR, LF; add FP to porridge	Akinsoji 1996, LI (BD-F), LI (BD-M)
							H: SE			FH: Edible	FH: Edible	Food: SE for balls in kunu (food)	Ayanbamiji 1996
								H: BA		MH: Blood tonic	MH: Blood	Blood tonic: BA decoction for blood formation	Akinsoji 1996, NCF n.d.
								H: LF		MH: Yellow fever	MH: Endemic & Epidemic Diseases	Yellow fever: 2-3 hf	Pellaumail 1998
								H: BA		MH: Gonorrhoea	MH: Fertility & Genitalia	Gonorrhoea: 2 hf	Pellaumail 1998
								H: LF		MH: Fever	MH: Fevers	Fever: boil LF	LI (BD-F), LI (BD-M)
								H: BA		MH: Stomach ache	MH: Gastrointestinal	Stomach ache: BA	Ayanbamiji 1996
								H: LF		MH: Vomiting	MH: Gastrointestinal	Vomiting: 2-3 hf	Pellaumail 1998
								H: BA		MH: Wound	MH: Skin	Wounds: take BA decoction	Akinsoji 1996
									H: ST	IH: Implements	IH: Timbers & Carving	Hoe handle: ST	Ayanbamiji 1996
									H: WP	IH: Shade tree	IH: Timbers & Carving	Shade tree	Akinsoji 1996
Rutaceae	*Teclea verdoorniana*				ST								
									C: TW LF TR	IC: Nest-tree	IC: Nest	1% of all nest-trees	Adanu 1998
Leguminosae - Papilionoideae	*Tephrosia platycarpa*				L								
							B: SE			FB: Edible	FB: Edible	Food	Warren p.c.
Leguminosae - Papilionoideae	*Tephrosia vogelii*	Fish-poison bean	Magimfa	Toke liddi	S								
									H: WO	IH: Firewood	IH: Firewood	Fuel: WO	Akinsoji 1996
									H: LF	IH: Fish poison	IH: Hunting & Fishing	Fish poison: LF	Akinsoji 1996
									H: ?	IH: Soil conditioner	IH: Soil conditioning	Soil conditioning	Akinsoji 1996
Combretaceae	*Terminalia spp.*		Baushe, Bauche, Sansami*	Boderlahi, Bodehi, Ku-lahi Gilohi	T	S LF							
							B: LY SE BA			FB: Edible	FB: Edible	Food	Warren p.c.
								H: RO		MH: Hypertension	MH: Blood	Hypertension: boil RO boil & drink	NCF n.d.
								H: BA		MH: Yellow fever	MH: Endemic & Epidemic Diseases	Yellow fever: 2 modos $ (bowls)	Pellaumail 1998
								H: BA		MH: Yellow fever	MH: Endemic & Epidemic Diseases	Yellow fever: soak BA in hot water, add calcium in water & drink	LI (FI-M)
								H: ?		MH: Malaria	MH: Fevers	Malaria	Martin 1996
								H: BA		MH: Fever	MH: Fevers	High fever & abonakeke: mix BA w. karlahi & konkehi, boil & bathe child	LI (GK-FH)
								H: BA LF		MH: Fever	MH: Fevers	Fever: boil BA & LF and bathe children or give to drink	LI (GK-F)
								H: BA		MH: Stomach ache	MH: Gastrointestinal	Stomach ache: mix w. konkehi, boil BA, bathe & enema	LI (DM-M)
								H: RO		MH: Diarrhoea	MH: Gastrointestinal	Diarrhoea	LI (MS-F)
								H: RO		MH: Dysentery	MH: Gastrointestinal	Dysentery	LI (MS-F)
								H: RO		MH: Jaundice	MH: Internal organs & Respiration	Jaundice: boil RO w. potash	NCF n.d.
								H: BA		MH: Teeth problem	MH: Oral health	Mouthwash: boil BA	LI (KM-F), LI (KM-M)

(continued)

Appendix (continued)

Family	Species	English	Hausa	Fulfulde	Habit	Habitat	Food	Medicine	Implement	Usage Description	Usage Category	Description in Original Source	Source
Combretaceae	*Terminalia avicennioides*		Baushe, Baotche	Boderlahi, Bodehi. Wane	T	S							
								H: RO BA		MH: Diarrhoea	MH: Gastrointestinal	Diarrhoea: RO & BA	Akinsoji 1996
									H: WO	IH: Implements	IH: Timbers & Carving	Implements: WO	Akinsoji 1996
Combretaceae	*Terminalia ivorensis*	Black afara			T	F							
									C: TW LF TR	IC: Nest-tree	IC: Nest	2% of all nest-trees GGH specimen	Adanu 1998
Combretaceae	*Terminalia laxiflora*		Farin baushe. Zinci	Ku ahi	T	LF MF							
							B: FL			FB: Edible	FB: Edible	Food	Warren p.c.
								A: BA		MA: Mastitis	MA: Female reproduction	Mastitis: pound BA, mix w. salt, potash or limestone, give as a lick	Martin 1996
								H: RO		MH: Hypertension	MH: Blood	Hypertension: RO decoction	Akinsoji 1996
								H: ST		MH: Cholera	MH: Endemic & Epidemic diseases	Cholera: soak ST in hot water (I think he means diarrhoea)	LI (S3-F)
								H: BA		MH: Diarrhoea	MH: Gastrointestinal	Diarrhoea: pound BA & soak in cold water. Drink for 4-5 days	LI (GK-F)
								H: RO		MH: Diarrhoea	MH: Gastrointestinal	Diarrhoea: 1 RO-1 hf	Pellaumail 1998
								H: RO		MH: Stomach upset	MH: Gastrointestinal	Stomach upset: 1 RO-1 hf	Pellaumail 1998
								H: BA		MH: Lung problems	MH: Internal organs & Respiration	Lung problems: 1 hf (w. red potash)	Pellaumail 1998
								H: BA		MH: Bilharzia	MH: Parasites	Bilharzia: 1 hf (w. red potash)	Pellaumail 1998
								H: BA		MH: Burns	MH: Skin	Burns: soak BA in cold water, apply to burns	Martin 1996
								H: BA		MH: Abonakeke	MH: Unknown	Abonakeke: boil BA, drink & bathe	LI (GK-FH)
									H: WO	IH: Roofing	IH: Crafts & Skills	Roofing: WO	Akinsoji 1996
									H: WO	IH: Implements	IH: Timbers & Carving	Implements: WO	Akinsoji 1996
Combretaceae	*Terminalia macroptera*	Cluster leaf	Baushe. Kwandari	Ku'ahi	T								
								H: BA		MH: Dysentery	MH: Gastrointestinal	Dysentery: BA	Akinsoji 1996
								H: LF BA		MH: Worms	MH: Parasites	Worm expeller: LF & BA for children	Ayanbamiji 1996
Combretaceae	*Terminalia schimperiana**	Tult	Baushe	Boderlahi, Bodehi	T								
								H: RO		MH: Constipation	MH: Gastrointestinal	RO decoction for constipation	Akinsoji 1996
Leguminosae - Mimosoideae	*Tetrapleura tetraptera*		Dawo	Wong	T	F							
							B: FR			FB: Edible	FB: Edible	Food	Warren p.c.
Marantaceae	*Thalia welwitschii*		Ganye. Fita		H							GGH specimen	
							B: ST			FB: Edible	FB: Edible	Food	Warren p.c.
Balanophoraceae	*Thonningia sanguinea*	Ground pineapple	Kubla										
							H: ST			FH: Seasoning	FH: Seasoning & Processing	Flavour: dry & pound rhizome & add to soup	Akinsoji 1996
								H: RO		MH: Worms	MH: Parasites	Worms: dry & pound rhizome & add to soup	Akinsoji 1996
Rubiaceae	*Tricalysia oligoneura**			Kale	T								
							B: FR			FB: Edible	FB: Edible	Food	Warren p.c.

Euphorbiaceae	*Uapaca spp.*		Wawan kurmi, Kafafago	Mambelko, Bakurehi	T								
							B: FR			FB: Edible	FB: Edible	Food	LI (GK-MA)
							C: FR			FC: Edible	FC: Edible	Food, dry to rainy season	LI (GK-MA)
							H: FR			FH: Edible	FH: Edible	Food	LI (GK-MA), LI (SB-F)
Euphorbiaceae	*Uapaca guineensis*	Red cedar	ditto	ditto	T	F							
							B: FR			FB: Edible	FB: Edible	Food	Warren p.c.
Euphorbiaceae	*Uapaca togoensis*		ditto	ditto	T	S							
							B: FR			FB: Edible	FB: Edible	Food	Warren p.c.
							H: FR			FH: Edible	FH: Edible	Food	Akinsoji 1996
									H: WO	IH: Firewood	IH: Firewood	Make charcoal: WO	Akinsoji 1996
Malvaceae	*Urena lobata*	Congo jute	Uwar magani	Uwar baganoji	L								
									H: BA	IH: Rope	IH: Crafts & Skills	Rope: BA	Akinsoji 1996
Compositae	*Vernonia ambigua*		Hakon diyam, Sarkin banja	Anyaka, Gam gam		S							
							H: LF			FH: Edible	FH: Edible	Food: collect LF & boil during rainy season. In wet areas during dry season	LI (BD-F), LI (BD-M), LI (GK-F), LI (GK-MA)
							H: WP			FH: Edible	FH: Edible	Food: dry WP & use powder for dala soup	LI (GK-MA)
								H: FL		MH: Measles	MH: Endemic & Epidemic Diseases	Meisi: mix purple FL w. shiwo laynde, boil, drink & batho	LI (GK-FH)
								H: RO		MH: Gonorrhoea	MH: Fertility & Genitalia	Gonorrhoea: boil & drink RO	LI (FI-M)
									H: LF	IH: Evil spirits	IH: Magico-religious	Evil spirit: burn 2x15 LF	Pellaumail 1998
Compositae	*Vernonia amygdalina*	Bitter leaf	Shuwaka	Shuwaka	ST	DC F S						GGH specimen	
							H: LF			FH: Edible	FH: Edible	Food: LF eaten as vegetable	Akinsoji 1996
							H: LF			FH: Edible	FH: Edible	Food: make dala soup	LI (GK-F)
								H/A: ? LF		MA / MH: Worms	MA / MH: Parasites	Worms: pound LF, add to water, filter, add to limestone & drink	Martin 1996
								H: LF		MH: Yellow fever	MH: Endemic & Epidemic Diseases	Yellow fever: 1-2 hf	Pellaumail 1998
								H: LF		MH: Lactation	MH: Female reproduction	Produce milk: mix w. ganye gwanda, make soup & give to lactating women	LI (GK-F)
								H: LF		MH: Fever	MH: Fevers	Fever: LF	Akinsoji 1996
								H: LF		MH: Constipation	MH: Gastrointestinal	Constipation: 1-2 hf	Pellaumail 1998
								H: LF		MH: Flatulence	MH: Gastrointestinal	Bloated stomach: drink or use as enema	LI (GK-F)
								H: LF		MH: Stomach ache	MH: Gastrointestinal	Stomach ache: LF	Akinsoji 1996
								H: TW		MH: Chewing stick	MH: Oral health	Chewing stick	Akinsoji 1996, Pellaumail 1998
								H: LF		MH: Worms	MH: Parasites	Worms: 1-2 hf: drink or give as enema to children	Pellaumail 1998, LI (CH-F), LI (GK-F), LI (SB-F)
								H: LF		MH: Skin conditions	MH: Skin	Skin problems: boil LF & rub onto skin	LI (GK-MA)
								H: RO LF		MH: Skin disease	MH: Skin	Skin disease: grind RO & LF (fresh) grind & rub affected part; batho in LF infusion	Akinsoji 1996, NCF n.d.
									H: WO	IH: Firewood	IH: Firewood	Firewood: use debarked RO	Akinsoji 1996
Sapotaceae	*Vitellaria paradoxa*	Shea butter tree	Kadanya	Karehi, Karelo, Kareli	T	S							

(continued)

Appendix (continued)

Family	Species	English	Hausa	Fulfulde	Habit	Habitat	Food	Medicine	Implement	Usage Description	Usage Category	Description in Original Source	Source
							A: FR			FA: Edible	FA: Edible	Edible: FR, SE produce edible oil	Ayanbamiji 1996
							B: FL FR			FB: Edible	FB: Edible	Food	Warren p.c., LI (GK-MA)
							H: ?			FH: Edible	FH: Edible	Food: mix w. kadanya to make oil	LI (GK-MA)
							H: FP			FH: Edible	FH: Edible	Food	Akinsoji 1996
							H: FR SE			FH: Edible	FH: Edible	Food: FR edible. SE produce edible oil	Ayanbamiji 1996, LI (9GK-F)
								A: ?		MA: Pacify	MA: Psychiatric diseases	Add salt for animals whose hair is standing on end	Martin 1996
								H: ?		MH: Bone weakness	MH: Bones	Bone weakness	NCF n.d.
								H: SE		MH: Sprains & fracture	MH: Bones	Treat sprains, fractures & dislocation: oil extracted from SE	Akinsoji 1996, Ayanbamiji 1996
								H: SE		MH: Rheumatism	MH: Bones	Rheumatism: SE oil	Ayanbamiji 1996
								H: ?		MH: Cough	MH: Fevers	Cough: add salt	Martin 1996
								H: SE		MH: Colds	MH: Fevers	Cold: 1-2 SE	Pellaumail 1998
								H: SE		MH: Colds	MH: Fevers	Cold: SE oil	Ayanbamiji 1996
								H: SE		MH: Stomach ache	MH: Gastrointestinal	Stomach ache (pregnant women): 1-2 SE	Pellaumail 1998
								H: BA		MH: General health	MH: Health promotion	BA is medicinal	Ayanbamiji 1996
								H: BA		MH: Toothache	MH: Oral health	Toothache: 1 hf	Pellaumail 1998
								H: BA		MH: Toothache	MH: Oral health	Toothache: boil BA & use as mouthwash	LI (GK-F)
								H: BA		MH: Worms	MH: Parasites	Worms in children: boil BA & use as enema	LI (GK-F)
								H: ?		MH: Mange	MH: Skin	Mange: add salt	Martin 1996
								H: BA		MH: Skin conditions	MH: Skin	Skin problems: pound BA & make oil	LI (GK-F)
								H: FR		MH: Skin softening	MH: Skin	Wound scarca: oil from FR to smooth wound scarca	NCF n.d.
								H: FR		MH: Skin conditions	MH: Skin	Skin problems: dry & pound FR to make oil	LI (GK-F)
								H: SE		MH: Skin softening	MH: Skin	Smoothen keloids on skin: rub on scars oil extracted from SE	Akinsoji 1996, Ayanbamiji 1996
								H: FR BA		MH: Abonakeke	MH: Unknown	Abonakeke: boil FR & BA & bathe	LI (GK-F)
									C: TW LF TR	IC: Nest-tree	IC: Nest	4% of all nest-trees	Adanu 1998
Verbenaceae	*Vitex doniana*	Black plum, West African plum	Dinya	Galbihi, Galbije	T	SF MF							
							A: LF			FA: Edible	FA: Edible	Cows eat LF	LI (FI-F)
							B: LY FR			FB: Edible	FB: Edible	Food	Warren p.c., LI (GK-MA)
							C: FR			FC: Edible	FC: Edible	Food	Fowler p.c., LI (GK-MA)
							H: FR LF			FH: Edible	FH: Edible	Food, LF for soup	Fowler p.c., LI (FI-F), LI (GK-F), LI (GK-MA), LI (MS-F), LI (SB-F)
							H: FR SE			FH: Edible	FH: Edible	Edible: FR & SE	Akinsoji 1996
								A: BA		MA: ?	MA: ?	Mix BA & limestone & drink. milk cow & inject w. own milk	Martin 1996
								A: BA		MA: Diarrhoea	MA: Gastrointestinal	Diarrhoea: soak BA into water & give cattle to drink	Ahmadu
								A: BA		MA: Dysentery	MA: Gastrointestinal	Dysentery: cook BA & give to calves	Martin 1996
								A: ?		MA: Liver	MA: Internal organs & Respiration	Liver problems	Martin 1996
								B: FR		MB: Worms	MB: Parasites	Worms: baboons eat to remove worms	LI (FI-F)

								H: BA		MH: Yellow fever	MH: Endemic & Epidemic Diseases	Yellow fever: BA	Akinsoji 1996
								H: ? LF		MH: Post-partum	MH: Female reproduction	Post-partum problems: drink or eat after birth to cleanse blood	Martin 1996, LI (FI-F), LI (SB-F)
								H: ?		MH: Dysentery	MH: Gastrointestinal	Dysentery	Martin 1996
								H: BA		MH: Stomach ache	MH: Gastrointestinal	Stomach ache: boil BA & drink	LI (GK-F)
								H: LF FR		MH: Dysentery	MH: Gastrointestinal	Dysentery: LF & FR	Akinsoji 1996, NCF n.d.
								H: SF		MH: Stomach problems	MH: Gastrointestinal	Stomach pain: cut SF, put in hot water & give children to drink	LI (SB-F)
								H: LF BA		MH: General health	MH: Health promotion	Boil BA & LF & give to children	LI (MS-F)
								H: BA		MH: Worms	MH: Parasites	Worms: mix BA w. Parkia, boil & give to children	LI (GK-MA)
								H: BA		MH: Skin disease	MH: Skin	4-5 hf to 1/2 basin for skin diseases (for bathing infants esp.)	Pellaumail 1998
								H: BA		MH: Skin conditions	MH: Skin	Skin problems: boil BA & bathe or drink	LI (GK-F)
									C: TW LF TR	IC: Nest-tree	IC: Nest	3% of all nest-trees	Adanu 1998
									H: WO	IH: Drums	IH: Crafts & Skills	Construction of drums: WO	Akinsoji 1996
									H: WO	IH: Firewood	IH: Firewood	Fuel: WO	Akinsoji 1996
Olacaceae	*Ximenia americana*	Wild olive, Blue sourplum	Tsada	Chabbule	T	S							
							B: FR			FB: Edible	FB: Edible	Food	Warren p.c., LI (GK-MA)
							C: FR			FC: Edible	FC: Edible	Food	LI (GK-MA)
							H: FR RO			FH: Beverage	FH: Beverages	Make simi drink w. RO	LI (GK-MA)
							H: FR			FH: Edible	FH: Edible	Food	Akinsoji 1996, LI (GK-FH)
								H: FR		MH: Pain	MH: Analgesic	Pain relief	LI (BD-F)
								H: RO		MH: Menstrual problems	MH: Female reproduction	Heavy menstrual flow: mix RO w. limestone & drink	LI (ST-MH)
								H: RO		MH: Gonorrhoea	MH: Fertility & Genitalia	Gonorrhoea	LI (GK-MA)
								H: SF		MH: Syphilis	MH: Fertility & Genitalia	Syphilis: squeeze JU from SF into patient's eye	Martin 1996
								H: RO		MH: Stomach ache	MH: Gastrointestinal	Stomach ache: RO	Akinsoji 1996
								H: LF		MH: Toothache	MH: Oral health	Toothache: LF decoction (gargled)	Akinsoji 1996, NCF n.d.
								H: BA		MH: Abonakoke	MH: Unknown	Boil BA & bathe for abonakoke	Gashaka FH
Annonaceae	*Xylopia spp.*	African pepper	Kimba	Kimbaro, Gosoji	T	F							
							B: FR			FB: Edible	FB: Edible	Food	LI (GK-MA)
							B: SF			FB: Edible	FB: Edible	Food	Warren p.c.
							C: FR			FC: Edible	FC: Edible	Food	LI (GK-MA)
							H: FR			FH: Edible	FH: Edible	Food: mix w. pepper & ginger & drink like tea or mix w. kunu	LI (GK-MA), LI (SB-F)
							H: LF			FH: Edible	FH: Edible	Food: make soup or cook w. meat	LI (BD-F), LI (BD-M), LI (SB-F)
								H: FR		MH: Post-partum	MH: Female reproduction	Cleanse after labour: add spices like pepper to cleanse women after birth	LI (BD-F), LI (BD-M)
								H: LF		MH: Labour	MH: Female reproduction	Facilitate labour: put LF in kunu w. pepper & wood ash to facilitate labour. For 2 weeks	LI (SB-F)
								H: BA		MH: Stomach problems	MH: Gastrointestinal	Stomach problems: boil BA	LI (BD-F), LI (BD-M)
									H: WO	IH: Roofing	IH: Crafts & Skills	Roofing	LI (GK-MA)

(continued)

Appendix (continued)

Family	Species	English	Hausa	Fulfulde	Habit	Habitat	Food	Medicine	Implement	Usage Description	Usage Category	Description in Original Source	Source
Rutaceae	*Zanthoxylum leprieurii*		Fasa kwari, Karya gatari*	Fasakorihi	T	F S							
								H: BA		MH: Toothache	MH: Oral health	Toothache: boil BA w. potash & wash tooth	NCF n.d.
									H: WO	IH: Carving	IH: Timbers & Carving	Carving: WO	Akinsoji 1996
									H: WO	IH: Implements	IH: Timbers & Carving	Pestle & drums: timber	Akinsoji 1996
Rutaceae	*Zanthoxylum zanthoxyloides*	Candlewood	Fasa kwari, Karya gatari*	Fasakorihi	T	F S							
							H: BA			FH: Seasoning	FH: Seasoning & Processing	Spice: BA (screened elsewhere as a cure for sickle-cell disease)	Akinsoji 1996
								H: RO		MH: Chewing stick	MH: Oral health	Chewing stick: RO	Akinsoji 1996
Gramineae	*Zea mays*	Maize	Masara	Butali	C								
							B: FR ST			FB: Edible	FB: Edible	Food	Warren p.c.
Rhamnaceae	*Ziziphus spina-christi*	Christ's thorn	Kurna	Kurnahi	T	S							
							H: FR			FH: Edible	FH: Edible	Edible: FR, potential oil extraction from SE	Akinsoji 1996
							H: SE			FH: Edible	FH: Edible	SE edible	Ayanbamiji 1996
								A: BA		MA: Fertility	MA: Fertility & Genitalia	Increase fertility: BA for early maturity & increased reproduction in female cattle	Akinsoji 1996, Ayanbamiji 1996
								H / A: ?		MH / MA: ? Fertility	MH / MA: Fertility & Genitalia	Fertility drug	Martin 1996
								H: LF		MH: Chicken pox	MH: Endemic & Epidemic diseases	Chicken pox: LF	Akinsoji 1996
								H: LF		MH: Skin infections	MH: Skin	Skin infections: LF	Akinsoji 1996
								H: BA		MH: Ciwon daji	MH: Unknown	Ciwon daji: BA & potash pound it & rub on the affected point	NCF n.d.
									H: ST WO	IH: Slates	IH: Crafts & Skills	Carve Koranic slates: ST WO	Akinsoji 1996, Ayanbamiji 1996
									H: FR BA	IH: Good luck	IH: Magico-religious	Good luck: FR & BA of durumi, gamji & complete vadiva	NCF n.d.

Chapter 6
Fulani of the Highlands: Costs and Benefits of Living in National Park Enclaves

David Bennett and Caroline Ross

Abstract Attempts to deal with human–animal conflicts are often beset with difficulties due to a lack of local consultation and appreciation of local knowledge systems. We address these issues by investigating human–wildlife conflict, specifically crop-raiding, as experienced by a population of Fulani living within the highland enclaves of Gashaka Gumti National Park. We classify sources of risk and perceived benefits experienced by the Fulani, who are traditionally pastoralists but have taken on a settled lifestyle. By distinguishing between risk incidence and severity of subjective risk perceptions we view crop-damage in the context of other costs and benefits associated with living in the enclaves and more specifically with living inside a national park. Attitudes toward specific crop depredating wildlife and domestic animals are also investigated, as well as reasons for recent changes in agricultural productivity and cattle herd dynamics. Results show that prosperity of livestock was considered the greatest benefit of living in the highland enclaves, whilst the secure environment afforded by the park authorities, contributing to the safety of family and cows, was identified as the greatest benefit of living inside the national park. The greatest perceived risk to local livelihoods associated with living in the enclaves was damage to crops, whilst the most serious problem with living in the park was prohibitions on the use of local resources. Animal damage by 16 species was rated as the risk that had the greatest negative impact on crop yields. The olive baboon was considered to pose the greatest threat to subsistence. Comparisons then made with actual recorded amounts of damage revealed a significant disparity. Why such a disparity should exist is complex. The Fulani live in the enclaves, not because they wish to farm, but because it is a good place to rear cattle. Adverse changes in crop productivity are perceived to have a direct negative impact on herd size, but the socio-economic status of individuals may well determine the

D. Bennett (✉)
Centre for Research in Evolutionary Anthropology, Roehampton University, London, SW15 4JD, UK
e-mail: c.ross@roehampton.ac.uk

V. Sommer and C. Ross (eds.), *Primates of Gashaka*, Developments in Primatology: Progress and Prospects 35, DOI 10.1007/978-1-4419-7403-7_6,

reasons for diversification from pastoralism into farming. This may influence the identification of particular sources of risk which, in turn, affects how an individual perceives the park authorities and the animals they are charged to protect.

Keywords Fulani • Human–wildlife conflict • Rapid rural appraisal • Risk • Benefit

Introduction

Farmer conflict with wildlife is not a new phenomenon in rural Africa, particularly amongst populations that live adjacent to remote or protected areas and habitats rich in wildlife. Domestically reared crops whether growing in the field or lying in store are an extremely cost efficient food in relation to the majority of wild foods available and animals find such a potential bonanza appealing. Crop damage and livestock depredation by wildlife can be a considerable drain on the economy of small scale subsistence farmers (Newmark *et al.* 1994, Tchamba 1996) whose coping strategies are hampered by an ever increasing scarcity of viable agricultural land and legislative hunting prohibitions (Naughton-Treves 1997). If such conflicts are not resolved it is possible that finally the persistence of depredating species may well depend upon the farmer's tolerance to the species' perceived threat. By engaging in poaching and habitat destruction, through illegal settlement and short-term resource use, the long-term survival prospects of rare and endangered species can be severely curtailed. Local farmers are unlikely to be sympathetic to aesthetic or ecological arguments promoting wildlife conservation when their very livelihoods are under threat (Balmford & Whitten 2003); and this threat can be very real.

However, many studies focusing on crop-raiding report a disparity between actual and perceived loss. The reasons for this disparity are often complex. It is possible that exaggerated and inaccurate accounts of wild animals damaging crops may be an imagined or contrived response to other more tangible conflicts set between people; individuals, local populations and possibly even the state (Gillingham & Lee 2003). So, whilst measurements of crop damage have an important role in determining loss, quantifiable measurements of damage alone may not provide an adequate representation of the actual impact of crop depredation upon affected populations (Bell 1984, Hill *et al.* 2002, Naughton-Treves & Treves 2005, Priston 2005). It is then important to investigate attitudes and perceptions of farmers towards species with which they share the environment, as well as measure damage to crops, in order to determine any mismatch between these perceptions and the probable economic impact of wild depredating species. Furthermore, it is important to distinguish human–wildlife conflict among the *many* risks and problems faced by the local population. Only by understanding why farmers prioritise among these risks can any appropriate management strategies be developed (Newmark *et al.* 1994, Naughton-Treves 1998).

In this context, the current study focuses on Fulani farmers living within Nigeria's Gashaka Gumti National Park (GGNP) in areas specifically demarcated for agriculture and livestock grazing, known as the enclaves.

The Fulani

The Fulani (sing. Pullo) are among the most widely distributed people of the African continent (*Fig. 6.1*). They are found throughout the Sahelian region from the Atlantic coast in the west to Ethiopia in the east. This dispersion across such a vast geographical space over a period of centuries has resulted in the development of diverse modes of subsistence necessary to meet the demands of particular ecological and historical factors (Frantz 1993). However, an economy of nomadic cattle-breeding is reputedly integral to Fulani spiritual and material culture (Zubko 1993), and it is claimed that even the sedentarised Fulani, many of whom have adopted a purely agrarian means of subsistence, maintain an affinity for cattle-breeding. The Fulani that inhabit the highland enclaves in Nigeria's north-eastern Gashaka Gumti National Park practice a semi-sedentary life-style in which households combine settlement, agriculture and cattle movement.

Fig. 6.1 Fulani life. (**a**) A Pullo herdsman at home. (**b**) Boy herding cattle. (**c**) The village of Selbe in the highland enclave of Tale, with surrounding grazed areas (Photos: David Bennett)

The Fulani have had a presence in the Gashaka region since a mid-19th century Jihad was extended to the regions of Koncha, Banyo (both now in Cameroon) and Gashaka, where a sub-emirate based on the principles of Islam was established (Nyuidze 2001). The historical narrative of the Fulani presence in the region permeates the social relations that the Fulani enjoy with each other and the non-Fulani, acting as a starting point from which conflict within the Fulani population of the park is experienced. Indeed, the three traditional chiefdoms within the Gashaka-Gumti region are still dominated by the Fulani. The Lamido (emir or chief) of Gashaka, (who was Alhaji Hamman Gabdo Muhammadu Sambo during the time of this study) is based in Serti and controls the southern area of the park. The traditional authority that the Lamido represents still retains a considerable degree of influence, political power, and respect within the region – a fact attested to by the ruling family's continued involvement in the affairs of local government and the management committee of GGNP. The majority of the people in the enclaves believe that it is only through the continuing influence of the present Lamido of Gashaka that the eviction of the local populations situated in the southern sector of the protected area have not taken place.

The Fulani now living in the enclaves were initially attracted to the region by the presence of significant tracts of montane grasslands subject to a regular high rainfall affording access to abundant pasture, coupled with a scarcity of tsetse fly responsible for the transmission of trypanosomes to cattle and other animals (Dunn 1995). Initial transhumance incursions by the Fulani in the 1940s were followed by later permanent settlement. When the game reserve was created in 1972 the Fulani were not evicted but participated in the creation of the park's enclaves.

Present problems voiced by the Fulani did not begin to materialise until the Nigerian Federal Government decreed the area a national park in 1991, undermining the authority of the traditional institutions that had previously managed the enclaves. Responsibility for the management of the park was then vested in the National Park Service (NPS) under the authority of the Federal Ministry of Agriculture and Natural Resources, along with a variety of other influential stakeholders, including the local government, national and international conservation NGOs and the traditional ruling authorities. The NPS then began action to evict two farming settlements now considered illegal and situated in close proximity to the main body of Fulani households (NPS / NCF / WWF 1998). Although such action was started in 1996, Tounga and Italal were not evicted until 2002 and 2003, respectively. The removal of these farming communities placed the resident Fulani in a difficult position. With no more access to cheap locally available agricultural produce the Fulani had to either import food at great expense into their highland homes from the lowland agricultural areas from within or outside the park, or dramatically increase the yield of their own household gardens. Thus, the Fulani are now forced to farm in order to sustain their families and keep their cattle herds. Loss of the cattle herds is unthinkable for the Fulani as this would be annihilate their culture and ethnic identity (Salzman 1980, Frantz 1982, Khazanov, 1984).

The Same Old Problem?

The negative impact of wildlife on local people's subsistence, as well as a shortage of quality farmland, has been highlighted by previous studies focusing on the attitudes that local people have toward GGNP and the problems they may face living in a protected area (Dunn 2000). Furthermore, local farmers claimed that many of the animals responsible for crop depredation are on the increase due to the creation of the national park and the prohibitions on hunting, trapping and poisoning. Andrew Dunn, a previous assistant project manager working for the Nigerian Conservation Foundation (NCF), points to the results of a survey performed in 1993, which estimated that crop damage by wildlife pests may account for as much as one third of the annual crop harvest (Dunn 2000). A later 2003 survey by a PhD researcher found that of all the 206 people interviewed within the park, 95.8 % claimed to have experienced wildlife depredation of their crops (Warren 2003). Such extreme amounts of damage would threaten the subsistence of all but the richest Fulani within the park. Indeed, economic pressures combined with the lack of roads, schools and clinics, have forced many Fulani to re-think the viability of their life within the enclaves as they are forced to sell livestock in order to meet their subsistence needs (NPS / NCF / WWF 1998).

Here we look at the experiences of the Fulani in recent years and the impact that living in the enclaves has on their lives.

Methods

Study Area and Data Collection

There are three quite distinct enclave types (*Fig. 6.2*): the highland enclaves (1650 – 1950 m asl.) of Nyumti, Tale, Shirgu, Hendu and Delam; sub-plateau enclaves (900 – 1200 m asl.) of Sabere and Filinga; and the lowland enclave (330 m asl.) of Gumti. The enclaves can only be reached on foot, by walking at least 4 – 5 h from the park border. The more remote settlements require hikes of at least 2 days to reach the nearest motorable road. Only minimal facilities exist for primary schooling. There are no hospitals, and the enclaves have neither television nor mobile phone coverage.

This study took place in the highland enclaves, and focused on 83 Fulani compounds spread amongst the surrounding hills and the small village of Selbe, which is also populated by some 20 households representing various West African ethnicities.

As part of a trans-disciplinary study into human–wildlife conflict, data were collected by DB from the heads of each Fulani household ($n = 83$), from Nov 04 – Apr 06 using Rapid Rural Appraisal (RRA) techniques (Chambers 1992). RRA enables outsiders to gain information and insight from rural people about rural conditions. It emphasises the importance of gaining rapport to learn from and utilise

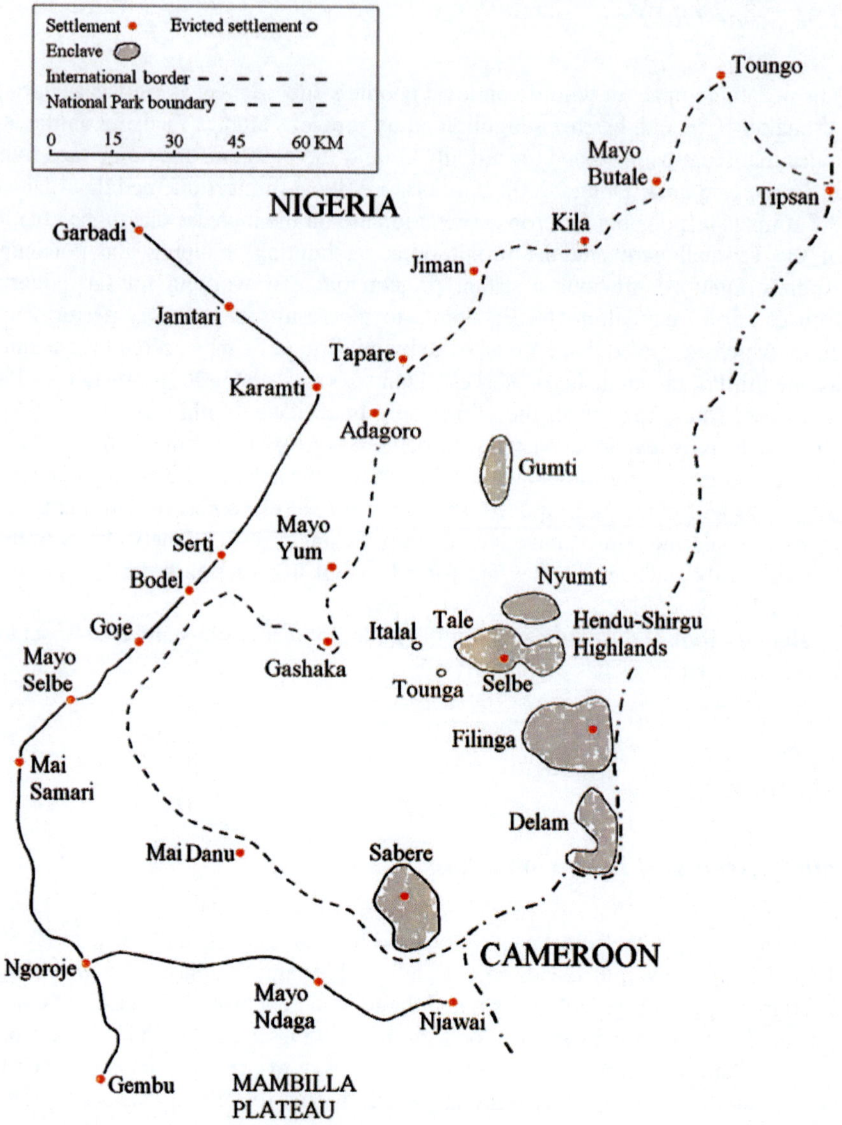

Fig. 6.2 Study area in Gashaka Gumti National Park / Nigeria

local knowledge systems to identify the problems, goals, and strategies of rural populations expressed in their own terms as opposed to those ideas defined by some outside perspective (Ghai & Vivian 1995). We follow Chambers' (1997) suggestion that the "R" should stand for "relaxed" rather than "rapid", and so minimise inaccurate or biased findings resulting from an unreflective application of RRA techniques. Relaxed techniques utilise non-formal data capturing, promoting the idea of an extended period of field residence combined with unhurried participant observation, and informal as well as formal conversations and interviews allowing the researcher to gain information and insight from rural people about rural conditions.

RRA was initially employed in Selbe (where DB lived for the duration of the study) to obtain an overview of problems the local population experienced in their daily lives. RRA discussions were held with residents of the highland enclaves in various group sizes. The group dynamic encourages a collective response and helps to pinpoint differences in opinion as well as areas of consensus. These discussions were always backed up by individual interviews and direct observation. Meetings were generally held in Fulani visitor huts, positioned at the entrance of the family compound (*sa'are*), where neighbours and fellow farmers would often gather to enjoy food and company. It was hoped that people would feel relaxed in this familiar environment and would find it more conducive to conversation than the researcher's hut, the bush or marketplace. Women were rarely present at such gatherings. Due to the strong hierarchical gender-structure of Fulani households it proved very difficult for DB and his field assistants (all male) to communicate with women in the presence of their husbands, who would simply answer for them. The husbands had no problem in granting permission to interview the older wives in the husband's absence, but younger wives were extremely shy and, as the husbands were sensitive to requests for their young bride's cooperation, the matter was never pushed. Overall, 119 male and 20 female respondents were interviewed. The mean average age of male respondents was 42.6 ± 14.9 years (range 23 – 86 years, median = 41) and the mean average age of female respondents was 41.9 ± 13.5 years (range 22 – 75 years, median = 40). All of the male respondents were heads of households with mean average cattle holdings of 56 ± 69.3 (range 0 – 333, median = 33). The females were all wives who expressed an interest in participating.

Baseline data were gathered regarding the demographic composition of the highland enclaves, natural resource use, local perceptions of the local wildlife, human–wildlife interactions, work habits, attitudes toward identified hegemonic structures, needs of the local population and portrayed perspectives upon any identified insecurities or problems. This information was the source for the questions used in the attitudinal semi-structured questionnaire administered to the local population (Feb – May 06). The combination of a formal structured survey together with qualitative research methods allows cross-checking and collation of information from different sources and facilitates research to be conducted on complex and sensitive issues (Bulmer & Warwick 1983, Nichols 1991). All interviews and discussions were conducted in Fulfulde with the help of a translator, Yakubu Ahmadu, a respected Gbaya (one of the main cultural groups of northern equatorial Africa) and son of the Mai'ungwa (village head) of Deu Mayo, a small settlement located just beyond the park boundary, but within the support zone of GGNP. Yakubu Ahmadu had previous experience working as a research assistant and translator in the region and was already known to the Fulani of the highland enclaves.

Measuring Crop Damage

To gain a perspective on the economic impact of crop-raiding on local farmers, damage was measured on selected farms over an entire growing season from Mar 05 – Apr 06. Farms ($n = 35$) were selected for practical reasons due to their proximity to Selbe village. Farm sizes ranged from 0.06 ha to 2.28 ha (mean 0.68 ha). Crop types and time of planting were recorded and the baseline seedling density of each crop was estimated (before any damage occurred). Density was measured by counting the number of seedlings/plants per square meter. This record was then used to estimate the starting total number of plants for each cultivated species for each field. Farms were then visited three times in a 20-day period, with visits no closer than 6 days apart.

In the highland enclaves montane grassland is cultivated primarily for the production of maize (*Zea mays*). However, a variety of crops are grown locally, including carbohydrate staples such as cassava, sweet potato, potato (*Solanum tuberosum*, known as "Irish potato" in West Africa), cocoyam, finger millet, legumes (e.g., beans) and groundnuts. Some vegetables and bananas are also widely cultivated. Avocado pear, guava, and cola nut trees are planted in the village of Selbe and some farmers are now growing eucalyptus trees for firewood and building poles. For most farmers there is only one growing season during the agricultural year. Maize is planted in April or May and harvested by the beginning of October. Potatoes, sweet and Irish, are planted between May and June and harvested in December; and recently in Nyumti farmers have taken to planting finger millet in July to be harvested in December. Cocoyam is planted in April or May and harvested after December when needed.

With the help of a local expert field assistant, records were obtained when damage was observed, noting crop type and age, nature of the damage (whether the crop was top-cut, dug up or totally destroyed, etc.), as well as the extent of the damage throughout the field. This was achieved in a number of ways depending on crop type. In the cases of maize, cocoyam, millet, cassava, beans, and bananas the numbers of ruined plants were individually counted. In the case of sweet and Irish potatoes the disturbed area of the crop stand in square metres was used to measure the damage. All instances of crop damage were catalogued to attempt an independent assessment of the animal responsible. Species were identified through damage type, bite marks, tracks, and spoor.

Damage measured on the focal farms was used to rank crop-damaging species, by creating a severity index and incidence index (see below). Damage to all variety of crops was included in the analysis. The approximate economic value of each crop type was used to calculate the relative cost to the farmer. Maize, as the most abundant and important crop, was used as the basic unit of damage against which other crop types were measured. Appropriate damage values were then assigned to other crop types relative to maize, dependent on the calculated cost of the approximate edible yield of a single plant. Data on crop damage measured over the entire growing season was used to estimate the total annual damage caused by each crop-raiding taxa. Using the assigned damage value per plant it was then possible to rank the potential risk of crop depredating species and so create a measured risk index.

Risk Analysis

For the purposes of this study, "risk" is defined as "uncertain consequences [...] in particular exposure to potentially unfavourable circumstances, or the possibility of incurring nontrivial loss" (Smith *et al.* 2000: 1946). Risk is then undesirable and differs from "uncertainty" in that although both reflect imperfect knowledge, uncertainty lacks any particular value judgement about the consequences. How risk is perceived is also a matter for conjecture. The distinction between objective and subjective risk has provided the foundation for many of the investigations into risk and uncertainty by social and behavioural sciences (Pidgeon *et al.* 1992). *Objective risk* refers to phenomena in the socio-ecological environment that may result in harmful effects. This may then be measured and probabilities of harm calculated. Objective risk is seen as a product of optimal value as described within the rational actor paradigm (RAP) of decision theory (Boholm 2003). RAP assumes that individual decision making is an act of optimising outcomes. For example, if a farmer experiences repetitive, yet trivial, crop damage from baboons and then looses over half his crop to cattle over the course of one night it would be rational to identify the cattle as posing the greater risk to subsistence. However, it is clear that individuals do not consistently make decisions in this way. *Subjective risk* therefore recognises that individual belief and opinion, as is often reported in human–animal conflict, may well deviate from any objective assessment (Renn & Rohrmann 2000). So the perceived risk posed by wild species (and in certain circumstances, domestic species) to local livelihoods may be coloured, not simply by amounts of actual damage, but by a symbolic threat, or existing social tensions / mechanisms (Knight 1999, 2003, Hill *et al.* 2002, Webber 2007). For example, farmers in Zanzibar were quick to blame red colobus monkeys for damaging coconuts in agricultural areas even though research showed that these primates had no significant negative impact on coconut harvest (Siex & Struhsaker 1999). The farmer's complaints were thought to be the result of a contentious debate with government officials over shares in the revenue collected from tourists wishing to see the monkeys (Siex & Struhsaker 1999).

Socioeconomic and socioecological factors such as education, ethnicity, religion, wealth, proximity to animal refuges and period of residency, etc. have all been reported as factors that influence risk perceptions of wildlife (Infield 1988, Heinen 1993, Nepal & Weber 1995, Zinn & Andelt, 1999, Knight 2000, Priston 2005, Webber 2007).

In order to classify, order, and analyse the types and relative importance of problems identified in the interviews, *risk maps* were produced (*sensu* Smith *et al.* 2000 and Quinn *et al.* 2003). However, the ordinal scheme of ranking was reversed and a value of 1 was given to the risk identified as the least severe and then 2 to the next risk identified as more severe etc. This allowed for any categorised risks unmentioned by a respondent, but mentioned by others, to be given a value of 0. Absence of response to a particular risk was considered to be important. Risks reported as equally severe by a respondent were given the same number and any subsequent ranked risks picked up the numbering, accounting for ties. Respondents were also asked to rank animals in this way, according to the perceived potential risk of each species to the farmer's crops. Each rank given by the respondent was then converted into a score: 1 for the least severe, 2 for the next rank, etc.

Risk maps use two measures of subjective risk, "incidence", i.e., the proportion of people affected, and "severity", i.e., how serious the risk is perceived to be. The measure of incidence, referred to as the incidence index (*Ij*), ranges from 0 (no one affected) to 1 (everybody affected). This then demonstrates how widespread the problem is within the population of the study. The measure of severity, referred to as the severity index (S_j) was then calculated for each problem as $S_j = 1 + (r - 1) / (n - 1)$, where r is the rank based on the order of response by the interviewee and n the total number of problems listed by the respondent (Quinn *et al.* 2003).

This sets the most serious risk at $S_j = 2.0$ and the least serious to $S_j = 1.0$. Although incidence and severity may be correlated, the incidence of risk does not always relate to the severity of the risk. Some things may pose a very small risk to many people, whereas others may be a very serious risk but only affect a small proportion of the population.

Participants were asked about the risks of living in the area (i.e., the highland enclaves) and about the specific risk of living within the national park. An overall risk index (R_j) was calculated by multiplying incidence index by severity index. This was done to create two risk indices, one relating to living in the area and one to living within the national park.

Benefit Analysis

Participants were asked about the benefits of living in the area (i.e., the highland enclaves) and about the specific benefits of living within the national park. In order to classify and analyse the types and relative importance of benefits associated with living in the region a benefit index was created. This was determined by applying the same method used for identifying subjective risk. A value of 1 was given to the benefit identified as least important and then 2 to the benefit identified as being more important etc. Again, this allowed for unmentioned benefits to be given a value of 0. The proportion of respondents who identify a benefit source provides a measure of incidence, ranging from 0 (no one affected) to 1 (everybody affected). An importance index and incidence index was then created from the rankings and numbers of responses, respectively. An overall benefit index (B_j) was then calculated by multiplying incidence index by importance index. This was done to create two benefit indices, one relating to living in the area and one to living within the national park.

Socioeconomic and Socioecological Variables

To investigate the possibility that socioeconomic, or socioecological variables have an influence on individual perceptions of risk or benefit, a large set of relevant variables, 11 in all, were collected: geographical isolation of the compound from others (distances measured using GPS); number of wives present in the household; number of children present in the household; literacy levels of respondents; number of bags of peeled maize obtained from 2005/2006 harvest (amounts based on estimates arrived at through discussions with farmers); percentage of planted crop harvested (again, amounts based on farmer's estimates); duration of tenancy within the enclaves; number of years farming in the enclaves; age of the head of the household; number of horses owned; number of cows owned. Generally data were self-reported by respondents and then, where possible, independently verified.

To help uncover any significant pattern or meaningful underlying variables within the data set, socioeconomic and socioecological independent variables were subject to Principal Components Analysis (PCA). Such a technique reduces the effects of multicollinearity. A principal component is defined as a linear combination of optimally weighted observed variables. "Linear combination" refers to the fact that adding together the scores on the observed variables included in the analysis creates scores on a component. "Optimally weighted" refers to the fact that observed variables are weighted such that the resulting components account for a maximal amount of variance in the data set. The number of components extracted in a principal component analysis is equal to the number of observed variables being analysed, however the first few usually explain most of the variance on the data and only these are retained to be used in further analyses. It is appropriate to use PCA when data have been obtained on a number of variables and there exists a possibility of redundancy amongst the variables. Redundancy here means that certain variables are correlated with one another, possibly because they are measuring the same construct. The presence of potential redundancy means that it should be possible to reduce the observed variables into a smaller number of principal components (surrogate variables) that will account for a significant amount of variance in the observed variables. This PCA analysis was done using all 11 socio-economic and socio-ecological variables.

Potential Pitfalls of Risk / Benefit Analysis

It is important to be aware of the possible interconnectedness of interviewee response when categorising risk using this method. For example, responses that cite the absence of a clinic and the scarcity of food can seem unrelated, but both represent a general lack of access to vital resources and particularly the need for a road into the highland enclaves. Such connections are revealed through the Rapid Rural Appraisal techniques applied here. However, to avoid arbitrary categorisation of cited risks, the respondent's definitions were retained unless they were clarified later by the respondent. For instance, when people mentioned that there were no proper farmers and there was no road, both problems were recorded separately even though they both influence food shortage. But, when scarcity of food is mentioned in the same sentence as being linked to the lack of good farms, then this was considered as a single risk. Some identified problems, such as the lack of a clinic; school or road, may seem to be structural problems associated with poverty rather than "risks". They were, however, genuine concerns voiced by the respondents and so are included (Smith *et al.* 2000).

Risk mapping has its limitations; it is a static representation, capturing attitudes that reflect an instant in time. This can be problematic if, for example, a farmer is interviewed the morning after baboons have just damaged his crop, or if park rangers have recently arrested the farmer for grazing cattle in the park, as such recently experienced events may influence his response. Furthermore, risk mapping works on the assumption that the first answer given is the one the respondent considers to be the most severe. The benefit of risk mapping lies in its ability to quickly identify who is experiencing what type of risk and where.

Results

Socioeconomic and Socioecological Variables

The PCA yielded 4 axes (eigenvalue ≥ 1), but only 2 were retained for further analysis (*Tab. 6.1*). The first axis explained 29.8 % of the total variance in the observed variables, above the percentage of variance expected due to random noise for 11 variables (expected is 27.4 % for axis 1). Furthermore, as there were three variables with significant loadings on the first axis and a further four variables with reasonably high loadings that share a similar conceptual meaning (i.e., appear to be measuring the same construct) it was considered appropriate to retain the component. The second axis explained 19.7 % of the total variance in the observed variables, also above the percentage of variance expected due to random noise for 11 variables (18.3 % for axis 2), and was also retained. The other axes failed to explain a significant percentage of variance for 11 variables and so were not used.

The first axis showed high loadings for the number of cows, horses, and bags of maize owned by a household. Furthermore, reasonable loadings were shown for the number of children, wives, isolation, and the percentage crop yield of the farm. The second axis showed a high loading for the number of years a farmer had spent farming and reasonable loadings for the age of the farmer and tenancy within the highland enclaves of the national park.

The high loadings on the first axis were judged to represent measures of economic wealth. Of course, wealth is a relativistic term that is difficult to measure. However, cows can be considered to be a good indicator (Barth 1964, Dercon &

Table 6.1 Axis loading for the two significant components of the principal components analysis on 11 variables of socioeconomic and socioecological data

	Component	
Variable	1	2
Cows	0.727*	0.100
Age	–0.355	0.660*
Number of years farming	–0.294	0.714*
Isolation	0.631	0.007
Number of children	0.536	0.617
Horses	0.731*	0.032
Bags of maize harvested	0.735*	0.055
Number of wives	0.502	0.571
Percentage crop yield	0.572	–0.339
Literacy levels	0.396	0.022
Tenancy within enclaves	–0.231	0.621*

The first component is thought to represent the relative wealth of a particular household and the second highland experience. Together these two components explain 57.6 % of the data. Asterisks indicate the relative high loadings that share a similar conceptual meaning within each component.

Krishnan 1996, Desta 1999, Little *et al.* 1999), although no attempt was made to check this reported measurement against actual livestock units. Horses, too, have a complex connection to status amongst the Fulani. They are useful for transport, but expensive to keep, requiring time and attention. It is not unusual for the Fulani traditional leaders to demand that a resident of the enclaves purchases and keeps a horse when his herd of cows has grown sufficiently large (usually around 80 head) to be split into two smaller sub-herds. It is not a surprise to find that the number of horses owned shares a significant positive correlation with the number of cows owned (Spearman's rank: $r_s = 0.437$, $n = 83$, $p < 0.001$). Similarly, bags of maize harvested is an absolute measure of maize production. A household that can produce many bags has either a large farm or can afford to employ workers to maximise farming efficiency, or both. Large farms cost more money to obtain and to work. Thus, the number of bags and cows owned are positively correlated (Spearman's rank: $r_s = 0.447$, $n = 83$, $p < 0.001$). With this in mind, the first axis will be used in further analyses as the surrogate variable "economic wealth".

The second axis retained may represent experience in years of living and working in the enclaves. There is a relatively high loading for the number of years farming and for age, as well as reasonable loadings for number of children and, to a lesser extent, the number of wives in a household. The relatively high loadings for the number of children and wives in both components reflect the close relationship that wealth and age have with the acquisition of wives and resulting progeny. Thus, this second axis will be used in further analysis as the surrogate variable "highland experience".

Life in the Enclaves: Pros and Cons

"What Is Bad About Living in the Highlands?"

Interviewees were asked about problems experienced while living within the enclaves. The greatest perceived risk was damage to crops, with several other factors also being identified (*Tab.* 6.2). A risk map illustrates the relationship between frequency and severity of the given responses (*Fig.* 6.3*a*).

Damage to crops has a higher risk index than other problems ($R_j = 0.57$), indicating that it was considered to be a major concern. This risk is found grouped with other more serious perceived risks, such as scarcity of food, restriction of access to resources and the lack of road. A lack of motorable road access may explain why the absence of any clinic has such a high severity index (S_j = 1.65), although far fewer respondents cited it as a problem (I_j = 0.17). Many local farmers referred to the problems of having to travel for hours or days on foot, or horseback, to receive medical aid outside of the park. Such an imposition would be greatly reduced if the highland enclaves could be accessed by road. Therefore, the sources of risk, "no road" and "no clinic" are probably interconnected.

Stepwise multiple linear regressions explored the relationship between sources of risk and surrogate independent variables of "economic wealth" and "highland experience" derived from the PCA analyses. In each regression the subject risk source was used as the dependent variable (log-transformed). However, only "no road" had a significant relationship with the surrogate variables (R^2_{adj} = 11.0 %, $F_{1,81} = 10.03$, $p = 0.002$), with higher levels of economic wealth found to be the only significant predictor of a respondent identifying this risk ($t_{81} = 3.17$, $p = 0.002$).

"What Is Bad About Living in a National Park?"

Questions were then asked specifically about living within a national park to see whether responses to questions about their dwelling place were different. Interviewees were initially asked if they knew what the enclaves were. If they had trouble with the actual word "enclave", then they were asked if there was anything distinct or special about the region in which they lived (n = 100). Of those asked, 37.0 % had no idea and assumed the land they lived on to be the same as anywhere else; 48.0 % reported that the enclave was a piece of land cut from the park where they could graze their cows and farm around their houses; 9.0 % thought it was land outside of the park; and 6.0 % thought it was the park. Interviewees (n = 83) were then asked to report problems living inside a national park to create another risk index.

The responses (*Tab.* 6.2) and risk map (*Fig.* 6.3*b*) indicate that restriction on the use of local resources was identified as the main risk, with the highest incidence index ($I_j = 0.60$), but the lowest severity index of ($S_j = 1.36$). The cited risk with the highest severity index is the lack of a road ($S_j = 1.65$). Crop damage has the second

Table 6.2 Perceived problems experienced with living inside the highland enclaves and the national park

Risk	Source	Risk index: living inside highland enclaves	Risk index: living in a national park
Damage to crops	Damage to crops, disease, wind and rain damage and not just damage caused by animals (wild or domestic)	0.573	–
Scarcity of food	Concern over lack of affordable food produce available in the enclaves	0.571	–
Restricted access to resources	Concern over prohibitions about use of park resources, especially inability to choose to farm where they want	0.432	0.819
No road	Lack of a road into highland enclaves, which hampers cheap import of food and means that medical care can only be reached on foot	0.427	0.338
No clinic	Lack of access to medical care which heightens the risk of personal injury and disease	0.278	0.121
Actions of park rangers	Economic costs incurred by local farmers and herders when arrested by park rangers, either legitimately as a result of transgressions against national park laws, or illegitimately through corrupt actions of park rangers	0.277	–
Livestock	Concerns over the viability of raising cattle in highland enclaves with particular regard to lack of vaccinations and access to grazing	0.196	0.482
Actions of leaders	Economic costs incurred by local farmers due to corrupt behaviour of traditional leaders (such as resident Ardo, Sarki Fulani and Lamido) and local government politicians	0.126	–
Lack of education	Lack of education due to the reluctance of teachers to stay in highland enclaves	0.033	–
Wildlife crop–damage	Damage caused by wild animals	–	0.325

Risk index ranked from most to least risk, calculated by multiplying the incidence index by the severity index. (a) Enclaves: of 83 interviewees, 18.0 % stated no problems living in the enclave and so gave no answer, 37.0 % identified a single problem, 45.0 % two or more resulting in more identified risks than actual respondents ($n = 152$). (b) Park: of 83 interviewees, 8.0 % stated no problems and so gave no answer, 53.0 % identified a single problem, 39.0 % identified two or more problems ($n = 116$).

highest severity index ($S_j = 1.50$), but a rather low incidence index ($I_j = 0.22$) resulting in crop damage ranked with a relatively overall low risk index ($R_j = 0.325$). Neither of the surrogate variables was found to be effective in predicting any of the responses.

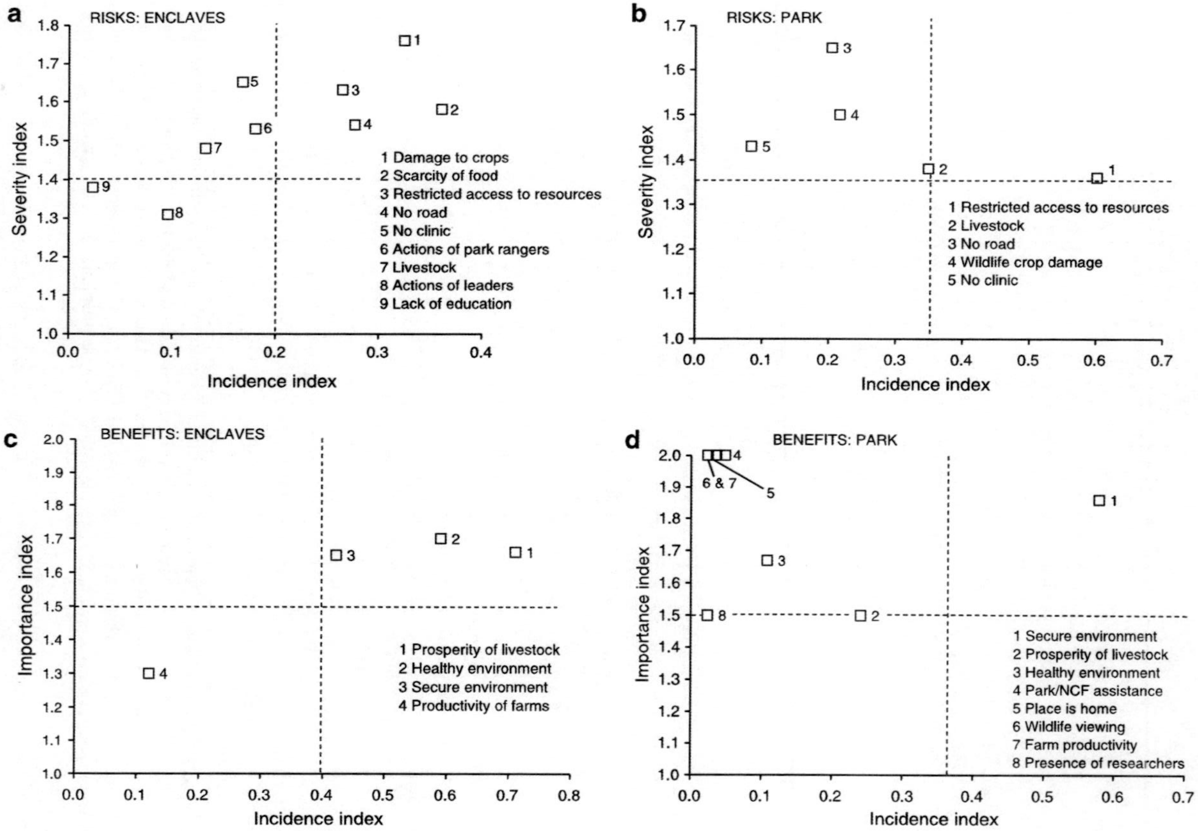

Fig. 6.3 Sources of *risk* (**a, b**) and *benefit* (**c, d**) identified with living within the highland enclaves and the national park ($n = 83$). For both risk and benefit maps the incidence index ranges from 0 (not mentioned by any informants) to 1 (mentioned by everyone). The severity index is measured in risk maps from 1 (least severe) to 2 (most severe). Risk index rank is added in parenthesis. In benefit maps, the importance index is measured from 1 (least beneficial) to 2 (most beneficial). Benefit index rank is added in parentheses

"What Is Good About Living in the Highlands?"

Responses (*Tab. 6.3*) and the relationship between frequency and importance of the given responses (*Fig. 6.3c*) reveal that "prosperity of livestock" is cited as the most significant benefit of living in the enclaves. This confirms the earlier assumption

Table 6.3 Perceived benefits experienced by those living inside the highland enclaves and the national park

Benefit	Source	Benefit index: living in a national park	Benefit index: living in the highland enclaves
Secure environment	Presence of park rangers and regulations that govern resource use within the park make enclaves a relatively safe haven from robbers, rustlers and unwanted intrusions of foreign cows	1.08	0.70
Prosperity of livestock	Benefits of raising cows in highland enclaves, due to the good grazing and healthy environment (particularly lack of tsetse fly)	0.36	1.18
Healthy environment	Generally held opinion that water in highland enclaves is good and there is less disease amongst the local human population and their livestock than in low lands outside enclaves	0.18	1.00
Park / NCF assistance	Refers to times past when Nigerian Conservation Foundation (NCF) assisted with vaccinations for cattle. Emergency assistance given to the sick through use of radio at Selbe and co-operation of park rangers. Upkeep of roads in and out of park (especially bridge over Mayo Kam in dry season)	0.10	–
Place is home	Many local farmers and herders simply say that this is their home and they know no other place.	0.07	–
Wildlife viewing	Opportunity to see wild animals	0.05	–
Productivity of farms	Perceived high quality of the land available for farming in the highland enclaves	0.05	0.16
Presence of researchers	Researchers bring gifts and co–operation in form of radio use and lifts in vehicles, in return for the co-operation of local population.	0.04	–

Benefit index ranked from most important to least calculated by multiplying the incidence index by the importance index. (a) Enclaves: of 83 interviewees, 22.0 % stated they enjoyed no benefit whatsoever, 38.5 % identified a single benefit, 39.5 % identified two or more ($n = 171$). (b) Park: of 83 interviewees, 18.0 % stated that they enjoyed no benefit whatsoever, 59.0 % identified a single benefit, 23.0 % two or more ($n = 171$).

(Dunn 1995) that the well-being of their cows is the main reason for the Fulani presence in the enclaves. However, the benefit with the highest rank of importance was "healthy environment", although it is cited less frequently than "prosperity of livestock". The benefit of a "secure environment" comes with a slightly lower rank of importance. The policing of the park means that there are few strangers in the enclaves and many of the Fulani can thus leave their cows unattended for long periods of time. This is unadvisable outside the park where rustling of animals and violence connected with disputes over cattle are common. Only a few individuals commented on the fertility of the land. This may be because the Fulani must necessarily farm close to their compounds high on the hillsides where the soil quality is poor, as they are prohibited from cultivating the more fertile river valleys by the park authorities.

Again, multiple linear regressions were undertaken using the surrogate variables derived from the PCA. Regressions using "productivity of farms" and "healthy environment" as the dependent variables gave no significant results. However, "secure environment" uncovered a weak, yet significant, relationship (R^2_{adj} = 6.2 %, $F_{1,81}$ = 5.32, p = 0.024), with "economic wealth" found to be the only significant predictor of a respondent identifying this benefit (t_{81} = 2.31, p = 0.024). Regression using "prosperity of livestock" also uncovered a significant relationship (R^2_{ad} = 4.8 %, $F_{1,81}$ = 4.06, p = 0.047), again with "economic wealth" the only significant predictor (t_{81} = 2.01, p = 0.047). Finally, regression using the response that "nothing" was good about living in the enclaves as the dependent variable again uncovered a significant relationship (R^2_{adj} = 4.9 %, $F_{1,81}$ = 4.14, p = 0.045), once again with "economic wealth" the only significant predictor of a respondent's negative response (t_{81} = – 2.03, p = 0.045). However, the poor fit of these models suggests that many other unexplained factors are probably involved in this decision process.

"What Is Good About Living in a National Park?"

Of several perceived benefits (*Tab. 6.3*), a secure environment is most importantly associated with living in the national park (*Fig. 6.3d*). However, it ranked as the third most important benefit from the question of "what is good about where you live?" Respondents identifying the benefit from both questions correlated significantly (Spearman rank correlation: r_s = 0.275, n = 83, p = 0.012). Whilst security of the person is important to local farmers and herders, who comment on the fact they do not need to lock their doors in the enclaves, the security of the herd is of paramount importance. So the prosperity of their livestock, which is ensured by the security offered by the park, is the most important. Otherwise the rank of benefits from living in the park is identical to those from living in the highlands.

Stepwise multiple linear regressions yielded a poor fit (R^2_{adj} = 4.7 %), but an overall significant relationship ($F_{1,81}$ = 4.02, p = 0.048) with "economic wealth" found to be a weak predictor of a respondent citing the benefit of raising cows inside the park (t_{81} = 2.00, p = 0.048).

Table 6.4 List of identified problem animals

Taxon	Animal	Damage to crops	Perceived as threat	Ranked risk
Carnivores	Civet	x	x	
	Dog	x	–	
	Jackal	x	x	
Ungulates	Aardvark	x	x	
	Bushbuck	x	x	
	Cattle	x	x	
	Donkey	x	x	5
	Horse	x	–	
	Red river hog	x	x	
	Sheep	x	x	
	Warthog	x	x	4
Rodents	Cane rat	x	x	
	Mouse	x	–	
	Porcupine	x	x	2
Primates	Olive baboon	x	x	1
	Tantalus monkey	x	x	
Birds	Bushfowl	x	x	3
	Harrier	–	x	
	Weaver bird	x	x	

The first column lists identified perpetrators of crop damage by independent measures, the second column those that have a perceived negative impact on local livelihoods.

Perceptions of Pest Severity

Whenever the issue of animals damaging crops was raised during interviews, respondents were asked to list any animals considered a threat to the productivity of their farms. Respondents identified 16 taxa that had a negative impact on their livelihoods (*Tab. 6.4*, *Fig. 6.4a*). The olive baboon is considered to pose the greatest threat ($R_j = 1.39$), closely followed by porcupines ($R_j = 1.38$). Warthog and red river hog are considered highly problematic with high severity indexes ($S_j = 1.52$ and 1.71 respectively), but they are not widespread and affect only a few farms.

Direct Measures of Crop Damage Compared with Perceived Risk

Damage measured on the focal farms was used, together with the assigned damage value per plant to rank the potential risk of crop depredating species and so create a measured incidence and severity risk index for each species. This could then be compared with the perceived risk (*Tab. 6.6*, *Fig. 6.4b*).

A number of key differences between measured and perceived risk exist for each type of animal (*Fig. 6.4.b*). For baboons, perceived severity and incidence indexes are higher than measured risk, suggesting that local farmers believe that baboons are a greater problem than the analysis of actual damage (proportional to other raiding species)

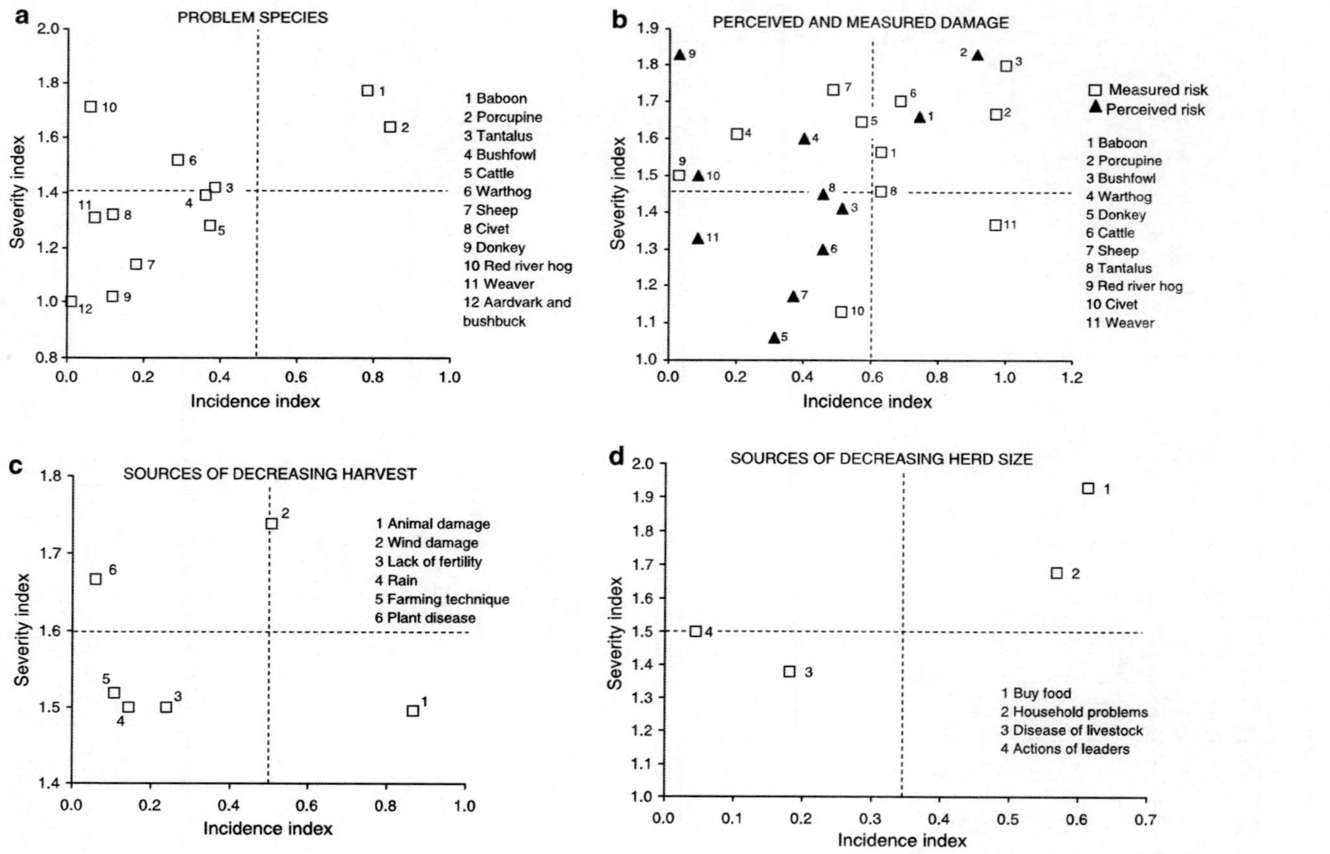

Fig. 6.4 (**a**) Perceptions of problem animals within the highland enclaves ($n = 80$). (**b**) Peceptions of problem animals and actual damage by those species to the focal farms ($n = 35$). (**c**) Sources of risk identified with decreasing harvests within the highland enclaves ($n = 83$). (**d**) Sources of risk identified with decreasing herd size ($n = 44$). Incidence index ranges from 0 (not mentioned by any informants) to 1 (mentioned by everyone). Severity index is measured from 1 (least severe) to 2 (most severe). Risk index rank is added in parentheses

Table 6.5 Assigned damage value for different crop types found in focal farms reflecting relative economic cost to the farmer in instances of damage

Crop type	Approx. yield per plant (g)	Approx. sale cost of plant yield (b)	Assigned damage value per plant
Maize	75	2.3	1
Potato (a)	200	6.0	2
Sweet potato	200	6.0	2
Cassava	200	12.0	5
Cocoyam	180	10.8	5
Groundnut	5	1.8	1
Bean	35	2.1	1

(a) Called "Irish potato" in West Africa
(b) In Nigerian currency (Naira): 250 = £1 = $2 (2007)

Table 6.6 Measured and perceived risk index for key raiding species calculated from focal farm data (n = 35). + indicates the highest factor; M = measured risk (independently measured by a researcher); P = perceived risk (according to the opinions of the farmers). Index ranges from 0 (no incidence) to 2 (universally most severe risk).

Species	Perceived risk index	Measured risk index	Variance
Baboon	1.23	0.98	+P
Porcupine	1.67	1.62	+P
Bushfowl	0.73	1.80	+M
Warthog	0.64	0.32	+P
Donkey	0.33	0.94	+M
Cow	0.59	1.17	+M
Sheep	0.43	0.84	+M
Tantalus monkey	0.66	0.92	+M
Red river hog	0.05	0.04	+P
Civet	0.13	0.58	+M
Weaver bird	0.11	1.33	+M

demonstrates. Warthogs also have a higher perceived incidence risk index than the actual risk. We can thus conclude that damage perpetrated on crops by both species is markedly less than local farmers believe. Although porcupines are perceived as highly destructive their damage is only marginally overestimated. However, the incidence risk for porcupines is underestimated.

In contrast, other taxa represent far more risk to crops than farmers would believe. Tantalus monkeys have a higher actual risk than perceived risk index, due to the higher rate of incidence of measured damage. Bushfowl and weaver birds also show differences between actual and perceived risk indexes. Measured and perceived severity indexes of weaver birds are approximately the same, whereas the measured incidence index for these small birds is far higher than the perceived incidence index. Thus it would seem that farmers often ignore weaver bird behaviour. The measured incidence index for bushfowl is also higher than the perceived incidence index. However, the actual severity index is also far above that of the perceived index. Most farms contain maize plants of varying ages, which is partly due to the consumption of seeds and emerging maize shoots by bushfowl and other depredating species early in the growing season. Farmers often replace plants damaged at such an early stage a few weeks later. This is done in the hope that the bushfowl's attention would, by then, be elsewhere, "thinking" the planting period to be over. It is possible that damage occurring relatively early in the growing season allows farmers to replace damaged seedlings. The potential for a species to perpetrate more lasting harm to livelihoods may therefore influence how farmers' judge a species (Hill 1997).

Civets have a higher rate of measured than perceived incidence but a far lower measured than perceived severity, i.e., farmers underestimate how often civets visit their farms but overestimate the damage they cause on each visit.

Not surprisingly, cows, sheep, and donkeys have a significantly higher actual risk than perceived risk indexes. Previous research reported that damage by domestic animals was rarely reported (Graham 1973, Hill 1997, Andama 2000, Hill *et al.* 2002). Possibly, domestic damage is not reported because farmers believe that the researcher is not interested in this effect, or because there are traditional practices in place to recompense the farmer for any damage to his crop from the owner of the raiding livestock. The Fulani frequently comment that they think little of the damage to their fields caused by domestic livestock, as they are likely to own the animals and so such damage is self-inflicted. Furthermore, the impact of livestock damage may be lessened by the comparative rarity of its occurrence, regardless of the substantial amounts of damage a field may suffer on a single raid by domestic livestock (Warren 2003).

Multiple linear regressions yielded a poor fit (R^2_{adj} = 5.3 %), but an overall significant relationship ($F_{1,81}$ = 4.51, p = 0.037) with "highland experience" found to be a weak predictor of a respondent identifying porcupine as a risk species (t_{81} = 2.12, p = 0.037). It may be that long established households suffer more attention from porcupine than more recently established households. Possibly the attentions of porcupine stretch back to before the formation of many of the farms when the Fulani grew cocoyam, banana, and cassava in small gardens around the house – crops commonly attacked by porcupine. On settling in the enclaves the Fulani tend not to move due to the complexity of land tenure and the permission needed from the local Fulani *Ardo* to farm and graze a herd on a particular piece of land. Again, the poor fit of this model suggests that the significant effects may well be obscured by many other unexplained factors.

The Maize Yield

To assess the actual cost of crop damage to the farmer, respondents were asked whether their crop of maize had increased, decreased or remained the same in relation to the previous growing season. Farmers in their first growing season were obviously unable to give an answer (5.0 %). The majority of respondents reported that their harvest had decreased from last year (88.0 %), with only a small minority reporting an increase (6.0 %) and even less stating that their harvest remained the same from year to year (4.0 %). Those whose crop yield had decreased were then asked the reasons why this had happened.

The relationships between these responses (*Fig. 6.4c*) reveal that animal damage by far the highest risk index ($R_j = 1.30$), having the highest incidence index ($I_j = 0.87$), but still a rather low severity index ($S_j = 1.50$). This implies that, although many farmers perceive crop depredation by animals as serious, it is often not seen as the main cause of decreasing harvests. However, when wind damage is an issue, it is perceived as the greatest contributor to poor yields of maize ($S_j = 1.74$).

A main theme surfacing during conversations was that unproductive harvests inevitably result in the need to sell cows and so over a period of years could destroy a herd. With this in mind, interviewees were also asked about changes in herd size over the past year. The majority reported that their herds had decreased (52.2 %), although many stated an increase (39.1 %) with a small minority reporting no change (8.7 %). Change in herd numbers correlated positively with reports of a change in harvest size from last year (Spearman rank correlation: $r_s = 0.265$, $n = 69$, $p = 0.028$). It is possible that unproductive harvests have an influential relationship with the selling of cattle and resulting diminished herd size.

Interviewees whose herd had decreased were then asked the reasons for such an occurrence. The relationships between these responses (*Fig. 6.4d*) indicate that the need to buy food had the highest risk index ($R_j = 1.18$) with both the highest incidence ($I = 0.61$) and severity indexes ($S = 1.93$). Such a high ranking supports the assertion that the lack of farm productivity can cause herds to become smaller. Household problems, such as structural repairs to buildings, medicines, and healthcare for the family and livestock etc., are perceived as the next highest risk to the stability of the herd ($R_j = 0.95$). However, neither of the surrogate variables predicted sources of risk associated with either decreasing harvests or decreasing herd size.

Discussion

Overall these results suggest that the Fulani look at a range of factors when considering the costs and benefits of living and farming within the national park. The status as protected land offers an environment relatively free of poachers and other trespassers but also results in isolation and a lack of access to facilities. They have limited powers to protect their crops and cattle from wild animals but at the same time they, and their livestock, enjoy the benefits of living in the highlands where tsetse fly and mosquito numbers are low. It is notable that the experience of the local population is not homogeneous, with wealth and degree of isolation influencing the farmers' experiences of farming in the enclaves. This study concurs with earlier work (Dunn 2000), in finding that local populations living within Gashaka Gumti National Park identify crop-raiding by animals as a major concern. Although crop-damage is seen as one of the most significant threats to their livelihoods, farmers' estimations of the importance of animal pests do not always match with the direct measures of crop-damage in this study. Together these findings suggest that farmers' perceptions of risk are likely to be influenced both by their own situation and by their attitudes towards the wildlife authorities, the park, and its various fauna.

Contextualising Crop-Raiding

Crop damage is seen by the Fulani as posing the greatest threat to their livelihoods. However, the heterogeneity of risk assessment is immediately apparent in the data relating to the question "What is bad about living in the highlands?", as no factor was cited as a risk by more than 36 % of the respondents. Furthermore, there is a possibility that many of the identified problems are interconnected and are simply different conceptions of the same source of risk.

Of course, some risks are idiosyncratic, but many reflect collectively experienced anxiety. The grouping of both "crop damage" and "no road" with "scarcity of food" in the more severe sector of the risk map might be expected as both the former risks are likely to contribute directly to the latter. Discussions with groups and individuals support this assumption. The ways in which respondents describe problems are then a matter of perception. The real problem may actually be scarcity of food. It is possible that the growing of crops is perceived by some as a viable solution to food scarcity, which would explain why crop damage is seen as the most serious risk faced by many, but not all, farmers. Indeed, many argue that if they could create larger fields in the more fertile areas of the highland enclaves, any damage to their crops would be relatively ineffectual in relation to their total possible yield and so food scarcity would be greatly alleviated. However, others argue that all problems, including food scarcity, would reduce if they could gain permission from the park authorities to build a road from Gashaka village up into the highlands. It is interesting to note that "economic wealth" is a predictor of a respondent identifying "no road" as a serious source of risk. The wealthier farmers in the enclave already have sizeable farms and a few of the richer individuals own other farms either in the Filinga basin or outside the park in the farmlands surrounding the town of Serti. It is possible that these farmers do not see farming as a viable solution to the problem of food scarcity, whereas the construction of a road would not only allow for the cheap importation of food, but would assist in the long-term development of the highland enclaves facilitating easier access to education, healthcare services and stimulate prospective business opportunities. However, road building is unlikely to be supported by the park authorities or other conservation organisations as it would also open the park to increased traffic and almost certainly result in more illegal settlement and poaching.

Despite the potential for conflict between the Fulani and the park authorities, certain positive attitudes toward the protected area exist. A "secure environment" was the most significant benefit associated with living within the national park and within the highlands themselves. Most pastoral populations require high labour to manage animals (Swift, 1986, Bayer 1990, Homewood & Rodgers 1991). In agropastoral enterprises labour bottlenecks form at the beginning and the end of the growing season (Eddy 1979). However, the level of security that surrounds the national park means that many Fulani do not have to spend all day with their cattle and can leave them for hours on end. This allows them more time to pursue other activities. The local population then enjoy the benefits of the protection offered by the park rangers, but conversely are threatened by the possibility of arrest should they be caught grazing their cattle outside of the enclaves and within the boundary of the park.

The responses to "what is good about living in the highlands?" support the premise that attitudes and expectations are divided on the basis of wealth, or cattle ownership. The most significant benefit was the "prosperity of livestock", or more specifically cattle. The surrogate variable "economic wealth" was found to be a predictor of a respondent citing "prosperity of livestock" as a benefit. This was expected as livestock is given as an indicator of economic wealth. Conversely, those respondents whose farms produced little or no bags of maize, and so considered to be relatively economically poor, were more likely to respond that nothing was good about living in the highlands.

Other problems such as access to clean water and the risk of disease (Smith *et al.* 1999, 2000, Quinn *et al.* 2003) are not a significant concern. The cool climate of the highland enclaves and the abundance of clear, clean mountain streams mean that there is relatively less disease of either humans or animals in the highland enclaves than in the hotter drier lowlands within or outside the park. More importantly there are very few tsetse flies (Dunn 1995, Wright 2003) and malaria-bearing mosquito due to the high altitude (Balls *et al.* 2004, Brooker *et al.* 2004). For this reason a "healthy environment" is seen as a significant benefit of living in the highlands.

Local Perceptions of Wildlife Species

Baboons are considered to pose the greatest threat to local subsistence, supporting previous questionnaire results (Dunn 1993, Kamaya 1996). The in-depth study of a smaller number of focal farms (where perceived risk was compared with measured damage) also showed that baboons were considered a highly destructive species but here ranked second after porcupines. Respondents in both the general population and those on focal farms agree in also placing warthogs, bushfowl, and tantalus monkeys in the top five wildlife pests.

The difference in perception and measured damage varies for these five taxa. Whereas baboons and warthogs appear to cause far less damage than farmers think, porcupine damage is estimated approximately correctly, and damage caused by tantalus monkeys and bushfowl is underestimated. It is probable that both the behaviour of animals and their place in Fulani culture contribute to this variation in perceptions.

Kingdon (1997) has argued that bush pigs are perceived as particularly detrimental due to their destructive rooting behaviour. Other studies have also found that farmers have a tendency to inflate risks posed by nocturnal species (Hill 1997,

2004, Hoare 1999, Naughton-Treves 2001). Nocturnal animals are also often less tolerated than diurnal species, possibly due to the lack of effective mitigating strategies (Hill 1997,2004, Hoare 1999, Naughton-Treves 2001). Nocturnal warthogs exhibit similar behaviour to bush pigs and so it is not surprising that they are perceived to be more of a threat than they actually are. Porcupines are also nocturnal and can be very destructive in a maize field. They lean on the mature maize stems flattening swathes of plants in order to feed on the ripened cobs (DB *pers. obs.*), yet interestingly there is little difference between the perceived and actual threat posed by these animals. Unlike warthogs, porcupines, are seen as magical creatures and are particularly difficult to hunt. It is said that, when chased, the porcupine simply hits the ground with its tail and the pursuing hunters fall over, speeding the creature's escape. They are also thought to have the power of divination and their tails are prized by those Fulani who know how to practice traditional medicine and construct magical charms. Such beliefs may weaken the formation of possible negative attitudes toward these animals.

Measures of crop-raiding demonstrate that both porcupine and baboon can cause considerable amounts of damage. However, although perceptions of porcupine damage appear to be based on actual loss, perceptions of baboons overestimate damage. The pervasive presence of large and highly visible groups of baboons may well divert the attention of farmers from the depredations of smaller perpetrators such as the weaver bird, possibly leading to the under-reporting of damage by these less conspicuous animals (Hill 1997). It is well known that large animals attract a disproportionate amount of blame for crop damage (Litsinger *et al.* 1982, Bell 1984, Parry & Campbell 1992, Naughton-Treves, 1997, De Boer & Baquette 1998, Gillingham & Lee 1999, Weladji & Tchamba 2003, Hill 2004) particularly in a region of low human density (Newmark *et al.* 1994), as they are visually intimidating and an obvious detectable threat. However, body size alone does not adequately explain why baboons are so vilified. Bushbuck, although timid, are far larger yet are not perceived as a threat despite having a measured risk index ($R_j = 0.40$) above that of warthog ($R_j = 0.32$). It is then possible that a combination of size and behaviour influence the perceived threat of a species.

Baboons are highly visible and mainly terrestrial. They approach farms in large groups and their unpredictable and sometimes seemingly human-like planned assaults on crops can be quite intimidating for affected households (Knight 1999, Hill *et al.* 2002, Hill 2005, Warren 2003, Warren *et al.* this volume [Ch. 8]). They are also difficult to drive from the field (Hill 1997, Warren 2003), often attacking dogs; women and children who dare to try and chase them (King & Lee 1987, Strum 1994, Hill 2000, Warren 2003). Such behaviour may well provoke distrust (Webber 2007). The tantalus monkeys' reputation is not so tarnished. They are more likely to make quick, and often independent, forays into a field of crops and so are perceived as opportunistic (Webber 2007).The reason civets have a higher perceived severity than actual measured severity may be due to their mainly nocturnal behaviour but it is more likely that is because of the destruction they cause as very efficient killers of chickens. It is possible that not many farmers are aware of the civets' depredating behaviour, but those that are may be prone to exaggerate the damage.

As in other areas (Naughton-Treves 1997, Knight 1999, Webber 2007), inhabitants of the enclaves commonly anthropomorphise wild animals. Analogous assessments of an animal's character and behaviour based on typically human attributes can also greatly influence farmer's attitude (Webber 2007). Baboons are frequently portrayed as cunning thieves by the Fulani. Herders describe how baboons lay motionless in the short grass near grazing goats or sheep and wait patiently for the herd to approach before leaping on unsuspecting lambs. Baboons are said to use decoys to distract any humans present whilst the rest of the troop enters the field from an alternative direction. Furthermore, the Fulani often comment on how mother baboons are too greedy to feed the young travelling on their backs. As pith eaters they chew the immature stems of maize plants causing extensive damage. By consuming plant parts that are unsuitable for human consumption baboons are often seen as "wasteful", needlessly destroying crops not ready to eat, but too far into the growing season to be replaced (Naughton-Treves 1996, 1997, Hill 1997, Knight 1999, Wheatley *et al.* 2002, Webber 2007). The highly destructive and observable foraging strategies of such species are then viewed negatively by local farmers (Webber 2007), or even conceived as a force of cultural violence (Hell 1996). There is a popular Nigerian axiom "monkey dey work, baboon dey chop" (Adejumo 2007). The word "chop" means to eat, take, or steal. This phrase is frequently used in Nigeria to represent the desirable and hard-working "us" (monkey) in opposition to the undesirable and scrounging "others" (baboon). For example, in newspaper articles discussing corruption, this phrase may well be used to contrast the hard-working public against corrupt officials who take bribes but do not work for their money. When wildlife that threatens a person's livelihood is identified as a "social other", it can be used to reinforce in-group boundaries. As John Knight states, "viewed in dynamic terms, the 'people' of people-wildlife conflict ceases to be a given or fixed category, but emerges in a complex process of conflict that may well span local, national and international levels" (Knight 2000: 22). It is possible then that crop-raiding species are considered problematic not simply for reasons of their economic cost, but also due to the transgression of the human settlement boundary (Douglas 1966) and their association with the park authorities that putatively prefer the continued presence and prosperity of the wild animals that inhabit the highland enclaves over and above that of the resident Fulani population.

Is Crop Damage a Real Problem?

The majority of farmers reported that their crop yield decreased from the previous year. Many Fulani complained that unusually strong winds close to the harvest had flattened the maize, breaking stems and making it easy for animals such as porcupine to eat the ripening cobs. Whilst devastating for many farms, the wind did not affect everyone and animal damage was still considered the most significant negative impact on crop yield. The high-risk index of "animal damage" is in no small part due to the extent of its reported incidence and in spite of its relatively low severity index.

However, comparisons between the farmers' perceived percentage of damaged plants and that of the independent measures, coupled with the high risk index associated with animal damage, suggests a substantial mismatch between local perceptions and its actual occurrence (Heinen 1993, Conover 1994, Naughton-Treves 1996, Sehkar 1998, Siex & Struhsaker 1999). This disparity between perceived and actual dimensions of crop depredation by animals should not simply lead us to discount local representations of animal damage (Wywialowski 1994, Gillingham & Lee 2003, Wang *et al.* 2006), but rather to understand why this may be the case.

Animal damage to crops can be defined as an external agricultural issue (Webber 2007), as can weather effects. Both are an outside influence that physically penetrates, often aggressively, into the agricultural domain. Issues of soil fertility and problems of farming techniques originate from within the farm boundary, i.e., they are internal issues that farmers are largely in control of. In this study, as in others, external issues were perceived as being more severe than internal issues (Tweheyo *et al.* 2005). Many Fulani farmers admit that they do not have the time to weed their fields, as they are too busy caring for their cows (Swift 1986). Diminishing soil fertility can also have a serious negative impact on crop yield (Yayock *et al.* 1988). Most farmers still failed to identify internal issues even though weeds have potentially the highest negative impact on global maize production (Oerke & Dehne 2004). External agricultural problems such as wind and animal damage are more difficult to control than internal issues. As individual risk assessments increase with exposure and perception and decrease with the capacity of an individual to instigate mitigating and coping strategies (Smith *et al.* 1999); individuals that lack sufficient strategies to deal with a particular potential hazard are likely to over-report that risk.

As with cultivation, a majority of respondents reported that their herds had decreased, although a considerable minority (40 %) stated the opposite. This correlated positively with reports of a decrease in harvest productivity from last year. It is then possible that unproductive harvests have an influential relationship with the selling of cattle. When asked for reasons why their herd size was decreasing, the practice of selling cows to "buy food" was indeed identified as the most significant cause. Another reason was identified as "household problems". This usually referred to unforeseen circumstances such as sickness in the household or amongst livestock. It is possible that some interviewees may have conceptually included the need to buy food in their answer. Respondents pointed out that unforeseen expense incurred through either sickness or essential maintenance to the farm or compound is often exacerbated by the isolation of the highland enclaves.

It's All About the Cows

Livestock are important as they can provide insurance against such possible income shocks (Dercon 1998, Desta *et al.* 1999). However, it seems difficult for poorer local herders to maintain decreasing herds over time. If a pastoralist's herd becomes too small the household might sedentarise and be ensnared in a poverty trap that is difficult to escape. The Fulani of the highland enclaves are already partially sedentary, but they insist that they are in the highland enclaves to raise cows and that to protect this livelihood they have turned to farming. While cultivation might be viewed as a practicable solution to risk (Campbell 1984, Smith 1998), it can also be seen as an unsustainable option that might even accentuate risk (Hogg 1987, 1988). If diversification into cultivation proves unproductive, livestock may still have to be sold to ensure subsistence, causing a potential reduction in herd size. It is possible that time, effort and financial resources are reapportioned from cattle to the cultivation of land further contributing to a reduction in herd size thus increasing risk. However, while risk is an important factor in the diversification of income sources, it may not be the only reason (Little *et al.* 2001). Much of the literature assumes that diversification is a strategy that always lowers exposure to risk and is often "scale-neutral" in that all members of a population have similar exposure to risk and coping strategies (see Bernstein *et al.* 1992, Dercon 1998, Ellis 1998, 2000). However, it is probable that the relationship between risk and diversification is far more complex and dynamic.

It is possible that the poorest households diversified into cultivation through necessity. For the relatively rich households, holding all their assets in the form of livestock can itself be considered a risk (Desta 1999) and so they are more likely to diversify for the sake of risk management, as an economic strategy for maximising herd capital (Barth 1964, Little, 1985, 1992, Dercon & Krishnan 1996, Homewood *et al.* 2006). Those who are experiencing a marked decrease in herd size might then have to farm to mitigate the loss of cattle. If farming fails to stem the loss, then exposure and perception of external influences that undermine their coping strategies (for example, crop-raiding) will cause their assessment of risk to become greatly exaggerated. If coping strategies fail then poorer individuals and households are likely to eventually lose all of their cattle and be pushed permanently out of pastoralism (Homewood *et al.* 2006). Indeed, over the time of the study three Fulani farmers were effectively forced to leave the enclaves because they no longer had enough animals upon which to survive. The perceived loss of these means of subsistence will result in negative attitudes toward protected areas (Parry & Campbell 1992, Newmark *et al.* 1993, DeBoer & Baquette 1998, Maikhuri *et al.* 2001, Hockings & Philips 1999, Stolton & Dudley 1999, Rao *et al.* 2000). Nevertheless, the Fulani presence in the highland enclaves is dependent on the continued co-operation of the park authorities.

Concerns over crop-raiding expressed by both the rich and poor Fulani suggests that it may be appropriate to define the inhabited space of the highland enclaves in terms of both wild animal and human populations (Knight 2000). Fields of crops, compounds, and villages within the enclaves represent an attempt to remove wildlife from the land, whilst the authority of the national park declares that wild animals should remain. Knight (2006) argues that this may result in resident human populations viewing wild animals as a competing rival demographic. Many Fulani believe that this unwanted competition is a direct result of the action, and inaction, of the park authorities. They maintain that the eviction of the illegal farming communities of Tounga and Italal, as well as the park's refusal to grant them permission to build a serviceable road into the highland enclaves, have forced many Fulani into adopting a complimentary agrarian mode of subsistence in a region replete with problematic crop-raiding species (Hill 1997, Naughton-Treves 1997). As such, human–wildlife conflict in the highland enclaves is dependent on perceptions of appropriate land-use. Opposing views held by the park authorities, conservation agencies and the local Fulani population on what the highland enclaves should be used for is a source of much of the underlying conflict in the region. Further antagonistic feelings are perpetuated by the widely held belief that it is only a matter of time before the Fulani themselves are evicted from the park.

Concerns then identified by the Fulani may be, in part, a means of coping with the insecurity of food acquisition and decreasing herd size as well as the insecurity of possible eviction. However, socioeconomic factors invariably influence how these problems are perceived and inform an individual's coping strategies, thus providing a discourse that enables people to express their problems with a perceived sense of legitimate morality and understanding (Croll & Parkin, 1992). It is then possible that the conflict and anger experienced by the Fulani of the highland enclaves does not result entirely from the animal depredation of their crops, but that these acts of destruction serve simply as a catalyst, uncovering feelings of anxiety and disaffection that were already present. However, it is important to recognise positive as well as negative attitudes toward protected areas when considering conservation strategies (Allendorf 2006).

Motivated by the continuing prosperity of their cattle, many Fulani have a vested interest in co-operating with the park authorities in the hope of ensuring their continuing tenancy. Moves to secure the status of the Fulani in the highland enclaves present an authentic opportunity for constructive dialogue between the Fulani and the park authorities for their mutual benefit. Furthermore, attempts to reduce the impact of food-insecurity would appreciably benefit the local population as a whole. We do not suggest that such efforts would instantly mitigate the existing conflicts that exist between the park authorities and local Fulani leadership, but it would help to promote a more constructive local attitude toward park policies. At present, there is little to suggest that the park authorities are overly enamoured with "community conservation" and the National Park Service (NPS) remains sceptical about the resident population of enclaves. They maintain that the Rinderpest epidemic, which swept through the park in the early 1980s, decimating the park's population of buffalo and antelopes, was initially introduced by the cattle of nomadic pastoralists encouraged to graze in the enclaves and the surrounding area by the Fulani leadership, and that continuing incursions into the enclaves by the *Mbororo'en* (nomadic Fulani) represent one of the greatest threats to the park's long-term sustainability (Dunn *et al.* 2000). There are also concerns about widespread degradation of habitat within the highland enclaves and areas of the national park caused by local residents (Adanu *et al.* this volume [Ch. 3]).

The Fulani recognise that their continued residency in the enclaves is dependent on their relationship with the park authorities and there exist sufficient traditional mechanisms to manage the use of the enclaves supplemented by rules and regulations provisionally agreed upon within the enclave agreements negotiated with the Nigerian Conservation Foundation in 1998 (although not yet ratified by the NPS). As Ostrom (1990) argues, how local people perceive their future participation in conserving commonly used resources impacts on their willingness to conserve a resource and ensure long-term use instead of simply exploiting a resource for short-term gain. The national park's management plan, finalised in 1998, recognised that strategic planning designed to conserve biodiversity would not succeed in the long-term unless "local communities" perceived such efforts as serving their economic and cultural interests (NPS / NCF / WWF 1998: 8 – 1). The majority of strategies outlined in the plan have yet to be implemented. Only through collaborative action with local people, leading to a favourable effect on local livelihoods and residential security, can the NPS expect any substantial support for the national parks' wider conservation objectives.

Acknowledgements

This project was conducted under a joint studentship from the Economic and Social Research Council and Natural Environment Research Council. Support for the project was generously given by the Nigerian Authorities, the National Park Service of Nigeria, the staff and management of Gashaka Gumti National Park, the Nigerian Conservation Foundation and all the staff at the *Gashaka Primate Project*. DB would also like to thank His Royal Highness Alhaji Hammangabdo Mohammadu Sambo, the Ardo'en and people of the highland enclaves, and particularly the people of Selbe for their support and indulgence. In addition, special thanks go to Yakubu Ahmadu and Ado for their help in the field. Thanks are also due to Volker Sommer, Garry Marvin, Claire Ozanne, and Amanda Webber for help and advice. We must thank Mary MacKenzie who provided help with interpreting GIS data and producing maps and Peter Shaw who gave invaluable statistical advice, and all the support staff at the Centre for Research in Evolutionary Anthropology who have helped with the logistics and administration of the project.

Literature

Adejumo, A. G. (2007). The sweet and sour aspect of integration as exemplified in Yoruba poetry. *Journal of Social Sciences* 14: 169–173.

Allendorf, T. D. (2006). Residents' attitudes toward three protected areas in southwestern Nepal. *Biodiversity and Conservation* 16: 2087–2102.

Andama, E. (2000). *Assessment of Wildlife Crop Damage around Mgahinga Gorilla National Park: A Particular Emphasis on Crop Depredation by Porcupine* (*Hystrix africaeaustralis*). Report to CARE-DTC.

Balls, M. J., Bodker, R., Thomas, C. J., Kisinza, W., Msangeni, H. A. & Lindsay, S. W. (2004). Effect of topography on the risk of malaria infection in the Usambara Mountains, Tanzania. *Transactions of the Royal Society of Tropical Medicine and Hygiene* 98(7): 400–408.

Barth, F. (1964). Capital investment and the social structure of a pastoral nomadic group in South Persia. In Firth, R. & Yamey, B. S. (eds.), *Capital, Savings and Credit in Peasant Societies*. London: Allen & Unwin, pp. 415–425.

Bayer, W. (1990). Behavioral compensation for grazing time by herded cattle in central Nigeria. *Applied Animal Behaviour Science* 27: 9–19.

Bell, R. H. V. (1984). The man animal interface: an assessment of crop damage and wildlife control. In Bell, R. H. V. & McCone-Caluzi, E. (ed.), *Conservation and Wildlife Management in Africa*. Malawi: US Peace Corps, pp. 387–486.

Bernstein, H., Crow, B. & Johnson, H. (1992). *Rural Livelihoods: Crisis and Responses*. London: Oxford University Press.

Boholm, A. (2003). The cultural nature of risk: can there be an anthropology of uncertainty? *Ethnos* 68: 159–178.

Brooker, S., Clarke, S., Njagi, J. K., Polack, S. & Mungo, B. (2004). Spatial clustering of malaria and associated risk factors during an epidemic in a highland area of western Kenya. *Tropical Medicine & International Health* 9: 757–766.

Bulmer, M. & Warwick, D. P. (1983). *Social Research in Developing Countries*. New York: John Wiley & Sons Ltd.

Campbell, D. J. (1984). Responses to drought among farmers and herders in Southern Kajiado District, Kenya. *Human Ecology* 12: 35–63.

Chambers, R. (1992). *Rural Appraisal: Rapid Relaxed and Participatory*. IDS Discussion Paper 311. Institute of Development Studies, University of Sussex, Brighton.

Chambers, R. (1997). *Whose Reality Counts? Putting the First Last*. London: Intermediate Technology.

Conover, M. R. (1994). Perceptions of grass-roots leaders of the agricultural community about wildlife damage on their farms and ranches. *Wildlife Society Bulletin* 22: 94–100.

Croll, E. & Parkin, D. (1992). Cultural understandings of the environment. In Croll, E. & Parkin, D. (eds.), *Bush Base, Forest Farm: Culture, Environment and Development*. London: Routledge, pp. 11–36.

De Boer, W. F. & Baquette, D. S. (1998). Natural resource use, crop damage and attitudes of rural people in the vicinity of Maputo Elephant Reserve, Mozambique. *Environmental Conservation* 25: 208–218.

Dercon, S. (1998). Wealth, risk and activity choice: cattle in Western Tanzania. *Journal of Development Economics* 55: 850–75.

Dercon, S. & Krishnan, P. (1996). Income portfolios in rural Ethiopia and Tanzania: choices and constraints. *Journal of Development Studies* 32: 850–75.

Desta, S., Coppock, D. L. & Barrett, C. B. (1999). Opportunities for asset diversification in a livestock system: the case of the pastoral Boran of Southern Ethiopia. In *Proceedings of the VI International Rangeland Congress*. Townsville, Australia.

Douglas, M. (1966). *Purity and Danger: An Analysis of the Concepts of Pollution and Taboo*. London: Routledge & Kegan Paul.

Dunn, A. (1993). *The Large Mammals of Gashaka-Gumti National Park, Nigeria: Line Transect Surveys from Forest and Savannah*. Report to WWF-UK, NCF & Federal Ministry of Agriculture, Water Resources and Rural Development / Nigeria.

Dunn, A. (2000). *Monitoring Local Attitudes and Perspectives: A Questionnaire Survey of the Support Zone and Enclaves, Gashaka Gumti National Park / Nigeria*. Report to NCF & WWF-UK.

Dunn, A. (1995). Gashaka Gumti, Nigeria – From Game Reserve to National Park. *RDFN Paper* 18.

Dunn, A., Mamza, J. U., Ananze, F. G. & Gawaisa, S. G. (2000). Sticking to the rules: working with local people to conserve biodiversity at Gashaka Gumti National Park, Nigeria. In *Promoting Partnerships: Managing Wildlife Resources in Central and West Africa*. IIED, Evaluating Eden Series No. 3.

Eddy, E. (1979). *Labor and Land Use on Mixed Farms in the Pastoral Zone of Niger* (Livestock Production and Marketing in the Entente States of West Africa, Monograph No. 3). Ann Arbor: University of Michigan Press.

Ellis, F. (1998). Household strategies and rural livelihood diversification. *Journal of Development Studies* 35: 1–38.

Ellis, F. (2000). *Rural Livelihoods and Diversification in Developing Countries*. Oxford: Oxford University Press.

Frantz, C. (1982). Settlement and migration among pastoral Fulbe in Nigeria and Cameroun. *Studies in Third World Societies* 17: 57–94.

Frantz, C. (1993). Are the Mboror'en boring, and Are the Fulbe finished? In Eguchi, P. K. & Azarya, V. (eds.), *Unity and Diversity of a People: The Search for Fulbe Identity. Senri Ethnological Studies* 35: 11–34.

Ghai, D. & Vivian, J. M. (1995). *Grassroots Environmental Action: People's Participation in Sustainable Development*. London: Routledge.

Gillingham, S. & Lee, P. C. (1999). The impact of wildlife-related benefits on the conservation attitudes of local people around the Selous Game Reserve, Tanzania. *Environmental Conservation* 26: 218–228.

Gillingham, S. & Lee, P. C. (2003). People and protected areas: a study of local perceptions of crop-damage conflict in an area bordering the Selous Game Reserve, Tanzania. *Oryx* 37: 316–325.

Graham, M. (1973). *A Natural Ecology*. Manchester: Manchester University Press.

Heinen, J. T. (1993). Park-people relations in Kosi Tappu Wildlife Reserve, Nepal: a socio-economic analysis. *Environmental Conservation* 20: 25–34.

Hell, B. (1996). Enraged hunters: the domain of the wild in north-western Europe. In Descola, P. & Palson, G. (eds.), *Nature and Society: Anthropological Perspectives*. London & New York: Routledge, pp. 205–218.

Hill, C. M. (1997). Crop-Raiding in wild vertebrates: the farmer's perspective in an agricultural community in western Uganda. *International Journal of Pest Management* 43: 77–84.

Hill, C. M. (2000). Conflict of interest between people and baboons: crop-raiding in Uganda. *International Journal of Primatology* 21: 299–315.

Hill, C. M. (2004). Farmers' perspectives of conflict at the wildlife-agriculture boundary: some lessons learned from African subsistence farmers. *Human Dimensions of Wildlife* 9: 279–286.

Hill, C. M. (2005). People, crops and primates: a conflict of interests. In Paterson, J. D. & Wallis, J. (ed.), *Commensalism and Conflict: the human-primate interface*. Norman, Oklahoma: American Society of Primatologists, pp. 40–59.

Hill, C., Osborn, F. & Plumptre, A. J. (2002). *Human–wildlife Conflict: Identifying the Problem and Possible Solutions*. Albertine Rift Technical Reports Series Vol. 1. Wildlife Conservation Society.

Hoare, R. E. (1999). Determinants of human-elephant conflict in a land-use mozaic. *Journal of Applied Ecology* 36: 689–700.

Hockings, M. & Phillips, A. (1999). *Parks 9*. Gland, Switzerland: IUCN.

Hogg, R. (1987). Development in Kenya: drought, desertification and food scarcity. *African Affairs* 86: 47–58.

Hogg, R. (1988). Water, harvesting and agricultural production in semi-arid Kenya. *Development and Change* 19: 69–87.

Holmes, C. M. (2003). The influence of protected area outreach on conservation attitudes and resource use patterns: a case study from western Tanzania. *Oryx* 37: 305–315.

Homewood, K. & Rodgers, W. A. (1991). *Maasailand Ecology: Pastoralists, Development and Wildlife Conservation in Ngorongoro, Tanzania*. Cambridge: Cambridge University Press.

Homewood, K., Coast, E., Kiruswa, S., Serneels, S., Thompson, M. & Trench, P. (2006). *Maasai pastoralists: diversification and poverty* [online]. London: LSE research online.

Infield, M. (1988). Attitudes of a rural community towards conservation and a local conservation area in Natal, South Africa. *Biological Conservation* 45: 21–46.

Kamaya, P. (1996). *A Preliminary Assessment of Crop Damage by Wild Animals in and Around Gashaka Gumti National Park*. Federal University Of Technology, Yola / Nigeria.

Khazanov, A. M. (1984). *Nomads and the Outside World. Cambridge*. London, New York, New Rochelle, Melbourne & Sydney: Cambridge University Press.

King, F. A. & Lee, P. C. (1987). A brief survey of human attitudes to a pest species of primate – *Cercopithecus aethiops*. *Primate Conservation* 8: 82–84.

Kingdon, J. (1997). *The Kingdon Field Guide to African Mammals*. San Diego: Academic Press.

Knight, J. (1999). Monkeys on the move: the natural symbolism of people-macaque conflict in Japan. *Journal of Asian Studies* 58: 622–647.

Knight, J. (2000). *Natural Enemies: People-Wildlife Conflicts in Anthropological Perspective*. London: Routledge.

Knight, J. (2003). *Waiting for Wolves in Japan*. Oxford: Oxford University Press.

Litsinger, J. A., Canapi, B. & Alviola, A. (1982). Farmers perceptions and control of rice pests in Solana, Cagayan Valley, a pre-green revolution area of the Phillippines. *Philippine Entomologist* 5: 373–383.

Little, P. D. (1985). Social differentiation and pastoralist sedentarization in Northern Kenya. *Africa* 55: 243–261.

Little, P. D. (1992). *The Elusive Granary: Herder, Farmer and State in Northern Kenya*. Cambridge: Cambridge University Press.

Little, P. D., Smith, K., Cellarius, B. A., Coppock, D. L. & Barrett, C. B. (2001). Avoiding disaster: diversification and risk management among East African herders. *Development and Change* 32: 387–419.

Maikhuri, R. K., Nautiyal, S., Rao, K. S. & Saxena, K. G. (2001). Conservation policy-people conflicts: a case study from Nanda Devi Biosphere Reserve (a World Heritage Site), India. *Forest Policy and Economics* 2: 355–365.

Naughton-Treves, L. & Treves, A. (2005). Sociological factors shaping local support for wildlife: crop raiding by elephants and other wildlife in Africa. In Woodroffe, R., Thirgood, F. & Rabinowitz, A. (ed.), *People and Wildlife: Conflict or Coexistence?* Cambridge: Cambridge University Press, pp. 252–277.

Naughton-Treves, L. (1996). *Uneasy Neighbours: Wildlife and Farmers Around Kibale National Park, Uganda*. PhD thesis, University Florida.

Naughton-Treves, L. (1997). Farming the forest edge: vulnerable places and people around Kibale National Park, Uganda. *The Geographical Review* 87: 27–46.

Naughton-Treves, L. (1998). Predicting patterns of crop damage around Kibale National Park, Uganda. *Conservation Biology* 12: 156–168.

Naughton-Treves, L. (2001). Farmers, wildlife and the forest fringe. In Weber, W., White, L. J. T., Vedder, A. & Naughton-Treves, L. (ed.), *African Rainforest Ecology and Conservation*. New Haven, Conn.: Yale University, pp. 369–384.

Nepal, S. K. & Weber, K. E. (1995). Prospects for coexistence: wildlife and local people. *Ambio* 24: 238–245.

Newmark, W. D., Leonard, N. L., Sariko, H. I. & Gamassa, D. M. (1993). Conservation attitudes of local people living adjacent to five protected areas in Tanzania. *Biological Conservation* 63: 177–183.

Newmark W. D., Manyanza, D. N., Gamassa, D. G. M. & Sariko, H. I. (1994). The conflict between wildlife and local people living adjacent to protected areas in Tanzania: human density as a predictor. *Conservation Biology* 9: 249–255.

Nichols, P. (1991). *Social Survey Methods: a Fieldguide for Development Workers*. Oxford: Oxfam.

NPS / NCF / WWF (1998). *Gashaka Gumti National Park: A Management Plan for Developing the Park and its Support Zone*. Nigeria National Park Service, Nigerian Conservation Foundation, World Wide Fund for Nature (UK).

Nyuidze, I. I. E. (2001). *Alhaji Hammangabdo Mohammadu Sambo Lamdo Gashaka: biography of a titan*. Yola / Nigeria: Paraclete.

Oerke, E. C. & Dehne, H. W. (2004). Safeguarding production – losses in major crops and the role of crop protection. *Crop Protection* 23: 275–285.

Ostrom, E. (1990). *Governing the Commons: The Evolution of Institutions for Collective Action*. Cambridge: Cambridge University Press.

Parry, D. & Campbell, B. (1992). Attitude of rural communities to animal wildlife and its utilization in Chobe enclave and Mbabe depression, Botswanna. *Environmental Conservation* 19: 245–252.

Pidgeon, N., Hood, C., Jones, D., Turner, B. & Gibson, R. (1992). Risk perception. In Royal Society (ed.), *Risk: Analysis, Perception and Management*. London: The Royal Society, pp. 98–134.

Priston, N. (2005). *Crop-Raiding by Macaca ochreata brunnescens in Sulawesi: Reality, Perceptions and Outcomes for Conservation*. PhD thesis, University of Cambridge.

Quinn, C. H., Hubby, M., Kiwasila, H. & Lovett, J. C. (2003). Local perceptions of risk to livelihood in semi-arid Tanzania. *Journal of Environmental Management* 68: 111–119.

Rao, K. S., Nautiyal, S., Maikhuri, R. K. & Saxena, K. G. (2000). Management conflicts in the Nanda Devi Bioshpere Reserve, India. *Mountain Research and Development* 20: 320–323.

Renn, O. & Rohrmann, B. (2000). *Cross-Cultural Risk Perception: A Survey of Empirical Studies*. Berlin: Springer.

Salzman, P. C. (1980). Introduction: processes of sedentarization as adaptation and response. In Salzman, P. C. (ed.), *When Nomads Settle*,. New York: Praeger, pp. ix-xviii.

Sehkar, N. U. (1998). Crop and livestock depredation caused by wild animals in protected areas: the case of Sariska Tiger Reserve, Rajasthan, India. *Environmental Conservation* 25: 160–171.

Siex, K. S. & Struhsaker, T. T. (1999). Colobus monkeys and coconuts: a study of perceived human–wildlife conflicts. *Journal of Applied Ecology* 36: 1009–1020.

Smith, K. (1998). Sedentarization and market integration: new opportunities for Rendille and Ariaal women of Northern Kenya. *Human Organization* 57: 459–468.

Smith, K., Barrett, C. B. & Box, P. W. (1999). *Participatory Risk Mapping for Targeting Research and Assistance: with an example from east African pastoralists*. GL-CRSP Pastoral Risk Management Project Technical Report No. 04 / 99. Logan: UT Utah State University.

Smith, K., Barrett, C. B. & Box, P. W. (2000). Participatory risk mapping for targeting research and assistance: with an example from east African pastoralists. *World Development* 28: 1945–1959.

Stolton, S. & Dudley, N. (1999). *A Preliminary Survey of Management Status and Threats in Forest Protected Areas*. PARKS: Protected Areas Programme. Gland Switzerland: IUCN.

Strum, S. C. (1994). Prospects for management of primate pests. *Behavioural Ecology and Sociobiology* 49: 295–306.

Swift, J. (1986). The economics of production and exchange in West African pastoral societies. In Adamu, M. & Kirk-Greene, A. H. M. (eds.), *Nomads and Pastoralists*. Manchester: Manchester University Press, pp. 175–190.

Tchamba, M. N. (1996). History and present status of the human elephant conflict in the Waza-Lagone Region, Cameroon, West Africa. *Biological Conservation* 75: 35–41.

Tweheyo, M., Hill, C. M. & Obua, J. (2005). Patterns of crop-raiding by primates around Budongo Forest Reserve, Uganda. *Wildlife Biology* 11: 237–247.

Wang, S. W., Lassoie, J. P. & Curtis, P. D. (2006). Farmer attitudes towards conservation in Jigme Singye Wangchuck National Park, Bhutan. *Environmental Conservation* 33: 1–9.

Warren, Y. (2003). *Olive Baboons (Papio cynocephalus anubis): Behaviour, Ecology And Human Conflict In Gashaka Gumti National Park, Nigeria*. PhD thesis, Roehampton University.

Webber, A. D. (2007). *Primate Crop Raiding in Uganda: Actual and Perceived Risks Around Budongo Forest Reserve*. PhD thesis, Oxford Brookes University.

Weladji, R. B. & Tchamba, M. N. (2003). Conflict between people and protected areas within the Benoue wildlife conservation Area, North Cameroon. *Oryx* 37: 72–79.

Wheatley, B. P., Stephenson, R., Kurashina, H. & Marsh-Kautz, K. G. (2002). A cultural primatological study of *Macaca fascicularis* on Ngeaur Island, Republic of Palau. In Fuentes, A. & Wolfe, L. D. (eds.), *Primates Face to Face*. Cambridge: Cambridge University Press, pp. 240–253.

Wright, M. (2003). *WWF Gashaka Gumti National Park Project Factsheet*. Godalming: WWF-UK.

Wywialowski, A. P. (1994). Agricultural producers' perceptions of wildlife caused losses. *Wildlife Society Bulletin* 22: 370–382.

Yayock J. Y., Lombin, G. & Owonubi, J. J. (1988). *Crop Science and Production in Warm Climates*. London: Macmillan.

Zinn, H. C. & Andelt, W. F. (1999). Attitudes of Fort Collins, Colorado, residents towards prairie dogs. *Wildlife Society Bulletin* 27: 1098–1106.

Zubko, G. V. (1993). Ethnic and cultural characteristics of the Fulbe. In Eguchi, P. K. & Azarya, V. (eds.), *Unity and Diversity of a People: The Search for Fulbe Identity. Senri Ethnological Studies* 35: 201–213.

Chapter 7
Pan-African Voyagers: The Phylogeography of Baboons

Dietmar Zinner, Umaru Buba, Stephen Nash, and Christian Roos

Abstract Baboons of the genus *Papio* have colonised wide areas of Africa and parts of the Arabian Peninsula. Traditionally, 5 phenotypically distinct morphotypes (species) are recognised: chacma baboons, yellow baboons, olive baboons, Guinea baboons and hamadryas baboons. We used mitochondrial DNA ("Brown" region) sequence data obtained mainly from faecal samples collected across the geographical range of baboons to reconstruct their phylogenetic relationships. Eight well-supported major haplogroups were detected, which reflect geographic populations. These disagree with the traditional classification of baboons into only 5 taxa. We found that West African olive and chacma baboons both comprise at least two deeply separated clades. In the case of chacma baboons, they correspond to recognised morphotypes (Cape chacma and grey-footed chacma). Our data also support a previously suggested distinction between yellow and Kinda baboons from central Zambia. Two other terminal clades from eastern Africa comprise either eastern olive and hamadryas baboons or eastern olive and yellow baboons. Southern yellow baboons cluster with grey-footed chacma baboons. Our data also indicate a possible mitochondrial overlap between Guinea baboons and a particular lineage of western olive baboons from Ivory Coast. These results support recent molecular studies, which detected several para- and polyphyletic mitochondrial clades in *Papio*, suggesting that the evolutionary history of baboons is even more complicated than previously thought. Thus, important roles might have been played by multiple phases of fragmentation, isolation, hybridisation, introgression, and nuclear swamping, hence, reticulation. These processes were most likely triggered by multiple cycles of expansion and retreat of savannah biomes during late Pliocene and Pleistocene glacial and inter-glacial periods. We also speculate on the likely dispersal pathways of these primates that may have led to their current distribution.

D. Zinner (✉)
Cognitive Ethology Laboratory, Deutsches Primatenzentrum, 37077 Göttingen, Germany
e-mail: dzinner@gwdg.de

V. Sommer and C. Ross (eds.), *Primates of Gashaka*, Developments in Primatology: Progress and Prospects 35, DOI 10.1007/978-1-4419-7403-7_7,

Keywords *Papio* spp. • Phylogeography • mtDNA • Evolution • Hybridisation • Dispersal

Introduction

Non-human primates have been extensively used as conceptual templates in attempts to understand the evolution of human sociality (Kinzey 1987, Cachel 2006). However, as many human traits evolved in savannah-like environments, our closest living relatives, chimpanzees and bonobos, are not necessarily the best models, as they often dwell in forests. Therefore, savannah-living African members of the Papionini tribe (e.g., baboons, geladas) serve as non-homologous models to explore the impact of ecological conditions on the evolution of human sociality (Strum & Mitchell 1987, Jolly 2001). But, extant geladas and baboons not only prefer open habitats, but also experienced an adaptive radiation in similar if not identical habitats in southern and eastern Africa corresponding to that of Plio-Pleistocene hominins (Jablonski 2002, Cachel 2006). Hominin-bearing sites in these areas also yield abundant fossils of papionins (Delson 1984, McKee & Keyser 1995). Nevertheless, although many of these are described as *Papio*, they may actually belong to the two closely related genera *Theropithecus* or *Parapapio* (Fleagle 1999). Baboons can also provide a model for the phylogeography of hominins with respect to the effect of gene-flow by occasional hybridisation between ecologically and adaptively distinct taxa (Jolly 2001, Arnold 2008, Zinner *et al.* 2009a). Both the baboon and the hominin lineages have been affected by recurrent and severe climatic changes in Africa, including expansion and contraction of savannahs during the late Pliocene and Pleistocene that parallel glacial and interglacial cycles in the Northern Hemisphere (Hamilton 1988, deMenocal 1995). Savannah dispersal routes opened only temporarily. Populations became isolated (vicariance by expansion of forest or desert), thus preventing gene-flow, or gene-flow became possible in (secondary) contact zones after a barrier had been removed. A valid reconstruction of the evolutionary history of baboons could thus also reveal much about the hominin lineage.

The basis for understanding the evolution of taxa, however, is an accurate phylogeny. That is, one has to know, or at least one has to have a well-supported hypothesis about the evolutionary relationships among various groups of organisms (e.g., species, populations). Describing, ordering and naming evolutionary units, hence their taxonomic classification (e.g., genera, species) is another important step. Adequate data also allow to date divergence events between lineages, calibrated with fossil records. These can ideally be correlated with environmental changes to identify mechanisms that drove evolution (Flagstad *et al.* 2001, Rohland *et al.* 2007).

As Robert M. May put it: "Without taxonomy to give shape to the bricks, and systematics to tell us how to put them together, the house of biological science is a meaningless jumble" (1990:130). Certainly when it comes to baboons, brick shape and the plan how to put them is all but clear. And, to make matters worse, "baboon systematics is a tangle", too (Groves 2001: 237).

Extant baboons of the genus *Papio* consist of clusters of parapatric populations with 5 traditionally recognised and phenotypically distinct morphotypes (*Fig. 7.1*): hamadryas, Guinea, olive, chacma, and yellow baboons (Jolly 1993, Kingdon 1997). They are distributed across sub-Saharan Africa (*Fig.* 7.2), excluding the West African and most parts of the central African rain forest. Baboons even colonised parts of the Arabian Peninsula. Beside the traditional 5 forms more than 25 geographic variants have been described mainly for East and South Africa (Hill 1970, Groves 2001). Of these, at least some deserve the same taxonomic level as the 5 basic types (Jolly 1993, 2001, Frost *et al.* 2003). A premier example is the Kinda baboon, which, because of its distinct external and cranial morphology, is increasingly recognised as its own taxon (Jolly 1993, Frost *et al.* 2003, Leigh 2006). Recent evidence from molecular genetics supports this assumption (Burrell 2008; Zinner *et al.* 2009a).

In contrast to eastern and southern Africa, baboons of western Africa have been neglected, with respect to both behavioural and ecological studies and morphological and phylogenetic analyses. This is particularly true for western olive baboons. For example, Hill (1967, 1970) lists 4–8 subspecies for chacma and yellow baboons, but only one throughout the vast range of olive baboons from Sudan west of the Nile to Mauretania. Nevertheless, *Papio nigeriae* has been previously described from the region of Ibi / Nigeria (Elliot 1909), *Papio choras* from northwest Nigeria (Ogilby 1843), and *Papio yokoensis* from Yoko in Cameroon (Matschie 1900). Groves (2001) followed Hill (1959, 1967, 1970) and classified all these taxa as synonyms of *Papio anubis*. Until recently, isolated populations of olive baboons survived in mountain areas of the Sahara desert, such as in Aïr and Tibesti (*Papio anubis tibestianus*, Bigourdan & Prunier 1937, Dekeyser & Derivot 1960, Monod 1963, Magin 1990). However, it remains questionable whether these populations still exist.

It is also not clear whether the type locality of the olive baboon (*Papio anubis*) is in West or East Africa (for a discussion see Hill 1959). There has been confusion about the correct name of olive baboons, because it seems that the name *Cynocephalus olivaceus* (I. Geoffroy St. Hilaire 1851) was given to a Guinea baboon and not to an olive baboon (Hill 1959). Probably because of this, baboons in a European zoo obtained from Ghana in 1940 were named *Papio papio* (Guinea baboons), despite the fact that only olive baboons occur in Ghana (Tahiri-Zagrët 1976).

Furthermore, little is known about contact zones and possible hybridisation (Tappen 1960, Jolly 1993, Groves 2001), with the exception of Awash, Ethiopia and Amboseli, Kenya, where olive baboons interbreed with hamadryas and yellow baboons, respectively. Interbreeding may also occur between western olive and Guinea baboons in Mauretania, Mali, Guinea and even in Ivory Coast (Tahiri-Zagrët 1976, Galat-Luong *et al.* 2006). Until recently, it was assumed that Kinda baboons and chacmas are parapatric or only marginally sympatric, and evidence for interbreeding was lacking (Machado 1969, Ansell 1978). However, Jolly (pers. comm.) observed respective intermediate forms in the contact zone between

Fig. 7.1 Baboon taxa. The descriptions of the 5 traditionally recognised baboon species and Kinda baboons follow Jolly (1993), Kingdon (1997) and Groves (2001). Drawings of male specimen by Stephen Nash (not to scale). For distribution, see *Fig. 7.2.* (**a**) Hamadryas or Sacred baboon (*Papio hamadryas*). Adult males are silvery-grey with large manes, reaching back to the rump and contrasting with the rather short body and limb hair. The mane bushes out to form white cheek ruffs, offset by darker, shorter crown hair. The tail tuft is white; hands and feet are dark. Females are plain olive-brown, with no mane. In males, the face and cushion-like skin around the ischial callosities are bright pinkish red. Kummer (1968) described an east-west gradient in male face colour, with darker faces in the west of the hamadryas range and more light-red faces in the east. The tail is held in a simple backward-pointing curve. Most likely, hamadryas baboons are named after the Hamadryads, nymphs of the Greek mythology whose lives began and ended with a particular tree. Nevertheless, hamadryas baboons live in open and dry areas and they rarely enter into forest. "Sacred baboon" refers to their significant role in the religion of Ancient Egypt, where these primates became an aspect of the sun god Re and the moon god Thoth. Hamadryas baboons inhabit the mountains and arid lowlands of eastern Eritrea and Ethiopia, extending into Djibouti and northern Somalia and north into the Red Sea Hills of Sudan. They are also found along the Red Sea coast and in the coastal mountains of the Arabian Peninsula, from Yemen north into Saudi Arabia up to Jeddah. The conservation status of hamadryas baboons is Least Concern (IUCN 2010). (**b**) Guinea baboon (*Papio papio*). The fur is reddish-brown, with a sharply defined mane in adult males similar to hamadryas baboons, but less full around the head. The face is blackish-pink, and the skin around the callosities is slightly lighter. The tail loops evenly up and back.

Fig. 7.1 (continued) Nostrils protrude beyond the end of the snout. The colour of females is similar to males, but females have no mane. They are named after Guinea, a historic term for the southern part of West Africa. Guinea baboons are confined to a relative small area between Senegal, Guinea, Guinea-Bissau and parts of Mauretania and Mali. The conservation status of Guinea baboons is Near Threatened (IUCN 2010). (**c**) Olive or Anubis baboon (*Papio anubis*). The fur is olive-brown or khaki in both sexes. Adult males have large manes, restricted to the foreparts and grading into shorter body hair, but not as sharply set off as in hamadryas or Guinea baboons. In some populations males have shell-shaped greyish cheek ruffs. The face and skin around the callosities are dark grey to black. The bare area on the rump is much smaller than in hamadryas or Guinea baboons. Nostrils project forward of the snout. One-fourth of the tail ascends before descending sharply as if broken. The name refers either to their olive or khaki-green coat colour or to the jackal-headed god Anubis of Egyptian mythology. Olive baboons inhabit most parts of the northern savannah belt, from Mali in the west to Eritrea and Ethiopia in the east, and southwards into Kenya and north-western Tanzania. They penetrated rain forest in the northern and eastern parts of the Congo forest, but not so in western Africa. Isolated populations may have persisted within the Sahara, e.g., in the Tibesti Massif. Hybridisation occurs with hamadryas baboons in Eritrea and Ethiopia and with yellow baboons in south-eastern Kenya and north-western Tanzania. Jolly (1993) distinguishes Heuglin's baboon (*P. anubis heuglini*) as a possible distinct morphotype. Several other morphotypes have been described (e.g., Elliot 1913, Hill 1970), but Jolly (1993) and also Groves (2001) dispute the existence of such "subspecies". The conservation status of olive baboons is Least Concern (IUCN 2010). (**d**) Chacma baboon (*Papio ursinus*). Chacmas are the largest baboons, but size varies among populations. Fur colour ranges dorsally from black or very dark brown to grizzled khaki or grey-buff, always paler ventrally and at the sides of the muzzle, and with paler patches between eyes and nostrils. Sexes are similar in colour. Males have no mane, but elongated hair tufts along the nape. The face and a small area around the callosities are black. The nostrils do not protrude forward above the lips. Like in olive baboons, one-fourth of the tail ascends before dropping sharply. The facial skeleton, unlike in other baboons, points downward as well as forward. Chacma baboons are found in large parts of southern Africa south of the Zambezi, except for the most arid parts such as the central Kalahari and parts of Namibia. In the west they occur north into southern Angola and south-western Zambia. The Merriam-Webster Online Dictionary states that chacma is a native name for this baboon, most likely of Khoikhoi origin. Several morphotypes have been described (e.g., Hill 1970). Jolly (1993) recognises the typical chacma (*P. u. ursinus*), the grey-footed chacma (*P. u. griseipes*) and the Transvaal chacma (*P. u. orientalis*). Groves (2001) recognises the first two taxa, but not the Transvaal chacma and instead distinguishes the Ruacana chacma (*P. u. ruacana*). The conservation status of chacma baboons is Least Concern (IUCN 2010). (**e**) Yellow baboon (*Papio cynocephalus*). Yellow baboons are generally more slender than other taxa (except Kinda baboons). Males and females are dorsally yellow to yellow-brown, contrasting with white underparts, inner surfaces of limbs, cheeks, and lateral patches on muzzle and fringing bands and feet. Males have no mane, or hardly any. In some populations males have a median nuchal crest of long hair or longer flank hairs, forming an inconspicuous fringe. The face and the (small) bare areas around callosities are black. The tail is usually carried as if broken as in chacma baboons, but held lower. The nostrils are set back from the lips. Their name refers to their yellowish coat. "Cynocephalus" derives from the Greek for "dog-head". Yellow baboons occur from the Zambezi north into eastern Zambia (Luangwa valley), Malawi and northern Mozambique and further to north-east and northern Tanzania, coastal Kenya, and for an unknown distance north into Somalia. As for olive or chacma baboons several more morphotypes have been described (e.g., Hill 1970). Jolly (1993) and Groves (2001) distinguish typical yellow baboons (*P. cynocephalus cynocephalus*), Ibean baboons (*P. c. ibeanus*) and Kinda baboons (*Papio [c.] kindae*). The Ibean yellow baboon is named after IBEA, an acronym of the short-lived Imperial British East Africa Company (Grubb 2006). According to Cotterill (2003) yellow and chacma baboons are separated by the Zambezi River. The conservation status of yellow baboons is Least Concern (IUCN 2008). (**f**) Kinda or Katanga baboon (*Papio [c.] kindae*). Kinda baboons (pronounced "keen-dah", not "kain-dah") are strikingly smaller in size than other baboon taxa. Their general colour is yellowish-brown, with lighter cheeks and belly. Adult individuals resemble older juveniles of yellow baboons. Uniquely among baboons, the neonatal coat appears to be whitish rather

(continued)

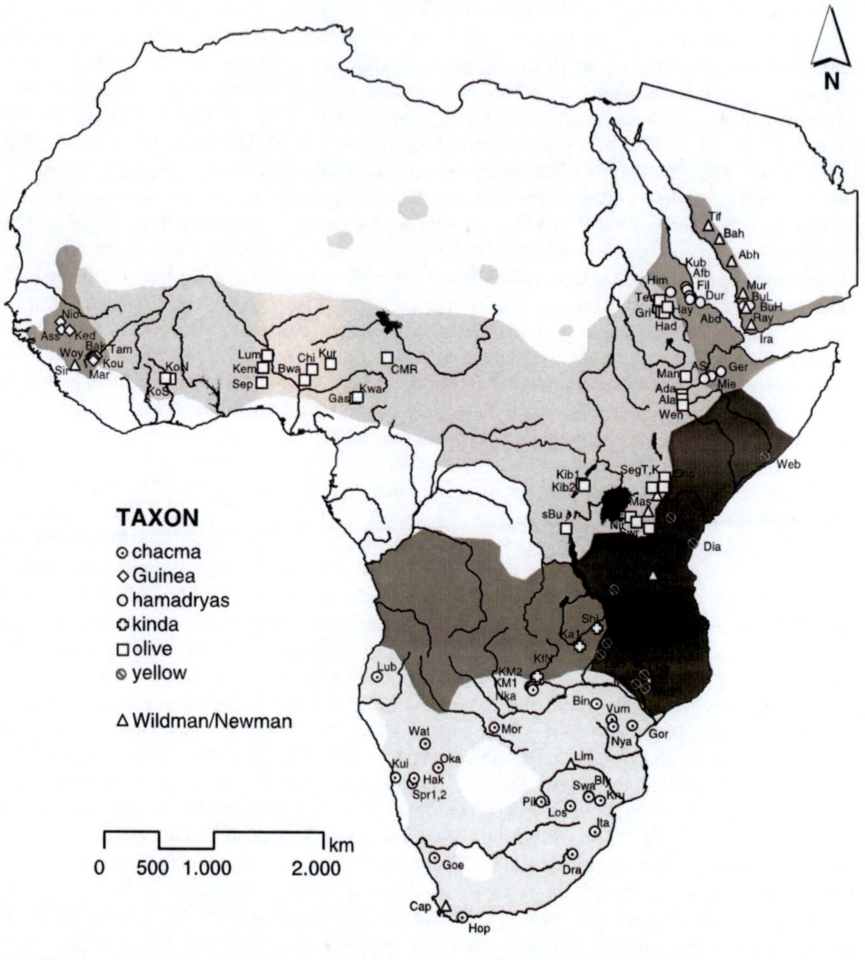

Fig. 7.2 Geographical distribution of traditionally recognised baboon species and sampling sites. Ranges are based on Kingdon (1997), and revised in accordance with Sarmiento (1997 – 1998) and Galat-Luong *et al.* (2006). The origins of samples from Newman *et al.* (2004) and Wildman *et al.* (2004) included in this study are indicated with triangles

Fig. 7.1 (continued) than black. Their tail is simply curved. Also typical for Kinda baboons seems to be a pink eye-ring, which Jane Phillips-Conroy and Clifford Jolly (pers. comm.) have found to be one of the most consistent field markers. The name was given by Lönnberg (1919) after the type locality near Kinda, a town in the southern DRC (Katanga). Kinda baboons occur on the upper Zambezi in central and south-western Zambia, in eastern Angola and south-eastern DRC. The ranges of this taxon and that of grey-footed baboons approach each other closely (e.g., in Kafue National Park, Zambia). Ansell (1978) found no evidence for hybridisation between Kinda and grey-footed baboons, however, recently Clifford Jolly (pers. comm.) observed individuals in the contact zone, which show an intermediate phenotype. Furthermore, Burrell (2008) found molecular evidence for hybridisation with yellow baboons where the ranges of both taxa meet, which fits with the observation of a broad cline in body size relating typical yellow and Kinda baboons. However, outside of this cline the Kinda morphotype appears to be stable throughout its range (Burrell 2008). The conservation status of Kinda baboons was not assessed (IUCN 2010)

Kinda baboons and grey-footed chacmas, indicating that hybridisation occurs. Also, there is clear evidence for interbreeding between Kinda and yellow baboons in eastern Zambia (Burrell 2008).

Thus, although baboons are among the best studied primates, their phylogeny and phylogeography are still unresolved, and numerous taxonomies are proposed (Tappen 1960, Roth 1965, Hill 1967, 1970, Jolly 1993, 1997–1998, Sarmiento 1997–1998) (*Fig. 7.3*).

Taxonomy attempts to order organism in hierarchically distinct groups according to character similarities and name them. These similarities should, of course, not reflect convergent evolution but common ancestry (synapomorphies, i.e., shared derived character-states). From this, one can infer evolutionary or genealogical relationships among taxa and establish a phylogeny that can be represented by an evolutionary tree. Terminal groups within a phylogeny form a clade or monophyletic group that includes all descendants and their common ancestor. Whether such terminal clades are equivalent to genera, species, or subspecies, i.e., which taxonomic

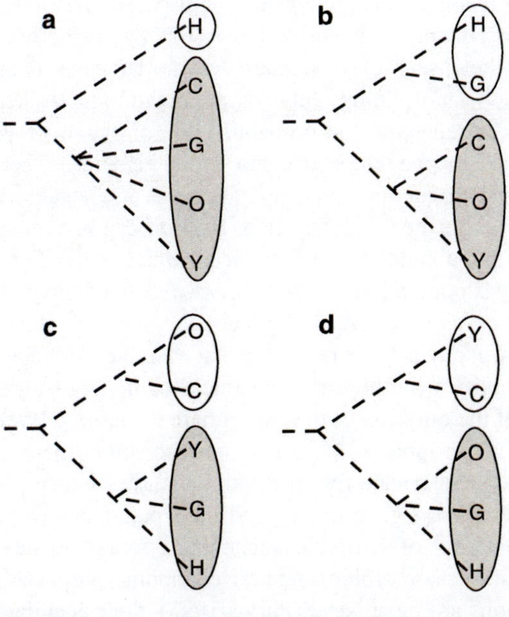

Fig. 7.3 Proposed taxonomies of baboons (H = hamadryas, O = olive, C = chacma, Y = yellow and G = Guinea). **(a)** The "socio-ecological" model (Buettner-Janusch 1966, Thorington & Groves 1970, Smuts *et al.* 1987) used differences in the social organisation of baboon taxa to place hamadryas baboons (with multi-level social organisation based on one-male units) as a sister taxon to all other taxa (savannah baboons, with multi-male multi-female organisation); **(b)** the "mantle" model (e.g., Hill 1967) distinguished taxa with a shoulder mane in adult males (hamadryas, Guinea) from those without (yellow, olive and chacma); **(c)** the "primitive hair" model (Ellermann *et al.* 1953; Kingdon 1971), differentiated olive and chacma baboons with their more agouti-like hair (possibly indicating a primitive trait) from yellow, hamadryas and Guinea baboons where this trait is absent; **(d)** the "north-south split" model (Jolly 1993) groups baboon taxa according to geographical range into a southern (chacma, yellow) and northern clade (olive, hamadryas, Guinea)

level they will be assigned to, is a matter of the degree of similarity, and also of the effect of the applied species concept.

Interbreeding occurs between some parapatric pairs of baboon taxa. Therefore, only a single baboon species would be recognised under a strict biological species concept (BSC; Mayr 1942), where barriers prevent successful reproduction among "good" species. In contrast, Groves (2001), Grubb *et al.* (2003) as well as Jolly (2007) recognise all 5 major forms as species, particularly because of the relationships between the forms where their ranges meet. Hybrid zones between olive and hamadryas baboons and between olive and yellow baboons were thought to be relatively restricted, thus fitting a "relaxed" BSC, where narrow hybrid zones are accepted. The phylogenetic species concept (PSC; Cracraft 1983), on the other hand, understands a species as "the smallest diagnosable cluster of individual organisms [identifiably distinct from other clusters] within which there is a parental pattern of ancestry and descent". These clusters represent monophyletic groups and the genus *Papio* could therefore include 5 or more species.

Jolly (1993), however, suggested that baboons constitute a single polytypic species, including 9 recognizable phylogenetic subspecies. His judgement is based on the discrepancies between "phenostructure" and "zygostructure" in baboons. The former concerns the observable characteristics of organisms at any level from social system (population), morphology (phenotype) down to DNA sequences (genome). Zygostructure is based upon the probability of formation of viable zygotes and includes all aspects of population structure in the geneticists' sense – such as rates of gene flow and cross-mating, intensity of positive and negative assortative mating by phenotype or avoidance of inbreeding. Baboon zygostructure would consist of 3 or 4 isolated population clusters between which gene flow is impeded: the Arabian hamadryas population, the possibly two isolated olive baboon populations in the Sahara desert and the rest of all other baboon populations of Africa.

Be it as it may, many authors recognise that even the most elaborate species concept will oversimplify a complex and dynamic evolutionary pattern (cf. Jolly 1993).

Irrespective of the question of the appropriate taxonomic level (species, subspecies) we need a well-supported phylogeny to reconstruct the evolution of a taxon. This is of particular relevance for co-evolution studies where phylogenies have to be reconciled between, e.g., primates and their pathogens (Switzer *et al.* 2005). Similarly, for patterns of heritable social behaviours we need to differentiate between ancestral (plesiomorphic) and derived (apomorphic) states. If, for example, hamadryas baboons are most basal (Purvis 1995), their peculiar multi-level social organisation with its one-male units (Kummer 1968) would be ancestral to the baboon clade, as opposed to the multi-male organisation of other baboon taxa (Melnick & Pearl 1987). However, the hamadryas system is most likely an autapomorphy, because hamadryas baboons are not basal in the baboon clade. Incorrect phylogenies will thus inevitably translate into incorrect inferences, not only about the evolution of "species", but also about respective physiological, morphological or behavioural traits.

The genus *Papio* probably originated in southern Africa, as evidenced by fossils (McKee 1992, Broadfield *et al.* 1994, Benefit 1999) and molecular studies

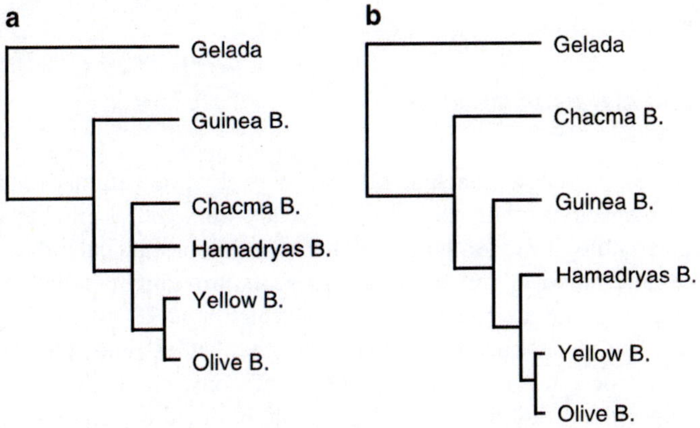

Fig. 7.4 Proposed molecular phylogenies of *Papio* taxa. Depending on studied loci and geographic provenance of baboon samples, different forms are proposed as most basal. (**a**) Guinea baboon (polymorphic protein loci, Williams-Blangero *et al.* 1990; mitochondrial COII gene, Disotell *et al.* 1992, Disotell 2000); (**b**) chacma baboon ("Brown" region, Newman *et al.* 2004). Williams-Blangero *et al.* (1990) emphasise their phylogeny applies only to the particular study sample of captive baboons. They argue that the clustering of olive and yellow baboons might be due to the fact that the geographic origin of the baboons was close to the contact zone of both taxa in Kenya, where gene flow was highly likely. This explanation stresses the importance of the exact provenance of samples in phylogenetic studies

(Newman *et al.* 2004, Zinner *et al.* 2009a). But apart from their origin, anything else about the evolution of baboons is still subject of debate and speculation (review in Newman *et al.* 2004). Even molecular genetics provides no simple solution (*Fig. 7.4*) given a discordance between mitochondrial phylogenies and known taxonomy (Hapke *et al.* 2001, Wildman *et al.* 2004, Newman *et al.* 2004, Burrell 2008, Zinner *et al.* 2009a, this study). Inconsistencies among various mitochondrial phylogenies, however, can largely be attributed to an incomplete taxon sampling in combination with paraphylies of the traditionally recognised baboon taxa.

Our contribution aims to determine the phylogenetic relationships within the genus *Papio* with special reference to the phylogenetic position of West African baboons. Our analysis is based on mitochondrial DNA sequences of the "Brown" region (Brown *et al.* 1982). We focused on this part of the mitochondrial genome in particular to build on a previous study by Newman *et al.* (2004) but expanded the sampling of regions and taxa to make our phylogenetic analysis more robust (Zwickl & Hillis 2002). Moreover, only 3 samples came from captive individuals, whereas the vast majority were of clear provenance. In addition, we included orthologous sequences available from GenBank if their geographic origin was documented.

This breadth enabled us to refine the phylogenetic resolution, in particular for West African baboon taxa, and seems to provide evidence for past genetic exchange between parapatric taxa.

Methods

Phylogenetic Reconstruction

Evolution is regarded as a branching process, whereby populations are altered over time. Their respective relationships are typically depicted as phylogenetic trees through palaeontological, morphological, behavioural, or molecular data. The latter have increased dramatically over the past 15 years providing detailed insights into the evolution of numerous organisms (e.g., Murphy *et al.* 2001a, 2001b, Ciccarelli *et al.* 2006). The reconstruction of molecular phylogenies relies on statistical and mathematical procedures such as maximum-parsimony, maximum-likelihood, and Bayesian algorithms, all of which depend on implicit or explicit models that describe the evolution of observed characters. In molecular phylogeny, these characters are usually aligned nucleotide or amino acid sequences. The robustness of the depicted relationships is tested by bootstrap, quartet-puzzling or posterior probability methods that indicate how reliable the obtained relationships among taxa are. Individuals or taxa represented by their specific DNA sequence (haplotype) that cluster together in a phylogenetic tree are called a clade or haplogroup. These are regarded as monophyletic if they contain all descendants of an ancestor and the ancestor itself. If not all descendants are included such a clade is called paraphyletic, and if such group does not contain the common ancestor at all, it is named polyphyletic.

Frequently, phylogenies based on molecular and other data differ to a certain degree. Even phylogenetic trees based on different genes are often incongruent (Avise 2000), with hybridisation gaining increasing acceptance as an explanation for this (Avise 2000, 2004, Funk & Omland 2003, Seehausen 2004). However, the implications of hybridisation for evolutionary processes are still unclear, at least in animals (e.g., Seehausen 2004, Mallet 2005, Arnold & Meyer 2006). Several studies assume that reticulate evolution or hybridisation can be a driving force for novel traits and diversification (Patterson *et al.* 2006, Koblmüller *et al.* 2007, Mallet 2007, 2008, McDonald *et al.* 2008; but see Barton 2001, 2006). For primates, natural hybridisation between species is now well documented (e.g., *Lepilemur* sp., Rumpler *et al.* 2008; *Alouatta* sp., Cortés-Ortiz *et al.* 2007; *Macaca* sp., Tosi *et al.* 2000; *Gorilla* sp., Thalmann *et al.* 2007), and may even lead to new species (e.g., *Macaca arctoides*, Tosi *et al.* 2000; *Macaca munzala*, Chakraborty *et al.* 2007; *Trachypithecus pileatus*, Osterholz *et al.* 2008; for review see Arnold & Meyer 2006, Arnold 2008).

Hybridisation creates offspring with a mixed genome that includes the mitochondrial genome of the mother and nuclear genome from both parents (except that all male offspring carry the father's Y chromosome). If hybrids are viable, but less fit than parent species, backcrossing with both parental species will possibly result in a narrow hybrid zone. A new species will possibly result, if hybrids have a selective advantage and if they prefer to breed amongst themselves instead of with members of one of the parent populations (assortative mating).

In addition to this bidirectional hybridisation, sex-specific, unidirectional hybridisation can occur when a limited number of females from one species transfer into the population of another species. If these immigrants produce daughters with a strong selective advantage over resident females, the invasive mitochondrial lineages would become fixed. At the same time, contributions of the associated nuclear lineages would decrease every generation because the hybrid females would breed with *resident* males (female introgression, mitochondrial capture).

Conversely, in male specific hybridisation, male immigrants would reproduce with *resident* females. If hybrid males have a selective advantage over resident males, the invasive Y chromosome would become fixed in an otherwise foreign genome (male introgression, Y chromosome capture). If, however, hybrid females backcross over generations preferentially with males of the *invasive* species one speaks of male introgression and nuclear swamping (*Fig. 7.5*). The nuclear genome of the invasive lineage would then replace that of the resident lineage until only footprints of the latter remain as either mitochondrial or Y chromosomal DNA – even after one of the two species has gone extinct in the former hybrid zone (Lehman *et al.* 1991).

Since in baboons male migration and female philopatry is predominant, male introgression and male-mediated nuclear swamping would be the more likely hybridisation mechanism. This process would be enhanced if the effective population size (N_e) of the resident species is small compared to the invasive species and if hybrids of the heterozygous sex (in baboon males with XY gonosomes) would have a selective disadvantage (Haldane's rule; Haldane 1922). Nuclear swamping and introgressive hybridisation are not distinguishable by genetic data alone and further information such as phenotype or time of lineage separation is required.

Data Sampling

Faecal material from 143 individuals representing all 5 *Papio* types and Kinda baboons (*Tab. 7.1*) was collected from free-ranging populations at 79 sites in Africa and the Arabian Peninsula. Three additional samples consisted of dry tissue from museum specimens (sample 404: *Papio cynocephalus*, north-east bank of Lake Rukwa, Tanzania, coll. no. 03-74959; sample 406: *Papio ursinus*, Kuisebthal, Walfishbay, Namibia, coll. no. 13-6524; both Humboldt Museum, Berlin / Germany) or of tissue preserved in ethanol (sample 507: *Papio* (*ruhei*) *cynocephalus*, 40 km north-west of Mogadishu at Webi Shebelli, Somalia, Zoologische Staatssammlung München / Germany). The geographic coordinates of the sampling sites were determined with GPS or, in case of the museum specimen, estimated from maps. Additional samples (no. 559) from northern Cameroon were provided by the Limbe Wildlife Centre in Cameroon and the zoo in Abidjan / Ivory Coast (no. 510 and 511). These specimens were phenotypically olive baboons and stem from free ranging populations of the respective countries. The provenance of the Cameroon sample was available, but the exact provenances of the Abidjan animals

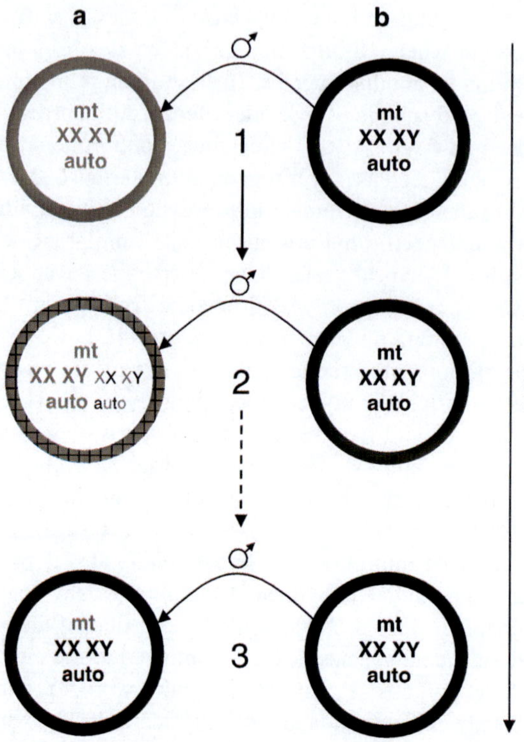

Fig. 7.5 Male introgression and nuclear swamping. Assuming that male migration and female philopatry is the predominant pattern in baboons, the most likely hybridisation scenario might be male introgression followed by nuclear swamping. The model assumes two parapatric populations. The grey population A is characterised by a "grey" mitochondrial genome (mt), "grey" gonosomes (XY) and "grey" autosomes (auto). The black population B has a respective "black" mitochondrial genome, gonosomes and autosomes. (1) Males from population B migrate into population A and produce hybrid offspring. These have all "grey" mt, but carry (statistically) 50 % "black" gonosomes and autosomes. Male offspring, carry exclusively "black" Y-chromosomes. Population A now contains individuals with pure "grey" and mixed "grey-and-black" genomes. If female hybrids of population A breed further with males from population B (2) and if this continues over generations (3), the frequency of "grey" gonosomes and autosomes will be significantly reduced. The "grey" nuclear genome (gonosomes and autosomes) will thus be "swamped" by the "black" nuclear genome until it is completely substituted. In contrast, the original "grey" mitochondrial genome remains in the population

are unknown. Our total baboon sample thus consists of 147 individual samples from 83 sites (plus 2 individuals from Abidjan zoo).

Our sampling pattern covered most of the baboon range (*Fig. 7.2*) with one major gap between Cameroon and Ethiopia, an area where only one baboon taxon should occur (*Papio a. anubis*, Hill 1970). Either fresh or dry faecal material was collected. Fresh samples were preserved in 75 % ethanol and dry samples simply in plastic tubes without any additive. Samples were stored at ambient temperature for up to 6 months before further processing. In addition, we included 20 "Brown"

Table 7.1 Sample localities of baboon faeces (geographic coordinates in decimal degrees), sample number (ID), haplotype designations (Hap) and GenBank accession numbers (Acc. No.). Additional sequences from localities where we were unable to obtain samples from were taken from GenBank. Sequences in *italic* represent unique haplotypes used in our phylogenetic analyses.

ID	Taxon (a)	Country (b)	Site (c)	Code	Longitude	Latitude	Hap	Acc. No. (d)
510	Pa	Ivory Coast	?	CD1	?	?	a01	EU885836
511	Pa	Ivory Coast	?	CD2	?	?	a02	EU885837
95	Pa	Ivory Coast	Komoé NP, north	KoN	–3.79000	8.80000	a03	EU885838
96	Pa	Ivory Coast	Komoé NP, north	KoN	–3.79000	8.80000	a03	EU885767
97	Pa	Ivory Coast	Komoé NP, north	KoN	–3.79000	8.80000	a03	EU885839
98	Pa	Ivory Coast	Komoé NP, north	KoN	–3.79000	8.80000	a03	EU885840
100	Pa	Ivory Coast	Komoé NP, north	KoN	–3.79000	8.80000	a03	EU885841
101	Pa	Ivory Coast	Komoé NP, south	KoS	–3.82000	8.74750	a03	EU885842
546	Pa	Nigeria	Lumma (Niger State)	Lum	4.26420	10.31810	a04	EU885769
550	Pa	Nigeria	Lumma (Niger State)	Lum	4.26420	10.31810	a05	EU885770
545	Pa	Nigeria	Bwari (Abuja)	Bwa	7.22030	8.81140	a06	EU885768
552	Pa	Nigeria	Kura (Plateau State)	Kur	9.26667	9.91667	a06	EU885777
555	Pa	Nigeria	Chigwa (Kaduna State)	Chi	7.81670	9.35000	a07	EU885776
549	Pa	Nigeria	Chigwa (Kaduna State)	Chi	7.81670	9.35000	a08	EU885775
553	Pa	Nigeria	Kemainja (Kwara State)	Kem	3.91670	9.71670	a09	EU885843
547	Pa	Nigeria	Kemainja (Kwara State)	Kem	3.91670	9.71670	a10	EU885771
138	Pa	Nigeria	Gashaka Gumti NP, Kwano	Kwa	11.58333	7.31667	a10	EU885844
144	Pa	Nigeria	Gashaka Gumti NP, Kwano	Kwa	11.58333	7.31667	a10	EU885845
147	Pa	Nigeria	Gashaka Gumti NP, Gashaka	Gas	11.50000	7.35000	a10	EU885846
169	Pa	Nigeria	Gashaka Gumti NP, Gashaka	Gas	11.50000	7.35000	a10	EU885847
EY01	Pa	Nigeria	Gashaka Gumti NP, Gashaka	Gas	11.50000	7.35000	a10	EU885773
EY10	Pa	Nigeria	Gashaka Gumti NP, Gashaka	Gas	11.50000	7.35000	a10	EU885774
548	Pa	Nigeria	Sepeteri (Oyo State)	Sep	3.65000	8.63330	a10	EU885772
559	Pa	Cameroon	Primate Rescue Center Limbe	CMR	13.93333	10.56667	a11	EU885778
512	Pa	DRC	Bukavu south	sBu	28.91092	–2.68258	a12	EU885780
513	Pa	DRC	Bukavu south	sBu	28.91092	–2.68258	a12	EU885848

(continued)

Table 7.1 (continued)

ID	Taxon (a)	Country (b)	Site (c)	Code	Longitude	Latitude	Hap	Acc. No. (d)
516	Pa	DRC	Bukavu south	sBu	28.91092	–2.68258	a12	EU885849
444	Pa	Uganda	Kibale Forest	Kib2	30.43333	0.51667	a13	EU885850
448	Pa	Uganda	Kibale Forest	Kib1	30.40000	0.48333	a13	EU885779
200	Pa	Eritrea	Haycota	Hay	37.06600	15.15695	a14	EU885851
77	Pa	Eritrea	Haycota	Hay	37.06600	15.15695	a15	EU885852
181	Pa	Eritrea	Tesseney	Tes	36.70142	15.14510	a15	EU885853
184	Pa	Eritrea	Ruba Griset	Gri	36.76018	14.88322	a16	EU885781
191	Pa	Eritrea	Ruba Griset	Gri	36.76018	14.88322	a17	EU885854
194	Pa	Eritrea	Ruba Hadejemi	Had	36.90710	14.35827	a18	EU885782
349	Pa	Ethiopia	Managasha NP	Man	38.58333	9.08333	a19	EU885783
332	Pa	Ethiopia	Adami Tulu	Ada	38.71493	7.82558	a20	EU885855
338	Pa	Ethiopia	Alambada	Ala	38.74768	7.50463	a21	EU885856
340	Pa	Ethiopia	Wendo Genet	Wen	38.64965	7.07127	a22	EU885857
527	Pa	Kenya	Segera Ranch	SegT	36.01581	0.25338	a23	EU885784
528	Pa	Kenya	Segera Ranch	SegK	36.75836	0.23566	a23	EU885858
530	Pa	Kenya	Segera Ranch	SegK	36.75836	0.23566	a23	EU885859
531	Pa	Kenya	Chololo Ranch	Cho	37.06000	0.91000	a23	EU885860
535	Pa	Kenya	Segera Ranch	SegK	36.75836	0.23566	a23	EU885861
295	Pa	Tanzania	Ngorongoro CA	Ngo	35.59039	–3.28206	a24	EU885862
351	Pa	Tanzania	Serengeti NP	Nir	34.80128	–2.42647	a24	EU885863
353	Pa	Tanzania	Serengeti NP	Swr	34.85236	–2.43100	a24	EU885786
537	Pc	Kenya	Amboseli NP	Amb	37.39000	–2.29000	a24	EU885790
538	Pc	Kenya	Amboseli NP	Amb	37.39000	–2.29000	a24	EU885864
W096	Pa	Kenya	Masai Mara NP	Mas	35.66667	–1.81667	a24	AY212096
W099	Pa	Kenya	Gilgil	Gil	36.35000	–0.58333	a24	AY212099
536	Pa	Kenya	Segera Ranch	SegT	36.01581	0.25338	a25	EU885785
*507**	Pc	Somalia	Webi Shebelli	Web	45.43333	2.42083	a25	EU885787
W097	Pa	Kenya	Gilgil	Gil	36.35000	–0.58333	a26	AY212097

W098	Pa	Kenya	Gilgil	Gil	36.35000	–0.58333	a27	AY212098
529	Pc	Kenya	Diani Beach	Dia	39.55000	–4.32000	c01	EU885788
533	Pc	Kenya	Diani Beach	Dia	39.55000	–4.32000	c01	EU885789
*404**	Pc	Tanzania	Lake Rukwa	Ruk	32.15517	–7.58297	c02	EU885791
288	Pc	Malawi	Mulanje Mt.	Mu2	35.51667	–15.95000	c03	EU885793
151	Pc	Malawi	Michiru Mts. CA	Mic	34.56667	–15.70000	c04	EU885792
286	Pc	Malawi	Liwonde NP	LNb	35.30000	–14.96667	c04	EU885865
570	Pc	Zambia	Luambe NP	Lua	32.14550	–12.45780	c05	EU885796
411	Pc	Zambia	South Luangwa NP	LuS	31.63793	–13.26840	c06	EU885795
408	Pc	Zambia	South Luangwa NP	LuS	31.63793	–13.26840	c07	EU885866
409	Pc	Zambia	South Luangwa NP	LuS	31.63793	–13.26840	c07	EU885794
W046	Pc	Tanzania	Mikumi NP	Mik	36.00000	–7.00000	c08	AY212046
W047	Pc	Tanzania	Mikumi NP	Mik	36.00000	–7.00000	c09	AY212047
W048	Pc	Tanzania	Mikumi NP	Mik	36.00000	–7.00000	c10	AY212048
209	Ph/Pa	Eritrea	Af Himbol	Him	37.39710	15.94505	h01	EU885867
74	Ph	Eritrea	Kubkub	Kub	38.63217	16.34482	h02	EU885801
391	Ph	Eritrea	Afabet	Afb	38.74958	16.12017	h03	EU885802
235	Ph	Eritrea	Filfil	Fil	38.94445	15.61442	h04	EU885868
232	Ph	Eritrea	Filfil	Fil	38.94445	15.61442	h05	EU885869
240	Ph	Eritrea	Durfo	Dur	38.96458	15.37370	h05	EU885870
37	Ph	Eritrea	Abdur	Abd	39.84585	15.12857	h06	EU885800
301	Ph	Ethiopia	Awash Station	ASt	40.17775	8.99268	h06	EU885803
300	Ph	Ethiopia	Awash Station	ASt	40.17775	8.99268	h07	EU885871
316	Ph	Ethiopia	Gerba Luku	Ger	41.53400	9.58740	h08	EU885872
319	Ph	Ethiopia	Gerba Luku	Ger	41.53400	9.58740	h09	EU885804
320	Ph	Ethiopia	Mieso	Mie	40.76408	9.20353	h10	EU885873
413	Ph	Yemen	Bura'a Forest, Hodaidah	BuH	43.41667	14.86667	h11	EU885874
414	Ph	Yemen	Bura'a Forest, Hodaidah	BuH	43.41667	14.86667	h11	EU885805
W073	Ph	Saudi Arabia	Abha	Abh	42.50523	18.21639	h11	AY212073

(continued)

Table 7.1 (continued)

ID	Taxon (a)	Country (b)	Site (c)	Code	Longitude	Latitude	Hap	Acc. No. (d)
W075	Ph	Saudi Arabia	Baha	Bah	41.46667	20.01667	h11	AY212075
W077	Ph	Saudi Arabia	Taif	Tif	40.41583	21.27028	h11	AY212077
415	Ph	Yemen	Bura'a Forest, Hodaidah	BuH	43.41667	14.86667	h12	EU885875
417	Ph	Yemen	Bura'a Forest, Hodaidah	BuL	43.86694	14.86722	h12	EU885876
W061	Ph	Yemen	Jabal Raymah	Ray	43.43333	14.66667	h13	AY212061
W062	Ph	Yemen	Jabal Iraf	Ira	44.25000	13.11667	h14	AY212062
W068	Ph	Yemen	Wadi Mur	Mur	43.41667	15.73333	h15	AY212068
W069	Ph	Yemen	Wadi Mur	Mur	43.41667	15.73333	h16	AY212069
569	Pk	Zambia	Shiwa N`gandu	Shi	31.73892	−11.19677	k01	EU885799
568	Pk	Zambia	Kasanka NP 1	Kal	30.25202	−12.59059	k02	EU885798
565	Pk	Zambia	Kafue NP, north	KfN	26.53577	−14.96779	k03	EU885797
518	Pp	Senegal	Kedougou	Ked	−12.12472	12.57556	p01	EU885808
523	Pp	Senegal	Kedougou	Ked	−12.12472	12.57556	p02	EU885809
257	Pp	Guinea	Haute Niger NP, Bakaria	Bak	−10.31542	10.54267	p03	EU885877
268	Pp	Guinea	Haute Niger NP, Mare	Mar	−10.33702	10.50143	p03	EU885878
276	Pp	Guinea	Haute Niger NP. Tambo	Tam	−10.29207	10.54283	p03	EU885879
519	Pp	Senegal	Kedougou	Ked	−12.12472	12.57556	p03	EU885880
520	Pp	Senegal	Kedougou	Ked	−12.12472	12.57556	p03	EU885881
521	Pp	Senegal	Kedougou	Ked	−12.12472	12.57556	p03	EU885882
522	Pp	Senegal	Kedougou	Ked	−12.12472	12.57556	p03	EU885883
525	Pp	Senegal	Mt. Assirik, NK NP	Ass	−12.76667	12.88333	p03	EU885884
526	Pp	Senegal	Mt. Assirik, NK NP	Ass	−12.76667	12.88333	p03	EU885806
556	Pp	Senegal	Niokolo Koba, NK NP	Nio	−12.72090	13.07467	p03	EU885807
W103	Pp	Senegal	?	SEN	?	?	p03	AY212103
W104	Pp	Sierra Leone	?	SIR	?	?	p03	AY212104
262	Pp	Guinea	Haute Niger NP, Mare	Mar	−10.33702	10.50143	p04	EU885885
368	Pp	Guinea	Woyumba, Haute Niger NP	Woy	−10.41442	10.50847	p05	EU885886
384	Pp	Guinea	Haute Niger NP, Kouroukorodgi	Kou	−10.07305	10.43605	p06	EU885887

252	Pp	Guinea	Haute Niger NP, Bakaria	Bak	−10.31542	10.54267	p07	EU885810
466	Pu	RSA	DeHoop NR	Hop	20.40658	−34.45621	u01	EU885888
472	Pu	Namibia	Hakos Guestfarm	Hak	16.36463	−23.23708	u01	EU885827
475	Pu	Namibia	Hakos Guestfarm	Hak	16.36463	−23.23708	u01	EU885889
482	Pu	Namibia	Namib Spreetshoogte NR	Spr1	16.20160	−23.73322	u01	EU885828
485	Pu	Namibia	Namib Spreetshoogte NR	Spr1	16.20160	−23.73322	u01	EU885890
483	Pu	Namibia	Namib Spreetshoogte NR	Spr2	16.20555	−23.64758	u01	EU885891
474	Pu	Namibia	Hakos Guestfarm	Hak	16.36463	−23.23708	u02	EU885892
463	Pu	RSA	DeHoop NR	Hop	20.40658	−34.45621	u03	EU885833
480	Pu	Namibia	Namib Spreetshoogte NR	Spr2	16.20555	−23.64758	u03	EU885893
484	Pu	Namibia	Namib Spreetshoogte NR	Spr2	16.20555	−23.64758	u04	EU885894
*406**	Pu	Namibia	Kuisebthal	Kui	14.78448	−23.20610	u05	EU885895
542	Pu	RSA	Giant's Castle GR	Dra	29.48333	−29.33333	u06	EU885830
543	Pu	RSA	Giant's Castle GR	Dra	29.48333	−29.33333	u06	EU885831
561	Pu	RSA	Goegap NR	Goe	18.03222	−29.69944	u07	EU885832
468	Pu	RSA	Loskop NR	Los	29.28162	−25.42147	u08	EU885825
560	Pu	Angola	Serra Leba, Lubango	Lub	13.24167	−15.14167	u09	EU885826
563	Pu	RSA	Ithala GR	Ita	31.26667	−27.53333	u10	EU885821
564	Pu	RSA	Ithala GR	Ita	31.26667	−27.53333	u10	EU885822
439	Pu	RSA	Blyde River, Swadini	Swa	30.79000	−24.68000	u11	EU885896
432	Pu	RSA	Blyde River, Blydepoort	Bly	30.78049	−24.66667	u12	EU885897
435	Pu	RSA	Blyde River, Blydepoort	Bly	30.78049	−24.66667	u12	EU885820
442	Pu	RSA	Blyde River, Swadini	Swa	30.79000	−24.68000	u12	EU885898
469	Pu	RSA	Loskop NR	Los	29.28162	−25.42147	u12	EU885899
470	Pu	RSA	Loskop NR	Los	29.28162	−25.42147	u12	EU885824
471	Pu	RSA	Loskop NR	Los	29.28162	−25.42147	u12	EU885900
493	Pu	RSA	Kruger NP	Kru	31.70000	−25.00000	u12	EU885901
494	Pu	RSA	Kruger NP	Kru	31.70000	−25.00000	u12	EU885902
422	Pu	RSA	Pilanesberg GR	Pil	26.87805	−25.11111	u13	EU885817

(continued)

Table 7.1 (continued)

ID	Taxon (a)	Country (b)	Site (c)	Code	Longitude	Latitude	Hap	Acc. No. (d)
425	Pu	RSA	Pilanesberg GR	Pil	26.87805	–25.11111	u13	EU885818
426	Pu	RSA	Pilanesberg GR	Pil	26.87805	–25.11111	u13	EU885903
427	Pu	RSA	Pilanesberg GR	Pil	26.87805	–25.11111	u13	EU885904
430	Pu	RSA	Pilanesberg GR	Pil	26.87805	–25.11111	u13	EU885905
476	Pu	Namibia	Waterberg Plateau	Wat	17.24221	–20.50450	u13	EU885906
477	Pu	Namibia	Waterberg Plateau	Wat	17.24221	–20.50450	u13	EU885907
478	Pu	Namibia	Waterberg Plateau	Wat	17.24221	–20.50450	u13	EU885819
479	Pu	Namibia	Waterberg Plateau	Wat	17.24221	–20.50450	u13	EU885908
557	Pu	Namibia	Okasewa Ranch	Oka	18.34910	–22.41203	u13	EU885909
558	Pu	Namibia	Okasewa Ranch	Oka	18.34910	–22.41203	u13	EU885823
486	Pu	Botswana	Moremi WR	Mor	23.00000	–19.18349	u14	EU885910
487	Pu	Botswana	Moremi WR	Mor	23.00000	–19.18349	u14	EU885911
488	Pu	Botswana	Moremi WR	Mor	23.00000	–19.18349	u14	EU885912
489	Pu	Botswana	Moremi WR	Mor	23.00000	–19.18349	u14	EU885913
490	Pu	Botswana	Moremi WR	Mor	23.00000	–19.18349	u14	EU885914
491	Pu	Botswana	Moremi WR	Mor	23.00000	–19.18349	u14	EU885915
492	Pu	Botswana	Moremi WR	Mor	23.00000	–19.18349	u14	EU885816
566	Pu	Zambia	middle Kafue NP, New Kalala	Nka	26.01077	–15.77360	u15	EU885811
567	Pu	Zambia	middle Kafue NP 1	KM1	25.99809	–15.79871	u15	EU885812
571	Pu	Zambia	middle Kafue NP 2	KM2	25.97031	–15.87178	u15	EU885916
495	Pu	Zimbabwe	Bindura	Bin	31.39892	–17.21172	u16	EU885917
501	Pu	Zimbabwe	Nyanga	Nya	32.81283	–19.05781	u17	EU885813
504	Pu	Zimbabwe	Vumba	Vum	32.66133	–18.52769	u18	EU885918
539	Pu	Mozambique	Gorongosa NP	Gor	34.36111	–18.97833	u19	EU885814
540	Pu	Mozambique	Gorongosa NP	Gor	34.36111	–18.97833	u19	EU885815
W057	Pu	RSA	Cape	Cap	?	?	u20	AY212057
W058	Pu	RSA	Cape	Cap	?	?	u21	AY212058

W059	Pu	RSA	Cape	Cap	?	?	u22	AY212059
W105	Pu	RSA	Limpopo	Lim	29.38333	–22.20000	u23	AY212105
459	Tgel	Zoo	outgroup Theropithecus gelada					EU885835

(a) Pa = olive baboon, Pc = yellow baboon, Ph = hamadryas baboon, Pp = Guinea baboon, Pu = chacma baboon

(b) DRC = Democratic Republic of Congo; RSA = Republic of South Africa

(c) CA = Conservation Area, GR = Game Reserve, NP = National Park, NR = Nature Reserve, WR = Wildlife Reserve; NK = Niokolo Koba National Park

(d) AY… accession numbers from Wildman *et al.* 2004 and Newman *et al.* 2004; *dry tissue or tissue preserved in ethanol from museum specimen region sequences deposited at GenBank from some sites where we were not able to collect material, increasing our total sample to 169 sequences (including 2 zoo samples) from 96 sites. Gelada (*Theropithecus gelada*) faecal samples were collected from zoo animals and were used as an outgroup.

Laboratory Procedures

DNA from tissue and faecal material was extracted using the DNeasy Blood & Tissue or QIAamp DNA Stool Mini Kits from QIAGEN. To prevent contamination, laboratory procedures followed standard protocols (Goossens *et al.* 2000, Karanth *et al.* 2005, Osterholz *et al.* 2008, Roos *et al.* 2008). Moreover, DNA extraction, polymerase chain reaction (PCR), PCR purification and sequencing was performed in separate laboratories and repeated randomly after several months, while always only one individual per taxon or location was tested. Finally, all PCR reactions were performed with negative (HPLC-purified water) controls.

The "Brown" region (896 bp), comprising 457 bp of the 3′ end of the NADH dehydrogenase subunit IV (ND4) gene, the tRNAs for histidine (His), serine (Ser), and leucine (Leu), and 239 bp of the 5′ end of the NADH dehydrogenase subunit V (ND5) gene, was amplified via two overlapping fragments according to PCR conditions and primers as described (Newman *et al.* 2004). The results of the PCR amplifications were checked on agarose gels. PCR products were cleaned with the QIAGEN PCR Purification Kit and subsequently sequenced on an ABI 3100-Avant sequencer using the BigDye Terminator Cycle Sequencing Kit (Applied Biosystems). All sequences were deposited at GenBank (for accession numbers see *Tab. 7.1*).

Statistical Methods

The final data set comprised 86 non-identical baboon sequences and a *Theropithecus gelada* sequence, used as an outgroup. Due to a 3-bp deletion in southern chacma baboons, the sequence length of the alignment was reduced to 893 bp. Phylogenetic trees were constructed on the basis of maximum-likelihood (ML) and Bayesian algorithms, using the programs GARLI v0.951 (Zwickl 2006) and MrBayes 3.1.2 (Ronquist & Huelsenbeck 2003), respectively. Reconstructions were performed with the TrN + G and TIM + I + G model of sequence evolution as they were selected as best-fitting models according to a hierarchical likelihood ratio test and under the Akaike information criterion with MODELTEST 3.06 (Posada & Crandall 1998), respectively. For ML trees, relative support of internal nodes was performed by 500 bootstrap replications. PAUP* v4.0b10 (Swofford 2002) was then used to calculate a majority-rule consensus tree in order to obtain bootstrap percentages. Bayesian posterior probabilities in MrBayes were estimated on the basis of two simultaneous, independent runs, which were allowed to proceed for 10000000 generations with chains sampled every 100 generations.

Results

Baboon faecal samples analysed for this study were collected over wide areas of the genus' geographical range. We amplified and sequenced the mitochondrial "Brown" region from 147 individuals from 83 localities plus two additional individuals from the zoo in Abidjan that were phenotypically olive baboons.

To exclude contamination of the data set with nuclear pseudogenes, coding regions of the fragment were checked for their potential to be correctly transcribed. Moreover, no inconsistent nucleotides were detected in overlapping regions of the studied locus and some of the observed haplotypes are identical with haplotypes published by Newman *et al.* (2004) and Wildman *et al.* (2004). Accordingly, pseudogenes are most likely not included in our data.

Among the 149 sequenced individuals, we detected 73 different haplotypes. We added 20 sequences from GenBank representing 16 different haplotypes of which 3 were also found among our haplotypes (*Tab. 7.1*). In total, 86 non-identical haplotypes became apparent, defined by 212 variable sites. In baboons from the southern part of the Republic of South Africa (RSA), southern and western Namibia and western Angola, a triplet deletion in the ND5 gene of the "Brown" region was detected (haplogroup 1 in *Fig. 7.6)*. This deletion constitutes an autapomorphy of this particular clade since it was not found in any other papionin taxon.

Identical haplotypes were mainly found in samples collected at the same locality (same social group or neighbouring groups), but we also detected identical sequences at sites > 1300 km apart (*Tab. 7.2*), with the largest distances between chacma baboon sites of RSA and Namibia. Furthermore, we detected identical haplotypes in different species, such as olive baboons from the Serengeti and central Kenya (haplotype a24; samples 295, 351, 353, W096, W099) and yellow baboons from the Amboseli region in southern Kenya (samples 537, 538) or olive baboons from central Kenya (haplotype a25; sample 536) and yellow baboons from the Webi Shebelli area in Somalia (sample 507). Distances between respective sites are 300 km and 1000 km.

Phylogenetic relationships among haplotypes were calculated with Bayesian and ML approaches, which both led to identical tree topologies and similar support values (*Fig. 7.6*). Several well supported monophyletic clades were detected, which do not correspond to the current taxonomic classification of baboons, thus making all taxa para- or polyphyletic. Instead, we found a strong geographical signal in the way that local populations cluster together irrespective of their taxonomic affiliations (*Fig. 7.7*).

In total, we detected 8 larger haplogroups (1 – 8 in *Fig. 7.6*), which cluster in a southern (1 – 3) and a northern (4 – 8) clade. However, monophylies of these clades are statistically only weakly supported. Furthermore, 5 additional lineages became obvious (a – e), mainly comprising only one or two haplotypes. These lineages will most likely turn into clades with further geographical sampling.

The major southern clade divides further into 3, well-supported clades: (1) southern chacma clade, representing haplotypes from the Drakensberg, the Cape,

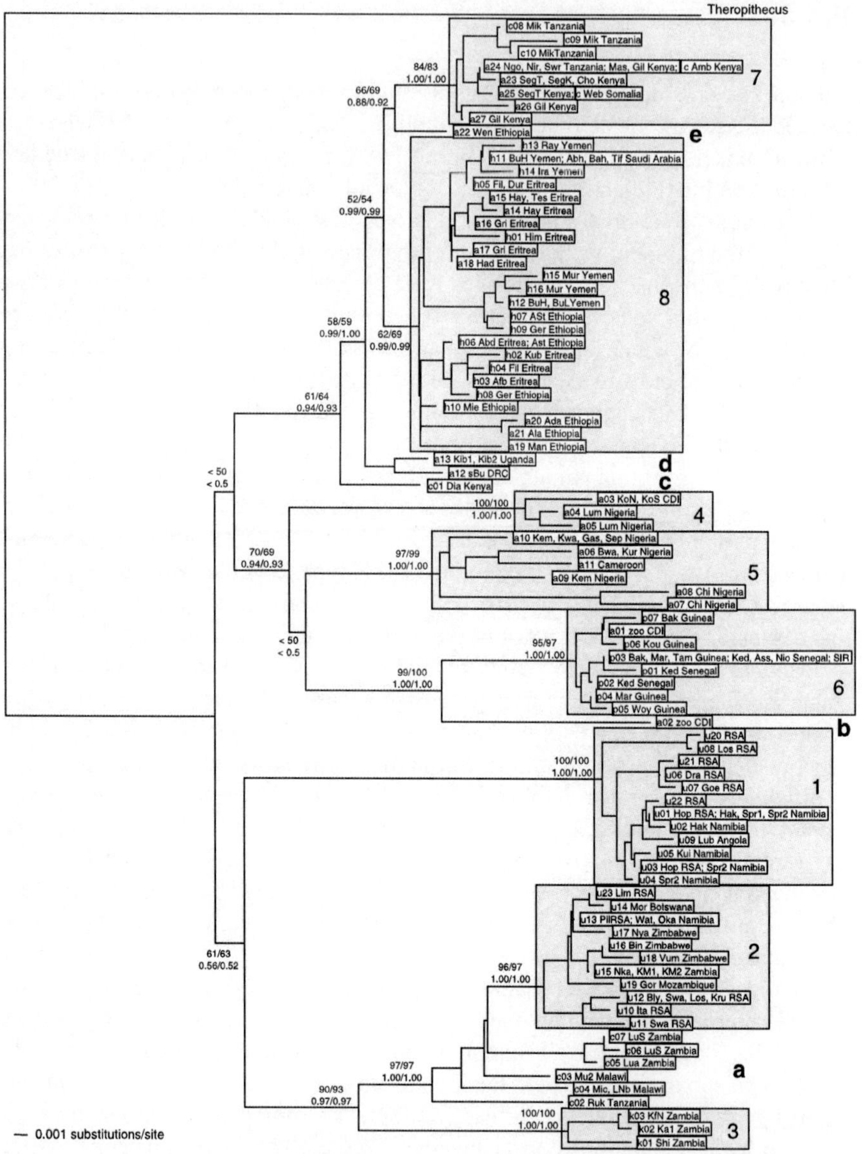

Fig. 7.6 Phylogenetic relationships among *Papio* based on 86 non-identical "Brown" region haplotypes. Numbers on branches indicate support for internal nodes, with the upper and lower numbers representing bootstrap and Bayesian posterior probability values, respectively. First and second numbers refer to values based on the TrN + G and TIM + I + G models, respectively. Numbers of terminal clades indicate haplogroups: (1) southern chacma baboons, (2) northern chacma baboons, (3) Kinda baboons, (4) olive baboons clade west 1, (5) olive baboons clade west 2, (6) Guinea baboons, (7) mixed clade of northern yellow baboons and eastern olive baboons, and (8) mixed clade of north-eastern olive baboons and hamadryas baboons. Several other lineages became also discernable: (**a**) a non-monophyletic group of southern yellow baboons, (**b**) a possible additional western olive baboon lineage, (**c**) possible lineages of coastal yellow baboons and (**d**) of central olive baboons, as well as (**e**) southern Ethiopian olive baboons

Table 7.2 Distances between sampling sites holding identical haplotypes (for abbreviation of haplotypes, sites and taxa, see *Table 7.1*)

Haplotype	Site 1	Site 2	Taxa (a)	Distance (km)
a24	Swr (Serengeti, Tanzania)	Amb (Amboseli, Kenya)	Pa – Pc	~ 300
a25	SegT (central Kenya)	Web (Webi Shebelli, Somalia)	Pa – Pc	~ 1000
h06	Abd (Abdur, Eritrea)	ASt (Awash Station, Ethiopia)	Ph – Ph	~ 690
h11	BuH (Bura'a Forest,Yemen)	Tif (Taif, Saudi Arabia)	Ph – Ph	~ 790
p03	Mar (Haute Niger NP, Guinea)	Nio (Niokolo Koba NP, Senegal)	Pp – Pp	~ 450
u01	Hop (DeHoop NR, RSA)	Hak (Hakos Guest Farm, Namibia)	Pu – Pu	~ 1300
u03	Hop (DeHoop NR, RSA)	Spr2 (Spreetshoogte NR, Namibia)	Pu – Pu	~ 1250
u13	Pil (Pilanesberg GR, RSA)	Wat (Waterberg Platea, Namibia)	Pu – Pu	~ 1100

(a) Key = *Table 7.1*

western RSA, southern and western Namibia and south-western Angola; (2) northern chacma baboon clade, which includes chacma haplotypes from Mozambique, northern RSA, Zimbabwe, Botswana, south-central Zambia and north-eastern Namibia; (3) clade of baboons that originated from central Zambia, which morphologically represent Kinda baboons.

The northern chacmas cluster together with yellow baboons from Malawi, eastern Zambia and south-eastern Tanzania (a), whereas the latter do not form a monophyletic clade.

The major northern clade divides into a western and an eastern clade. The western clade consists of one clade of olive baboon haplotypes found in Ivory Coast and western Nigeria (4), a second clade of olive baboon haplotypes mainly from central and eastern Nigeria (including baboons from Gashaka Gumti National Park) and northern Cameroon (5), and a clade of Guinea baboon haplotypes covering the area from Guinea to Senegal (6). One of two haplotypes (haplotype a01) from morphologically olive baboons from the zoo in Abidjan / Ivory Coast clusters with Guinea baboons and most likely represents a hybrid between olive and Guinea baboons. The second haplotype (a02) represents a distinct lineage (lineage b; cf. *Fig. 7.6*), which, however, clusters significantly with clade 6. The relationships within haplogroup 5 suggest an internal geographical structure, with haplotypes from Chigwa / Nigeria showing a relative large genetic distance to other members of the haplogroup.

The eastern clade is composed of two major haplogroups (7 and 8), which both represent mixtures of different taxa. However, the first divergence within the eastern clade concerns yellow baboons from the Kenyan coast (lineage c) followed by an olive baboon lineage (lineage d) from western Uganda and eastern Democratic

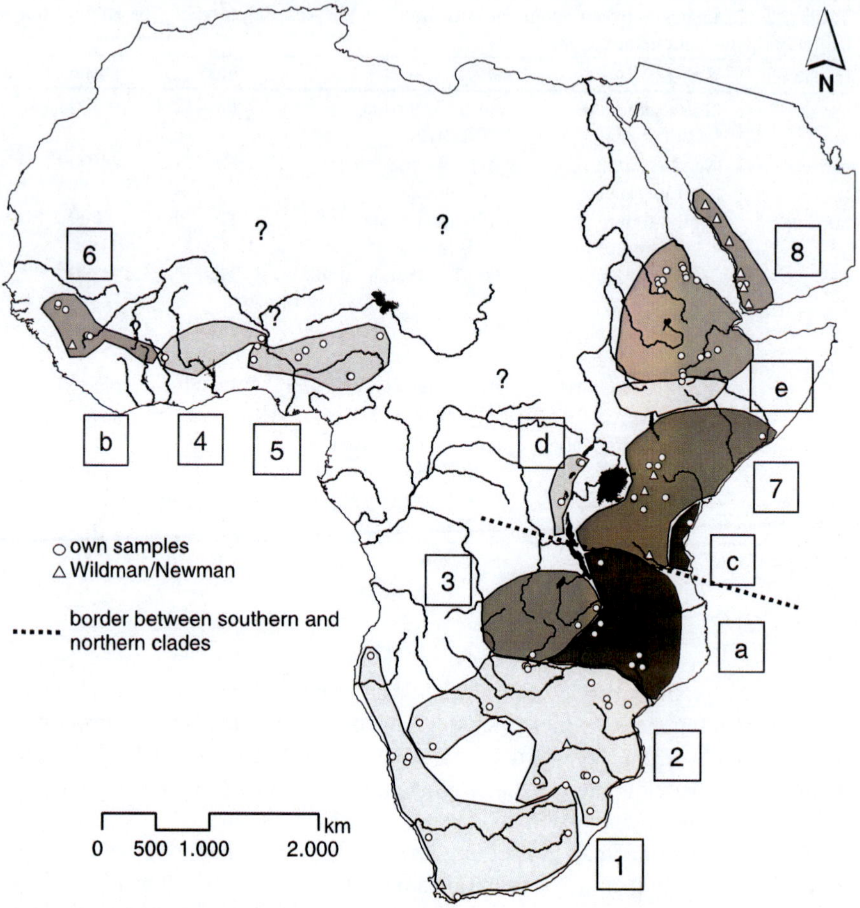

Fig. 7.7 Geographical distribution of terminal haplogroups and lineages. Circles and triangles represent sample locations, shaded areas approximate ranges of haplogroups (for numbers, see *Fig. 7.6*). Question marks indicate important future sampling areas (contact zone between olive and Guinea baboons; contact zone between olive baboon haplogroups in Nigeria; isolated olive baboon populations from the Sahara desert; olive baboons from Central African Republic, Congo and north-eastern DRC)

Republic of Congo (DRC), before further lineages diverged from each other. The major clade (7) consists of yellow and olive baboon haplotypes from northern Tanzania, Kenya, and Somalia, while clade (8) represents a mixture of olive and hamadryas baboon haplotypes from Ethiopia, Eritrea, and the Arabian peninsula. Interestingly, Arabian and African hamadryas haplotypes do not form distinct lineages. Moreover, haplotype a22 found in southern Ethiopia (Wendo Genet) is geographically close to the range of clade 8, but constitutes a distinct lineage (lineage e) and clusters with haplogroup 7.

Discussion

Our phylogenetic reconstruction suggests that baboon populations can be diagnosed through mitochondrial DNA and sorted into several coherent and reasonably well-supported geographic groups. These, however, do not match the traditionally recognised baboon taxa. This becomes particularly obvious when looking at the deep splits between several haplogroups of olive, yellow, and chacma baboons and the mixture of taxa within certain mitochondrial clades, which suggests multiple para- or polyphylies in the genus (cf. *Fig. 7.6*).

However, it has to be remembered that our results are based only on maternally inherited mitochondrial DNA and that the evolutionary history of baboons will remain incomplete, unless phylogenies based on nuclear genes are incorporated. These are apparently responsible for morphological and behavioural characteristics of a taxon and might therefore be more in concordance with traditionally recognised baboon morphotypes. The discrepancy between mitochondrial haplogroups and baboon morphotypes thus most likely reflects a discordance between mitochondrial and nuclear phylogenies and indicates, that events of hybridisation, introgression and nuclear swamping (cf. *Fig. 7.5*) might have shaped the evolution of baboons.

Taxonomy

Whether taxonomic units in the genus *Papio* should be classified as species or subspecies has long been discussed (reviews in Jolly 1993, Groves 2001, Grubb *et al.* 2003). Our study exemplifies that the use of mitochondrial DNA can also not solve that issue unequivocally.

If one would regard terminal clades in our phylogeny as "species", morphotypes would have to be split into different species along the lines of haplogroups. This would apply, for example, to olive baboon haplogroups, which are not only similar in morphology but also in behaviour and ecology. Alternatively, one would have to lump different morphotypes into the same species. For example, hamadryas and eastern olive baboons would be grouped together, although they differ not only in morphology but also in their social systems.

We nevertheless believe it is currently best to apply a phylogenetic species concept and refer to baboon allotaxa as "species", in the light of both morphological data (e.g., Frost *et al.* 2003) and genetic data (Hapke *et al.* 2001, Newman *et al.* 2004, Wildman *et al.* 2004, Burrell 2008, Zinner *et al.* 2008, 2009a, this study). The "species" view is also adopted by Groves (2001), Grubb *et al.* (2003), Jolly (2007) and Burrell (2008). Further studies that include various nuclear DNA loci will be required to establish a more robust taxonomy of baboons.

Nevertheless, the available genetic data already provide important insights into the evolutionary history of baboons and can be employed to identify possible taxonomic units.

Accordingly, for southern Africa, we identified 3 monophyletic haplogroups and the polyphyletic southern yellow baboons. This suggests that 2 forms of chacma baboons exist (Jolly 1993, 2001), probably Cape chacmas (*P. u. ursinus*) and grey-footed chacmas (*P. u. griseipes*), whereas a further separation of *P. u. ruacana* described from Namibia is not warranted. Ranges of grey-footed and Cape chacmas seem to overlap in the Loskop Nature Reserve (northern RSA). Of 8 sequenced Loskop individuals, 7 belonged to the northern clade (grey-footed) and 1 to the southern haplogroup (Cape). These clades probably also overlap in northern Namibia, where we identified them in relatively short geographical distances from each other.

Our data also indicate that Kinda baboons should be treated as a separate taxon. They are distinct from other baboons as they are smaller, have a less expressed sexual dimorphism, are set apart by cranial morphometric analyses (Freedman 1963, Frost *et al.* 2003) and also display behavioural peculiarities (Phillips-Conroy 2009). Samples from south-central Zambia that match the Kinda baboon morphotype constitute a well-supported monophyletic clade, as confirmed by Burrell (2008).

In contrast, southern yellow baboons are not monophyletic (lineage a). They could be partly identical with *P. cynocephalus strepitus*, which supposedly occur in southern Tanzania and Malawi (Hill 1967,1970), and *P. cynocephalus jubilaeus*, the "dwarf chacma" from the Luangwa Valley in eastern Zambia (Hill 1967, 1970). More geographical sampling would be needed to resolve the distribution and complex phylogeny of baboon forms in this region.

The situation in eastern and north-eastern Africa is similarly complicated, due to possible ongoing introgressive hybridisation between olive and northern yellow baboons (Alberts & Altmann 2001, Storz *et al.* 2002, Newman *et al.* 2004, Tung *et al.* 2008, Zinner *et al.* 2008, 2009a) and olive and hamadryas baboons (Shotake *et al.* 1977, Shotake 1981, Hapke *et al.* 2001, Wildman *et al.* 2004). It is currently not clear whether the various forms of yellow and olive baboons in this region deserve species or subspecies status. Similarly, it is not clear whether it is justified to separate olive baboon in western Africa. We indicate this uncertainty by placing the species name in parentheses. Northern yellow baboons from the south Kenyan coast (lineage c) and most likely also from the north Tanzanian coast (C. Groves pers. comm.) would then be referred to as *P. (cynocephalus) cynocephalus* (Hill 1967, 1970), whereas yellow baboons from the mixed olive and yellow baboon clade would be referred to as *P. (c.) ibeanus* (Hill 1967, 1970). Central olive baboons from the DRC and Uganda (lineage d) are probably best referred to as *P. (anubis) tesselatum* (Hill 1967, 1970). Several candidate names exist for olive baboons in different regions from Eritrea to northern Tanzania, i.e., *P. (a.) heuglini*, *P. (a.) doguera*, *P. (a.) furax*, *P. (a.) neumanni* (Hill 1967, 1970). If one or more of these names would be appropriate remains to be seen.

Hill (1967, 1970) synonymised all West African olive baboons and listed only one subspecies. Whether our western olive baboon haplogroups correspond to previously described forms such as *Papio (a.) nigeriae* (Elliot 1909), *P. (a.) choras* (Ogilby 1843), or *P. (a.) yokoensis* (Matschie 1900) needs likewise to be explored.

Guinea baboons, on the other hand, constitute a fairly well-justified taxonomic unit. However, the inclusion of the two "olive baboons" from the Abidjan zoo / Ivory Coast, in our analysis shows that Guinea baboons may have been also subject to introgressive hybridisation with western olive baboons. At least one of the zoo baboons clusters tightly with Guinea baboons from Haute Niger National Park, Guinea, suggesting that "olive baboons" can carry Guinea baboon mitochondria. The other zoo baboon probably constitutes a lineage by itself (lineage b), with some genetic distance to the Guinea baboon haplogroup, but even more to olive baboons from the Komoé National Park / Ivory Coast. Since the exact provenance of both zoo baboons is unknown, sampling in the contact zone between Guinea and western olive baboons is necessary to further elucidate the relationships between the two taxa.

Conservation

In general, baboons are not listed as threatened by IUCN (2010). Only Guinea baboons have recently been upgraded from the category "least concern" to "near threatened". However, challenges to conservation will increase if a taxon once considered as wide-ranging is split up, if and when local populations are recognised as species or subspecies. Thus, if one accepts the various baboon haplogroups as evolutionary units worthy to be conserved in their own right, one would be faced with the likely consequence that several would have to be classified as threatened, given small ranges or small populations. Possible candidates include the coastal yellow baboons (lineage c) or the western olive baboon clade 4. In any case, the overall conservation status of olive baboons in western Africa was last assessed 30 years ago (Balzamo *et al.* 1980) and has likely deteriorated since then.

Phylogeography

The phylogenetic patterns within *Papio* suggest a complicated biogeographic history, most likely triggered by multiple cycles of expansion and retreat of savannah biomes during late Pliocene and Pleistocene glacial and inter-glacial periods (*Fig. 7.8;* Turner 1999, Hewitt 2000, deMenocal 2004). A relatively warm and humid climate with lowland rain forest in most parts of western and eastern Africa predominated during the late Miocene (11 – 5 million years ago [mya]; Hamilton 1988). However, the climate became progressively cooler and drier after 2.8 mya. The mid-Pliocene vegetation shifted from closed canopy forest to open savannah vegetation in large parts of Africa, and arid-adapted flora and fauna expanded as glaciation in the north proceeded (deMenocal 1995). The last 2.5 million years saw about 20 glacial cycles with major forest expansions and retreats in Africa (Hamilton 1982, Hamilton & Taylor 1991) as evidenced by pollen core data and the current distribution of forest organisms (Sowunmi 1981, Hamilton 1981, 1988,

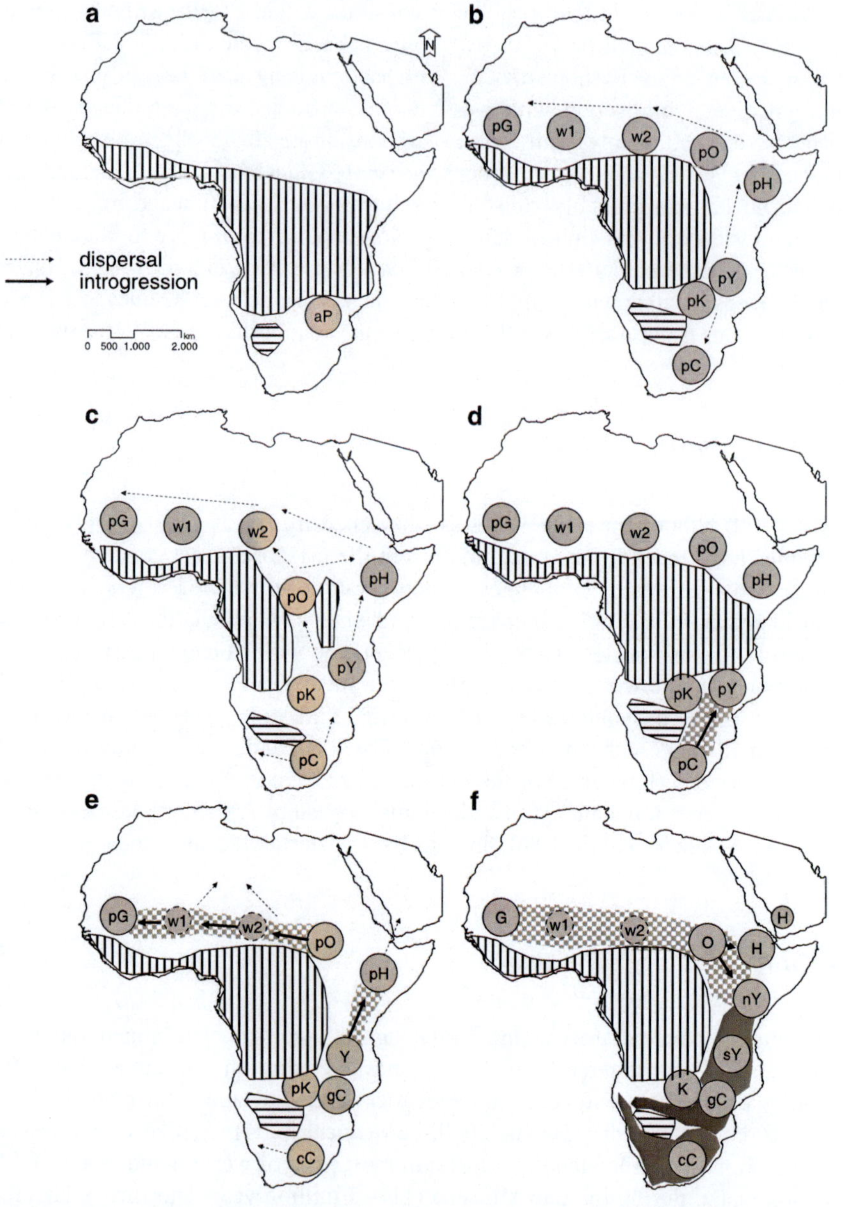

Fig. 7.8 Dispersal hypothesis of *Papio* baboons. Vertical stripes indicate forest and forest refuges, horizontal stripes the Kalahari desert. Illustrated are possible dispersal scenarios and introgression events between local populations. Circles depict nuclei of possible local baboon populations: aP = ancestral *Papio*, C = chacma, cC = Cape chacma, gC = grey-footed chacma, K = Kinda, Y = yellow, sY = southern yellow, nY = northern yellow, H = hamadryas, O = olive, G = Guinea, w1 = western clade 1, w2 = western clade 2 (p = proto). (**a**) Fossil evidence supports a southern African origin of *Papio* 3.8 – 4.4 mya (aP). Dispersal to the north was prevented by the equatorial forest belt. Dispersal to the south was most likely possible. (**b**) During glacial periods in the late

Hamilton & Taylor 1991, Grubb 1982, Tosi 2008). Glacial period environments were characterised by dry, tropical scrub and grassland with limited gallery forest along drainages. Thus, with the retreat of dense forest into high altitude and large river refugia, savannah corridors opened and provided pioneering and recurrent dispersal possibilities for open-country mammals such as baboons. Refugia and dispersal routes were further affected by various events in the Pliocene and early Pleistocene related to the formation of the Great Rift Valley, intensive volcanic activity, e.g., of Mt. Kilimanjaro (Nonnotte *et al.* 2008), the development of the Congo Basin (Coetzee 1993; Myers-Thompson 2003) and dramatic changes in river drainage systems, such as the Zambezi and Niger rivers or Lake Mega-Chad (Goudie 2005, Leblanc *et al.* 2006 a, b).

Baboons were certainly impacted by these dynamics in multiple ways. At times, new dispersal opportunities opened up while once suitable habitats vanished; some populations became fragmented whereas others reconnected. Multiple phases of isolation, hybridisation, and introgression might have occurred at different times and in different regions, leading to speciation and subsequent reticulation.

Fig. 7.8 (continued) Pliocene and early Pleistocene a savannah corridor opened in eastern Africa towards the equator, connecting the southern and northern savannah belt, and baboons migrated to the north. Several local populations evolved into proto-species. The Kalahari Desert most likely expanded in arid periods possibly isolating most of the southern population. (**c**) Alternatively or in combination with the East African savannah corridor a second dispersal route might have opened east of the Rift Mountains. Whichever route the baboons took, after they reached the northern savannah belt, they further dispersed into West and north-east Africa. (**d**) During subsequent humid inter-glacial periods northern and southern populations became separated by the vicariance effect of the extended forest belt, again forming a broad band of unsuitable habitat. Proto-chacma or chacma baboons moved north and introgressed with the parapatric proto-yellow or yellow baboons, thus replacing the yellow nuclear genome but retaining the yellow mitochondria. The hybrid population evolved into grey-footed chacmas (gC) with a chacma like nuclear genome and morphology but a mitochondrial genome corresponding to yellow baboons. (**e**) During recurred arid phases the savannah corridor opened again and yellow baboons possibly went north and introgressed with proto-hamadryas baboons (pH) or a now extinct local population, closely related to hamadryas. The result of this introgression are northern yellow baboons (nY), carrying the yellow baboon nuclear genome but mitochondria closely related to those of hamadryas baboons. Hamadryas baboons possibly invaded the Arabian Peninsula during glacial phases when sea level was low. During humid phases the Sahara desert was also invaded by baboons. (**e** and **f**) Olive or proto-olive baboons started a second dispersal wave into western and eastern Africa and connected with Guinea baboons in the west and hamadryas and yellow baboons in the east, where male introgression progressed. Local populations that once existed in western (w1 and w2) and possible also in eastern Africa were completely swamped with the olive baboon nuclear genome. Only their former mitochondrial genomes remained as vestiges. However, it remains questionable, whether one of the detected olive baboon lineages (w1, d; *Fig. 7.6, Fig. 7.7*) represent the original olive baboons or whether this population is not yet detected and persists in the so far unsampled regions of northern DRC, CAR or south-eastern Sudan. (**f**) The hybridisation process still continues in Ethiopia and Eritrea and south-eastern Kenya. In southern Africa, Cape and grey-footed chacmas increased their ranges into Namibia. Neighbouring populations came into contact again with ongoing hybridisation. (Smaller lineages such as a – d in *Fig. 7.6* are not depicted.)

Our findings are largely consistent with the scenario of the "north-south split" hypothesis (Jolly 1993). Accordingly, baboons originated and dispersed in southern Africa before spreading towards the equator and further north. Haplotypes are therefore divided into "north tropical" and "south tropical" subclades, with a geographical boundary in central Tanzania. Southern haplotypes are now distributed all over southern Africa into central Tanzania, and are seen in Cape chacmas, grey-footed chacmas, Kinda, and some yellow baboons. The "north tropical" branch expanded through a narrow East African corridor, and gave rise to species currently distributed from Senegal to Arabia and south to northern Tanzania. These haplotypes are seen in Guinea, hamadryas, and olive baboons as well as in yellow baboons from Somalia, Kenya, and northern Tanzania. Baboons might have also moved north through a second savannah corridor in eastern DRC west of the Pleistocene forest refuge of the Rift Mountains (Bonnefille *et al.* 1990, Hamilton & Taylor 1991, Maley 1996, Nichol 1999, Plana 2004), but this possibility needs further exploration.

The divergence of the *Papio* lineage (Zinner *et al.* 2009a) accompanied by dispersal from southern Africa to the north (~2.1 mya) and the further division into several distinct southern (~1.8 mya) and northern lineages (~1.9 mya) coincides with savannah expansions and major radiations of hominins and antelopes (Bovidae) (Vrba 1999). A similar pattern of an early north-south division and the subsequent division of the northern clade into an eastern and western clade has also been reconstructed for other large African savannah mammals, such as hartebeest (*Alcelaphus*; Arctander *et al.* 1999, Flagstad *et al.* 2001), topi and wildebeest (*Damaliscus lunatus* and *Connochaetes taurinus*; Arctander *et al.* 1999), roan antelope (*Hippotragus equinus*; Alpers *et al.* 2004), warthog (*Phacochoerus*; Muwanika *et al.* 2003), giraffe (*Giraffa*; Hassanin *et al.* 2007) and lion (*Panthera leo*; Barnett *et al.* 2006). Within several of these taxa paraphyletic relationships were detected, similar to the patterns found in baboons (Arctander *et al.* 1999).

We agree with Jolly (1993) about a current boundary in central Tanzania (cf. *Fig. 7.7*) as baboons from Mikumi National Park (included in Newman *et al.* 2004) belong to the northern clade and those from Lake Rukwa (included in our study) to the southern. Recently, Tom Butynski and Yvonne de Jong (pers. comm.) found that yellow baboons of Mahale National Park on the eastern shores of Lake Tanganyika closely resemble Kinda baboons, rendering the picture for yellow baboons even more complex.

Jolly (1993) further suggested that extant *Papio* baboons do not belong to a simple, closely related clade, but that they have a complicated mitochondrial relationship (Newman *et al.* 2004, Wildman *et al.* 2004). For example, olive baboons from Ethiopia and Eritrea carry mitochondria derived from a neighbouring hamadryas or proto-hamadryas stock. Yellow baboons of Zambia carry mitochondria only distantly related to those of northern Tanzanian and Kenyan yellow baboons, and chacma baboons from Botswana and Zambia are mitochondrially very distant from those living in South Africa.

Our data revealed even more dramatic discrepancies, likely based on introgressive hybridisation and nuclear swamping. Thus, southern baboons were found to be split into Cape chacmas and a combined clade of grey-footed chacmas, Kinda and yellow baboons (*Fig. 7.8d*). Male Cape chacmas or proto-chacmas probably invaded a part of the southern yellow baboon population, which slowly turned into phenotypical chacmas while retaining yellow baboon mitochondria.

Recurrent extensions of the Kalahari Desert during the Pleistocene (Cooke 1963) or the dynamics of the southern African drainage system (Cotterill 2003) might have affected the gene flow among southern populations. Currently, both grey-footed and Cape chacma baboons have small overlapping ranges in north-eastern Namibia and around Johannesburg in RSA. Gene flow might also occur between Kinda and yellow baboons in a narrow contact zone in eastern Zambia (Burrell 2008) and between grey-footed chacmas and Kinda baboons. Clifford Jolly (pers. comm.) indeed noticed some intermediate morphotypes between the two taxa where they meet.

The already complicated evolutionary history of southern baboons will almost certainly turn out to be even more complex as distribution data accumulate (Tom Butynski pers. comm.). Interestingly, the recently described genus *Rungwecebus* (Davenport *et al.* 2006, Olson *et al.* 2008) also carries mitochondria of the southern yellow baboon haplogroup (Zinner *et al.* 2009b). Similarly, male introgression and nuclear swamping can explain the large genetic distance between the southern yellow baboons and their northern conspecifics. In a second dispersal to the north, yellow baboons thus encountered proto-hamadryas, hamadryas or a now extinct hamadryas relative, possibly in eastern and north-eastern Africa (*Fig. 7.8e*). The result is a population of morphologically yellow baboons that carry mitochondria of the invaded taxon, very distant from the mitochondria of southern yellow baboons, but similar to the hamadryas mitochondria.

Jolly (1993) assumed that northern baboons differentiated into hamadryas baboons in the east and Guinea baboons in the west. He provides circumstantial evidence that the olive baboon phenotype, which combines features of both northern and southern phenotypic clades, could have been formed by hybridisation of a northern and a southern population in north-central Africa. It subsequently propagated mainly, if not entirely, by male dispersal into populations of neighbouring taxa to the west and east.

We are currently unable to support or refute this scenario. We also have to remain agnostic about whether olive baboons originate from one of the West African populations (clades 4 or 5) or the central olive baboons (lineage d) or an as yet unidentified population in the unsampled region between Cameroon and the Nile. What seems to be clear is that olive baboons dispersed in a second wave into West and East Africa and hybridised with local baboon populations (*Fig. 7.8e*). Proto-olive baboons most likely adapted to forests during moister interglacial periods. Indeed, olive baboons still occupy the most humid and most forested habitats of all baboons (Gautier-Hion *et al.* 1999, Kamilar 2006, Ross *et al.* this volume [Ch. 9]). During suitable conditions, olive baboons dispersed into West and East Africa. Here, due to higher competitive abilities, male olive baboons fertilised females from local populations with subsequent introgressive hybridisation and nuclear swamping. Currently, olive baboons hybridise with Guinea baboons in the west and hamadryas and northern yellow baboons in the east. On their way west, olive baboons most likely "swallowed" several local baboon forms, such as w1 and w2, which are still traceable in western olive baboons through their mitochondrial genome (clade 4 and 5; *Fig. 7.8e*). However, this dispersal scenario remains speculative, due to a lack of data.

The phylogeographic process in the western part of the baboon range was possibly influenced by recurrent expansion of unsuitable forest or desert habitats, creating various isolated savannah refuges for baboons at certain periods which were subsequently reconnected (Nichol 1999). Rivers are generally only incomplete barriers to baboon dispersal because of historical and seasonal variation in water levels and historical shifts in drainage patterns. Nevertheless, the dramatic dynamics of Lake Mega-Chad with expansion, shrinkage and final disappearance in the late Quaternary, have most likely influenced the dispersal and distribution of West African baboons. The spill-way of this mega-lake was what is now the Mayo Kebbi River forming an outlet to the Gulf of Guinea, via the Rivers Benue and Niger. Moreover, the northward flow of the Niger into the Sahara was probably blocked by dune construction during arid phases of the Pleistocene (Goudie 2005). In wetter phases, a lake formed that spilled over the Tosaye sill to join the lower Niger system. The Niger drainage catchment was once far more extensive. Drainage would run from the south-western slopes of the Ennedi massif into Lake Mega-Chad and then flow over into the Benue and Niger. This not only affected baboon distribution but also that of other larger mammals including West African giraffes. Surprisingly, the giraffes of Niger seem to be closer related to East African than to neighbouring central African populations (Hassanin *et al.* 2007). It is therefore hypothesised that East and central African populations became separated and that East African giraffes moved into North Africa and from there back south into West Africa, separated from central African populations by Lake Mega-Chad (Hassanin *et al.* 2007).A clearer picture emerges for eastern baboon populations (hamadryas, eastern olive and northern yellow baboons). Our data suggest a similar phylogeographic scenario as found by Wildman *et al.* (2004) and Newman *et al.* (2004). Accordingly, olive baboons from Eritrea and Ethiopia cluster with hamadryas baboons, and those from Kenya and northern Tanzania cluster with neighbouring northern yellow baboons from Kenya and Somalia (possibly Ibean baboons, Zinner *et al.* 2008). Since olive baboons from East and north-east Africa carry either hamadryas or yellow baboon mitochondria, Jolly (1993) argued that both hamadryas and northern yellow baboons have been subject to male introgression and nuclear swamping by olive baboons, and that this lead to morphologically olive baboons with hamadryas or yellow baboon mitochondria (*Fig. 7.8f*). Evidence for a gradual introgression of male olive baboons into northern yellow baboon populations has been found in Amboseli (Alberts & Altmann 2001, Tung *et al.* 2008). In addition to previous studies, we detected a second lineage of yellow baboons at the southern Kenyan coast (lineage c) and an olive baboon haplogroup from western Uganda and eastern DRC (lineage d), which are both basal to all eastern and north-eastern haplotypes. If olive baboons invaded the populations of yellow and hamadryas baboons from west to east, as hypothesised by Jolly (2009) and Wildman *et al.* (2004), these two haplogroups may represent populations that have not been subject to hybridisation between olive and yellow baboons.

Conclusions

Based on mitochondrial data, the present study demonstrates multiple para- and polyphylies in the genus *Papio* and a general disagreement between morphology-based baboon taxonomy and mitochondrial DNA data. The geographic pattern of haplogroup distribution suggests a complex evolutionary history of baboons and emphasises the role of recurrent cycles of climate change during the late Pliocene and Pleistocene in Africa. Periodical isolation of populations and subsequent partial or complete range overlap and hybridisation had possibly a strong impact on baboon diversification and speciation. Hybridisation is recently discussed as an important mechanism driving speciation in primates (Arnold & Meyer 2006, Arnold 2008). Introgressive hybridisation and nuclear swamping, up to the complete "swallowing" of local taxa by introgressing taxa, might have occurred several times during baboon evolution. Baboon evolution is therefore reticulate. Introgressive hybridisation is a promising hypothesis to explain inconsistencies between mitochondrial DNA-based relationships and proposed taxonomic classifications – much more than incomplete lineage sorting, i.e., the unequal distribution of different ancestral haplotypes into progeny lineages. To clarify relationships among baboon taxa, to estimate the impact of hybridisation on their evolution and to test hypotheses of possible male introgression in various populations, we will need nuclear sequence data from autosomal and gonosomal (XY) loci. However, at least for parts of the Y chromosome, only minimal variation has been detected between certain types of baboons (Lawson Handley *et al.* 2006).

A comprehensive assessment of baboon phylogeography will require sampling in all parts of the baboon range. We will attempt to close important sampling gaps in future studies (cf. *Fig. 7.7*). In western Africa, this would be the contact zone between Guinea and western olive baboons and isolated baboon populations in the Sahara Desert. Discerning the haplogroup distribution in western and north-western Nigeria and in the baboon range between Cameroon, Congo, DRC, Central African Republic, and western Sudan would likewise be a matter of priority.

Acknowledgements We thank the following individuals and institutions who assisted in the collection of samples or provided permits: Christos Astaras, Augu Basabose, Olaf Behlert, Dawit Berhane, Rebecca Deleu, Marion East, Elodie Ey, Julia Fischer, Sery Gonedelé Bi, Kurt Hammerschmidt, Heribert Hofer, Marie-Claude Huynen, Cuthbert Katsvanga, Christina Keller, Britta Kunz, Klaus Lorenz, Markus Metz, Karim Nasher, Ryne A. Palombit, Steve Pechey, Susanne Rensing, Christelle Scheid, Carsten Schradin, Kerry Slater, Simone Teelen, Dagmar Thierer, Ymke Warren, Kerstin Wilhelm; Direction Parcs Nationaux / Senegal; Ministère de l'Agriculture et des Ressources Animales / Ivory Coast; *Gashaka Primate Project* / Nigeria; The Limbe Wildlife Center / Cameroon; Ministry of Agriculture / Eritrea; Ethiopian Wildlife Conservation Organization; Institute of Primate Research / Kenya; The Spotted Hyaena Project,

IZW / Tanzania; Bindura University / Zimbabwe; ZAWA / Zambia; Ministry of Environment and Tourism / Namibia; DeHoop Baboon Project / RSA; Humboldt Museum, Berlin / Germany; Zoologische Staatssammlung München / Germany; Zoo Duisburg / Germany. – We are grateful to Christina Oberdieck and Christiane Schwarz for their excellent laboratory work and to Colin Groves, Tom Butynski and Clifford Jolly for sharing their thoughts on baboon systematics and phylogeography. – DZ was supported by a Grant of the German Science Foundation (DFG ZI 548 / 3-1). – Parts of this chapter were published earlier in *BMC Evolutionary Biology*.

Literature

Alberts, S. C. & Altmann, J. (2001). Immigration and hybridization patterns of yellow and anubis baboons in and around Amboseli, Kenya. *American Journal of Primatology* 53: 139–154.

Alpers, D. L., Van Vuuren, B. J., Arctander, P. & Robinson, T. J. (2004). Population genetics of the roan antelope (*Hippotragus equinus*) with suggestions for conservation. *Molecular Ecology* 13: 1771–1784.

Ansell, W. F. H. (1978). *The Mammals of Zambia*. Chilanga, Zambia: The National Parks & Wildlife Service.

Arctander, P., Johansen, C. & Coutellec-Vreto, M. A. (1999). Phylogeography of three closely related African bovids (tribe Alcelaphini). *Molecular Biology and Evolution* 16: 1724–1739.

Arnold, M. L. (2008). *Reticulate Evolution and Humans. Origins and Ecology*. Oxford: Oxford University Press.

Arnold, M. L. & Meyer, A. (2006). Natural hybridization in primates: One evolutionary mechanism. *Zoology* 109: 261–276.

Avise, J. C. (2000). *Phylogeography –The History and Formation of Species*. Cambridge, MA: Harvard University Press.

Avise, J. C. (2004). *Molecular Markers, Natural History, and Evolution*. Sunderland, MA: Sinauer Associates.

Balzamo, E., Seri, B. & Kouako, K. (1980). Répartition géographique de *Papio anubis* dans le nord-est et le centre-sud de Côte d'Ivoire. *Primates* 21: 524–537.

Barnett, R., Yamaguchi, N., Barnes, I. & Cooper, A. (2006). The origin, current diversity and future conservation of the modern lion (*Panthera leo*). *Proceedings of the Royal Society of London Series B-Biological Sciences* 273: 2119–2125.

Barton, N. H. (2001). The role of hybridization in evolution. *Molecular Ecology* 10: 551–568.

Barton, N. H. (2006). How did the human species form? *Current Biology* 16: R647–R650.

Benefit, B. (1999). Biogeography, dietary specialization, and the diversification of African plio-pleistocene monkeys. In Bromage, T. G. & Schrenk, F. (eds.), *African Biogeography, Climate Change, and Human Evolution*. Oxford: Oxford University Press, pp. 172–188.

Bigourdan, J. & Prunier, R. (1937). *Les Mammifères Sauvages de l' Ouest Africain et Leur Milieu*. Montrouge (Paris): Jean de Rudder.

Bonnefille, R., Roeland, J. C. & Guiot, J. (1990). Temperature and rainfall estimates for the past 40,000 years in equatorial Africa. *Nature* 346: 347–349.

Broadfield, D. C., Delson, E. & Atsalis, S. (1994). Cercopithecid fossils from the later Pleistocene of Taung, South Africa. *American Journal of Physical Anthropology Suppl.* 18: 59–60.

Brown, W. M., Prager, E. M., Wang, A. & Wilson, A. C. (1982). Mitochondrial DNA sequences of primates: Tempo and mode of evolution. *Journal of Molecular Evolution* 18: 225–239.

Buettner-Janusch, J. (1966). A problem in evolutionary systematics: Nomenclature and classification of baboons, genus *Papio*. *Folia Primatologica* 4: 288–306.

Burrell, A. (2008). *Phylogenetics and Population Genetics of Central African Baboons*. PhD thesis, New York University.

Cachel, S. (2006). *Primate and Human Evolution*. Cambridge: Cambridge University Press.

Chakraborty, D., Ramakrishnan, U., Panor, J., Mishra, C. & Sinha, A. (2007). Phylogenetic relationships and morphometric affinities of the Arunachal macaque (*Macaca munzala*), a newly described primate from Arunachal Pradesh, northeastern India. *Molecular Phylogenetics and Evolution* 44: 838–849.

Ciccarelli, F. D., Doerks, T., von Mering, C., Creevey, C. J., Snel, B. & Bork, P. (2006). Toward automatic reconstruction of a highly resolved tree of life. *Science* 311: 1283–1287.

Coetzee, J. (1993). African flora since the terminal Jurassic. In Goldblatt, P. (ed.), *Biological Relationships Between Africa and South America*. New Haven, CT: Yale University Press, pp. 37–61.

Cooke, H. B. S. (1963). Pleistocene mammal faunas of Africa, with particular reference to southern Africa. In Howell, F. C. & Bourlière, F. (eds.), *African Ecology and Human Evolution*. New York: Wenner-Gren Foundation for Anthropological Research, pp. 65–116.

Cortes-Ortiz, L., Duda, T. F. Jr., Canales-Espinosa, D., Garcia-Orduna, F., Rodriguez-Luna, E. & Bermingham, E. (2007). Hybridization in large-bodied New World primates. *Genetics* 176: 2421–2425.

Cotterill, F. P. D. (2003). Geomorphological influences on vicariant evolution in some African mammals in the Zambezi Basin: some lessons for conservation. In Plowman, A. (ed.), *Ecology and Conservation of Mini-antelope: Proceedings of an International Symposium on Duiker and Dwarf Antelope in Africa*. Fürth / Germany: Filander Verlag, pp. 11–58.

Cracraft, C. (1983). Species concepts and speciation analysis. *Current Ornithology* 1: 159–187.

Davenport, T. R. B., Stanley, W. T., Sargis, E. J., De Luca, D. W., Mpunga, N. E., Machaga, S. J. & Olson, L. E. (2006). A new genus of African monkey, *Rungwecebus*: Morphology, ecology, and molecular phylogenetics. *Science* 312: 1378–1381.

Dekeyser, P. L. & Derivot, J. (1960). Sur de nouveaux spécimens de Cynocéphales du Tibesti (Mission Carl et Petit, 1957). *Bulletin de l'I. F. A. N.* 22 ser A: 1453–1456.

Delson, E. (1984). Cercopithecid biochronology of the African Plio-Pleistocene: Correlations among eastern and southern hominid-bearing localities. *Courier Forschungsinstitut Senckenberg* 69: 199–218.

deMenocal, P. B. (1995). Plio-Pleistocene African climate. *Science* 270: 53.

deMenocal, P. B. (2004). African climate change and faunal evolution during the Pliocene-Pleistocene. *Earth and Planetary Science Letters* 220: 3–24.

Disotell, T. R. (2000). The molecular systematics of the Cercopithecidae. In Whitehead, P. F. & Jolly, C. J. (eds.), *Old World Monkeys*. Cambridge: Cambridge University Press, pp. 29–56.

Disotell, T. R., Honeycutt, R. L. & Ruvolo, M. (1992). Mitochondrial DNA phylogeny of the Old-World monkey tribe Papionini. *Molecular Biology and Evolution* 9: 1–13.

Ellerman, J. R., Morrison-Scott, T. C. S. & Hayman, R. W. (1953). *Southern African Mammals, 1758 to 1951: A Reclassification*. London: British Museum (Nat. Hist.).

Elliot, D. G. (1909). Descriptions of apparently new species and subspecies of monkeys of the genera *Callicebus*, *Lagothrix*, *Papio*, *Pithecus*, *Cercopithecus*, *Erythrocebus* and *Presbytis*. *The Annals and Magazine of Natural History* 8: 244–274.

Flagstad, O., Syvertsen, P. O., Stenseth, N. C. & Jakobsen, K. S. (2001). Environmental change and rates of evolution: The phylogeographic pattern within the hartebeest complex as related to climatic variation. *Proceedings of the Royal Society of London Series B-Biological Sciences* 268: 667–677.

Fleagle, J. (1999). *Primate Adaptation and Evolution*. New York: Academic Press.

Frost, S. R., Marcus, L. F., Bookstein, F. L., Reddy, D. P. & Delson, E. (2003). Cranial allometry, phylogeography, and systematics of large-bodied papionins (Primates: Cercopithecinae) inferred from geometric morphometric analysis of landmark data. *Anatomical Records* 275A: 1048–1072.

Funk, D. J. & Omland, K. E. (2003). Species-level paraphyly and polyphyly: frequency, causes, and consequences, with insights from animal mitochondrial DNA. *Annual Review of Ecology, Evolution, and Systematics* 34: 397–423.

Galat-Luong, A., Galat, G. & Hagell, S. (2006). The social and ecological flexibility of Guinea baboons: implications for Guinea baboon social organization and male strategies. In Swedell,

L. & Leigh, S. R. (eds.), *Reproduction and Fitness in Baboons. Behavioral, Ecological, and Life History Perspectives*. New York: Springer, pp. 105–121.

Gautier-Hion, A., Colyn, M. & Gautier, J. P. (1999). *Histoire Naturelle des Primates d'Afrique Centrale*. Libreville, Gabon: ECOFAC.

Geoffroy St. Hilaire, I. (1851). *Catalogue méthodique de la collection des mammifères de la collection des oiseaux et des collections annexes*. Paris: Muséum d'Histoire Naturelle de Paris (p. 34).

Goossens, B., Chikhi, L., Utami, S. S., De Ruiter, J. R. & Bruford, M. W. (2000). A multi-samples, multi-extracts approach for microsatellite analysis of faecel samples in an arboreal ape. *Conservation Genetics* 1: 157–162.

Goudie, A. S. (2005). The drainage of Africa since the Cretaceous. *Geomorphology* 67: 437–456.

Groves, C. P. (2001). *Primate Taxonomy*. Washington, DC: Smithsonian Institution Press.

Grubb, P. (1982). Refuges and dispersal in the speciation of African forest mammals. In Prance, G. T. (ed.), *Biological Diversification in the Tropics*. New York: Columbia University Press, pp. 537–553.

Grubb, P. (2006). English common names for subspecies and species of African primates. *Primate Conservation* 20: 65–73.

Grubb, P., Butynski, T. M., Oates, J. F., Bearder, S. K., Disotell, T. R., Groves, C. P. & Struhsaker, T. T. (2003). Assessment of the diversity of African primates. *International Journal of Primatology* 24: 1301–1357.

Haldane, J. B. S. (1922). Sex ratio and unisexual sterility in hybrid animals. *Journal of Genetics* 12: 101–109.

Hamilton, A. C. (1981). The quaternary history of African forests: its relevance to conservation. *African Journal of Ecology* 19: 1–6.

Hamilton, A. C. (1982). *Environmental History of East Africa: A Study of the Quaternary*. London: Academic Press.

Hamilton, A. C. (1988). Guenon evolution and forest history. In Gautier-Hion, A., Bourliere, F., Gautier J. P. & Kingdon, J. (eds.), *A Primate Radiation. Evolutionary Biology of the African Guenons*. Cambridge: Cambridge University Press, pp. 13–34.

Hamilton, A. C. & Taylor, D. (1991). History of climate and forests in tropical Africa during the last 8 million years. *Climatic Change* 19: 65–78.

Hapke, A., Zinner, D. & Zischler, H. (2001). Mitochondrial DNA variation in Eritrean hamadryas baboons (*Papio hamadryas hamadryas*): Lifehistory influences population genetic structure. *Behavioural Ecology and Sociobiology* 50: 483–492.

Hassanin, A., Ropiquet, A., Gourmand, A.L., Chardonnet, B. & Rigoulet, J. (2007). Mitochondrial DNA variability in *Giraffa camelopardalis*: consequences for taxonomy, phylogeography and conservation of giraffes in West and central Africa. *Comptes Rendus Biologies* 330: 265–274.

Hewitt, G. (2000). The genetic legacy of the Quaternary ice ages. *Nature* 405: 907–913.

Hill, W. C. O. (1959). The correct name of the olive baboon. *Journal of Mammalogy* 40: 143–145.

Hill, W. C. O. (1967). Taxonomy of the baboon. In Vagtborg, H. (ed.), *The Baboon in Medical Research II*. Austin, TX: University of Texas Press, pp. 3–11.

Hill, W. C. O. (1970). *Primates: Comparative Anatomy and Taxonomy VIII Cynopithecinae: Papio, Mandrillus, Theropithecus*. Edinburgh: Edinburgh University Press.

IUCN 2010. *The IUCN Red List of Threatened Species. Version 2010, 3*. <http://www.iucnredlist.org>.

Jablonski, N. G. (2002). Fossil Old World monkeys: The late Neogene radiation. In Hartwig W. C. (ed.), *The Primate Fossil Record*. Cambridge: Cambridge University Press, pp. 255–299.

Jolly, C. J. (1993). Species, subspecies and baboon systematics. In Kimbel, W. H. & Martin, L. B. (eds.), *Species, Species Concepts, and Primate Evolution*. New York: Plenum Press, pp. 67–107.

Jolly, C. J. (1997–1998). Baboon nomenclature. *African Primates* 3: 47–48.

Jolly, C. J. (2001). A proper study for mankind: Analogies from the papionin monkeys and their implications for human evolution. *Yearbook of Physical Anthropology* 44: 177–204.

Jolly, C. J. (2007). Baboons, mandrills, and mangabeys: Afro-Papionin socioecology in a phylogenetic perspective. In Campbell, C. J., Fuentes, A., MacKinnon, K. C., Panger, M. & Bearder, S. K. (eds.), *Primates in Perspective*. New York: Oxford University Press, pp. 240–251.

Jolly, C. J. (2009). Fifty years of looking at human evolution – backward, forward, and sideways. *Current Anthropology* 50: 187–199.

Jolly, C. J. (in press). Philopatry at the frontier: a demographically-driven origin of multi-level societies in baboons (genus *Papio*).

Kamilar, J. M. (2006). Geographic variation in savanna baboon (*Papio*) ecology and its evolutionary implications. In Lehmann, S. M. & Fleagle, J. G. (eds.), *Primate Biogeography*. New York: Springer, pp. 169–200.

Karanth, K. P., Delefosse, T., Rakotosamimanana, B., Parsons, T. J. & Yoder, A. D. (2005). Ancient DNA from giant extinct lemurs confirms single origin of Malagasy primates. *Proceedings of the National Academy of Sciences, USA* 102: 5090–5095.

Kingdon, J. (1971). *East African Mammals. An Atlas of Evolution in Africa. Vol. 1–7*. London: Academic Press.

Kingdon, J. (1997). *The Kingdon Field Guide to African Mammals*. San Diego: Academic Press.

Kinzey, W. G. (ed.) (1987). *The Evolution of Human Behaviour: Primate Models*. New York: New York State University Press.

Koblmüller, S., Duftner, N., Sefc, K. M., Aibara, M., Stipacek, M., Blanc, M., Egger, B. & Sturmbauer, C. (2007). Reticulate phylogeny of gastropod-shell-breeding cichlids from Lake Tanganyika –the result of repeated introgressive hybridization. *BMC Evolutionary Biology* 7: 7.

Kummer, H. (1968). *Social Organization of Hamadryas Baboons. A Field Study*. Basel: Karger.

Lawson Handley, L. J., Hammond, R. L., Emaresi, G., Reber, A. & Perrin, N. (2006). Low Y chromosome variation in Saudi-Arabian hamadryas baboons (*Papio hamadryas hamadryas*). *Heredity* 96: 298–303.

Leblanc, M. J., Leduc, C. Stagnitti, F., van Oevelen, P. J., Jones, C., Mofor, L. A., Razack, M. & Favreau, G. (2006a). Evidence for Megalake Chad, north-central Africa, during the late Quaternary from satellite data. *Palaeogeography, Palaeoclimatology, Palaeoecology* 230: 230–242.

Leblanc, M. J., Favreau, G., Maley, J., Nazoumou, Y., Leduc, C., Stagnitti, F., van Oevelen, P. J., Delclaux, F. & Lemoalle, J. (2006b). Reconstruction of Megalake Chad using Shuttle Radar Topographic Mission data. *Palaeogeography, Palaeoclimatology, Palaeoecology* 239: 16–27.

Lehmann, N., Eisenhawer, A., Hansen, K., Mech, L. D., Peterson, R. O., Gogan, P. J. P. & Wayne, R. K. (1991). Introgression of coyote mitochondrial DNA into sympatric North American gray wolf populations. *Evolution* 45: 104–119.

Leigh, S. R. (2006). Cranial ontogeny of *Papio* baboons (*Papio hamadryas*). *American Journal of Physical Anthropology* 130: 71–84.

Lönnberg, E. (1919). Contributions to the knowledge about the monkeys of Belgian Congo. *Revue Zoologique Africaine* 7: 107–154.

Machado de Barros, A. (1969). Mammíferos de Angola ainda não citados ou pouco conhecidos. *Publições culturais da Companhia de Diamantes de Angola* 46: 93–232.

Magin, C. (1990). *The status of wildlife populations in the Aïr and Ténéré National Nature Reserve 1988–1990*. Niamey, Niger: Ser. Rapp. Tech. No. 14. IUCN / WWF.

Maley, J. (1996). The African rain-forest –main characteristics of changes in vegetation and climate from the Upper Cretaceous to the Quaternary. In Alexander, I. J., Swaine, M. D. & Watling, R. (eds.), *Essays on the Ecology of the Guinea-Congo Rain Forest. Proceedings of the Royal Society of Edinburgh* 104B: 31–73.

Mallet, J. (2005). Hybridization as an invasion of the genome. *Trends in Ecology and Evolution* 20: 229–237.

Mallet, J. (2007). Hybrid speciation. *Nature* 446: 279–283.

Mallet, J. (2008). Hybridization, ecological races, and the nature of species: empirical evidence for the ease of speciation. *Philosophical Transactions of the Royal Society B* 363: 2971–2986.

Matschie, P. (1900). Einige Säugethiere aus dem Hinterlande von Kamerun. *Sitzungsbericht der Gesellschaft naturforschender Freunde zu Berlin, 20. März 1900*: 89–91.

May, R. M. (1990). Taxonomy as destiny. *Nature* 347: 129–130.

Mayr, E. (1942). *Systematics and the Origin of Species*. New York: Columbia University Press.

McDonald, D. B., Parchman, T. L., Bower, M. R., Hubert, W. A. & Rahel, F. J. (2008). An introduced and a native vertebrate hybridize to form a genetic bridge to a second native species. *Proceedings of the National Academy of Sciences, USA* 105: 10837–10842.

McKee, J. K. (1992). The southern African origin of the genus *Papio. South African Journal of Medicine* 82: 193.

McKee, J. K. & Keyser, A. W. (1995). Craniodental remains of *Papio angusticeps* from the haasgat cave site, South Africa. *International Journal of Primatology* 16: 823–841.

Melnick, D. J. & Pearl, M. C. (1987). Cercopithecines in multimale groups: genetic diversity and population structure. In Smuts, B. B., Cheney, D. L., Seyfarth, R. M., Wrangham, R. W. & Struhsaker, T. T. (eds.), *Primate Societies*. Chicago: The University of Chicago Press, pp. 121–134.

Monod, T. (1963). The late Tertiary and Pleistocene in the Sahara and adjacent southerly regions. In Howell, C. F. & Bourliere, F. (eds.), *African Ecology and Human Evolution*. New York: Aldine, pp. 117–229.

Murphy, W. J., Eizirik, E., Johnson, W. E., Zhang, Y. P., Ryder, O. A. & O'Brien, S. J. (2001a). Molecular phylogenetics and the origins of placental mammals. *Nature* 409: 614–618.

Murphy, W. E., Eizirik, E., O'Brien, S. J., Madsen, O., Scally, M., Douady, C. J., Teeling, E., Ryder, O. A., Stanhope, M. J., de Jong, W. W. & Springer, M. S. (2001b). Resolution of the early placental mammalian radiation using Bayesian phylogenetics. *Science* 294: 2348–2351.

Muwanika, V. B., Nyakaana, S., Siegismund, H. R. & Arctander, P. (2003). Phylogeography and population structure of the common warthog (*Phacochoerus africanus*) inferred from variation in mitochondrial DNA sequences and microsatellite loci. *Heredity* 91: 361–372.

Myers-Thompson, J. A. (2003). A model of the biogeographical journey from Proto-pan to *Pan paniscus*. *Primates* 44: 191–197.

Newman, T. K., Jolly, C. J. & Rogers, J. (2004). Mitochondrial phylogeny and systematics of baboons (*Papio*). *American Journal of Physical Anthropology* 124: 17–27.

Nichol, J. E. (1999). Geomorphological evidence and Pleistocene refugia in Africa. *Geographical Journal* 165: 79–89.

Nonnotte, P., Guillou, H., Le Gall, B., Benoit, M., Cotton, J. & Scaillet, S. (2008). New K-Ar age determinations of Kilimanjaro volcano in the North Tanzanian diverging rift, East Africa. *Journal of Volcanology and Geothermal Research* 173: 99–112.

Ogilby, W. (1843). On two new species of baboon. *Proceedings of the Zoological Society of London*: 10–12.

Olson, L. E., Sargis, E. J., Stanley, W. T., Hildebrandt, K. B. P. & Davenport, T. R. B. (2008). Additional molecular evidence strongly supports the distinction between the recently described African primate *Rungwecebus kipunji* (Cercopithecidae, Papionini) and *Lophocebus*. *Molecular Phylogenetics and Evolution* 48: 789–794.

Osterholz, M., Walter, L. & Roos, C. (2008). Phylogenetic position of the langur genera *Semnopithecus* and *Trachypithecus* among Asian colobines, and genus affiliations of their species groups. *BMC Evolutionary Biology* 8: 58.

Patterson, N., Richter, D. J., Gnerre, S., Lander, E. S. & Reich, D. (2006). Genetic evidence for complex speciation of humans and chimpanzees. *Nature* 441: 1103–1108.

Phillips-Conroy, J. E. (2009). Genetic and behavioral observations of "Kinda" baboons (*Papio cynocephalus kindae*) in Zambia. *American Journal of Physical Anthropology* (Suppl. 48) 138: 211.

Plana, V. (2004). Mechanisms and tempo of evolution in the African Guineo-Congolian rainforest. *Philosophical Transactions of the Royal Society of London B: Biological Sciences* 359: 1585–1594.

Posada, D. & Crandall, K. A. (1998). Modeltest: testing the model of DNA substitution. *Bioinformatics* 14: 817–818.

Purvis, A. (1995). A composite estimate of primate phylogeny. *Philosophical Transactions of the Royal Society of London B: Biological Sciences* 348: 405–421.

Rohland, N., Malaspinas, A. S. Pollack, J. L., Slatkin, M., Matheus, P. & Hofreiter, M. (2007). Proboscidean mitogenomics: Chronology and mode of elephant evolution using mastodon as outgroup. *PLoS Biology* 5: e207.

Ronquist, F. & Huelsenbeck, J. P. (2003). MrBayes 3: Bayesian phylogenetic inference under mixed models. *Bioinformatics* 19: 1572–1574.

Roos, C., Nadler, T. & Walter, L. (2008). Mitochondrial phylogeny, taxonomy and biogeography of the silvered langur species group (*Trachypithecus cristatus*). *Molecular Phylogenetics and Evolution* 47: 629–636.

Roth, W. T. (1965). The taxonomy of the baboon and its position in the order of primates. In Vagtborg, H. (ed.), *The Baboon in Medical Research I*. Austin, TX: University of Texas Press, pp. 3–16.

Rumpler, Y., Warter, S., Hauwy, M., Fausser, J. L., Roos, C. & Zinner, D. (2008). Comparing chromosomal and mitochondrial phylogenies of sportive lemurs (Genus *Lepilemur*, Primates). *Chromosome Research* 16: 1143–1158.

Sarmiento, E. (1997). Current problems with Papio taxanomies. *African Primates* 3: 48–52.

Seehausen, O. (2004). Hybridization and adaptive radiation. *Trends in Ecology and Evolution* 19: 198–207.

Shotake, T. (1981). Population genetical study of natural hybridisation between *Papio anubis* und *Papio hamadryas*. *Primates* 22: 285–308.

Shotake, T., Nozawa, K. & Tanabe, Y. (1977). Blood protein variations in baboons. I. Gene exchange and genetic distance between *Papio anubis*, *Papio hamadryas* and their hybrid. *Japanese Journal of Genetics* 52: 223–237.

Smuts, B. B., Cheney, D. L., Seyfarth, R. M., Wrangham, R. W. & Struhsaker, T. T. (1987). *Primate Societies*. Chicago: The University of Chicago Press.

Sowunmi, M. A. (1981). Aspects of late Quaternary vegetational changes in West Africa. *Journal of Biogeography* 8: 457–474.

Storz, J. F., Beaumont, M. A. & Alberts, S. C. (2002). Genetic evidence for long-term population decline in a savannah-dwelling primate: Inferences from a hierarchical Bayesian model. *Molecular Biology and Evolution* 19: 1981–1990.

Strum, S. C., Mitchell, W. (1987). Baboon models and muddles. In Kinzey, W. G. (ed.), *The Evolution of Human Behavior: Primate Models*. New York: New York State University Press, pp. 87–104.

Switzer, W. M., Salemi, M., Shanmugam, V., Gao, F., Cong, M., Kuiken, C., Bhullar, V., Beer, B. E., Vallet, D., Gautier-Hion, A., Tooze, Z., Villinger, F., Holmes, E. C. & Heneine, W. (2005). Ancient co-speciation of simian foamy viruses and primates. *Nature* 434: 376–380.

Swofford, D. (2002). *PAUP Phylogenetic analysis using parsimony (and other methods). Vers. 4.* Sunderland, MA: Sinauer Associates.

Tahiri-Zagrët, C. (1976). Les cercopithecidae de Côte d'Ivoire. *Bulletin de l'I. F. A. N. Ser A* 38: 206–230.

Tappen, N. (1960). Problems of distribution and adaption of the African monkeys. *Current Anthropology* 1: 91–120.

Thalmann, O., Fischer, A., Poinar, H. N., Pääbo, S. & Vigilant, L. (2007). The complex evolutionary history of gorillas: insights from genomic data. *Molecular Biology and Evolution* 24: 146–158.

Thorington, R. & Groves, C. (1970). An annotated classification of the Cercopithecoidea. In Napier, J. R. & Napier, P. H. (eds.), *Old World Monkeys*. New York: Academic Press, pp. 629–647.

Tosi, A. J. (2008). Forest monkeys and Pleistocene refugia: a phylogeographic window onto the disjunct distribution of the *Chlorocebus lhoesti* species group. *Zoological Journal of the Linnean Society* 154: 408–418.

Tosi, A. J., Morales, J. C. & Melnick, D. J. (2000). Comparison of Y chromosome and mtDNA phylogenies leads to unique inferences of macaque evolutionary history. *Molecular Phylogenetics and Evolution* 17: 133–144.

Tung, J., Charpentier, M. J. E., Garfield, D. A., Altmann, J. & Alberts, S. C. (2008). Genetic evidence reveals temporal change in hybridization patterns in a wild baboon population. *Molecular Ecology* 17: 1998–2011.

Turner, A. (1999). Evolution in African Plio-Pleistocene mammalian fauna: Correlation and causation. In Bromage, T. G. & Schrenk, F. (eds.), *African Biogeography, Climate Change & Human Evolution*. Oxford: Oxford University Press, pp. 76–87.

Vrba, E. (1999). Habitat theory in relation to the evolution in African neogene biota and hominids. In Bromage, T. G. & Schrenk, F. (eds.), *African Biogeography, Climate Change & Human Evolution*. Oxford: Oxford University Press, pp. 19–34.

Wildman, D. E., Bergman, T. J., al-Aghbari, A., Sterner, K. N., Newman, T., Phillips-Conroy, J. E., Jolly, C. J. & Disotell, T. R. (2004). Mitochondrial evidence for the origin of hamadryas baboons. *Molecular Phylogenetics and Evolution* 32: 287–296.

Williams-Blangero, S., Vandenberg, J. L., Blangero, J., Konisberg, L. & Dyke, B. (1990). Genetic differentiation between baboon subspecies: Relevance for biomedical research. *American Journal of Primatology* 20: 67–80.

Zinner, D., Kraft, R. & Roos, C. (2008). The phylogenetic position of *Papio ruhei* – a unique baboon taxon from Somalia? *Zoologischer Garten NF* 77: 303–311.

Zinner, D., Groeneveld, L. F., Keller, C. & Roos, C. (2009a). Mitochondrial phylogeography of baboons (*Papio* spp.) – Indication for introgressive hybridization? *BMC Evolutionary Biology* 9: 83.

Zinner, D., Arnold, M. L. & Roos, C. (2009b). Is the new primate genus *Rungwecebus* a baboon? *PLoSONE* 4: e4859.

Zwickl, D. J. (2006). *Genetic Algorithm Approaches for the Phylogenetic Analysis of Large Biological Sequence Data Sets under the Maximum Likelihood Criterion*. PhD thesis, The University of Texas.

Zwickl, D. J., Hillis, D. M. (2002). Increased taxon sampling greatly reduces phylogenetic error. *Systematic Biology* 51: 588–598.

Chapter 8
Crop-raiding and Commensalism in Olive Baboons: The Costs and Benefits of Living with Humans

Ymke Warren, James P. Higham, Ann M. Maclarnon, and Caroline Ross

Abstract We investigated the causes and consequences of crop-raiding for the ecology and life-history of two troops of olive baboons studied in Nigeria's Gashaka Gumti National Park over 8 years. Kwano troop feeds entirely on wild foods whilst the Gamgam troop regularly consumes crops grown within its home-range. Crop-raiding provides both energetic and reproductive advantages as Gamgam troop spent less time travelling and feeding and more time resting and socialising. The crop-raiding troop has also shorter inter-birth intervals and lower infant mortality. Costs to crop-raiding due to chasing and attacks by farmers are outweighed by the benefits of increased access to high-quality foods, a reduced susceptibility to pathogen loads, and a consequently increased reproductive output.

Keywords Human-wildlife conflict • Crop-raiding • Food-enhancement • Forest baboons

Introduction

> "Oh baboon,
> I greet you, possessor of hard-skinned swollen buttocks. [...]
> He who, after raiding a farm, returns to his perch,
> his cheek pouches hang down like Dahomean's pocket.
> Possessor of eyes shy like bride's,
> seeing the farmers' wives on their husband's farms. [...]
> Gentleman on the tree-top, whose fine figure intoxicates him like liquor.
> Ladoogi whose mouth is protuberant and longish like a grinning rod.
> Whose jaws are like wooden spoons,
> and whose chest looks as if it has a wooden bar in it.

C. Ross (✉)
Centre for Research in Evolutionary Anthropology,
Roehampton University, London, SW15 4JD, UK
e-mail: c.ross@roehampton.ac.uk

V. Sommer and C. Ross (eds.), *Primates of Gashaka*, Developments in Primatology: Progress and Prospects 35, DOI 10.1007/978-1-4419-7403-7_8,

> Whose eyes are deep-set as it goes a-raiding farms…
> He whom his mother gazed and gazed upon and burst out weeping.
> Saying her child's handsomeness would be the ruin of him.
> (*Singing:*) A baboon I saw on my forest farm, as it was munching away.
> Stout it was, munching away."
>
> (Yoruba hunting poem; cited in Jolly 1999)

Humans and non-human primates have interacted throughout history (Sponsel *et al.* 2002), with wild primates living alongside human communities in a wide variety of contexts. There are many consequences of these interactions for both primates and humans; one of the most important for primates is access to human foods, and hence improved nutrition. Such "food-enhancement" can come about in several ways, including by crop-raiding and rubbish-raiding, where primates take resources for themselves, or by provisioning, where primates are purposely given food by humans such as pilgrims, worshippers, animal lovers and tourists (Paterson & Wallis 2005).

Provisioned and rubbish-raiding primates may be thought of as commensal species, benefiting from the association without harming the human population. In contrast, there are situations where primates may cause damage and cost their human neighbours a great deal – such as crop-raiding, from farms, gardens and plantations, which can result in farmers sustaining significant economic loss. This behaviour is particularly common when cultivated areas are close to forest edges (Naughton-Treves 1997, Hill 2005). To date, over 70 species from almost all primate families have been recorded to raid crops (Wolfheim 1983, Cowlishaw & Dunbar 2000, Warren 2003). Of these, cercopithecine species such as macaques in Asia and baboons in Africa are some of the most frequent raiders (Wolfheim 1983, Lee & Priston 2005, Ross & Hill 2006). A number of characteristics make them well adapted to crop-raiding as they are generalist and opportunistic feeders, relatively terrestrial, large-bodied, have broad habitat tolerance, and possess cheek pouches to transport food away from fields, and angry farmers, to places of safety.

We report on our observations of crop-raiding olive baboons at Gashaka Gumti National Park, Nigeria (*Fig. 8.1*) and set our results into the wider context of studies of food-enhanced primates, focusing on the costs and benefits of living alongside humans to the animals.

Only two other studies have so far investigated the behaviour and ecology of crop-raiding baboons from direct observation, both in Kenya (Gilgil: Forthman-Quick & Demment 1988, Strum 1994; Roka: Maples *et al.* 1976). There is also a long-term study report on rubbish-raiding baboons at Amboseli, Kenya (Altmann & Muruthi 1988, Bronikowski & Altmann 1996). Others mention the phenomenon in passing (Dunbar & Dunbar 1974b, Davidge 1978, Anderson 1981a), or rely on reports from farmers, or measured crop damage (Else 1991, Hill 1997, Naughton-Treves 1997). Collectively, these studies conclude that baboons crop-raid because, despite risk of injury or death from control methods, it is a very efficient foraging strategy as crops are spatially and temporally predictable, which reduces searching time and processing (Forthman-Quick 1986, Altmann & Muruthi 1988, Strum 1994).

In addition, it is true for baboons and other crop-raiding taxa that, relative to wild foods, crops contain elevated concentrations of easily digestible carbohydrates, fewer

Fig. 8.1 Olive baboons of the crop-raiding Gamgam troop at Gashaka Gumti National Park. (**a**) Lactating female (photo: Alejandra Pascual-Garrido); (**b**) feasting on maize harvested by local farmers (photo: YW)

digestion inhibitors, such as fibre, and fewer toxins (Iwamoto 1973, 1974, Waterman 1984, Fa 1986, Forthman-Quick & Demment 1988, Strum 1994, Naughton-Treves *et al.* 1998). Moreover, food-enhanced animals tend to alter their activity budgets noticeably, as they tend to spend less time foraging and travelling, and hence spend more time resting and in social activity, than their completely wild foraging counterparts.

There are also life-history benefits to food-enhancement (e.g., baboons: Anderson 1981a, Altmann 1986, Strum 1991; vervets: Lee *et al.* 1986; Japanese macaques: Sugiyama & Oshawa 1982, Mori *et al.* 1997), although few studies have focused solely on crop-raiding. Benefits include lower infant and / or adult mortality (Altmann 1986, Lee *et al.* 1986, Fa & Southwick 1988, Lyles & Dobson 1988, Asquith 1989, Samuels & Altmann 1991), accelerated growth and higher final weight (Strum 1991) as well as better body condition with more subcutaneous fat and fewer ectoparasites (Eley *et al.* 1986). An additional benefit of crop-raiding may relate to the ability of a taxon to colonise new areas as anthropogenic influence alters natural habitats. Some authors have suggested that the opening up of forests by slash and burn agriculture (Tappen 1965), or crop planting (Greenhood 1971), may have enabled baboons to expand their range into areas from which they would otherwise be excluded.

Potential negative consequences of living close to humans and their livestock may include an increased risk of parasite transmission (Muller-Graf *et al.* 1997), although access to better nutrition may result in improved overall health, and so higher parasite resistance. However, another study that investigated the relationship between parasite load and rubbish-feeding found that the differences did not appear to be explained by variation in the degree of human contact (Hahn *et al.* 2003). Living and feeding close to humans may also have other effects. For example, rubbish-feeding baboons in Amboseli have high levels of antibiotic resistant gastro-intestinal bacteria (Rolland *et al.* 1985), and poor peri-odontal health (Phillips-Conroy *et al.* 1993) compared to nearby wild-feeding animals. Animals feeding on rubbish and crops also face increased rates of competition and aggression (Lee *et al.* 1986, Warren 2008), increased exposure to human diseases (Dittus 1974, Rolland *et al.* 1985, Routman *et al.* 1985, Lee *et al.* 1986, Altmann & Muruthi 1988, Malik & Johnson 1994, Strum 1994), and loss of learned foraging strategies, which could result in greater mortality during years of food shortage (Power 1986, Samuels & Altmann 1991). Crop-raiders have to cope with particularly serious hazards if farmers resort to prevention techniques, including shooting, poisoning and trapping (Lee & Priston 2005), which can lead to injury or death (Dittus 1974, Forthman-Quick & Demment 1988, Eley *et al.* 1989, Richard *et al.* 1989) and a reduction in population size and distribution (Malik & Johnson 1994, Wheatley & Putra 1994). It should also be stressed that primate crop-raiding may lead to negative attitudes to wildlife among humans local to the affected area. In the past decades, various authors have stressed the importance of analysing such human-wildlife conflict, to consider how best to maintain or increase the quality of life of farmers, and to encourage local support for conservation initiatives (Harcourt *et al.* 1986, Ezealor & Giles 1997, Hill 1997, Naughton-Treves 1997, Udaya Sekhar 1998, Hill *et al.* 2002, Wheatley *et al.* 2002, Weladji & Tchamba 2003, Woodroffe *et al.* 2005).

Collectively, previous research illustrated how difficult it can be to understand fully the consequences of crop-raiding for raiding animals, as effects cover diverse areas of animal ecology, and stray into the human-wildlife interface.

Our study in Nigeria focuses on the ecology of a crop-raiding troop of olive baboons in comparison with a wild-feeding troop living nearby. We look at activity budgets, life-history variables and parasite load to compare the potential energetic and reproductive costs and benefits of living a crop-raiding life.

Methods

Our focal troops of olive baboons range at the edge of Gashaka Gumti National Park (GGNP) located in north-eastern Nigeria (06°55′ – 08°05′ N and 11°11′ – 12°13′ E), which shares its eastern border with the Republic of Cameroon (Dunn 1999). Research has focused on various aspects such as socioecology (Warren 2003), vocal communication (Ey & Fischer this volume [Ch. 10], MacGregor Inglis *et al.* 2008), parasitology (Weyher 2004, Weyher *et al.* 2006), endocrinology (Higham 2006, Higham *et al.* 2007, 2008 a, b), and life-history theory (Higham *et al.* 2009). Here, we summarise findings about the impact of crop-raiding on baboon ecology, behaviour, and reproduction (see also Warren 2003, 2008, Warren *et al.* 2007, Higham *et al.* 2009).

Focal Species

Baboons (genus *Papio*) are Old World Monkeys of the family Cercopithecidae (cheek pouched monkeys). Baboons in Nigeria belong to the olive baboon type (*Papio anubis*; Zinner *et al.* this volume [Ch. 7]).

Except for hamadryas baboons, other baboon taxa such as Guinea baboons, yellow baboons, chacma baboons, as well as olive baboons have a social structure based on female philopatry and male emigration. Troop sizes are highly variable but always composed of several females, one or more adult males plus infants, juveniles and subadults. Females first give birth at between 6 – 7 years of age and then reproduce every 2 – 3 years (Hill *et al.* 2000, Leigh & Bernstein 2006). Infants are cared for almost entirely by the mother although other individuals may interact, assist, and guard infants at times. Weaning is usually completed between 1 and 2 years of age but infants may continue to associate preferentially with their mothers even after the birth of the next infant.

Focal Troops and Study Site

The first baboon study group is *Gamgam (G) troop* (in previous publications sometimes referred to as "Gashaka troop" or CR [crop-raiding] troop). The second is *Kwano (K) troop* (previously sometimes referred to as NR [non-raiding] troop). The troops range about 10 km apart with G troop's range around 1 km outside the national park boundary, and K troop's range inside the park (*Fig. 8.2*).

The vegetation in both study areas is a mosaic of Southern Guinea savannah-woodland, grassland, lowland forest and riverine gallery forest and resembles that throughout the southern sector of the national park (see also descriptions in Dunn 1999, Adanu *et al.* this volume [Ch. 3], Koutsioni & Sommer this volume [Ch. 5]). However, K

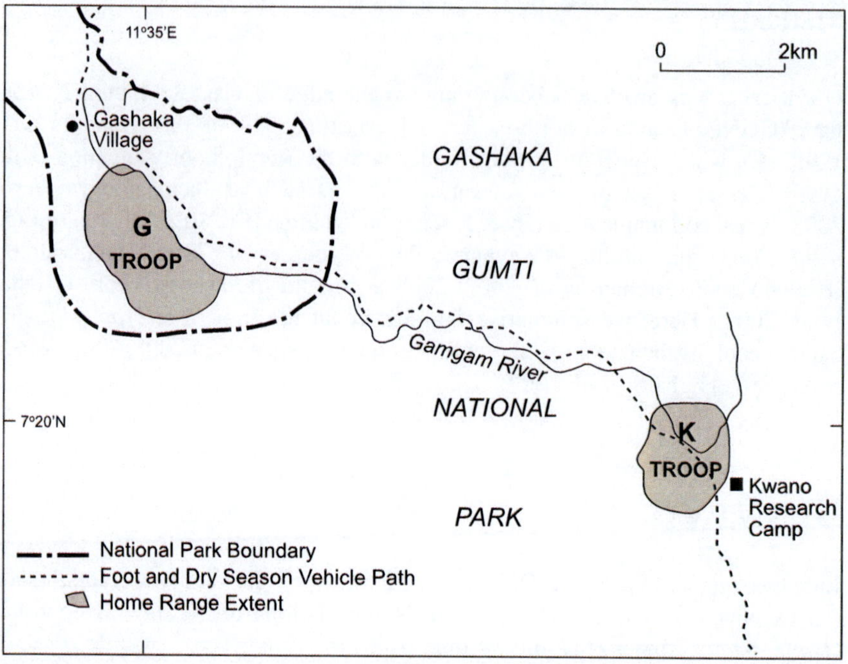

Fig. 8.2 Baboon study sites, indicating home-ranges of the crop-raiding Gamgam (G) troop near the village of Gashaka, and the wild-feeding Kwano (K) troop further inside the national park

troop's home-range has substantially more lowland and gallery forest than grassland, whereas G troop's range has a higher proportion of grassland and is more open.

The savannah-woodland includes undersized trees and is dominated by coarse tall grasses in the wet season. During the dry season, much of the grass dies off, and the woodland becomes bare and open. In addition, grasses are burnt from Dec – Mar, especially in those areas of G troop's range close to Gashaka village. Moreover, the range of G troop is grazed during some years by cattle of pastoralist Fulani. The Gamgam river is lined with elephant grass (*Panicum maximum*), which is cleared in some areas for farming, but large swathes still persist adjacent to fields. Crops are grown along the alluvial flats of the river in both wet and dry seasons and include maize, rice, sweet potatoes, cassava, okra, groundnuts, peppers, green leaf vegetables, pawpaw, and banana. These fields were regularly raided by baboons. Some trees such as mango (*Mangifera indica*), oil palm (*Elaeis guineensis)* and the fan palm (*Borassus aethiopum*) are sparsely distributed within G troop's range, and produce fruits that are eaten by both humans and baboons.

The home-ranges differ in altitude, with G troop found at approximately 320 m, and K troop at 583 m. The climate is highly seasonal, with hardly any rain or no rain at all during the pronounced dry season (mid Nov – mid Mar). There are minor, but noticeable differences between the study sites, as averaged from data collected at weather stations of the *Gashaka Primate Project* at Kwano and Gashaka from 2000 – 2008:

- mean maximum / minimum temperature (°C): Kwano 31.9 / 20.9, Gamgam 32.5 / 21.0;
- coolest / hottest day (°C): Kwano 14 / 39, Gashaka 12 / 42;
- rainy days / month (%): Kwano 40, Gashaka 35;
- rain (mm / year): Kwano 1973, Gashaka 1897.

The Kwano site is thus slightly cooler and wetter, as it receives 4.1 % (77 mm) more rain than the Gamgam site.

Troop sizes (*Tab. 8.1*) varied over the study period 2000 – 2008, and averaged 28.4 individuals for K troop (range 26 – 35) and 19.3 individuals for G troop (range 14 – 23). The overall average is 23.9 animals. This figure seems to be representative for the general population, as it corresponds well with census data on 22 other baboon troops in the Kwano, Gashaka, and nearby Selbe areas of GGNP collected from Mar – Jul 06 by field assistant Halidu Ilyasu. From this, we calculated an overall troop size in the wider Gashaka area of 22.7 individuals, including an average of 5 females (Higham *et al.* 2009).

Data Collection

The main field researchers (YW, JH) were supported by local field assistants (mainly Bobbo Buba and Halidu Ilyasu).

Female Reproductive Parameters

Life-history data refer to the period Mar 01 – Feb 06. In addition, gestation and cycle length were estimated via hormonal data from female faecal samples collected between Apr 04 – Apr 05 during a field study of JH. Female reproductive state was recorded as: *lactating*: suckling an infant; *pregnant*: not cycling and showing the pregnancy signs (Altmann 1973); *SO*: cycling with no swelling; *S1*: cycling with small swelling of the ano-genital area (AGA); *S2*: cycling with a medium-large swelling of the AGA and a small swelling of the para-callosal skin (PCS); *S3*: cycling with a large swelling of the AGA and full swelling of the PCS.

Table 8.1 Troop sizes of troop G (crop-raiding) and troop K (wild-feeding)

	K troop							G troop						
Period	AM	AF	SM	SF	J	I	Total	AM	AF	SM	SF	J	I	Total
2000 – 2001	6	7	0	2	5	8	28	4	4	0	0	3	3	14
2002 – 2003	6	7	1	1	4	8	27							
2003 – 2005	6	6	2	5	3	4	26	4	4	0	1	3	7	19
2005 – 2006	3	7	2	3	6	5	26	3	5	0	0	6	7	21
2008	6	10	3	2	8	6	35	1	5	2	0	13	2	23

AM = adult male, AF = adult female, SM = subadult male, SF = subadult female, J = juvenile, I = infant.

Data were used to estimate life-history parameters including gestation length, post-partum amenorrhea and inter-birth intervals with and without survival of the preceding infant as well as birth seasonality and infant survivorship (Altmann *et al.* 1977, Bentley-Condit & Smith 1977, Altmann 1980).

Independent two-tailed sample *t*-tests compared parameters between K and G troops. There is less confidence in some of the data for G troop as, unlike for K troop they are not continuous over the whole period. Data of uncertain accuracy are excluded. Sample sizes for G troop are therefore lower, but they are, in general, not large for both troops. Although this is partly due to the short length of study compared with some sites, it also reflects relatively small troop sizes at GGNP compared with many other sites.

Endoparasites

Three faecal samples per individual were collected by Anna Weyher from all adults in G troop (6 males, 4 females) and K troop (6 males, 7 females) between Apr – May 04 and analysed at Roehampton University (Weyher *et al.* 2006). An ocular micrometer was used to identify eggs, cysts, and larvae of helminth and protozoan species, and to obtain an average parasite count. Comparisons of parasitic richness (total number of parasitic taxa that each troop harboured), prevalence, output, and load were made between the troops. Mean parasite counts were normalised via log $(x + 1)$ transformation before further statistical analysis. As tests were carried out for each of the 9 identified taxa and for total number of helminths and protozoa recovered, type-I error rate was controlled by calculating table-wide critical significance levels, using a sequential Bonferroni adjustment (Rice 1989) for a level of significance at 0.05 and the (k) value equal to 11.

Activity Budget and Ranging

Data were collected by YW between Mar – Aug 01 and Nov 01 – Apr 02, for 10 h / day, for 8 days / troop each month, with early (06:00) and late starts (08:00) on alternate days. The troops were followed in alternate weeks so that each was observed for 4 days at the start and at the end of each month. A Psion™ workabout handheld computer and Observer™ software (Noldus Information Technology 1996) was used, programmed to record age-sex class and activity state with one or two modifiers.

Behavioural observations totalled 549 h for G troop, and 454 h for K troop and were based on troop scan samples at 2.5-min intervals, with mutually exclusive behavioural categories (Altmann 1974). As it was uncommon to see all troop members at all times, a focal individual was followed and data scored for all animals in view of the observer. The number of records collected in each hour was not the same for either troop across all hours of the day. Budgets were therefore adjusted for between-troop comparisons, by calculating the number of records for each activity per hour. The mean of these hourly proportions was used to produce daily activity budgets. These were then used as the sample unit for statistical analysis and to compare the troops. For this, observations were paired in that the first day for G troop in May was paired with the first day for K troop in the same month, the second day for G troop in May with the second for K troop in May, and so on. In order to meet assumptions of parametric statistics, proportional data were arcsine-transformed using the formula: $X' = \text{arcsine} \sqrt{p}$, where X' is the transformed value and p is the observed proportion (Krebs 1999). A Bonferroni correction was used to adjust the p value for repeated correlations.

Home-range use was noted at half-hourly intervals, when troop location was recorded with a GPS (Garmin 12®), and habitat type noted in the Observer™ program. GPS points were used to calculate the distance travelled during each half-hour interval. The resulting time period averages were summed to produce an estimated day journey length (Bronikowski & Altmann 1996). The study troops were compared by using paired t-tests on the mean half-hour travel distance for each month.

Crop-raiding

Crop-raiding was recorded by YW whenever it was observed between Mar – Aug 01 and Nov 01 – Apr 02. As a relatively rare event, it was noted *ad libitum* and continuously, as well as at the 2.5-min scans. The attempt, success, failure or otherwise of the raid was documented, along with the type of crop consumed, baboon behaviour, and farmers' response. The presence or absence of crops was noted on a weekly basis, to allow estimation of the number of days when crops were available. Due to animosity from some farmers towards both baboons and sometimes also researchers, crop-raiders were not always followed into fields and detailed data are thus sometimes lacking.

The following description of a crop-raiding day in Feb 2 recorded by YW illustrates the interplay between the raiding strategies of the monkeys and the reactions of farmers: "I set out with field assistant Bobbo Buba before dawn to look for G troop, but the troop is not in the expected sleeping trees. After much searching, it is located at 09:00 and sampling of focal animal adult male A begins. The troop spends the first hour of observations resting and foraging at intervals. At 10:00, they reach the maize stores in the central part of their range, just north of a farming area. Some animals pick up the last remaining scraps of maize that have been left after the farmers had cleared out the stores because of the baboons' earlier raids. – An infant sits and gazes at the store and then at the swaying fields of green maize across the river. An adult female looks in the same direction. Suddenly, we hear the shouts of a farmer and a chase. Adult male M appears, splashing across the river with a female farmer in dusty pursuit; he has a cob of maize in his mouth as he runs. The rest of the troop already moves fast in the opposite direction as he approaches and passes us. The farmer stops at the water's edge and shouts angrily. Within 5 min the troop is back scavenging around the store and adult male M is chomping on his fresh maize cob. – Fifteen minutes later, adult male M followed by an adult female heads down the steep sand bank towards the river. They cross a small piece of farmland, which is empty of crops because cattle have eaten them earlier in the season, but a small patch of green vegetables has been planted and these are avidly eaten. We see adult male B crossing back to our side of the river, empty-handed and empty-mouthed. A farmer (male this time) approaches along the path on the way to his farm. Seeing the baboons down by the river he shouts and runs at them with his spear raised. The baboons scatter and bark as they run through the savannah-woodland. The farmer chases the baboons far into the trees. This particular villager is one of the few who would do this, others normally stop after 50 m. The troop rests for a while and then begins to travel east, foraging as they go. Around 12:30 they rest and have a social session, with bouts of mutual grooming. At 14:30 they move off again, foraging here and there as they go. — At 16:00, the baboons are heading towards a farm but appear to hear or see something and veer away, continuing to move in the same direction but circling the farm. Then we hear the shouts of guards; some of the troop have entered the farm. As the baboons move through a narrow piece of gallery forest, stones are catapulted into the trees and rain down. None of these missiles seem to hit their targets. Despite the sounds of drumming and shouting from the farm adult male M heads in that direction, 'barking' as he goes. The young boys guarding are clearly visible, shooting stones from their catapults. They chase the baboons who skirt the edges of the maize field. Adult male M grabs a cob and runs off even as a small boy is running up behind him. Adult male W escapes from the maize field and straight up a tree. He seems cornered by two boys, but somehow manages to climb through the trees fast enough to gain ground and run off without being captured. A juvenile male is sitting high up a tree in the gallery forest about 30 m away, eating a maize cob and watching the progress of the guards. Using their catapults, the boys shoot at him to no avail as he continues to eat. — The boys turn and walk back, laughing and talking. Even as they are turning, adult male A is moving into the field, picking two cobs and running back to the gallery forest. With a final meal in their stomachs the baboons settle down for the night in trees very close to the farm – a good place to sleep for raiding again tomorrow!"

Results

The Ecology of Crop-Raiding

Members of Gamgam (G) troop invaded fields regularly when crops were available during the study period, i.e., from Jun 01 – Mar 02. A total of 45 raids were recorded on 32 days (57 % of observation days). Of these, 31 (69 %) resulted in the baboons successfully gaining access to crops, 4 (9 %) were thwarted by farmers, 6 (13 %) terminated by the baboons moving away, and in 4 cases (9 %) the outcome was unknown. When baboons were chased away by farmers they often attempted to raid other fields nearby.

Successful raids were concentrated on maize (the most commonly grown crop) at all stages of production, with 45 % of raids on maize fields at various growth stages, 4 % on maize remnants in harvested fields and 13 % on maize stores.

Animals also raided other crops including cassava, sweet potato, banana, and greenleaf. There were 54 observations of baboons foraging on different types of crop. The number of observed crop types eaten is greater than the total number of raids as during any one event, more than one type of crop could be eaten. Fresh maize from fields was most often raided, followed by stored maize and cassava scraps left behind in fields after farmers had harvested.

Over the total observation period that included raids (n = 549 h), they occurred at a rate of 0.08 / h, and an average of 2 % of all time was spent foraging on crops, equivalent to just 9 % of foraging time. However, the nutritional benefits for each animal could be high at times when crops were present. From observed damage to maize fields it was estimated that baboons, including focal troops and others with overlapping ranges, could obtain 2 – 7 cobs per day when maize was available.

Raiding strategies employed by the baboons included various watchful behaviours before entering fields and use of cheek pouches to transport food to "safe" locations. One clear vigilance behaviour that took place prior to entering fields, involved resting in trees 50 m from fields looking over them. The troop undertook this activity for a mean daily proportion of 2. 0 (± 2.7) % of raiding days (cf., the poem at the beginning of this chapter).

Activity Budget

Overall, the raiding G troop spent significantly more time resting and in social activity, and significantly less time foraging and travelling than the wild-feeding K troop (*Tab. 8.2*). A comparison including only the 7 months in which crop-raiding occurred (Jun – Aug 01 and Dec 01 – Mar 02) still revealed a significant difference ($p < 0.05$) for forage, rest and social activity, but not for time spent travelling (independent sample t-test, $t = -1.728$, df = 107, $p = 0.087$).

Ranging Behaviour

Day journey length (DJL) was significantly longer for G troop (3.1 km) compared to K troop (2.4 km; $t = 2.629$, $n = 10$, $p = 0.027$). This difference existed despite the fact that home-range sizes of the study troops were almost identical at about 1.5

Table 8.2 Activity budgets in crop-raiding baboon troop G (549 h of observation) and wild-feeding troop K (454 h of observation) between Mar 01 – Apr 02

Activity	G troop	K troop	t	df	p
Forage	27	31	–2.545	152	0.012
Rest	35	28	3.862	152	<0.001
Social	11	9	2.887	152	0.004
Travel	25	30	–3.247	152	0.001

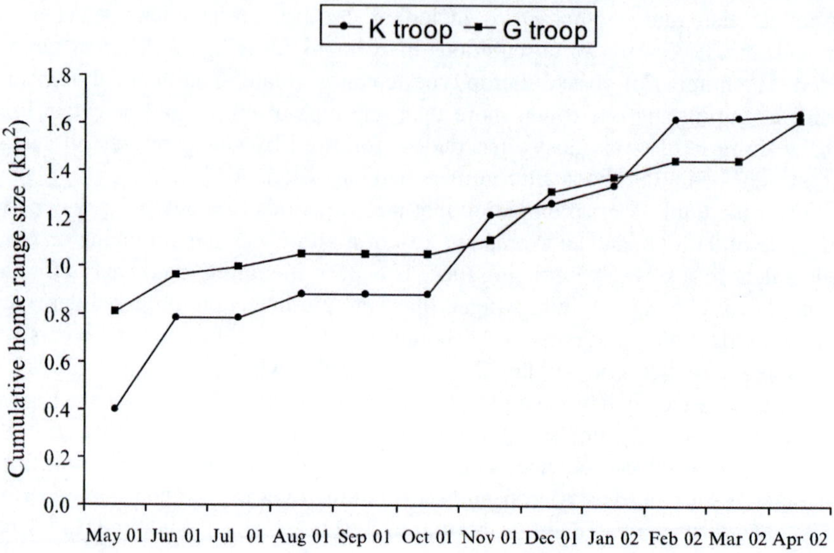

Fig. 8.3 Home-range sizes of baboon troops G and K, estimated through cumulative plotting of newly entered areas

km². However, the range of G troop is likely to be underestimated as the cumulative plot of monthly home-range size did not reach an asymptote (*Fig. 8.3*). This is at least partly due to the fact that no location points were recorded when the troop crossed the river to crop-raid, in contrast with the curve for K troop's home-range that reached an asymptote and is therefore likely to be fairly accurate.

Life-history

Female reproductive parameters (from Higham *et al.* 2009). are provided in *Tab. 8.3*. Accordingly, inter-birth intervals were significantly shorter in G than in K troop ($t_4 = -6.0$, $p = 0.004$) although, post-partum amenorrhea was not significantly shorter ($t_9 = 0.8$, $p = 0.429$). Instead, G troop had a greatly reduced duration of cycling compared with K troop ($t_5 = -6.3$, $p = 0.002$). Neither gestation length ($t_3 = -1.1$, $p = 0.367$) nor cycle length ($t_5 = 0.2$, $p = 0.988$) differed between the troops.

Infant mortality in K troop was high compared with G troop (*Fig. 8.4*), where the only known death of an infant younger than a year was that of a twin. Most infant deaths in K troop were clustered in time, with 4 infants dying during the peak of the rainy season in 2004, a year when annual rainfall was 18 % higher than the average for the Kwano range. In addition, 4 adult females disappeared in K troop over the 5-year study period, including one adult female during the 2004 heavy rains. In all cases, females were seen struggling with disease or illness before disappearance. Females in K troop in general are often seen exhibiting disease symptoms during

Table 8.3 Estimates of female reproductive parameters in troops G and K (after Higham *et al.* 2009)

Parameter	Troop	Mean	SD	Range	*n*
Post-partum amenorrhea (PPA) (months)	K	11.3	4.8	7 – 17	7
	G	9.3	1	8 – 10	4
Amenorrhea following infant death (months)	K	1.4	1.5	0 – 4	8
	G	—	—	—	—
Duration of cycling (DC) (months)	K	12.4	2.3	10 – 16	5
	G	1.5	0.7	1 – 2	2
Duration of cycling following infant death (months)	K	3.4	2.6	1 – 8	7
	G	—	—	—	—
Gestation length (GL) (days)	K	185.3	4	180 – 188	4
	G	188 – 192	n/a	n/a	1
Cycle length (days)	K	42.3	3.5	39 – 46	3
	G	42.3	2.3	41 – 45	3
Observed inter-birth interval (months)	K	25	2.8	23 – 27	2
	G	16	1.2	15 – 17	4
Inter-birth interval following infant death (months)	K	13.8	5.1	8 – 21	5
	G	—	—	—	—
Inter-birth interval (months) (a)	K	29.9	n/a	n/a	n/a
	G	16.5	n/a	n/a	n/a
Infant mortality (proportion dying < 12 mo)	K	0.470	n/a	n/a	19
	G	0.063	n/a	n/a	16

(a) Calculated from mean of (PPA + DC + GL)
"–" = parameter could not be estimated. Sample sizes are incidences, not females.

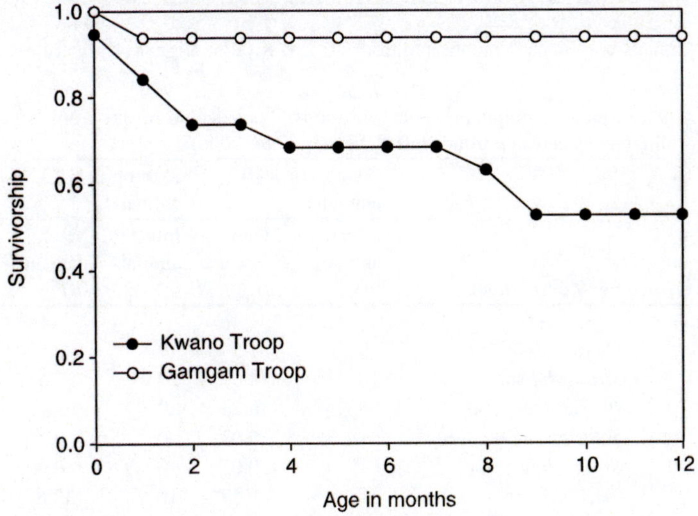

Fig. 8.4 Survivorship curve for infants of troops G and K from 0 – 12 months (after Higham *et al.* 2009)

the heavy rainfalls from August to October. In contrast, no adult females died or disappeared in G troop over the study period, and they often appeared to be in excellent health, with good body condition. One adult male death in G troop in 2005 was a direct consequence of crop-raiding after a prime adult male was attacked by a farmer with a machete, inflicting a groin wound that later proved fatal.

In K troop, births occurred in all months except April, with little evidence of seasonality (*Fig. 8.5*). G troop on the other hand, may exhibit a birth peak 6 months – the approximate length of gestation – after wet season crop-raiding.

Parasitology

Most individuals in both troop G and K displayed low infection levels, with only a few baboons having high intensities (Weyher *et al.* 2006). Seven helminth and 2 protozoan parasites were recorded in baboon faeces, although not all could be identified to genus level (*Tab. 8.4*). Overall, G troop had higher parasitic species richness, mostly due to

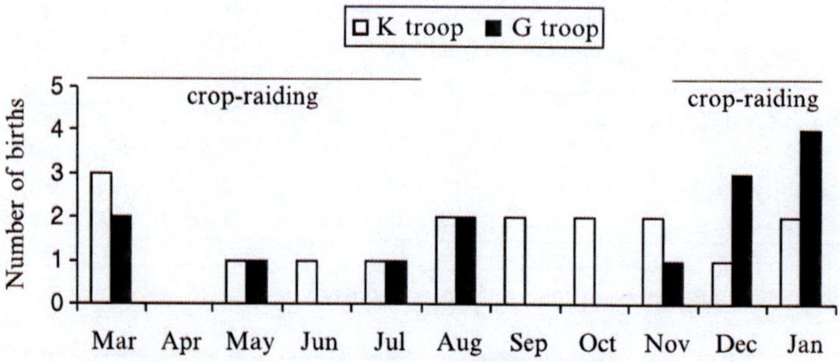

Fig. 8.5 Infants born in each month into troops G and K (after Higham *et al.* 2009)

Table 8.4 Mean parasite outputs recorded in troops G and K and results from 2-way ANOVA testing for differences between troops (after Weyher *et al.* 2006)

		G troop (n = 10 animals)		K troop (n = 13 animals)		
Parasite type	Egg / cyst output	Infected animals (n)	Mean parasites (n)	Infected animals (n)	Mean parasites (n)	ANOVA p value
Helminth						
	Dicrocoelium spp.	9	1.53	9	0.87	0.190
	Strongylid spp.	10	14.07	10	17.51	0.518
	Physaloptera spp.	9	6.23	9	63.23	0.001
	Schistosoma mansoni	1	0.03	0	0.00	0.156
	Strongyloides spp.	1	0.03	0	0.00	0.156
	Trichuris spp.	10	9.57	10	32.69	0.001
	Unidentified nematode	1	0.07	5	2.08	0.129
	Total helminth load	10	31.50	13	116.38	0.001
Protozoa						
	Amoeba spp.	10	32.30	10	20.57	0.198
	Balantidium coli	10	520.87	10	43.41	0.003
	Total protozoan load	10	533.16	13	63.98	0.008

one female that was parasitised by both the bilharzia-causing *Schistosoma mansoni* and a type of *Strongyloides* not recorded in K troop. On the other hand, the mean output of helminth worms was significantly higher for K troop (2-way ANOVA, $F = 16.496$, df $= 1$, $p = 0.001$), particularly because of the loads of 2 taxa, *Physaloptera* sp. (F = 15.235, df = 1, $p = 0.001$) and *Trichuris* sp. ($F = 15.713$, df = 1, $p = 0.001$). Although G troop had a higher protozoan load than K troop this difference was not significant after sequential Bonferroni correction (2-way ANOVA, $F = 8.917$, $p = 0.008$). G troop was significantly more parasitised with *Balantidium coli* (F = 11.787, df = 1, p = 0.003).

Discussion

Our research on olive baboons in Nigeria's Gashaka region revealed considerable effects of crop-raiding behaviour on their ecology, including how the animals spend their time, on patterns of mortality and reproduction, and on parasite dynamics. We first discuss these different aspects of animal ecology separately, and then examine how crop-raiding may shape the lives of these baboons.

Crop-raiding, Activity Budget, and Ranging

The rate of observed crop-raiding by G troop seems low compared to some other populations such as those of yellow baboons at Roka / Kenya, with 1.8 raids / h of observation (Maples *et al.* 1976). However, the study troop there was much larger, and the area of available crops much greater, which may explain some of the difference. Taking an alternative measure, G troop entered farmland to raid on 57 % of observation days. This is comparable to chacma baboons in Suikerbosrand, who raided on 52 % of observation days (Anderson 1981a).

Activity budgets of G troop differ significantly from those of K troop, and the trends are similar to other populations of wild and food-enhanced primates (*Tab. 8.5*). Overall, food-enhanced animals spend more time resting than foraging or travelling, and the least amount of time in social activity, such that typically rest > forage > travel > social activity. Wild-feeding counterparts usually forage for the largest part of their time budget, with social activity still the least common, such that typically forage > travel > rest > social activity. Differences between G and K troops were consistent with these patterns but not as great (cf. *Tab. 8.2*).

Overall, the following differences in % activity time emerged between food enhanced and wild foraging baboons (WFB) in other populations, compared to differences between the crop-raiding G-troop and wild-feeding K troop, respectively:

- foraging, WFB 20 – 43 % more, K troop 4 % more;
- rest, WFB 17 – 36 % less, K troop 7 % less;
- social activity, WFB 4 – 10 % less, K troop 2 % less;
- travel, WFB 5 – 6 % more, K troop 4 % more.

Table 8.5 Comparison of activity budgets (% daytime) of food-enhanced and wild-feeding primates

Taxon	Country / Site / Remark	Food	Forage	Rest	Social	Travel	Source
Baboons							
	Ethiopia, Awash valley	W	31	31	12	25	Nagel 1973, Dunbar 1992, Dunbar & Dunbar 1974
	Ethiopia, Bole valley	W/C	21	35	12	25	Nagel 1973, Dunbar 1992, Dunbar & Dunbar 1975
	Kenya Gilgil	W	51	7	8	30	Eley *et al.* 1989
	Kenya Gilgil	W/C/R	29	24	18	27	Eley *et al.* 1990
	Kenya, Amboseli	W	48	16	9	27	Bronikowski & Altmann 1996
	Kenya, Amboseli	W/R	24	44	13	19	Bronikowski & Altmann 1996
Savannah (tantalus / vervet) monkeys							
	Senegal, Parc National du Niokolo-Koba	W	50	45	12	nd	Harrison 1983
	Kenya, Amboseli, dry season	W	35 – 45	~29	~5	~27	Lee *et al.* 1986

(continued)

At a proximate level, these trends are probably brought about because crops are a more clumped resource compared to wild foods, which are more scattered. Wild foods thus require more travel time to locate as well as more processing time, so that less time is available for resting and socialising (see below).

One possible explanation for the smaller differences in activity budgets between K and G troops compared to wild-foraging and food-enhanced baboons elsewhere is that habitat quality might be better around Gashaka. Tropical forests (which make up large patches at Gashaka) have the highest net primary productivity per unit area per year of any known terrestrial community (Krebs 1994). However, most East African baboon study sites are located in savannah grassland, where baboons spend 30 – 60 % of time foraging, whereas those inhabiting East African forests show similar levels (20 – 30 %) to those seen in the present study (Nagel 1973, Paterson 1976, Bercovitch 1983, Depew 1983, Rasmussen 1985, Dunbar 1992b).

Group sizes also tend to be smaller in forest-living animals. As a consequence, within-troop competition over food is less severe. Hence, it may be that forest baboons are already in productive environments, and as such crop-raiding may make a smaller difference to forest baboons than to baboons in less productive habitats.

Some support for this comes from a comparison of the Ethiopian olive baboon populations at Awash and Bole whose habitats are partially forested and where crop-raiding decreases foraging time by only 10 %, whereas greater reductions are seen in Gilgil (23 %) and Amboseli (24 %) (cf. *Tab.* 8.5). It is also notable that at both Gashaka and Bole, baboons supplement their diet mainly by crop-raiding, whereas animals in Gilgil both crop- and rubbish-raid, and those at Amboseli

Table 8.5 (continued)

Taxon	Country / Site / Remark	Food	Forage	Rest	Social	Travel	Source
	Kenya, Amboseli, dry season	W/R/H	20 (50)	43	20	16	Lee *et al.* 1986
	Uganda, Entebbe	W/C/H	24 (50)	44	11	14	Saj *et al.* 1999
Barbary macaques							
	Morocco, Moyen Atlas mountains	W	49	14	11	23	Fa 1986
	Gibraltar, Rock of Gibraltar	H	8	39	11	38	Fa 1986
Rhesus macaques							
	Tughlaqabad, India, forest,	W	40	28	(2)	26	Seth & Seth 1985
	Tughlaqabad, India, temples	W/H	27	35	(11)	19	Seth & Seth 1986
	Tughlaqabad, India, urban	W/C	16	46	(14)	17	Seth & Seth 1987
	Tughlaqabad, India	W/C/H	8 (10)	30	(12)	19	Malik 1986
Hanuman langurs							
	Nepal, Himalayas, Nepal, breeding group	W	41	39	(4)	6	Sugiyama 1976
	India, Jodhpur, all male group	W/C/H	25 (18)	64	(4)	4	Little & Sommer, 2002, Sommer pers. com.
	India, Jodhpur, all male group	W/H	29 (2)	58	(4)	4	Little & Sommer, 2002, Sommer pers. com.
	India, Jodhpur, breeding group	W/H	20(24)	59	(9)	9	Little & Sommer, 2002, Sommer pers. com.
	India, Kanha Tiger Reserve, breeding group	W	26	42	(6)	13	Newton 1992

Food: W = wild foraging, C = crop-raiding, R = rubbish-raiding, H = provisioned by humans. *Forage*: Figures in brackets indicate % of all foraging time spent feeding on human sources only as opposed to wild resources. *Social*: Definitions vary, and include sometimes all social contacts, sometimes only allo-grooming; figures in brackets indicate time spent allo-grooming only (for langurs, grooming includes a small proportion of auto-grooming).

rubbish-raid. The costs of feeding on rubbish are likely to be lower than those of crop-raiding, as animals are less likely to be chased away from the food source, and hence need to spend less time in vigilance and flight behaviours.

Previous studies suggest that day journey length (DJL) is affected by food availability (Henzi *et al.* 1992) and hence might be expected to be reduced in crop- and rubbish-raiding animals (Altmann & Muruthi 1988). This pattern holds true for primates generally but some exceptions exist and some differences are small (cf. *Tab. 8.4*). One likely reason why G troop does not show a more substantially reduced travel time, is that crop-raiding necessitates travel in itself, in the form of both initial travel to fields, and in flight behaviour when chased from cultivated areas. This may also help explain why DJL is higher in G troop (3.1 km) than K troop (2.4 km). As K troop still invested more time in travel, this leads to the logical conclusion that G troop travelled faster – a plausible assumption, given that raiding often requires swift exits! Differences in travel time may also be related to differences in habitat type. Forest-living baboons travel over less distance (Harding 1976), whereas poor quality and or arid environments require a longer DJL (Kummer 1968, Anderson 1981b, Sigg & Stolba 1981). As such, the lower amount of forest and greater expanse of open woodland in G troop's range might contribute to their longer DJL compared to K troop.

Finally, G troop spent more time in social activity than K troop. This pattern is consistent with other food-enhanced primate groups. The amount of time that primates spend socialising may be limited by how much individuals need to spend servicing social relationships (Dunbar 1992a, Henzi *et al.* 1997). As a consequence, members of smaller groups may have to invest less time per capita in socialising. The relatively small difference between G and K troop (2 % more socialising) might therefore be more significant as G troop was smaller through most of the study period.

Watching agricultural areas from prominent elevated positions was observed in 38 % of G troop's successful raids. Similar sentinel behaviour as a prelude to raiding is also reported for yellow baboons in Roka, Kenya (Maples *et al.* 1976). "Watching" farms, has also been noted in other baboon studies (Dunbar & Nathan 1972, Else 1991). The behaviour seems unlikely to have evolved specifically to crop-raid, but may reflect a pre-adaptation observed in non-food-enhanced baboons. For example, in Honnet Nature Reserve, South Africa, where several troops of chacma baboons shared waterholes, an approaching troop male would spend several hours in a tree scanning before approaching to drink (Stoltz & Saayman 1970). Presumably, this allows avoidance of predators, other troops that might act aggressively, or competition with other males.

Life-history

G troop has significantly shorter inter-birth intervals than K troop, largely due to a reduced duration of cycling, unlike other food enhancement studies which have not detected a reduced number of cycles to conception when compared with wild-feeding animals. Garcia *et al.* (2006) showed that captive olive baboons exhibited inter-birth intervals that were 60 % shorter than wild baboons, this being due to a very short period of post-partum amenorrhea but not a reduction in mean duration of cycling. Differences in inter-birth intervals between G and K troops do not appear to be reflected in differences in gestation or cycle lengths, perhaps unsurprisingly as these variables are known to be relatively invariant (Martin & MacLarnon 1985).

Moreover, infant mortality in K troop was much higher than in G troop. A number of the deaths were clustered, as both infant and adult female mortality events in

K troop occurred during the rainy season. Thus, high rainfall may be a problem for K troop baboons at certain times of the year. The absence of these same mortality events in G troop offers the possibility that the improved nutrition obtained by crop-raiding could provide some sort of buffer against adverse environmental conditions.

As crops are seasonal, food availability is more seasonal for G troop than for K troop and as such we might expect births to be more seasonal in G troop than in K. Although our data are rather limited, this may indeed be the case, with a possible birth peak approximately 6 months after wet season crop-raiding for G troop. Other baboon studies have also shown at least some reproductive seasonality, again often linked to patterns of food availability (Lycett *et al.* 1999, Cheney 2004, Beehner & Whitten 2004).

Parasites

In terms of parasitology, there may be both costs and benefits of crop-raiding to animals (Weyher *et al.* 2006). The overall difference between G and K troops were due to output levels, with 2 parasite taxa being more common in K troop and 1 more common in G troop. Although, parasite richness was higher for G troop, this was due to 2 taxa that were recorded in only 1 female of G troop. K troop had a significantly higher total helminth load than the crop-raiding troop, with higher output of the identified helminth parasites (*Physaloptera* sp., and *Trichuris* sp.). Nutrition of a host is an important factor affecting parasite load, with higher load being linked to better host health and nutrition (Dobson 1985, Bundy & Golden 1987, Nesheim 1993, Coop & Holmes 1996, Coop & Kyriazakis 1999, Ezenwa 2004). Better host health increases the effectiveness of the immune response to parasites and allows animals to cope better with parasitic infections once established (Hausfater & Sutherland 1984).

The significant differences in the output of *Physaloptera* sp. may also be influenced by diet, as animals become infected with this parasite by consuming intermediate hosts such as beetles, cockroaches, and crickets (Ivens *et al.* 1978, Muller-Graf 1994, Foreyt 2001). Detailed analysis on baboon insectivory may help identify the intermediate host of the *Physaloptera* parasite.

On the other hand, G troop had a significantly higher output rate of the protozoan parasite *Balantidium coli* cyst than K troop (Weyher *et al.* 2006). This is a parasite of both humans and nonhuman primates (Ash & Orthiel 1997). The higher levels of *Balantidium coli* in G troop contrast with lower levels of helminth parasites, and this difference may be explained both by G troop's higher exposure to human contact and by the reproductive biology of this parasite. Like most protozoa (and unlike parasitic helminths) *Balantidium coli* reproduces within the host and thus a single infection event can be followed by rapid host reproduction and lead to high parasite loads (Wakelin 1996). A host diet rich in starch allows a high growth of *Balantidium coli* (Noble *et al.* 1989). G troop's diet includes starch-rich crops such as maize, guinea corn, and cassava (Warren 2003) and this may explain the difference in output between the troops.

One female in G troop was found to be infected with *Schistosoma mansoni*, but it is possible that more individuals were infected with the parasite, as it is difficult to detect in faecal matter (Hall 1982). This parasite is commonly found in baboons (McGrew *et al.* 1989, Muriuki *et al.* 1989, Ghandour *et al.* 1995, Muller-Graf *et. al.* 1997, Murray *et. al.* 2000) and in humans (Muller-Graf 1997). Both the Gamgam river and the human populations using its water are also known to be infected with *Schistosoma*. This raises the possibility of transmission back and forth between humans and baboons, something that has been suggested at Gombe, Tanzania where baboons may have been infected from humans (Muller-Graf 1997).

Thus, although the better health of G troop allows it to resist some parasites (even those that may be also found in humans), a crop-raiding life may lead to higher levels of at least some other parasites that may be transmitted back and forth from humans and baboons.

Crop-raiding and Baboon Ecology at Gashaka: Why Crop-raid?

The lives of the crop-raiding members of G troop differ both qualitatively and quantitatively from the lives of wild foraging K troop members, which experience much less anthropogenic exposure.

There are some clear disadvantages to crop-raiding, although these may not always be easy to quantify. Farmers typically respond to animals near or in their fields by throwing stones, chasing and shouting. Baboons, as cercopithecine monkeys, have an advantage over many other species as they can use cheek-pouches to store food when chased and flee to eat elsewhere undisturbed. G troop animals were frequently observed to do this, or to carry food away in their hands. This mitigates the costs of being able to make only short forays into fields when farmers are present. There are also instances of baboons being killed or injured by villagers; in fact, before the national park was created, and for a while thereafter, baboons and other crop-raiding monkeys near Gashaka were regularly shot (Barnwell this volume [Ch. 1]). Despite this, G troop members are persistent in their attempts to gain access to crops. For example as described above, on one occasion the troop was chased away from one end of a field, only to circle around the angry guards and begin raiding again at the opposite end of the field. The continued prevalence of crop-raiding, despite harassment from farmers, suggests that the costs of the activity must outweigh the benefits.

One might argue that differences between G and K troops in certain aspects of their ecology (troop size, activity budgets, DJL) are only marginally affected by crop-raiding. Nevertheless, differences in life-history are clearly more marked, as the shorter inter-birth interval and low infant mortality in G troop translates into a higher reproductive rate. For individual females, these life-history differences could lead to significantly more offspring over their lifetimes compared to sympatric females with no access to crops. Over a 10 year period, an average female in K troop can be expected to produce 6.1 infants with only 3.2 surviving to more than

12 months, whereas an average G troop female would produce 7.6 infants of which 7.1 would survive to at least one year of age. This two-fold reproductive advantage to crop-raiding helps to explain why animals might be selected to pursue a crop-raiding life despite its apparent risks of injury and possibly increased susceptibility to at least some parasites. Female mortality would need to be greatly elevated to outweigh the increased reproductive output due to raiding. In fact our limited evidence suggests the opposite, with mortality rates for K troop adult females at 0.8 year and no deaths observed at all for G troop adult females. We do not have data on age at first reproduction in females but it seems likely that the food enhanced animals will mature more rapidly than their wild-feeding neighbours (Lyles & Dobson 1988). This would add another reproductive advantage to crop-raiding.

Environmental differences between G and K troops' home-ranges, related anthropogenic influence such as land clearance, burning practices, crop planting, and differential exposure to humans, can have thus surprisingly beneficial consequences for baboons (Tappen 1965, Greenhood 1971). Anthropogenic habitats such as those found in G troop's range are becoming more common across West and Central Africa, as forests continue to be degraded and, in some cases, replaced by agriculture. Although these changes may have detrimental effects for many forest primate species, they may allow baboons, a primate with extraordinary ecological flexibility, to flourish and breed more rapidly than in closed forests. Baboons may be one of the few primate species to benefit from human habitat disturbance, as their long history of commensalism and crop-raiding continues to benefit the species, although often to the detriment of their human neighbours.

Acknowledgements Field work was made possible with a permit by the Nigerian National Park Service to the *Gashaka Primate Project*, which receives core funding from the Chester Zoo Nigeria Biodiversity Programme. NCF / WWF-UK provided logistical support. Local field assistants Bobbo Buba, Halidu Ilyasu, Buba Bello, Nuhu Husseini, Haruna Ahmadu, and volunteers Helen Cross and Alejandra Pascual-Garrido helped collect data. Stuart Semple gave invaluable advice and help to many aspects of this study. At Roehampton University, Mary MacKenzie, Balbir Singh and Martin Evans provided technical assistance. Funding for YW and JH came from Roehampton University, with additional funding for JH from The Leakey Trust and for YW from the Primate Society of Great Britain, NUMICO Research B.V., British Federation of Women Graduates and the International Federation of Animal Welfare.

Literature

Altmann, J. (1974). Observational study of behaviour: sampling methods. *Behaviour* 49: 227–267.

Altmann, J. (1980). *Baboon Mothers and Infants*. London: Harvard University Press.

Altmann, J. (1986). Adolescent pregnancies in non-human primates: An ecological and developmental perspective. In Lancaster, J. & Hamburg, B. (eds.), *School-Age Pregnancy and Parenthood*. Chicago: Aldine Press, pp. 247–262.

Altmann, J. & Muruthi, P. (1988). Differences in daily life between semiprovisioned and wild-feeding baboons. *American Journal of Primatology* 15: 213–221.

Altmann, J., Altmann, S. A., Hausfater, G. & McCuskey, S. A. (1977). Life history of yellow baboons: Physical development, reproductive parameters and infant mortality. *Primates* 18: 315–330.

Altmann, S. A. (1973). The pregnancy sign in savannah baboons. *Journal of Zoo Animal Medicine* 4: 8–12.
Anderson, C. M. (1981a). Intertroop relations of chacma baboons (*Papio ursinus*). *International Journal of Primatology* 2: 285–310.
Anderson, C. M. (1981b). Subtrooping in a chacma baboon (*Papio ursinus*) population. *Primates* 22: 445–458.
Asquith, P. J. (1989). Provisioning and the study of free-ranging primates: History, effects and prospects. *Yearbook of Physical Anthropology* 4: 288–308.
Beehner, J. C. & Whitten, P. L. (2004). Modifications of a field method for fecal steroid analysis in baboons. *Physiology & Behavior* 82: 269–277.
Bentley-Condit, V. & Smith, E. (1997). Female reproductive parameters of Tana River yellow baboons. *International Journal of Primatology* 18: 581–595.
Bronikowski, A. M. & Altmann, J. (1996). Foraging in a variable environment: Weather patterns and the behavioural ecology of baboons. *Behavioural Ecology and Sociobiology* 39: 11–25.
Coop, R. & Holmes, P. (1996). Nutrition and parasite interaction. *International Journal of Parasitology* 26: 951–962.
Cowlishaw, G. & Dunbar, R. (2000). *Primate Conservation Biology*. Chicago: The University of Chicago Press.
Davidge, C. (1978). Ecology of baboons (*Papio ursinus*) at Cape Point. *Zoologica Africana* 13: 329–50.
Depew, L. A. (1983). *Ecology and Behaviour of Baboons* (*Papio anubis*) *in the Shai Hills Game Production Reserve, Ghana*. MSc thesis, Cape Coast University.
Dittus, W. P. J. (1974). The socioecological basis for conservation of the toque monkey (*Macaca sinica*) of Sri Lanka (Ceylon). In Prince Rainer III of Monaco, H. S. H. & Bourne, G. H. (eds.), *Primate Conservation*. New York: Academic, pp. 238–267.
Dunbar, R. I. M. (1992a). Neocortex size as a constraint on group size in primates. *Journal of Human Evolution* 20: 469–493.
Dunbar, R. I. M. (1992b). Time – a hidden constraint on the behavioural ecology of baboons. *Behavioural Ecology and Sociobiology* 31: 35–49.
Dunbar, R. I. M. & Dunbar, P. (1974a). On hybridization between *Theropithecus gelada* and *Papio anubis* in the wild. *Journal of Human Evolution* 3: 187–192.
Dunbar, R. I. M. & Dunbar, P. (1974b). Ecological relations and niche separation between sympatric terrestrial primates in Ethiopia. *Folia Primatologica* 21: 36–61.
Dunbar, R. I. M. & Nathan, M. F. (1972). Social organisation of Guinea baboon. *Folia Primatologica* 17: 321–334.
Dunn, A. (1999). *Gashaka Gumti National Park: a Guidebook*. Lagos: Gashaka Gumti National Park, NCF/WWF-UK.
Eley, R. M., Strum, S. C. & Muchemi, G. (1986). Nutritional effects upon physical condition of three free-ranging troops of baboons in Gilgil, Kenya. *Primate Report* 14: 136.
Eley, R. M., Strum, S. C., Muchemi, G. & Reid, G. D. F. (1989). Nutrition, body condition, activity patterns and parasitism of free-ranging olive baboons (Papio anubis) in Kenya. *American Journal of Primatology* 18: 209–220.
Else, J. G. (1991). Non-human primates as pests. In Box, H. O. (ed.), *Primate Responses to Environmental Change*. London: Chapman & Hall, pp. 155–156.
Ezealor, A. U. & Giles, R. H. (1997). Vertebrate pests of the Sahelian wetland agro-ecosystem: perceptions and attitudes of the indigenes and potential management strategies. *International Journal of Pest Management* 43: 97–104.
Ezenwa, V. (2004). Interactions among host diet, nutritional status and gastrointestinal parasite infection in wild bovids. *International Journal of Parasitology* 34: 535–524.
Fa, J. E. & Southwick, C. H. (1988). *Ecology and Behaviour of Food-enhanced Primate Groups*. New York: Alan R. Liss Inc.
Fa, J. E. (1986). *Use of Time and Resources by Provisioned Troops of Monkeys: Social Behaviour, Time and Energy in the Barbary Macaque* (Macaca sylvanus *L.) at Gibraltar*. Contributions to Primatology 23. Basel: Karger.

Forthman-Quick, D. L. (1986). Activity budgets and the consumption of human foods in two troops of baboons (*Papio anubis*) at Gilgil, Kenya. In Else, J. G. & Lee, P. C. (eds.), *Primate Ecology and Conservation*. New York: Cambridge University Press, pp. 221–228.

Forthman-Quick, D. L. & Demment, D. (1988). Dynamics of exploitation: Differential energetic adaptations of two troops of baboons to recent human contact. In Fa, J. E. & Southwick, C. H. (eds.), *Ecology and Behaviour of Food-Enhanced Primate Groups*. New York: Alan R. Liss, pp. 25–51.

Garcia, C., Lee, P. & Rosetta, L. (2006). Dominance and reproductive rates in captive female olive baboons, *Papio anubis*. *American Journal of Physical Anthropology* 131: 64–72.

Ghandour, A., Zahid, N., Banaja, A., Kamal, K. & Boug, A. (1995). Zoonotic intestinal parasites of hamadryas baboons *Papio hamadryas* in the western and northern regions of Saudi Arabia. *Journal of Tropical Medicine and Hygiene* 98: 431–439.

Greenhood, W. F. (1971). *Diets and Ecology of Crop-Raiding Baboons*. MSc thesis, University of Florida.

Hahn, N. E., Proulx, D., Muruthi, P. M., Alberts, S. & Altmann, J. (2003). Gastrointestinal parasites in free-ranging Kenyan baboons (*Papio cynocephalus* and *P. anubis*). *International Journal of Primatology* 24: 271–279.

Harcourt, A. H., Pennington, H. & Weber, A. W. (1986). Public attitudes to wildlife and conservation in the Third World. *Oryx* 20: 152–154.

Harding, R. S. O. (1976). Ranging patterns of a troop of baboons (*Papio anubis*) in Kenya. *Folia Primatologica* 25: 143–185.

Harrison, M. J. S. (1983). Age and sex differences in the diet and feeding strategies of the green monkey *Cercopithecus sabaeus*. *Animal Behaviour* 31: 969–977.

Hausfater, G. & Sutherland, R. (1984). Little things that tick off baboons. *Natural History* 2: 54–61.

Henzi, S. P., Byrne, R. W. & Whiten, A. (1992). Patterns of movement by baboons in the Drakensberg Mountains: primary responses to the environment. *International Journal of Primatology* 13: 601–628.

Henzi, S. P., Lycett, J. E. & Weingrill, T. (1997). Cohort size and the allocation of social effort by female mountain baboons. *Animal Behaviour* 54: 1235–1243.

Higham, J. P. H. (2006). *The Reproductive Ecology of Female Olive Baboons* (Papio hamadryas anubis) *at Gashaka Gumti National Park, Nigeria*. PhD thesis, Roehampton University, London, UK.

Higham, J. P., Ross, C., Warren, Y., Heistermann, M. & MacLarnon, A. M. (2007). Reduced reproductive function in wild olive baboons (*Papio hamadryas anubis*) related to natural consumption of the African black plum (*Vitex doniana*). *Hormones and Behavior* 52: 384–390.

Higham, J. P., MacLarnon, A. M., Ross, C., Heistermann, M. & Semple, S. (2008a). Baboon sexual swellings: information content of size and color. *Hormones and Behavior* 53: 452–456.

Higham, J. P., Semple, S., MacLarnon, A. M., Heistermann, M. & Ross, C. (2008b). The timing of ovulation with respect to sexual swelling detumescence in wild olive baboons. *Primates* 49: 295–299.

Higham, J. P., Warren, Y., Adanu, J., Umaru, B.N., MacLarnon, A. M., Sommer, V. & Ross, C. (2009). Living on the edge: life-history of olive baboons at Gashaka-Gumti National Park, Nigeria. *American Journal of Primatology* 71: 293–304.

Hill, C. M. (1997). Crop-Raiding in wild vertebrates: the farmer's perspective in an agricultural community in western Uganda. *International Journal of Pest Management* 43: 77–84.

Hill, C. M. (2005). People, crops, and primates: a conflict of interests. In Paterson, J. & Wallis, J. (eds.), *Commensalism and Conflict: The Primate-Human Interface*. Norman, Oklahoma: The American Society of Primatology, pp. 41–59.

Hill, R. A., Lycett, J. E. & Dunbar, R. I. M. (2000). Ecological and social determinants of birth intervals in baboons. *Behavioral Ecology* 11: 560–564.

Hill, C. M., Osborn, F. & Plumptre, A. J. (2002). *Human-Wildlife Conflict: Identifying the Problem and Possible Solutions*. Albertine Rift Technical Reports Series Vol. 1. Wildlife Conservation Society.

Iwamoto, T. (1973). A brief report on feeding activity of Japanese monkey. *Annual Report Primate Research Institute Kyoto University* 3: 25–26.

Iwamoto, T. (1974). A bioeconomic study of a provisional troop of Japanese monkeys (*Macaca fuscata*) at Koshima Islet, Miyazaki. *Primates* 15: 241–262.

Jolly, A. (1999). *Lucy's Legacy: Sex and Intelligence in Human Evolution*. London: Harvard University Press.

Krebs, C. J. (1994). *Ecology: The Experimental Analysis of Distribution and Abundance*. New York: Harper & Row.

Krebs, C. J. (1999). *Ecological Methodology*. New York: Addison-Wesley Educational Printers, Inc.

Kummer, H. (1968). *Social Organization of Hamadryas Baboons. A Field Study*. Basel: Karger.

Lee, P. & Priston, N. (2005). Human attitudes to primates: Perceptions of pests, conflict and consequences for conservation. In Paterson, J. & Wallis, J. (eds.), *Commensalism and Conflict: The Primate-Human Interface*. Norman, OK: *Am. Soc. Primatol.* pp. 1–23.

Lee, P., Brennan, E., Else, J. & Altmann, J. (1986). Ecology and behaviour of vervet monkeys in a tourist lodge habitat. In Else, J. G. & Lee, P. C. (eds.), *Primate Ecology and Conservation*. Cambridge: Cambridge University Press, pp. 229–235.

Leigh, S. R. & Bernstein, R. M. (2006). Ontogeny, life history, and maternal investment in baboons. In Swedell, L. & Leigh, S. R. (eds.), *Reproduction and Fitness in Baboons. Behavioral, Ecological, and Life History Perspectives*. New York: Springer, pp. 225–255.

Little, K. A. & Sommer, V. (2002). Changes of enclosure in langur monkeys: Implications for the evaluation of environmental enrichment. *Zoo Biology* 21: 549–559.

Lyles, A. M. & Dobson, A. P. (1988). Dynamics of provisioned and unprovisioned primate populations. In Fa, J. E. & Southwick, C. H. (eds.), *Ecology and Behavior of Food-Enhanced Primate Groups*. New York: Alan R. Liss, pp. 167–198.

Lycett, J., Weingrill, T. & Henzi, S. (1999). Birth patterns in the Drakensberg Mountain baboons (*Papio cynocephalus ursinus*). *South African Journal of Science* 95: 354–357.

MacGregor Inglis, D., Ross C. & Semple, S. (2008). Multi-modal communication in olive baboons (*Papio hamadryas anubis*). (Poster abstracts No. 168, XXII Congress, *The International Primatological Society*, 03 – 08 Aug 2008, Edinburgh, UK).

Malik, I. (1986). Time budgets and activity patterns in free-ranging rhesus monkeys. In Else, J. G. & Lee, P. C. (eds.), *Primate Ecology and Conservation*. Cambridge: Cambridge University Press, pp. 105–114.

Malik, I. & Johnson, R. L. (1994). Commensal rhesus in India: The need and cost of translocation. *Revue d'Ecologie – La Terre Et La Vie* 49: 233–243.

Maples, W., Maples, M., Greenwood, W. & Walek, M. (1976). Adaptations of crop-raiding baboons in Kenya. *American Journal of Physical Anthropology* 45: 309–316.

Martin, R. D. & MacLarnon, A. M. (1985). Gestation period, neonatal size and maternal investment in placental mammals. *Nature* 313: 220–223.

McGrew, W., Tutin, C., Collins, D. & File, S. (1989). Intestinal parasites of sympatric *Pan troglodytes* and *Papio* spp. at two sites: Gombe (Tanzania) and Mt. Assirik (Senegal). *American Journal of Primatology* 17: 147–155.

Mori, A., Yamaguchi, N., Watanabe, K. & Shimizu, K. (1997). Sexual maturation of female Japanese macaques under poor nutritional conditions and food-enhanced perineal swelling in the Koshima troop. *International Journal of Primatology* 18: 553–579.

Muller-Graf, C., Collins, D., Packer, C. & Woolhouse, M. (1997). Schistosoma mansoni infection in a natural population of olive baboons (*Papio cynocephalus anubis*) in Gombe Stream National Park, Tanzania. *Parasitology* 115: 621–627.

Muriuki, S., Murugu, R., Munene, E., Karere, G. & Chai, D. (1989). Some gastro-intestinal parasites of zoonotic (public health) importance commonly observed in old world non-human primates in Kenya. *Acta Tropica* 71: 73–82.

Murray, S., Stem, C., Boudreau, B. & Goodall, J. (2000). Intestinal parasites of baboons (*Papio cynocephalus anubis*) and chimpanzees (*Pan troglodytes*) in Gombe National Park. *Journal of Zoo and Wildlife Medicine* 31: 176–178.

Nagel, U. (1973). A comparison of anubis baboons, hamadryas baboons and their hybrids at a species border in Ethiopia. *Folia Primatologica* 19: 104–165.

Naughton-Treves, L. (1997). Farming the forest edge: vulnerable places and people around Kibale National Park, Uganda. *The Geographical Review* 87: 27–46.

Naughton-Treves, L., Treves, A., Chapman, A. C. & Wrangham, R. (1998). Temporal patterns of crop-raiding by primates: linking food availability in croplands and adjacent forest. *Journal of Applied Ecology* 35: 596–606.

Nesheim, M. (1993). Human nutrition needs and parasitic infections. *Parasitology* 107: S7–S18.

Newton, P. (1992). Feeding and ranging patterns of forest Hanuman langurs (*Presbytis entellus*). *International Journal of Primatology* 13: 245–285.

Noldus Information Technology (1996). *The Observer System for Collection and Analysis of Data. Reference Manual (Version 3. 0).* Wageningen: Noldus Information Technology.

Paterson, J. D. (1976). *Varying Ecology and Differential Adaption of the Ugandan Baboon, Papio cynocephalus anubis, with Special Reference to Forest Conditions and Behaviour Models for Early Hominids.* PhD thesis, University of Toronto.

Paterson, J. D. & Wallis, J. (2005). *Commensalism and Conflict: The Human-Primate Interface.* Norman, Oklahoma: American Society of Primatologists.

Phillips-Conroy, J., Hildebolt, C., Altmann, J., Jolly, C. & Muruthi, P. (1993). Peri-odontal health in free-ranging baboons of Ethiopia and Kenya. *American Journal of Physical Anthropology* 90: 359–371.

Power, M. (1986). The foraging adaptation of chimpanzees and the recent behaviours of the provisioned apes in Gombe and Mahale National Parks, Tanzania. *Human Evolution* 1: 251–266.

Rasmussen, K. R. L. (1985). Changes in the activity budgets of yellow baboons (*Papio cynocephalus*) during sexual consortships. *Behavioral Ecology & Sociobiology* 17: 161–70.

Richard, A. F., Goldsmith, S. J. & Dewar, R. E. (1989). Weed macaques: the evolutionary implications of macaque feeding ecology. *International Journal of Primatology* 10: 369–594.

Rolland, R. M., Hausfater, G., Marshall, B. & Levy, S. B. (1985). Antibiotic-resistant bacteria in wild primates: Increase prevalence in baboons feeding on human refuse. *Applied Environmental Microbiology* 49: 791–794.

Ross, C. & Hill, C. M. (2006). Non-human primates as pests: variation in primate species and human cultures. *International Journal of Primatology* 27: 398.

Routman, E., Miller, R., Phillips-Conroy, J. & Hart, D. (1985). Antibiotic resistance and population structure in Escherichia coli from free-ranging African yellow baboons. *Applied Environmental Microbiology* 50: 749–754.

Saj, T., Sicotte, P. & Paterson, J. (1999). The influence of human food consumption on the time budget of vervet monkeys. *International Journal of Primatology* 20: 977–994.

Samuels, A. & Altmann, J. (1991). Baboons of the Amboseli basin: Demographic stability and change. *International Journal of Primatology* 12: 1–9.

Seth, P. K. & Seth, S. (1985). Ecology and feeding behavior of the free ranging rhesus monkeys in India. *Indian Anthropologist* 15: 51–62.

Sigg, H. & Stolba, A. (1981). Home range and daily march in a hamadryas baboon troop. *Folia Primatologica* 36: 40–75.

Sponsel, L. E., Ruttanadakul, N. & Natadecha-Sponsel, P. (2002). Monkey business? The conservation implications of macaque ethnoprimatology in southern Thailand. In Fuentes, A. & Wolfe, L. D. (eds.), *Primates Face to Face*. Cambridge: Cambridge University Press, pp. 288–309.

Strum, S. C. (1991). Weight and age in olive baboons. *American Journal of Primatology* 25: 219–237.

Strum, S. C. (1994). Prospects for management of primate pests. *Behavioural Ecology and Sociobiology* 49: 295–306.

Sugiyama, Y. (1975). Characteristics of the ecology of the Himalayan langurs. *Journal of Human Evolution* 5: 249–77.

Sugiyama, Y. & Oshawa, H. (1982). Population dynamics of Japanese monkeys with special reference to artificial feeding. *Folia Primatologica* 39: 238–263.

Tappen, N. C. (1965). The taxonomy of the baboon and its position in the order of primates (comments on W. Ted Roth). In Vagtborg, H. (ed.), *The Baboon in Medical Research.* Austin: University of Texas Press.

Udaya Sekhar, N. (1998). Crop and livestock depredation caused by wild animals in protected areas: the case of Sariska Tiger Reserve, Rajasthan, India. *Environmental Conservation* 25: 160–171.

Warren, Y. (2003). *Olive Baboons (Papio cynocephalus anubis): Behaviour, Ecology and Human Conflict in Gashaka Gumti National Park, Nigeria*. PhD thesis, Roehampton University.

Warren, Y. (2008). Crop-raiding baboons (*Papio anubis*) and defensive farmers: A West African perspective. *West African Journal of Applied Ecology* 14: 33–46.

Warren, Y., Buba, B. & Ross, C. (2007). Patterns of crop-raiding by wild and domestic animals near Gashaka Gumti National Park, Nigeria. *International Journal of Pest Management* 53: 207–216.

Waterman, P. G. (1984). Food acquisition and processing as a function of plant chemistry. In Chivers, D., Wood, B. & Bilsborough, A. (eds.), *Food Acquisition and Processing in Primates*. New York: Plenum Press, pp. 177–211.

Weladji, R. B. & Tchamba, M. N. (2003). Conflict between people and protected areas within the Benoue wildlife conservation Area, North Cameroon. *Oryx* 37: 72–79.

Weyher, A. H. (2004). *A Comparison of Gastrointestinal Parasites in a Crop-Raiding and a Forest-dwelling Troop of Olive Baboons* (Papio cynocephalus anubis) *in Gashaka Gumti National Park, Nigeria*. MRes thesis, Roehampton University, London, UK.

Weyher, A. H., Ross, C. & Semple, S. (2006). Gastrointestinal parasites in crop raiding and wild foraging *Papio anubis* in Nigeria. *International Journal of Primatology* 27: 1519–1534.

Wheatley, B. P. & Putra, D. K. H. (1994). The effects of tourism on conservation at the monkey forest in Ubud, Bali. *Revue d'Ecologie – La Terre Et La Vie* 49: 245–257.

Wheatley, B. P., Stephenson, R., Kurashina, H. & Marsh-Kautz, K. G. (2002). A cultural primatological study of *Macaca fascicularis* on Ngeaur Island, Republic of Palau. In Fuentes, A. & Wolfe, L. D. (eds.), *Primates Face to Face*. Cambridge: Cambridge University Press, pp. 240–253.

Wolfheim, J. H. (1983). *Primates of the World, Distribution, Abundance and Conservation*. Seattle & London: University of Washington Press.

Woodroffe, R., Thirgood, S. & Rabinowitz, A. (2005). *People and Wildlife: Conflict or Coexistence*? Cambridge: Cambridge University Press.

Chapter 9
How Different Are Gashaka's Baboons? Forest and Open Country Populations Compared

Caroline Ross, Ymke Warren, Ann M. Maclarnon, and James P. Higham

Abstract Our long-term study of baboons in Gashaka Gumti National Park (GGNP) is one of very few that looks at West African baboons, and is also unusual because it focuses on animals living in a forested environment with high annual rainfall. Here, we present data on troop size, activity budgets, ranging behaviour, diet, and life-history, and compare these data with those from other baboon study sites. The troop sizes at GGNP are significantly smaller than many other baboon populations, and this may be linked to low predation pressure at the site. The high productivity and floral diversity at GGNP allow animals to avoid eating difficult to process subterranean foods and feed on preferred foods such as fruits and seeds. The diet of wild-feeding animals at the site is diverse, and includes a high proportion of fruit and low proportion of leaves, subterranean items, and animal foods when compared with other baboon populations. Gestation length, cycle length, and length of post-partum amenorrhea for the wild-feeding K troop are typical of baboons but the inter-birth interval in this population is significantly longer than that found in most other baboon populations due to a long duration of cycling. The high rainfall at GGNP may increase disease risk, forcing wild-feeding mothers to invest heavily in their offspring, resulting in a long period of cycling before successful pregnancy and hence a long inter-birth interval.

Keywords Baboons • Life-history • Reproduction • Mortality

Introduction

Comparative biology is a powerful tool that allows us insight into the processes of evolution and adaptation. The comparative method is much used for comparisons between higher taxonomic levels such as genera but its use intra-specifically or within

C. Ross (✉)
Centre for Research in Evolutionary Anthropology, Roehampton University, London SW15 4JD, UK
e-mail: c.ross@roehampton.ac.uk

V. Sommer and C. Ross (eds.), *Primates of Gashaka*, Developments in Primatology: Progress and Prospects 35, DOI 10.1007/978-1-4419-7403-7_9,

the same genus is less common. This is partly because there is often limited variation within genera or species, and partly because few taxa are studied well enough to give us sufficient information for inter-population comparisons.

However, there are some questions that can be investigated only by using intra-generic or intra-specific data, for example, when investigating the flexibility of a generalist taxon, which lives in a range of environments. Such adaptable primate groups include the Asian macaques and hanuman langurs and, in Africa, the baboon, a large-bodied highly adaptable monkey that is widely distributed across sub-Saharan Africa and into the Arabian peninsula.

Since the 1960s, studies of wild baboons have investigated ecology and social organisation, both as a subject of study in itself, and as a model of the evolution of human behaviour (DeVore & Hall 1965, Dunbar 1976, 1983, Jolly 2001, Elton 2006). These early studies were some of the first to realise that the wide distribution of baboons gives us unparalleled insight into the way in which ecological adaptability can be mirrored by a similar flexibility in social organisation, something that continues to drive many baboon research programmes today (Jolly 2001, Alberts & Altmann 2006). Yet, as Hall (1965) noted, a complete understanding of baboon ecology is not possible without knowledge of baboons in a range of environments: "The baboon [...] being probably the most widely distributed of any species of nonhuman primate in Africa, is adaptable to a considerable range of habitats. So far these habitats have been adequately sampled from only Kenya and Southern Africa [...] no systematic data on ecology and behaviour of the species being available from any area south of the Sahara in West Africa" (p. 29).

Despite a flurry of field studies since the pioneering work on baboons began, there are still large gaps in our knowledge. Baboons have been well studied in savannah grasslands and other relatively open habitats, particularly in southern and east Africa. However, few data are published from West Africa, and none from large areas of Central Africa. Our understanding of how baboons adapt to forest habitats is therefore relatively limited. West African forest-living animals are also of interest because they appear to be living in a habitat that is marginal for the taxon. In areas of west-central Africa with high rainfall, baboons are not found, with their terrestrial niche being filled by mandrills or drills. Kingdon (1997) suggests that this indicates that the wetter forests are difficult for baboons to colonise, possibly because of their susceptibility to disease, and because of competition from forest-adapted primate species.

It is partly the flexible feeding behaviour of baboons that allows them to exist in a wide range of habitats, although this is combined with dietary selectivity (Whiten *et al.* 1991). Studies show considerable variation in both the types of dietary items eaten and dietary diversity (Aldrich-Blake *et al.* 1971, Stoltz & Saayman 1970, Post 1982, Byrne *et al.* 1987; Norton *et al.* 1987, Barton 1989, Whiten *et al.* 1991, Altmann 1998, Bronikowski & Altmann 1996, Hill 1999). Similarly, flexibility in troop size and composition, activity budget and life-history, may also be key factors that have allowed baboons to attain a geographical and habitat spread that is extraordinary amongst non-human primates.

Possible correlates and constraints of baboon flexibility have recently been investigated using comparative databases (Dunbar 1992, 1996, Hill *et al.* 2000,

Fig. 9.1 Olive baboons of a wild-feeding troop at Gashaka Gumti National Park (photo: David MacGregor Inglis)

Hill & Dunbar 2002, Barrett *et al.* 2006). All draw on the plethora of available studies, which encompass considerable geographical and ecological diversity. Once again, all these models have largely been based on open-country populations from southern and eastern Africa. Our study extends this approach by adding data from a population of olive baboons that penetrate the forests of West Africa in Nigeria's Gashaka Gumti National Park (GGNP) (*Fig. 9.1*). This is an outlier in geography, habitat, and climate, and hence offers a fascinating opportunity to test theories and models that predict variation in activity budgets, troop size, diet and life-history parameters in baboons.

For this, we describe troop size, activity budget, ranging behaviour (day journey length and home-range), diet, and life-history parameters of GGNP baboons and compare these values with those of other baboon populations (*Fig. 9.2*). In particular, we investigate whether previously published models based on variables such as climate successfully predict the socioecology of baboons in north-eastern Nigeria.

Methods

Study Species, Focal Troops, and Study Site

Baboons (genus *Papio*) are Old World Monkeys of the family Cercopithecidae (cheek pouched monkeys) widely distributed across Africa and into the Arabian

Fig. 9.2 Location of 33 baboon study sites. Approximate position of sites 1 – 31 taken from Dunbar (1992). 1. Badi / Senegal; 2. Mt. Assirik / Senegal; 3. Shai Hills / Ghana; 4. Bole Valley / Ethiopia; 5. Mulu / Ethiopia; 6. Metahara / Ethiopia; 7. Awash Falls / Ethiopia; 8. Awash Station / Ethiopia; 9. Erer-Gota / Ethiopia; 10. Ishasaha / Uganda; 11. Murchison / Uganda; 12. Budongo Forest / Uganda; 13. Cholol / Kenya; 14. Laikipia / Kenya; 14. Gilgil / Kenya; 16. Nairobi NP / Kenya; 17. Manyara / Tanzania; 18. Serengeti National Park (NP) / Tanzania; 19. Amboselia / Tanzania; 20. Gombe / Tanzania; 21. Ruaha / Tanzania; 22. Mikumi / Tanzania; 23. Kariba / Zimbabwe; 24. Honnet / S. Africa; 25. Suikerbosrand / S. Africa; 26. Giant's Castle / S. Africa; 27. Drakensberg / S. Africa; 28. Mt. Zebra NP / S. Africa; 29. Cape Point / S. Africa; 30. Okavango / Botswana; 31. Kiuseb / Namibia; 32. De Hoop / S. Africa (Hill *et al.* 2003); 32. Comoé NP / (Kunz & Linsenmair 2008); 33. Gashaka Gumti NP / Nigeria (this study)

peninsula. Various morphotypes are typically distinguished, including hamadryas, Guinea, yellow, chacma, Kinda and olive baboons (Zinner *et al.* this volume [Ch. 7]). Except for hamadryas, baboon taxa have a social structure based on female philopatry and male emigration.

Baboons in Nigeria belong to the olive baboon type (*Papio anubis*). We conducted research at Gashaka Gumti National Park, located in north-eastern Nigeria (06°55′ – 08°05′ N and 11°11′ – 12°13′ E; Dunn 1999), with a focus on two habituated troops studied since 2001 (see, e.g., Warren *et al.* this volume [Ch. 8]; Ey & Fisher this volume [Ch. 10]; Higham *et al.* 2009).

The first study group is Gamgam (G) troop (in previous publications sometimes referred to as Gashaka troop or CR [crop-raiding] troop). The second is Kwano (K) troop (previously sometimes referred to as NR [non-raiding] troop). The troops range about 10 km apart with G troop's range around 1 km outside the national park boundary, and K troop's range inside the park. The vegetation in both study areas is a mosaic of Southern Guinea savannah-woodland, grassland, lowland forest, and riverine gallery forest. However, K troop's home-range has substantially more lowland and gallery forest than grassland, whereas G troop's range has a higher proportion of grassland and is more open.

The climate data used in the testing of predictive models all come from data collected at Kwano research station and Gashaka village from 2000 – 2008 (Sommer & Ross this volume [Ch. 1]). Data on baboon socioecology come from both study troops, but only those for the wild-feeding K troop are used in most comparisons with other wild-feeding baboon populations. When discussing data that come from both troops, and those in the surrounding area, we refer to "GGNP baboons", otherwise, we state specifically whether troops K or G are being discussed.

Data Collection and Analyses

Data were collected by the primary researchers (YW and JH), who were supported by local field assistants (mainly Bobbo Buba and Halidu Ilyasu). Students and volunteers from Nigerian and European universities also contributed to the collection of supporting data.

Troop Sizes

Censusing of troops (total $n = 22$) by an experienced field assistant (Halidu Ilyasu) took place from Mar – Jul 06 around the areas of Kwano ($n = 5$), Gashaka ($n = 7$) and Selbe ($n = 10$). Three counts of individuals in age-sex classes in each troop were taken, each separated by a minimum of 30 min. Home-range areas and distinguishing features of individuals in troops were noted to prevent accidental re-sampling. We compare the average GGNP troop size with other published troop sizes using a one-sample t-test (SPSS 15).

Dunbar (1992, 1996) published models of baboon behaviour that predict maximum ecologically tolerable and minimum ecologically permissible troop sizes for areas with a range of different climates. Following the methods of Higham *et al.* (2009) and using more up-to-date climate data, we use Dunbar's (1996) published group-size tables to investigate whether the troop sizes measured in our census are within his suggested range. We use the points in the table that most closely approximate the mean climate at both troop sites, a mean annual rainfall of about 1900 mm (cf. Kwano = 1973 mm; Gamgam = 1897 mm), and a mean annual temperature of 25 °C (cf. Kwano: 26.4 C, Gamgam 26.8 °C).

Activity Budget and Ranging

Data were collected by YW between Mar – Aug 01, and Nov 01 – Apr 02 for 10 h / day, for 8 days each month, with early (06:00) and late (08:00) starts on alternate days (details of scan sampling methods used for collecting activity budget and ranging data in Warren *et al.* this volume [Ch. 8]). Activity was defined as *social* (including allo-grooming, aggression, play and sexual behaviour), *rest* (both vigilant and non-vigilant), *travel* and *foraging* (picking, processing, and handling foodstuffs, including sweeping through the leaf litter and turning over rocks in search of invertebrates). Foraging time included putting items in cheek pouches, but the analyses shown here are from direct foraging observations only, as it was impossible to identify foods eaten from cheek pouches.

It is worth noting that activity budgets from other studies on the same troops (Higham 2006; Ey & Fischer this volume [Ch. 10]) give different results from those of YW, with higher values for foraging and lower values for travel and resting. This may partly be due to increased habituation levels in later years of the research, with animals spending more time moving away from observers in earlier years. However, because only YW's data provide a representative sample of all individuals, and as they include both wet and dry seasons, we have used only these data for all further comparisons.

Previous inter-population comparisons of primates, and specifically baboons, show that mean troop size is positively correlated with day journey length (DJL) and home-range (HR) size (Sharman & Dunbar 1982, Stacey 1986, Barton *et al.* 1992, Hill 1999). We conducted similar analyses restricting our sample to wild-feeding baboon populations only, including previously published data and values from K troop. We used regression analyses, with troop size as the independent variable and either DJL or HR as the dependent variable. All variables were normally distributed.

Diet

The daily schedule given above was also used by YW to collect dietary data. Although G troop regularly included crop foods in its diet, we focus here on the use of wild foods only. We discuss the effects of feeding on crops in more detail in Warren *et al.* (2007, this volume [Ch. 8]). During foraging (defined above), YW recorded age / sex class of actor; food type (animal, tree, grass, herb, shrub, lianas, vine, or other); species eaten and part consumed (flower, fruit, bean / seed, leaf, roots, bark etc).

For comparisons with other populations, diet was split into categories used by previous authors: i.e., fruit / seeds (including beans), leaves, flowers, and underground and animal matter. Food item was identified by both species and part consumed, (e.g., leaves and fruits of one plant species were considered as two separate food items). Food items that made up more than 1% of all forage observations overall were designated as *core foods*,and food items that took more than 1% of forage observations in any one month were designated as *monthly staples*.

Hill & Dunbar (2002) found that percentage time spent feeding correlated with the percentage of some dietary foods. They also suggest that inter-population variability in baboon diet can be predicted, at least partly, by variation in climate and geography, with rainfall, temperature, and latitude variables predicting the amounts of leaves, fruit, and subterranean items in the diet. We use their figures and equations, and long-term GGNP climate data, to predict the diet of GGNP baboons and compare these predictions with observed values.

Diet and Food Availability

Plant food availability of the four main food categories (fruit, bean / seed, flower and leaves) was measured from Mar 01 – Apr 02; we could not measure animal food availability. The densities of fruit, bean / seed, flower and leaves of 30 wild plant species, all known to be a part of the baboons' diet, were measured using 10 (50 × 50 m) quadrats, randomly placed within the home-ranges of both troops: 5 in Guinea savannah-woodland and 5 in lowland and gallery forest. Five specimens of each of the 30 commonly eaten food species were monitored in both troops' ranges, and the presence or absence of fruit, bean / seed, flower and leaf noted twice a month. Total number of food trees in each troop's home-range was estimated as the mean number of each wild food species per m^2 in each habitat type multiplied by the area of the habitat type in m^2.

A monthly index of food availability (FAI) for fruit, bean / seed, flower and leaf was calculated, using data from 24 food tree species for G troop and 23 for K troop. FAI was calculated each month using FAI = $\sum_{i=1}^{N} p_i d_i s_i$ (p_i is the percent of the i-th food tree species with food, d_i is the density of the i-th tree species in the home-range of the troop, and s_i is the mean DBH of the i-th species in cm). We used Spearman's rank correlations to test for a relationship between the FAI of each food class and its percentage consumption for both troops. As these required multiple correlations of related data ($n = 4$ for each troop) we used a Bonferroni correction to adjust the significance value for these correlations, so that $p < 0.0125$ was used as the adjusted threshold p value.

Reproductive Parameters

Life-history data presented here were collected from Dec 00 – Feb 06 (Higham *et al.* 2009). The long-term data are continuous for K troop but there was no detailed data collection from G troop from May 02 – Nov 03. Life-history parameters have been estimated for both troops using both long-term data records and analysis of faecal hormones (Higham *et al.* 2009, Warren *et al.* this volume [Ch. 8]). Here, we focus on the life-history of the wild-feeding K troop and compare this to other wild-feeding baboon populations.

Hormonal data were used to investigate the relationship between sexual swellings and ovulation and to calculate cycle length and gestation length (Higham *et al.* 2007, 2008a, b). Faecal samples were collected from Apr 04 – Apr 05 from females during focal follows, according to protocols in Hodges & Heistermann (2003), aiming to obtain samples from individuals every 1 – 2 days. Fresh samples, with no urine contamination, were collected from the ground in the morning, usually before 07:30. Following *in situ* homogenisation of the faecal bolus using a stick, we collected approximately 2 g (wet weight) of sample and placed it into 10 ml (> 95 %) of ethanol, pre-measured into screw-top plastic bottles (Azlon 7BWH0030N). Laboratory analyses were carried out at Roehampton University, where faecal extracts were analysed for oestrogen and progesterone metabolites using Enzyme-Immuno-Assay (EIA) for total oestrogens (Etotal) (Patzl *et al.* 1998) and 5ß-pregnane-3α, 20α-diol (PdG) (Heistermann *et al.* 1995). Ovulation was detected, and ovulation date estimated, from profiles of PdG and E total, according to methods described in Higham *et al.* (2008b).

Data on menstrual cyclicity were taken from cycles where length could be accurately measured from one ovulation to the next. These measured cycles did not occur during periods of reproductive suppression caused by feeding on African black plum (*Vitex doniana*), from Aug – Oct and Jan – Feb (Higham *et al.* 2007). Fruit and leaves of this plant contain high concentrations of progestogen-like compounds, their ingestion by females results in suppression of ovulation, sexual swellings, and hence reproductive behaviour. Inter-swelling intervals are also extended at this time, but we do not have sufficient accurate hormonal ovulation data to enable these cycle lengths to be measured from ovulation date to ovulation date.

Following Higham *et al.* (2009), we compare inter-birth intervals from the wild-feeding K troop with a data set published by Barrett *et al.* (2006), using a one-sample *t*-test (SPSS 15). We also compare a greater range of life-history variables with previously published values from 4 sites that have presented similar, directly comparable data: Amboseli / Kenya (Altmann *et al.* 1977, Altmann 1980); Mchelelo / Kenya (Bentley-Condit & Smith 1997); Gilgil / Kenya (Smuts & Nicholson 1988) and Erer / Ethiopia (Sigg *et al.* 1982). Finally, we use our data to test previously published comparative baboon models for inter-birth intervals from Hill *et al.* (2000) and Barrett *et al.* (2006).

Results

Comparative socioecological data between baboons of Gashaka Gumti National Park (GGNP) and other populations throughout Africa are compiled in *Tab. 9.1*. Expected and observed values for GGNP parameters as predicted by previous models are given in *Tab. 9.2*.

Troop Size

The mean size of K troop was 28.4 (range 26 – 35), including 6 – 10 females and 3 – 6 males, and the mean for G troop was 19.3 (range 14 – 23), including 4 – 5 females and 1 – 4 males (2 peripheral males which were included in the 6 males sampled by

Table 9.1 Socioecological parameters at long-term baboon study sites across Africa. Values for activity budgets and feeding times rounded to whole numbers. Data on group size, day journey length and interval were taken from Sharman & Dunbar (1982) and Hill *et al.* 2000. Values for activity budgets and feeding times are rounded to whole numbers. Data on group size, day journey length, and interval were taken from Sharman and Dunbar (1982) and Hill *et al.* (2000)

							Activity budget (% total time)				Diet (% feeding time)					
Baboon taxon	Country	Site	Day journey length (km)	Home range (km²)	Group size	Inter-birth interval (months)	Forage	Travel	Rest	Social (grooming)	Leaves	Flowers	Fruit & seeds	Under-ground	Animal	Source
Olive																
	Nigeria	Gashaka (K troop)	2.4	1.4	28.4	29.9	31	30	28	9 (3)	6	5	63	1	8	Warren 2003, Higham 2006, Higham *et al.* 2009
	Nigeria	Gashaka (G troop)	3.1	1.4	19.3	16.5	27	25	35	11 (4)	4	5	54	6	14	Warren 2003, Higham 2006, Higham *et al.* 2009
	Ghana	Shai Hills	1.3	0.7	38.8	30.3	20	18	62	22	8	5	59	17	0	Dunbar 1988, 1992
	Ivory Coast	Comoé		9.8	15.3	20.5	42	26	19	14	5	6	35	>0 (a)	1	Kunz & Linsenmair 2008
	Ethiopia	Mulu	1.1		22.0		41	25	22	15						Dunbar 1992
	Ethiopia	Awash (Metahara)	5.8	2.6	47.8											Dunbar 1992
	Ethiopia	Awash river	5.3		55.8											Nagel 1973
	Uganda	Budongo	3.8		43.0		59	18	6	17						Dunbar 1992, Hill 1999
	Uganda	Ishasha	2.4		45.0											Dunbar 1992
	Uganda	Ishasha	7.5		50.5											Sharman & Dunbar 1982
	Kenya	Gilgil	4.6	20.1	47.0	26.5	51	30	7	8 (5)	29	3	34	27	2	Nicolson 1982, Harding 1976, Eley *et al.* 1989
	Kenya	Laikipia	6.6	43.8	102.0	26.9	40	36	15	8	26	30	24	17	0	Barton 1989, Barton *et al.* 1993
	Tanzania	Gombe	2.4	1.1	23.5	25.0	26	19	30	11	14	2	49	7	13	Dunbar 1992, Barton *et al.* 1993

(continued)

Table 9.1 (continued)

Baboon taxon	Country	Site	Day journey length (km)	Home range (km^2)	Group size	Inter-birth interval (months)	Activity budget (% total time)				Diet (% feeding time)					Source
							Forage	Travel	Rest	Social (grooming)	Leaves	Flowers	Fruit & seeds	Under-ground	Animal	
Yellow																
	Kenya	Amboseli	5.0	34.0	59.2	21.6	46	25	20	8	19	4	31	29	1	Altmann 1980, Post 1978, 1981, Stacey 1986
	Tanzania	Mikumi	3.4	43.0	120.0	21.0	37	26	25	6	13	3	51	22	5	Rhine *et al.* 1989, Dunbar 1992, Barrett *et al.* 2006
	Tanzania	Ruaha	2.0		72.0	26.3	47	24	17	5	19	1	16	52	9	Dunbar 1988, 1992
Guinea																
	Senegal	Badi					26	9	47	16						Dunbar 1992
	Senegal	Mt. Assirik	7.9	29.0	85.0		23	37	21	19	8	9	74	4	3	Dunbar 1992
Chacma																
	South Africa	Cape Point	7.9	37.0	85.0	27.7	34	29	26	11	16	13	45	25	1	Davidge 1978a, 1978b
	South Africa	De Hoop		15.6		21.8	40	31	16	13	13	5	35	42	5	Hill 1999, Barrett *et al.* 2006
	South Africa	Drakensberg	4.6	17.5	11.8	38.5	40	31	16	13	26	14	3	53	4	Whiten *et al.* 1987, 1991, Henzi *et al.* 1992, Lycett *et al.* 1998,
	South Africa	Honnett Reserve	9.0	18.1	47.2											Stolz & Saayman 1970
	South Africa	Mkuzi	4.5		47.3	20.0	36	30	21	12	6	3	90	1	1	Hill 1999, Barrett *et al.* 2006

	Botswana	Moremi, Okovango Delta			75.0	24.3										Cheney *et al.* (2006)
Hamadryas																
	Ethiopia	Erer-Gota	8.9	28.0	85.0	24.0					28	22	45	2	0	Kummer 1968, Sigg *et al.* 1982
	Ethiopia	Awash	6.1		54.0											Dunbar 1992
Mean			4.8	18.9	53.1	25.1	37	26	24	13	15	8	44	20	4	
Maximum			1.1	0.7	11.8	16.5	20	9	6	5	4	1	3	1	0	
Minimum			9.0	43.8	120.0	38.5	59	37	62	22	29	30	90	53	14	

(a) Grass = 29.1 %, including subterranean parts, blades and seeds

Table 9.2 Comparative models with values used in predictive equations

Parameter	Predictive model (a)	K troop		G troop		Source
		Predicted	Observed	Predicted	Observed	
Activity budget (% total time)						
Forage (F)	$\ln(F) = 6.866 + 4.077 \ln(Z) - 0.75 \ln(T) - 0.39 \ln(V) + 0.155 \ln(DJL)$	27.7	31.2	27.6	26.7	Dunbar 1996
Rest (R)	$\ln(R) = 0.97 - 7.923 \ln(Z) + 0.601 \ln(V)$	22.1	27.8	23.8	34.5	Dunbar 1996
Travel (M)	$\ln(M) = 2.201 + 0.1631 \ln(N) + 0.219 \ln(V)$	21.3	29.6	21.1	25.5	Dunbar 1992
Social (S)	$\ln(S) = -2.275 + 1.32 \ln(N) - 0.0445 (\ln(N))^2$	4.0	8.5	3.8	11.0	Dunbar 1996
Diet (% feeding time)						
Fruit (Fr)	$Fr = 30.55 + 2.63\,T - 0.023\,A + 4.23\,PPI$	50.7	62.8	55.3	54.3	Hill & Dunbar 2002
Subterranean items (Sub)	$Sub = 164.45 - 3.06\,T - 97.07\,P_s$	5.3	0.6	6.2	5.7	Hill & Dunbar 2002
Leaves (Le)	$Le = 90.86 - 2.69\,T - 0.83\,L$	10.3	5.6	9.6	3.9	Hill & Dunbar 2002
Flowers	None	-	4.9	-	4.8	Hill & Dunbar 2002
Animal	None	-	8.1	-	13.6	Hill & Dunbar 2002
Day journey length (DJL)	$\ln(DJL) = 1.344 + 0.784 \ln(N) - 0.4731 \ln(P)$	1.2	2.4	1.2	3.1	Dunbar 1996
Inter-birth interval (IBI)	$\ln(IBI) = 20.736 - 2.914 \ln(F) + 0.553 (\ln(F))^2 - 8.953 \ln(T) + 1.452 (\ln(T))^2$	38.3	25.0	47.5	16.0	Hill *et al.* 2000

(a) Parameter:	K troop	G troop
Z = Simpson's index for evenness of rainfall over the year ($Z = 1 - \text{sum}\ \pi$ for all months)	0.86	0.85
V = months / yr with < 50 mm rain	5	5
DJL = day journey length (km)	2.4	3.1
N = troop size (mean in area)	22.3	21.2
T = mean ambient temperature (mean of average monthly temperatures) (°C)	26.4	26.8
P = mean annual rainfall (mm)	1973	1897
Y = number females (given as F in original paper)	5.4	4.5
A = altitude	500	300
L = latitude	11.5	11.5
PPI = plant productivity index, i.e., months / year receiving more rainfall (mm) than twice annual temperature (°C)	7	7
P_s = Shannon's index of rainfall diversity	0.81	0.79
$P > 2t$ = measure of habitat productivity, i.e., number of months / year with precipitation (mm) greater than twice the mean annual temperature (°C)	7	7

Weyher *et al.* (2006) did not join the group). Troops counted in the 2006 census show an average of 22.7 animals per troop (range 10 – 32) with no significant difference between troop sizes in the three areas (Kwano 22.3, $n = 7$; Gashaka 21.2, $n = 5$; Selbe 19.4, $n = 10$; ANOVA $p > 0.05$).

GGNP troop sizes are at the lower end of the range predicted by Dunbar (1996). His model suggests the climate at GGNP would result in troop sizes of a minimum of 17 and a maximum of 71.

Mean GGNP troop size is significantly smaller than those of non-GGNP baboon populations ($t = 9.47$, df = 21, $p < 0.0001$) and smaller than for 14 other study sites used by Hill *et al.* (2000) in his comparative analyses ($t = 6.22$, df = 13, $p < 0.001$).

Activity Budgets and Ranging

The model used to predict overall activity budget from Dunbar (1992, 1996) gives values that are close to those observed at GGNP, although it does underestimate the amount of social and feeding time.

K troop has a day journey length (DJL) of 2.4 km and G troop 3.1 km, and both troops have a home-range size (HR) of 1.4 km^2.

A significant positive relationship is found between DJL and troop size (T) for wild-feeding baboon populations, including K troop (so that: DJL = $0.04\,T + 2.72$, $r = 0.424$, $p < 0.05$, $n = 22$). If the outlying point for Mikumi baboons (troop size = 120, DJL 3.4 km) is omitted, this relationship becomes even stronger (DJL = $0.06\,T + 1.62$, $r = 0.598$, $p < 0.05$, $n = 21$).

DJL for K troop is not unusual for its troop size (*Fig. 9.3*). However, DJL for G troop (not used in the regression analyses) is, perhaps surprisingly for a food-enhanced group, slightly high for its troop size. Both troops are found to have longer travel distances than the 1.2 km predicted by Dunbar (1996). HR is predicted well by troop size in the sample of wild-feeding baboon populations, including K troop (HR = $0.39\,T - 1.80$, $r = 0.835$, $p < 0.01$, $n = 22$). The HR of both K and G troop are both relatively low for their troop sizes (*Fig. 9.4*).

Diet

Feeding time for both GGNP study troops on flowers and for fruits / seeds falls within the upper range of wild-feeding baboons at other sites. However, compared with most other populations, K troop spent relatively little time foraging on underground items and leaves.

Hill & Dunbar (2002) showed that percentage time feeding was negatively related to percentage fruit in the diet and positively related to percentage subterranean items. A comparison with other baboon populations that have similar feeding time (Hill & Dunbar 2002) suggests that K troop feeds on approximately the expected amount of fruit / seeds but a lower amount of subterranean items than expected.

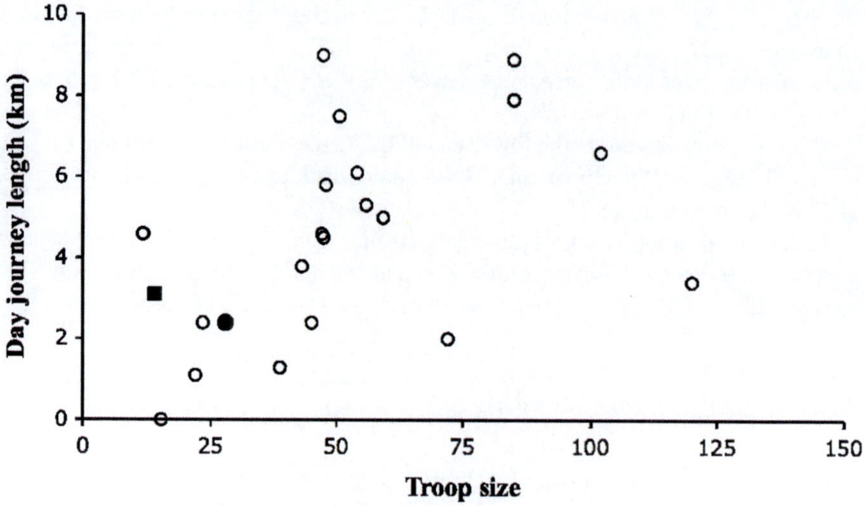

Fig. 9.3 Day journey length versus troop size for 25 wild-feeding baboon populations. Data from *Tab. 9.1.* K troop = closed circle, G troop = closed square

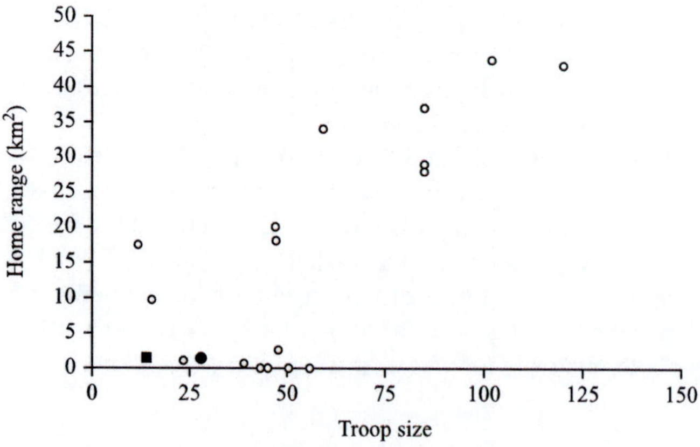

Fig. 9.4 Home-range size versus troop size for 25 wild-feeding baboon populations. Data from *Tab. 9.1.* K troop = closed circle, G troop = closed square

Similarly, the model of Hill & Dunbar (2002) does not accurately predict the diet of the wild-feeding K troop from climate and other environmental variables, as they ate a higher than expected proportion of fruit and a lower than expected proportion of leaves and underground food items. When compared with other wild-feeding populations, G troop spent relatively more time feeding on animals (mostly invertebrates) and little time eating leaves. Hill & Dunbar's (2002) predicted diet for G troop is close to that observed for fruit and underground items, but not leaves, although G troop's consumption of underground items would be far lower if consumption of crops (sweet potato and cassava) were not included in these data. However, all these findings need to be viewed with caution as G troop could not always be followed onto farms. Thus our estimation of the amount of crops eaten is likely to be underestimated and this will also influence estimation of proportions of other foods eaten.

Overall, the two focal troops consumed a total of 211 identified food items from 165 species, including different items from some species, e.g., seeds, bean pulp and flower bracts from the locust bean tree (*Tab. 9.3*). K troop consumed a total of 119 identified food items and had 18 core foods, whereas G troop had a more diverse diet including 149 food items and 26 core foods. Some foods that did not contribute greatly to the annual diet were nevertheless eaten frequently in some months, giving

Table 9.3 Core foods (≥ 1 % of total forage time) eaten by Gashaka study troops K and G. If more than one part of the same species is eaten, these are listed separately. Items given in descending order, according to time spent foraging by K troop. Taxon: Not all food items could be identified to the species level. In particular, "Gramineae" includes all grass species, except for the two that could be identified (*P. maximum*, *R. exaltata*). "Insects" includes all insects except Lepidoptera and Scarabaeidae. Parts: A = animal, F = fruit, I = insect, L = leaf, LC = leaf with caterpillars, N = nectar, S = seed, SD = seed from foraging in cattle dung, O = cocoon, R = root, T = stem. Consumption: np = food item not present in troop's range; p = food item present but < 1 % forage time or not observed eaten

Taxon	Part	% consumption (K troop)	% consumption (G troop)	Rank (K troop)	Rank (G troop)
Elaeis guineensis (oil palm)	F	19.4	2.8	1	8
Erythrophleum suaveolens (poison wood tree)	S	15	0.6	2	p
Uapaca togoensis	F	5.5	4.7	3	4
Cynometra megalophylla	LC	3.1	0	4	p
Aquatic invertebrates	A	2.7	0.2	5	p
Parkia biglobosa (locust bean tree)	S	2.5	2.6	6	10
Piliostigma thonningii	S	2.3	5.5	7	3
Daniellia oliveri	N	2.1	3.1	8	7
Cynometra megalophylla	L	1.7	0	9	p
Elaeis guineensis (oil palm)	S	1.7	2	10	13
Insects (unidentified)	A	1.3	0	11	p
Ficus ovata (fig)	F	1.3	np	12	–
Gramineae (unidentified grass)	T	1.3	1.1	13	22
Landolphia macrantha	F	1.2	2.3	14	11
Gramineae (unidentified grass)	L	1.1	1.3	15	18
Ficus polita (fig)	F	1	np	16	–
Zea mays (maize)	S	np	5.8	–	1
Piliostigma thonningii	SD	np	3.5	–	5
Scarabaeidae larvae	A	np	1.9	–	14
Borassus aethiopum (palm)	F	np	1.8	–	16
Manihot esculenta (cassava)	R	np	1.7	–	17
Ipomoea batatas (sweet potato)	R	np	1.2	–	19
Mucuna poggei	R	np	1.1	–	21

(continued)

Table 9.3 (continued)

Taxon	Part	% consumption (K troop)	% consumption (G troop)	Rank (K troop)	Rank (G troop)
Panicum maximum (elephant grass)	T	np	1	–	24
Lepidoptera cocoon	A	0	5.5	p	2
Landolphia spp.	F	0.2	3.3	p	6
Mangifera indica (mango)	F	0.7	2.7	p	9
Rottboelia exaltata (grass)	S	0	2.3	p	12
Lepidoptera caterpillar	A	0.8	1.9	p	15
Gramineae (unidentified grass)	S	0.1	1.1	p	20
Prosopis africana	S	0.2	1.1	p	23
Total % diet of foods ≥ 1 % forage time (N)		65.2 (16)	62.1 (24)		
Total % diet of foods < 1 % forage time (N)		12.5 (81)	37.9 (97)		
Unidentified items % diet		34.8	24.3		

much higher numbers of monthly staple foods. K troop had a total 45 monthly staple foods (5 – 15 per month) and G troop a total 61 (6 – 16 per month).

K troop included oil palm fruits as a staple food in every month, making this the most frequently consumed food item, with poison wood tree seeds and aquatic invertebrates also being important, both being monthly staples for 7 months. The aquatic invertebrates found in rivers (usually by turning over stones) made up more than 2 % of foraging time for K troop.

For G troop, maize seed was the most frequently consumed item and oil palm fruits were monthly staples in 6 months. Although some other wild foods were eaten by G troop in 8 or more months of the study (invertebrates, *Ficus sur* fruits, and grass stems and leaves), none were monthly staples. However, maize was eaten by G troop in all 7 months when it was available (Jun – Aug 01 and Dec 01 – Mar 02), and was a monthly staple in 4 months (Aug 01, Jan – Mar 02). Cassava was a monthly staple in Dec 01, Jan 02, and Feb 02, and sweet potato in Dec 01, Jan 02, and Mar 02.

There was limited support for a significant relationship between food consumed and food availability, whereas no significant relationship was detected between the availability of fruits or beans / seeds, and the time spent foraging on these items for either troop. Only K troop had a significant positive correlation between leaf eating and leaf availability after application of the Bonferroni correction ($n = 10$, $r_s = 0.758$, $p < 0.001$) and there was a significant positive correlation between flower eating and flower availability in both troops (G troop $n = 10$, $r_s = 0.708$, $p < 0.001$; K troop, $n = 10$, $r_s = 0.680$, $p < 0.001$).

Reproductive Parameters

Estimates for gestation length, cycle length and length of post-partum amenorrhea for the wild-feeding K troop are within, or very close to, the range found for other baboon populations (Tab. 9.4). However, the inter-birth interval (IBI) in this population is

Table 9.4 Life-history parameters: A comparison between Gashaka study troops and other field sites. All values are averages, given in months (after Higham *et al.* 2009)

Parameter	Definition	Gashaka K troop	Gashaka G troop	Amboseli	Mchelelo	Gilgil	Erer
Cycle length	Time between successive ovulation dates as measured via faecal hormones	1.39	1.39	1.1	1.3	1.4	
Gestation length (G)	Time between probable conceptive ovulation to birth date	6.1	6.3 (a)	5.8	6.0	5.9	
Post-partum amenorrhea (PPA)	Birth until first day of sexual swelling	11.3	9.3	12.0	14.6	13.4	14.0
Duration of cycling (DC)	Duration of cycling from end of PPA to conception	12.4	1.5	4.0	4.9	5.6	
Inter-birth interval	Sum of G, PPA and DC	29.9	16.5	22.0	24.5	24.9	
Observed inter-birth interval (IBI)	IBI between consecutive births when first infant survived until weaning	25	16	21.0	27.9	25.0	24.0
Source		*Higham et al.* 2009	Higham *et al.* 2009	Altmann *et al.* 1977, Altmann 1980	Bentley-Condit & Smith 1997	Smuts & Nicolson 1989	Sigg *et al.* 1982

(a) Based on one gestation estimate of 188–192 days

markedly longer than found in most other baboon populations, and it is significantly longer than the sample of 11 sites given in Barrett *et al.* (2006; $t = -3.0$, df = 10, $p = 0.013$). This is because the K troop females spend longer cycling without successfully conceiving than do other populations for which we have data (Higham *et al.* 2009).

The regression model of Hill *et al.* (2000) predicts an IBI that is considerably different from the values observed for either troop, with an expected value of 38.3 months for K troop as compared with the observed value of 29.9 months. Even more inaccurately predicted is an expected value of 47.5 months rather than the observed 16 months for G troop.

The relationship between IBI and habitat productivity suggested by Barrett *et al.* (2006) is likewise not supported by our data. When Kwano is added to their dataset, there is no significant relationship between the variables ($r^2 = 0.025$, $F_{1,9} = 0.230$, $p = 0.643$). This remains insignificant even when the outlying Drakensberg population is removed from the analysis ($r^2 = 0.239$, $F_{1,8} = 2.511$, $p = 0.152$).

Discussion

Here we focus on a comparison of the socioecology of GGNP baboons with populations throughout Africa. Differences between GGNP study troops are discussed in Warren *et al.* (this volume [Ch. 8]).

Troop Size

The mean troop size of GGNP baboons is lower than the values for all but one of the 25 populations compiled in Sharman & Dunbar (1982) and Hill & Dunbar (2002). These reviews record only 3 areas where there are fewer than 30 animals / troop: olive baboons in Bole / Ethiopia and Gombe / Tanzania and chacma baboons in Drakensberg / South African Republic. More recently, one other West African population of olive baboons in Comoé / Ivory Coast was found to have a small mean troop size (15.3 individuals; Kunz & Linsenmair 2008).

There are a number of reasons why we find small troops at GGNP. The models developed by Dunbar (1992, 1996) predict that populations experiencing a high rainfall and high temperature are unlikely to be able to form very large troops. However, troop sizes at GGNP are far below the maximum tolerable mean of 71 these models would predict for the GGNP climate. Kunz & Linsenmair (2008) relate the similarly small troop sizes at Comoé to a combination of low baboon population density and low predation pressure.

Accurate measures of population density for GGNP do not exist, but we can estimate this in the Kwano area from home-range size (2.4 km), mean troop size (22.3), and a home-range overlap of 36 % (Warren 2003). This method is likely to overestimate population density, by looking only at a limited area where animals are known to occur. Nevertheless, our census and foot surveys throughout the southern sector of the park (Adanu *et al.* this volume [Ch. 3]) confirm that baboons are indeed ubiquitous. The figures for K troop thus give a (rough) density estimate of 21.6 individuals / km. This value is similar to Gombe / Tanzania, a forested East African site (Kunz & Linsenmair 2008) but far higher than the 1.2 per km^2 seen at Comoé / Ivory Coast. This finding suggests that low population density is not the cause of small troop sizes at GGNP.

Another reason for the small GGNP troop sizes of baboons might be a relative paucity of large predators. Leopards do occur at least in the K troop home-range (Adanu *et al.* this volume [Ch. 3]). Nevertheless, we have no evidence that mammalian carnivores prey on baboons or other monkeys. Currently, there is also no evidence that sympatrically occurring chimpanzees hunt baboons in this area.

A further possibility is that small troop sizes are caused by forest-living and fruit distribution. For example, some resources such as oil palm fruit are highly clumped in small dispersed patches, which would make it difficult for large groups of animals to exploit them. A forested habitat may also impose high costs of maintaining a large social group, as visibility is low.

High rainfall may also reduce the time available for social activity, as animals are usually inactive during periods of heavy rain and thus are constrained to spend much of the remaining time foraging. This limitation on social time will mean that the number of social relationships that can be maintained is also limited, leading to smaller groups.

Activity Budgets

The activity budgets of K troop are generally not unusual when compared with other baboons. This suggests that the forest habitat of GGNP does not require any major adjustments in daily activity.

Some differences exist between the observed time budgets and those predicted by Dunbar (1992, 1996), but these are generally quite small. The most consistent discrepancy is seen in social activity, which is higher than predicted for small troops. Nevertheless, only social time shows a difference above expected activity time approaching 1 standard deviation (> 0.9 SD for K troop) or more (> 1.4 SD for G troop). Perhaps, a relatively low predation risk allows the animals to spend less time on vigilance behaviour when resting, and hence more time on servicing social relationships.

Ranging

The small troop sizes of GGNP baboons lead us to expect the short DJL and a small HR that we indeed see in this population. Relative to troop size, the DJL of wild-feeding K troop is about as expected, but HR sizes are smaller than predicted.

Small HR is not a feature of all West African olive baboon populations, as those in the Ivory Coast have a home-range of 9.75 km^2, despite their smaller troops (Kunz & Linsenmair 2008). However, these Ivory Coast animals live in a savannah habitat with just over half the rainfall of GGNP (1053 mm vs. 1973 mm at Kwano). Other forest baboon populations also have small home-ranges for their troop size, supporting the idea that this is caused by densely distributed, high quality food (Dunbar 1988). Open country habitats have comparably lower food density and a scarcity of sweet fleshy fruits; the fruits that occur are often also small and may be high in fibre (Clutton-Brock 1977, Post 1981, Barton 1989, Johnson 1990, Ganzhorn 1999, Hunt 2002). Accordingly, the area required to support baboons increases on a continuum from forest to riverine forest to woodland to savannah / desert. Other studies also find larger range sizes in habitats of lower quality for baboons (Dunbar 1988, Byrne 1990, Henzi *et al.* 1992) and for a variety of other primates (e.g., howler monkeys, Estrada 1984; Barbary macaques, Mehlman 1989; rhesus macaques, Jiang *et al.* 1991; mountain gorillas, McNeilage 2001).

Plentiful drinking water and sleeping sites may also allow a small HR. The high rainfall and forested habitat in GGNP mean that water is never more than 2 km away for either troop, and is usually much nearer. Similarly, sleeping trees are readily available throughout both troops' home-ranges.

Diet

It is for good reason that baboons have been described as eclectic omnivores (DeVore & Hall 1965, Altmann 1998). Many sites report long lists of foods. At Mikumi / Tanzania, up to 6 parts of more than 183 species were consumed by baboons over a 5-yr period (Norton *et al.* 1987). In Laikipia / Kenya, 108 food items from 76 plant species were eaten (Barton 1989). Yearling baboons at Amboseli / Kenya ate a total of 277 plant foods (Altmann 1998). The GGNP troops, which consume 211 identified food items from 165 different species, fall within these reported ranges.

Although baboon diets are diverse, baboons are selective feeders with a relatively small number of food items usually making up a high proportion of the total diet. Although the total number of items eaten is not high in comparison to other baboon populations, GGNP baboons do have a more varied diet than most, with 19 identified foods for K troop and 23 for G troop making up 60 % of the foraging time. This is very high in comparison to other sites. In Amboseli / Kenya, only 5 species constituted more than 60 % of feeding time (Post 1982) and 10 foods accounted for 67 – 76 % (Stacey 1986). In the austral winter at De Hoop / South Africa, 9 food items comprised 92 – 97 % of daily feeding time of chacma baboons (Hill 1999). In Laikipia / Kenya, the 13 most common foods accounted for 68% of feeding time. Finally, at the Drakensberg / South Africa, the 8 most common foods accounted for 68% of feeding time (Whiten *et al.* 1991).

The length of studies will influence the total recorded number of foods (Barton 1989). Nevertheless, the number of items making up the majority of foraging time of GGNP baboons is high compared to studies of similar or greater length. The high floral biodiversity within this Nigerian forest area (Mutke *et al.* 2001) provides a greater range of different foods than found in other baboon habitats, thus leading to a wider choice of core foods.

Although some of Hill & Dunbar's (2002) predictive models are not supported by our results, we do find that the relatively low amount of time spent feeding is correlated with a high proportion of fruit in the diet. This supports the idea that, when possible, baboons will select foods dense in protein and / or energy, like seeds and fruit, which will result in decreased foraging times (Barton & Whiten 1994, Altmann 1998, Hill & Dunbar 2002).

The low amounts of subterranean food and leaves, and high amounts of fruit and seeds, in the GGNP baboon diet may be explained by the unusually high biodiversity and productivity of their habitat. A negative relationship between rainfall and the proportion of underground items in the diet across baboon populations is known from previous studies (Barton 1989, Hill & Dunbar 2002). Nevertheless, the amount of underground items and leaves eaten by GGNP baboons is far lower than expected. This is probably because fruit, seeds or beans are available in all months of the year (Warren 2003, Sommer *et al.* this volume [Ch. 12]), and animals are not constrained to switch to alternative foods in some months.

This supports previous suggestions that, only where food availability is lower in drier habitats, baboons are likely to utilise underground items as "fall-back" foods in periods of food shortages (Post 1978, Barton 1989, Whiten *et al.* 1987). At GGNP such foods are rarely eaten, as foraging on subterranean foods is more costly than feeding on more easily obtained alternatives – except when, as for G troop, they are easily obtainable subterranean crop items.

Demography and Life-history

The gestation length in K troop is not significantly different from other populations (cf. *Tab. 9.2*). This supports previous work suggesting that gestation length is a relatively conservative life-history variable that does not fluctuate significantly in response to environmental conditions Martin & MacLarnon 1985).

Still, K troop has a slightly longer gestation period than estimated at other sites. However, this value is an estimate based on hormonal data whereas previous studies followed Hausfater (1975) in using the day of swelling detumescence (D-day) to estimate the date of ovulation and hence gestation length (Altmann *et al.* 1977, Bentley-Conduit & Smith 1997, Smuts & Nicolson 1899). Higham *et al.* (2008b) found that at GGNP ovulation never occurred on D-day but always before this time. The use of D-day to estimate the day of ovulation may thus lead to spurious inferences about sexual behaviour around this time and will slightly underestimate gestation length and overestimate its variability.

K troop baboons also have a cycle length that is not significantly different from other populations. However, their life histories are clearly distinguished by their inter-birth intervals. The IBI is longer than for any other population of olive baboons reported and even longer than for any *Papio* population, bar chacma baboons in Drakensberg / South Africa. This extended IBI is primarily due to females having several non-conceptive cycles before finally conceiving. Other constituents of IBI (gestation length and post-partum amenorrhea) are not extended.

We have detected no evidence for anoestrus swelling periods in our studies (Higham 2006). This suggests that females are either failing to conceive despite ovulation, and/or that they are experiencing early miscarriage following conception. Although we do not presently know the reason why wild-feeding GGNP baboons fail to conceive quickly once they resume cycling, it is clear that this has a cost to their future reproductive output.

The model of Hill *et al.* (2000) using temperature and number of females in the troop fails to predict K troop IBI. An alternative insight may come from another unusual population. A short study of baboon in Shai Hills / Ghana reports a long IBI of 30 months (Dunbar 1992). Apart from this, Drakensberg is the only long-term baboon study that, with an IBI of 38.5 months, exceeds K troop (Lycett *et al.* 1998, Barrett *et al.* 2006). Drakensberg baboons live in a harsh seasonal habitat, which prevents infants feeding independently at an early age, thus resulting in slow infant growth and prolonged suckling (Lycett *et al.* 1998). Unusually for baboons, seasonality at this site also acts as a constraint, meaning that Drakensberg mothers cannot start to wean their infants at 6 – 7 months of age as seen in other populations. Lycett *et al.* (1998) argue that a prolonged suckling period is possible because negligible predation pressure reduces the need for vigilance and hence allows mothers both to forage themselves and to feed their infants without interruption. They suggest that predation in other baboon populations leads to a constrained time budget that forces mothers to "decide" whether to allocate time to their own foraging or to suckling their infants. The resulting high levels of prolonged maternal investment in Drakensberg are linked with high infant survival rates and a lack of overt mother-infant conflict over access to the nipple (Lycett *et al.* 1998, Barrett *et al.* 2006).

Barrett *et al.* (2006) further propose that less seasonal but more productive habitats may "tempt" infants into early independence, with the risk that, if conditions change unpredictability, infants may not survive. In such habitats the risky transition to independence leads to both high infant mortality and longer IBIs, as mothers seek to mitigate the dangers of weaning by increased investment.

Barrett *et al.* (2006) link IBI to habitat productivity index ($P > 2t$). By this measure, K troop has a moderately seasonal environment but not the IBI of about 2 years that would be predicted. Thus, seasonality and climate alone cannot explain the similarity of life histories in GGNP and Drakensberg baboons. However, they may both experience low predation pressure. Although mortality levels at GGNP are much higher than those recorded at Drakensberg (Higham *et al.* 2009), these are probably not significantly related to predation.

Possibly, heavy rainfall and consequent disease risk for K troop lead to a high risk of infant mortality and females combat this by investing heavily in their infants, thus leading to longer IBIs. Although disease risk is often thought of as an extrinsic mortality cause (Barrett *et al.* 2006), good health may mitigate this risk.

The very short birth intervals of G troop do suggest that IBI may be reduced in certain circumstances (see Warren *et al.* this volume [Ch. 8]). Probably, crop-raiding leads to better nutrition and good body condition in G troop females, thus decreasing disease risk. Mothers can then flexibly reduce maternal investment without compromising infant survival. Testing this hypothesis requires data on suckling rates, weaning ages, weaning weights, and survival rates of K and G troop infants, which are currently not available.

Conclusions

Baboons in Gashaka Gumti National Park are living in an extreme environment when compared with most other previously studied populations. High rainfall, low predation pressure and availability of fruit all year round lead to their having small troop sizes, a diet that is low in subterranean foods and leaves, and relatively small home-range sizes. Although previously published models do predict the activity budgets of these individuals reasonably well, observed values for inter-birth interval and some dietary components are quite different from those expected. These models are based mainly on data from open-country eastern and southern African populations, and it seems that they cannot necessarily be extrapolated to predict socioecology across the whole geographical range of baboons.

The high productivity of Nigerian forests, combined with pronounced seasonality in rainfall, may lead to selection pressures very different from those at other baboon habitats. Low predation levels may also influence reproductive strategies, as low extrinsic mortality risk will select for females who invest heavily in their offspring. Wild-feeding Nigerian baboons have long inter-birth intervals, indicating a need for high maternal investment that prevents a quick return to pregnancy when postpartum amenorrhea ends.

A comparison between the GGNP study troops suggest that differences found between these West African forest animals and their counterparts elsewhere in Africa are primarily due to an ability to respond flexibly to environmental challenges. Further research into this population will lead to a more complete understanding of how forest-living affects the evolution of flexibility in adaptable primates such as baboons and humans.

Acknowledgements The Nigerian National Park Service granted a permit to the *Gashaka Primate Project*, which receives its core funding from the Chester Zoo Nigeria Biodiversity Programme. NCF / WWF-UK provided logistical support. Bobbo Buba, Halidu Ilyusu, Buba Bello, Nuhu Husseini, Haruna Ahmadu, Elodie Ey, Helen Cross and Alejandra Pascual-Garrido helped collect data in the field. At Roehampton University, we were supported by Stuart Semple, Mary MacKenzie, Balbir Singh, and Martin Evans. Funding was provided by Roehampton University, as well as The Leakey Trust, Primate Society of Great Britain, NUMICO Research B.V., British Federation of Women Graduates and the International Federation of Animal Welfare.

Literature

Alberts, S. & Altmann, J. (2006). The evolutionary past and the research future: environmental variation and life history flexibility in a primate lineage. In Swedell, L. & Leigh, S. R. (eds.), *Reproduction and Fitness in Baboons. Behavioral, Ecological, and Life History Perspectives*. New York: Springer, pp. 277–303.

Aldrich-Blake, F. P. G., Bunn, T. K., Dunbar, R. I. M., & Headley, P.M. (1971). Observations on baboons, *Papio anubis*, in an arid region in Ethiopia. *Folia Primatologica* 15: 1–35.

Altmann, J. (1974). Observational study of behaviour: sampling methods. *Behaviour* 49: 227–267.

Altmann, J. (1980). *Baboon Mothers and Infants*. London: Harvard University Press.

Altmann, J., Hausfater, G. & Altmann S. A. (1988). Determinants of reproductive status in savannah baboons. In Clutton-Brock, T. H. (ed.), *Reproductive Success*. Chicago: Chicago University Press, pp. 403–418.

Altmann, J., Altmann, S. A., Hausfater, G. & McCuskey, S. A. (1977). Life history of yellow baboons: Physical development, reproductive parameters and infant mortality. *Primates* 18: 315–330.

Altmann, S. A. (1998). *Foraging for Survival*. Chicago: The University of Chicago Press.

Barrett, L., Henzi, P. & Lycett J. E. (2006). Whose life is it anyway? Maternal investment, developmental trajectories, and life history strategies in baboons. In Swedell, L. & Leigh, S. R. (eds.), *Reproduction and Fitness in Baboons. Behavioral, Ecological, and Life History Perspectives*. New York: Springer, pp. 199–244.

Barton, R. A. (1989). *Foraging Strategies, Diet and Competition in Olive Baboons (Papio anubis)*. PhD thesis, University of St. Andrews.

Barton, R. A., Whiten, A., Byrne, R. W., & English, M. E (1993). Chemical composition of baboon plant foods: implications for the interpretation of intra- and inter-specific structure in baboons. *Behavioral Ecology & Sociobiology* 38: 321–329.

Barton, R, A., Whiten, A., Strum, S. C., Byrne, R.W., & Simpson, A. J. (1992). Habitat use and resource availability in baboons. *Animal Behaviour* 43: 831–844.

Bentley-Condit, V. & Smith, E. (1997). Female reproductive parameters of Tana River yellow baboons. *International Journal of Primatology* 18: 581–595.

Bronikowski, A. M. & Altmann, J. (1996). Foraging in a variable environment: Weather patterns and the behavioural ecology of baboons. *Behavioural Ecology and Sociobiology* 39: 11–25.

Byrne, R. W. (1990). Baboon societies as responses to ecological constraints. Introduction. In de Mello, M. T., Whiten, A. & Byrne, R. W. (eds.), *Baboons: Behaviour and Ecology, Use and Care*. (Selected Proceedings, XIIth Congress, International Primatological Society). Brasilia: University of Brasilia Press, pp. 41–42.

Byrne, R. W., Whiten, A., & Henzi, S. P. (1990). Measuring the food constraints on mountain baboons. In de Mello, M. T., Whiten, A. & Byrne, R. W. (eds.), *Baboons: Behaviour and Ecology, Use and Care*. (Selected Proceedings, XIIth Congress, International Primatological Society). Brasilia: University of Brasilia Press, pp. 105–122.

Cheney, D. L., Seyfarth, R. M., Fischer, J., Beehner, J. C., Bergman T. J., Johnson, S. E., Kitchen, D. M., Palombit, R. A., Rendall, D. & Silk, J. B. (2006). Reproduction, mortality and female reproduction in chacma baboons of the Okavango Delta, Botswana. In Swedell, L. & Leigh, S. R. (eds.), *Reproduction and Fitness in Baboons. Behavioral, Ecological, and Life History Perspectives*. New York: Springer, pp. 147–176.

Clutton-Brock, T. H. (1977). Methodology and measurement. In Clutton-Brock, T. H. (ed.), *Primate Ecology: Studies of Feeding and Ranging Behaviour in Lemurs, Monkeys and Apes*. New York: Academic Press, pp. 585–590.

Davidge, C. (1978a). Activity patterns of chacma baboons (*Papio ursinus*) at Cape Point. *Zoologica Africana* 13: 143–155.

Davidge, C. (1978b). Ecology of baboons (*Papio ursinus*) at Cape Point. *Zoologica Africana* 13: 143–155.

DeVore, I, & Hall, K. R. I. (1965). Baboon Ecology. In DeVore, I. (ed.), *Primate Behavior, Field Studies of Monkeys and Apes*. New York: Holt Rinehart & Winston, pp. 20–65.

Dunbar, R. I. M. (1983). Theropithecines and hominids: Contrasting solutions to the same ecological problem. *Journal of Human Evolution* 12: 647–658.

Dunbar, R. I. M. (1988). *Primate Social Systems: Studies in Behavioural Adaptation*. London: Croom Helm.

Dunbar, R. I. M. (1992). Time a hidden constraint on the behavioural ecology of baboons. *Behavioral Ecology and Sociobiology* 31: 35–49.

Dunbar, R. I. M. (1996). Determinants of group size in primates: a general model. In Runciman, W. G., Maynard Smith, J. & Dunbar, R. I. M. (eds.), *Evolution of Social Behaviour Patterns in Primates and Man*. Oxford: Oxford University Press, pp. 33–57.

Dunbar, R. I. M. & Sharman, M. (1983). Female competition for access to males affects birth rate in baboons. *Behavioral Ecology and Sociobiology*. 13: 157–159.

Eley, R. M., Strum, S. C., Muchemi, G. & Reid, G. D. F. (1989). Nutrition, body condition, activity patterns and parasitism of free-ranging olive baboons (*Papio anubis*) in Kenya. *American Journal of Primatology* 18: 209–220.

Elton, S. (2006). Forty years on and still going strong: the use of hominin-cercopithecid comparisons in palaeoanthropology. *Journal of the Royal Anthropological Institute* 12: 19–38.

Estrada, A. (1984). Resource use by howler monkeys (*Alouatta palliata*) in the rain forest of Los Tuxtlas, Veracruz, Mexico. *International Journal of Primatology* 5: 105–31.

Ganzhorn, J. U. (1999). Body mass, competition and the structure of primate communities. In Fleagle, J. G., Janson C. & Reed, K. E. (eds.), *Primate Communities*. Cambridge: Cambridge University Press, pp. 141–157.

Hall, K. R. L. (1965). Experimentation and quantification in the study of baboon behavior in its natural habitat. In Vagtborg, H. (ed.), *The Baboon in Medical Research*. Austin: University of Texas Press, pp. 29–42.

Harding, R. S. O. (1976). Ranging patterns of a troop of baboons (*Papio anubis*) in Kenya. *Folia Primatologica* 25: 143–185.

Hausfater, G. (1975). Dominance and reproduction in baboons (*Papio cynocephalus*). *Contributions to Primatology 7*. Basel:Karger.

Heistermann, M., Finke, M. & Hodges J. K. (1995). Assessment of female reproductive status in the Hanuman langur (*Presbytis entellus*) by measurement of urinary and fecal steroid excretion patterns. *American Journal of Primatology* 37: 275–284.

Henzi, S. P., Byrne, R. W. & Whiten, A (1992). Patterns of movement by baboons in the Drakensberg Mountains: primary responses to the environment. *International Journal of Primatology* 13: 601–628.

Higham, J. P. (2006). *The Reproductive Ecology of Female Olive Baboons (Papio hamadryas anubis) at Gashaka-Gumti National Park, Nigeria*. PhD thesis, Roehampton University.

Higham, J. P., Ross, C., Warren, Y., Heistermann, M. & MacLarnon, A. M. (2007). Reduced reproductive function in wild olive baboons (*Papio hamadryas anubis*) related to natural consumption of the African black plum (*Vitex doniana*). *Hormones and Behavior* 52: 384–390.

Higham, J. P., MacLarnon, A. M., Ross, C., Heistermann, M. & Semple, S. (2008a). Baboon sexual swellings: information content of size and color. *Hormones & Behavior* 53: 452–456.

Higham, J. P., Semple, S., MacLarnon, A. M., Heistermann, M. & Ross, C. (2008b). The timing of ovulation with respect to sexual swelling detumescence in wild olive baboons. *Primates* 49: 295–299.

Higham, J. P., Warren, Y., Adanu, J., Umaru, B., MacLarnon, A., Sommer, V. & Ross C. (2009). Life on the edge: Life-history of olive baboons at Gashaka-Gumti National Park, Nigeria. *American Journal of Primatology* 71: 293–304.

Hill, R. A. (1999). *Ecological and Demographic Determinants of Time Budgets in Baboons: Implications for Cross-Populational Models of Baboon Socioecology*. PhD thesis, University of Liverpool.

Hill, R. A. & Dunbar, R. I. M. (2002). Climatic determinants of diet and foraging behaviour in baboons. *Evolutionary Anthropology* 16: 579–593.

Hill, R. A., Lycett, J. E. & Dunbar, R. I. M. (2000). Ecological and social determinants of birth intervals in baboons. *Behavioral Ecology* 11: 560–564.

Hodges, J. K. & Heistermann, M. (2003). Field endocrinology: Monitoring hormonal changes in free-ranging primates. In Setchell, J. M. & Curtis, D. J. (eds.), *Field and Laboratory Methods in Primatology. A Practical Guide*. Cambridge: Cambridge University Press, pp. 282–294.

Hunt, K. D. & McGrew, W. C. (2002). Chimpanzees in the dry areas of Assirik, Senegal and Semliki Wildlife Reserve, Uganda. In Boesch, C., Hohmann, G. & Marchant, L. F. (eds.), *Behavioural Diversity In Chimpanzees and Bonobos*. Cambridge: Cambridge University Press, pp. 35–51.

Jiang, H., Lui, Z., Zhang, Z. & Southwick, C. (1991). Population ecology of rhesus monkeys (*Macaca mulatta*) at Nanwan Nature Reserve, Hainan, China. *American Journal of Primatology* 25: 207–217.

Johnson, R. B. (1990). The effects of plant chemistry on food selection in adult male yellow baboons (*Papio cynocephalus*). In de Mello, M. T., Whiten, A. & Byrne, R.W. (eds.), *Selected Proceedings of the XIIth Congress of the International Primatological Society*. Brasilia: University of Brasilia Press, pp. 23–29.

Jolly, C. J. (2001). A proper study for mankind: Analogies from the papionin monkeys and their implications for human evolution. *Yearbook of Physical Anthropology* 44: 177–204.

Kingdon, J. (1997). *The Kingdon Field Guide to African Mammals*. San Diego: Academic Press.

Kummer, H. (1968). *Social Organization of Hamadryas Baboons. A Field Study*. Basel: Karger.

Kunz, B. K., Linsenmair, K. E. (2008). The disregarded west: diet and behavioural ecology of olive baboons in the Ivory Coast. *Folia Primatologica* 79: 31–51.

Lycett, J. E., Henzi, P. & Barrett, L. (1998). Maternal investment in mountain baboons and the hypothesis of reduced care. *Behavioral Ecology & Sociobiology* 42: 29–56.

Martin, R. D. & MacLarnon, A. M. (1985). Gestation period, neonatal size and maternal investment in placental mammals. *Nature* 313: 220–223.

McNeilage, A. (2001). Diet and habitat use of two mountain gorilla groups in contrasting habitats in the Virungas. In Robbins, M. M., Sicotte, P. & Stewart, K. J. (eds.), *Mountain Gorillas: Three Decades of Research at Karisoke*. Cambridge: Cambridge University Press, pp. 266–292.

Mehlman, P. J. (1989). Comparative density, demography and ranging behaviour of Barbary macaques (*Macaca sylvanus*) in marginal and prime conifer habitats. *International Journal of Primatology* 10: 269–292.

Mutke, J., Kier, G., Braun, G., Schultz C. & Barthlot, W. (2001). Patterns of African Vascular Plant Diversity: A GIS Based Analysis. *Systematics and Geography of Plants*. 71: 1125–1136.

Nagel, U. (1973). A comparison of anubis baboons, hamadryas baboons and their hybrids at a species border in Ethiopia. *Folia Primatologica* 19: 104–165.

Norton, G.W., Rhine, R. J., Wynn, G. W. & Wynn, R. D. (1987). Baboon diet: a five-year study of stability and variability in the plant feeding and habitat of the yellow baboons (*Papio cynocephalus*) of Mikumi National Park, Tanzania. *Folia Primatologica* 48: 78–120.

Patzl, M., Schwarzenberger, F., Osmann, C., Bamberg, E. & Bartmann W. (1998). Monitoring ovarian cycle and pregnancy in the giant anteater (*Myrmecophaga tridactyla*) by fecal progestagen and oestrogen analysis. *Animal Reproduction Science* 53: 209–219.

Post, D. G. (1978). *Feeding and Ranging Behaviour of the Yellow Baboon (Papio cynocephalus)*. PhD thesis, Yale University.

Post, D. G. (1981). Activity patterns of yellow baboons (*Papio cynocephalus*) in the Amboseli National Park, Kenya. *Animal Behaviour* 29: 357–374.

Rhine, R. J., Norton, G. W., & Wynn, G. W. (1989). Plant feeding of yellow baboons (*Papio cynocephalus*) in Mikumi National Park, Tanzania, and the relationship between seasonal feeding and immature survival. *International Journal of Primatology* 10: 319–42.

Sharman, M. J. & Dunbar, R. I. M. (1982). Observer bias in selection of study group in baboon field studies. *Primates* 23: 567–573.

Sigg, H., Stolba, A., Abegglen, J- J., Dasser, V. (1982). Life history of hamadryas baboons: Physical development, infant mortality, reproductive parameters and family relationships. *Primates* 23: 473–487.

Smuts, B. B. (1983). Special relationships between adult male and adult female olive baboons: Selective advantages. In Hinde, R. A. (ed.), *Primate Social Relationships: An Integrated Approach.* Sunderland, Massachusetts: Sinauer Associates, pp. 262–266.

Smuts, B. B. & Nicolson, N. (1989). Reproduction in wild female olive baboons. *American Journal of Primatology* 19: 229–246.

Stacey, P. (1986). Group size and feeding efficiency of yellow baboons. *Behavioral Ecology & Sociobiology* 18: 175–187.

Warren, Y. (2003). *Olive Baboons (Papio cynocephalus anubis): Behaviour, Ecology and Human Conflict in Gashaka Gumti National Park, Nigeria.* PhD thesis, Roehampton University.

Warren, Y., Buba, B. & Ross, C. (2007). Patterns of crop-raiding by wild and domestic animals near Gashaka Gumti National Park, Nigeria. *International Journal of Pest Management* 53: 207–216.

Weyher, A. H., Ross, C. & Semple, S. (2006). Gastrointestinal parasites in crop raiding and wild foraging *Papio anubis* in Nigeria. *International Journal of Primatology* 27: 1519–1534.

Whiten, A., Byrne, R. W., Barton, R. A., Waterman, P.G. & Henzi, S. P. (1991). Dietary and foraging strategies of baboons. *Philosophical Transactions of the Royal Society London B* 334: 187– 197.

Whiten, A., Altmann, S. A., Barton, R. A., Byrne, R. W., Bulger, J., Hamilton, W. J., Henzi, S. P, Johnson, R.B., King, B.J., Lochhead, D., Norton, G. W. & Waterman, P. W. (1987). The Nature and Development of Foraging Strategies in Baboons. *International Journal of Primatology* 8: 413.

Chapter 10
Keeping in Contact: Flexibility in Calls of Olive Baboons

Elodie Ey and Julia Fischer

Abstract Vocal production learning, and in particular the ability to imitate other people's vocalisations, is a key foundation of human speech. From an evolutionary perspective, it is puzzling that our closest living relatives, monkeys and apes, show so little flexibility in terms of altering the structure of their vocalisations. Instead, non-human primates typically have a restricted vocal repertoire consisting of different call types with varying degrees of variability within and between types. This raises two questions: first, which factors affect the overall morphology of vocal repertoires, and second, within the rather fixed system of a given vocal repertoire, which aspects of vocal production may reveal a certain degree of plasticity. Baboons lend themselves for an investigation of these questions because they show a high degree of flexibility with respect to social structure and habitats they range in. Moreover, their phylogenetic relatedness is well understood. In this study, we investigated whether olive baboons from two troops ranging in Gashaka Gumti National Park / Nigeria, adjust their calls in relation to the habitat. We compare the results to findings from other sites (olive baboons from Uganda, chacma baboons from Botswana), to explore variation within and between populations. We focus on contact calls used over short distances (grunts) and long distances (clear barks) and tested whether usage (grunt and bark rates) and call structure differed in relation to the habitat. We expected a larger degree of flexibility in call usage compared to variation in call structure and predicted that subjects would call more frequently when the visibility is poor. If individuals are in fact able to modify the structure of their calls, theory predicts that they should optimise the propagation distance by using longer calls in forest compared to woodland-savannah, with a lower frequency and energy concentrated in lower frequencies.

Indeed, the baboons uttered significantly longer grunts in forest than in open woodland, suggesting some degree of intra-individual short-term flexibility. Contrary to our expectations, grunt usage did not vary with the habitat type, perhaps

E. Ey (✉)
Department of Cognitive Ethology,
German Primate Centre, 37077, Göttingen, Germany
e-mail: elodie.ey@dpz.gwdg.de

V. Sommer and C. Ross (eds.), *Primates of Gashaka*, Developments in Primatology: Progress and Prospects 35, DOI 10.1007/978-1-4419-7403-7_10,

because a large proportion were used in social contexts, during infant handling and friendly approaches. Since these calls are given at close range, visibility was not affected. Our observation also showed little variation in grunt rate between the troops. However, compared to some other populations and taxa, Nigerian baboons grunted rarely, so that the lack of variation in relation to habitat may simply reflect a floor effect. Clear barks were given mostly as single calls by immature individuals and adult females when separated from the group or particular group members, while resting alone and travelling. Bark rate varied between troops, but not in relation to habitat type. Due to small sample size, bark structure was not analysed. Overall, our findings highlight similarities between baboon taxa in call contexts, as well as variation in responses to changing environmental conditions. Probably, other factors than the environment – such as interaction rates, for instance – may affect call usage. Future studies will need to integrate data from multiple baboon taxa to establish a better picture of the interplay between different factors that govern variation in call usage and structure.

Keywords Vocal communication • Contact calls • Olive baboon • Habitat

Introduction

The structure of non-human primate vocalisations is considered to be largely innate, as evidenced by the ontogeny of vocal production and the neurobiological foundations of vocal control (reviewed in Seyfarth & Cheney 1997, Janik & Slater 2000, Fischer 2002, Hammerschmidt & Fischer 2008). Given that humans are adept vocal imitators and speech production is to a large degree based on learning, this restriction in vocal communication of non-human primates – and indeed most terrestrial mammal – is puzzling and unexpected. Current research aims to improve our understanding of the genetic basis underlying vocal production to shed light on constraints that apparently play a role in the evolution of communication (e.g., Jamain *et al.* 2008, Marcus & Fisher 2003). In addition, field studies of vocal behaviour are indispensable to identify the selective pressures that affect its structure and usage. Key factors influencing signal usage and design are the social system (for instance via the dominance style or reproductive skew), as well as the environment. Because the largely innate structure of calls phylogenetic descent may explain a large degree of variation between primate taxa.

Baboons are particularly suitable for an integrated investigation of different factors such as social system, habitat, and phylogenetic descent. First, they are widely distributed across Africa (de Vore & Hall 1965, Kingdon 1997, Sarmiento 1997, Zinner *et al.* 2009). Chacma baboons (*Papio ursinus*) live in the south of Africa; yellow baboons (*P. cynocephalus*) range north of them. East and Central Africa and a large part of West Africa are occupied by olive baboons (*P. anubis*). Guinea baboons (*P. papio*) can be found at the extreme West Africa, and hamadryas baboons (*P. hamadryas*) range near the Arabic peninsula (Kingdon 1997,

Zinner *et al.* 2009). Recent investigations shed additional light on phylogenetic relatedness between taxa and, to some extent, populations (Zinner *et al.* 2009 this volume [Ch. 7]). Second, baboons form various social systems (Kingdon 1997; summarised in Maestripieri *et al.* 2007). For example, hamadryas baboons live in one-male-multi-female groups (harems), which gather into much larger units (Kummer 1968). Chacma (e.g., Hall 1963), yellow (e.g., Semple *et al.* 2002) and olive baboons (de Vore & Hall 1965, Rowell 1966) live in multi-male-multi-female groups. Guinea baboons, the least investigated taxon, seem to live in complex fission-fusion multi-male-multi-female groups (Dunbar & Nathan 1972, Byrne 1981). Some authors suggested that the core of Guinea baboon societies may consist of one-male units (Maestripieri *et al.* 2007), reflecting a harem structure as in hamadryas baboons. Recent observations of a group of wild Guinea baboons in Senegal, however, suggest otherwise (JF. D. Zinner pers. obs.). Social systems can also be intermediate between different forms in hybridisation zones (e.g., Bergman & Beehner 2004). Variations in social systems are believed to emerge from the interplay of ecological factors, such as climate, resource availability and predation risk (see Henzi & Barrett 2003, for review). Third, baboons range across various habitat types. Hamadryas baboons occur in semi-desert areas (Kummer 1968), chacma baboons in often flooded grassland interspersed with patches of woodland (e.g., Cheney *et al.* 2004, Fischer *et al.* 2001a), woodlands (e.g., Ron *et al.* 1996), mountainous areas (Hall 1963) and up to desert edges (e.g., Davies & Cowlishaw 1996). Guinea baboons live in savannah-woodlands and bushy areas (Dunbar & Nathan 1972) as well as woodland-savannah with stripes of gallery forest (Byrne 1981). Olive baboons range in diverse environments: open grassland (de Vore & Hall 1965, Harding 1976), gallery forest and grassland (Rowell 1966), savannah (Strum 1987), grassland clearings and moist semi-deciduous tropical forest (Rahn 2008), and in the Gashaka area of Nigeria in woodland-savannah, lowland rainforest and riverine forest.

Long-term behavioural data from Amboseli (e.g., Altmann & Altmann 1970, Altmann 1980), Gilgil and Chololo (e.g., Harding 1976, Strum 1987), Gombe (Ransom 1981), de Hoop (e.g., Hill *et al.* 2003) and Moremi (e.g., Cheney *et al.* 2004) allow for multi-faceted comparisons between populations. However, most of the current knowledge on baboon vocal communication refers to chacma baboons. This taxon has been investigated with respect to both the *usage*, i.e., contexts in which calls are given, rate of calling, age and sex classes of callers (e.g., Cheney *et al.* 1995, Cheney *et al.* 1996, Palombit *et al.* 1999, Rendall *et al.* 1999, Fischer *et al.* 2001b), and the *structure*, i.e., acoustic features of vocal signals, their variations with caller characteristics and contexts (e.g., O Connell & Cowlishaw 1994, Owren *et al.* 1997, Fischer *et al.* 2001a, 2002, 2004, Rendall *et al.* 2004). In contrast, vocal communication is poorly investigated in other baboon taxa. For instance, the last comprehensive studies on olive baboons were published more than 20 years ago (Hall & de Vore 1965, Rowell 1966, Ransom 1981). Otherwise, only a few abstracts on the usage and function of grunts and copulation calls are available (Gilmore 1978, 1979, 1983a, 1983b), as well as a recent study on friendship between males and lactating females that uses playbacks of screams to elicit support (Lemasson *et al.* 2008). Apart from chacma baboons, investigations of vocalisations generally lack modern techniques of acoustic analyses (but see Semple *et al.* (2002) for a study on copulation calls in yellow baboons; Pfefferle & Fischer 2006 for a study on grunts of hamadryas baboons). In sum, little is currently known about the variability in vocal behaviour (or lack thereof), which might exist among baboon taxa.

Generally, environmental conditions alter sound during propagation (reviewed in Wiley & Richards 1978). Strategies used to adapt vocal behaviour were widely investigated in birds and to a lesser extent in anurans and in primates, mostly by comparing species or populations (reviewed in Ey & Fischer 2009). However, intra-individual vocal plasticity in response to the environment has never been documented in wild primates. Whereas chacma baboons occupy mostly open habitats, olive baboons (*Fig. 10.1a*) can be found in open savannah as well as riverine forest, with all intermediates possible. Olive baboons are thus a useful model to examine how range might affect short-term variations in the usage and structure of vocal signals. These large, heavily-built primates live in multi-male-multi-female groups, and have a diversified diet that is mostly herbivorous and frugivorous but also includes animal proteins such as ants, grasshoppers, or crabs (de Vore & Hall 1965, Rowell 1966; see also Warren *et al.* this volume [Ch. 8]).

To document variability between baboon taxa and flexibility in response to environmental factors, contact calls are particularly suitable, since equivalent calls are found in all taxa. Grunts (*Fig. 10.2a*) are used for short-distance communication. These are short, quiet, low-pitched, and harmonically rich calls uttered during affiliative interactions or during resting and feeding and before a group movement (Ransom 1981, Owren *et al.* 1997, Rendall *et al.* 1999, 2004). Clear barks (*Fig. 10.2b*) – also termed contact barks – are used for long-distance communication. These are loud and frequently uttered by individuals separated from the rest of the group or from particular individuals (Ransom 1981, Rendall *et al.* 2000, Fischer *et al.* 2001a, 2002). In males, these calls are acoustically similar to contest "wahoos" but are given at a lower rate and not in the same contexts (Fischer *et al.* 2002).

We investigated the flexibility in vocal communication in two troops of olive baboons in Gashaka Gumti National Park / Nigeria. Both ranged in forest and woodland-savannah. This allowed intra-individual comparisons of call rates and acoustic features between different habitats. We relate our findings first to those of olive baboons in Uganda, and secondly to chacma baboons in Botswana, to evaluate flexibility within and between different taxa. We focus on the usage of calls, the contexts, the emission rate and its variation between sexes, and acoustic structure. General acoustic features are described and compared between sexes and between contexts. Here, we concentrate on frequently used acoustic variables, namely call duration, fundamental frequency, and peak frequency (*Fig. 10.2a*; see methods for definitions).

The usage of vocal signals is flexible in primates and learning plays an important role in understanding meaning and context (reviewed in Seyfarth & Cheney 1997). For instance, several studies succeeded in conditioning primates of various species

Fig. 10.1 (**a**) Olive baboons: An adult female from the Gamgam troop / Nigeria, eats flowers of *Daniellia oliveri*, while carrying her infant (photo: EE). (**b**) An adult female chacma baboon from Botswana approaches a mother and her infant in a friendly manner (photo: JF)

to utter vocalisations at a given signal (reviewed in Pierce 1985). Another example is the case of young vervet monkeys, which gradually become more specific in their usage of different alarm calls (reviewed in Cheney & Seyfarth 1990). In contrast, acoustic structure is generally considered to be innate (e.g., Jürgens 2002, Hammerschmidt & Fischer 2008). For instance, squirrel monkeys do not need to listen to conspecifics vocalising or to have auditory feedback from their own vocalisations to develop their normal vocal repertoire (Winter *et al.* 1973, Hammerschmidt *et al.* 2001). Nevertheless, some degree of plasticity exists in the acoustic structure as a response to different factors (e.g., learning: Sutton *et al.* 1973; distance to

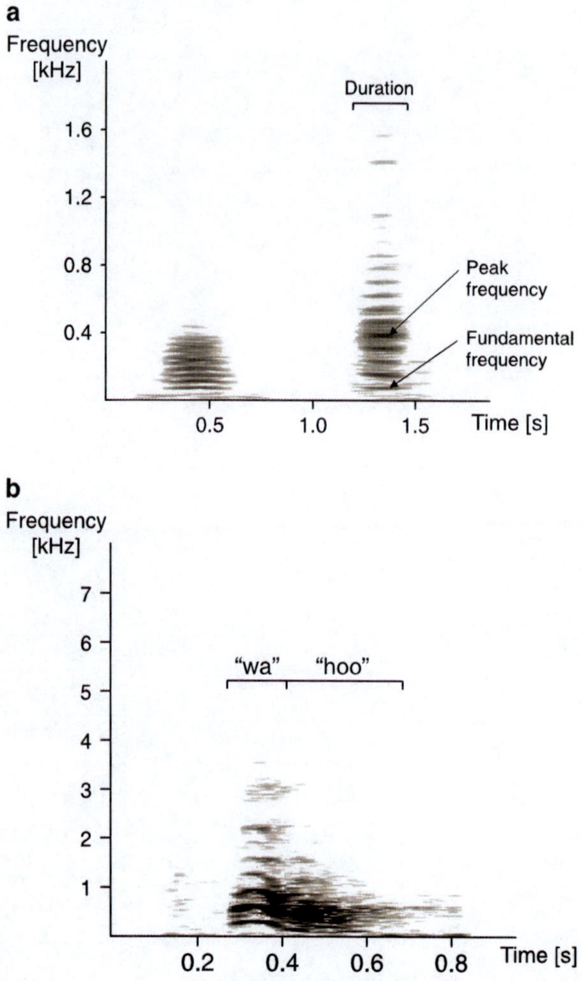

Fig. 10.2 Spectrograms of contact calls in olive baboons. (**a**) Grunt (left: adult male; right: adult female). (**b**) Clear bark, also termed contact bark (adult female), which consists of a tonal, relatively high-pitched first part ("wa"), and a quieter, noisier and lower-pitched second one ("hoo")

conspecifics: Oda 1996, Sugiura 2007; noise level: Brumm *et al.* 2004, Egnor & Hauser 2006).

We thus predict that call usage should show a higher flexibility according to habitat type and possibly also a higher level of variability between groups than call structure. Moreover, vocal signals in closed habitats are expected to be given at a higher rate to increase the likelihood of signal detection, unless predation pressure suggests otherwise (Waser & Waser 1977). Further, in areas of poor visibility, calling may be the most effective means of locating other group members and maintaining group cohesion. Thus, call rates are expected to be higher when visibility is poor. In terms of the acoustic structure, theory predicts that signals used in closed environments should have a longer duration, a lower fundamental frequency, and energy concentrated in lower frequencies than in open habitat to optimise signal propagation (Nemeth *et al.* 2006, "acoustic adaptation hypothesis": Morton 1975, reviewed in Ey & Fischer 2009). While we generally expect little acoustic variation in acoustic structure, potential changes should be in the direction of lengthening the call duration, lowering the fundamental frequency and a general shift of energy towards lower frequencies.

Methods

Study Site and Subjects

Data were collected by EE in the southern sector of Gashaka Gumti National Park, eastern Nigeria, near the border with Cameroon. This park covers ~6700 km^2, and altitudes range from 300 – 2400 m. The Gashaka sector is drained by many rivers and is mostly hilly, with steep mountains up to 900 m high from the base. The area is located in the Guinea savannah zone, but nevertheless includes a mosaic of vegetation types, such as lowland forest and woodlands of the southern Guinea savannah type (hereafter woodland-savannah), which is regularly burnt. As a result, woody species in woodland-savannah are more dispersed and smaller than in forests. Grass may grow up to 2 m or even higher when trees are more scattered (Dunn 1993, Harcourt & Ellerton 1995, Akinsoji 1996, Chapman & Chapman 2001).

Two troops of olive baboons were studied over a total of 10 months (Nov – Dec 05, Feb – May 06, Nov 06 – Apr 07), with observations distributed over two dry seasons since audio-recordings are difficult during the rainy season (Jul – Oct). The *Gamgam troop* range is located near Gashaka village, just outside the park. This troop supplements its diet by raiding crops, which leads to regular chases by farmers (Warren *et al.* this volume [Ch. 8]). Most of its home-range consists of woodland-savannah, with narrow bands of forest along seasonal streams. The *Kwano troop* lives within the park. Lowland forest largely covers its range, which also includes patches of woodland-savannah. This troop has little interaction with human beings, except for field assistants and researchers (see Warren 2003, Warren *et al.* this volume [Ch. 8]). Forest with low visibility was present in both home-ranges, as well as open woodland-savannah with dry or burnt grass. When the grass is high during and shortly after the rains, visibility is as low as in the forest. Data presented here exclude this situation to allow for a clearer distinction between "closed" and "open" habitats. Baboons range in both forest and woodland-savannah during the day. Both groups have been fully habituated to researchers since 2000 (Warren *et al.* this volume [Ch. 8]). Adults and subadults could be observed from a 2 – 6 m distance. Infants and juveniles could usually not be approached to < 5 m. The size and composition of the troops varied slightly over the observation period, with 20 – 21 individuals in the Gamgam troop and 24 – 29 in the Kwano troop.

Phylogenetic reconstruction based on mitochondrial DNA extracted from faecal samples indicates that the troops share the same haplotypes, suggesting a common matrilineal ancestry (Zinner *et al.* this volume [Ch. 7]).

Data Collection

Each day, one focal animal was followed from 06:00 to 12:00 by EE and a field assistant. Focal observations rotated between 6 animals in the Gamgam troop (1 adult male, 5 adult females) and 13 in the Kwano troop (3 adult males, 10 adult females [1 subadult female was considered adult since she became pregnant in early 2007]). Every 15 min, a scan recorded the habitat type (forest, woodland-savannah), activity of the focal animal, and number of group members within 10 m. Non-social contexts included foraging (actively foraging or eating from a food source), resting (sitting, standing or lying), and travelling (walking, running, climbing in trees). In these contexts, focal animals were clearly not directly interacting with other group members and no other individual was within 2 m, except dependent infants, who were usually with their mothers. Social contexts included friendly approaches (the outcome was sitting near the other one, passing, sitting alone as the other left, or interacting with the other), infant handling (by a non-mother or mother), resting (sitting, standing or lying), grooming / being groomed, embracing, presenting or receiving a presentation, mounting / being mounted, and aggressing / being aggressed. In social contexts, there was always at least one individual within 2 m around the focal animal. It should be noted that activity scans of the focal animals recorded mostly long-lasting states. The behaviours "approach", "embrace", "present", and "mount" did not appear in scans, because they are short events, but they were noted in the contexts of grunts.

All grunts were recorded, as well as those from other individuals whenever caller identity and context could be determined. Clear barks, uttered more rarely, were recorded *ad libitum* from all group members. Time, caller identity, context, habitat type, and height of the caller were noted. Audio-recordings were made using a Marantz PMD660 solid-state recorder (44100 Hz sampling frequency, 16-bit resolution, mono format) and a Sennheiser directional microphone (K6 power module and ME66 recording head, with a Rycote softie windscreen). Audio-recordings were interrupted when conditions were adverse, for instance when the troop stayed in a river bed, as background noise from the water covered the frequency range of grunts and therefore the signal-to-noise ratio was not strong enough.

A total of 6124 and 4064 grunts were recorded in 370 and 455 h of observation in the Gamgam and Kwano troops, respectively. Among these, 3016 resp. 2141 grunts in Gamgam resp. Kwano troop were recorded from the focal animal in respective focal follows. From these data, we examined grunt contexts and rates. For the acoustic analyses, 1853 (Gamgam) and 1044 (Kwano) grunts were of sufficient quality (here, grunts from non-focal animals were included). Only grunts recorded from a distance of 3 – 12 m were used for the acoustic analyses, since calls are distorted over longer distances (Fischer *et al.* 2002). This data set was further reduced to obtain an even distribution between callers, contexts and habitat types. Thus, 2 – 10 grunts per animal per habitat type (same number in forest and woodland-savannah) and per context (social / non-social) were selected from 4 males and 8 females, leading to a set of 330 grunts (social context: 97 grunts per habitat; non-social contexts: 68 grunts per habitat). For a comparison of grunts given in more specific contexts (social: infant handling; non-social: resting alone; see Owren *et al.* 1997, Meise 2008 for the choice of these specific contexts), we used recordings from 6 females (7 – 48 grunts per female) in both habitat types and in both contexts (infant handling: 59 grunts; resting alone: 92 grunts). Concerning clear barks, 131 and 368 calls were recorded from the Gamgam and Kwano troops, respectively. Among these, 31 and 89 recordings were of sufficient quality, but, as for grunts, the data set was further reduced to obtain a more even distribution. Here, 28 clear barks from 11 adult females (1 – 5 barks per female) constituted the final set for acoustic analyses.

An "individual" grunt rate was calculated by counting all single grunts (i.e., the units in bouts) uttered by the focal animal over the total recording time and over the time spent in each habitat type. Since clear barks were given much more rarely – and usually not by the focal animal – a "group" bark rate was estimated: all clear barks within earshot (even from unidentified callers) were counted and related to daily observation time (Gamgam: 57 days, Kwano: 66 days). This represents an estimation of how many calls a baboon might hear, since the main energy of barks (0.3 – 10 kHz) falls within the hearing range of baboons (*P. cynocephalus*: 0.04 – 40 kHz) and humans (0.03 – 17.6 kHz; Heffner 1998, Heffner 2004). To analyse the effect of habitat on bark rate, we only used days with data for both forest and woodland-savannah (Gamgam: 42 days, Kwano: 60 days) and calculated rates over the time spent in each habitat type.

Acoustic Analyses

Definition of Acoustic Variables

The mechanisms of sound production involve inter-costal muscles which contract, so that the volume of the rib cage is reduced. Air is then forced from the lungs in the trachea. The amount of expulsed air and its speed determine the duration of a sound. The air flow passes through the vocal folds, which begin to oscillate. The fundamental frequency is that at which the vocal folds are vibrating and depends on their tension, mass, and elasticity. During tonal sound production, the source signal is comprised of the fundamental frequency and its multiple integers (harmonics). The sound waves then pass through the vocal tract (between the glottis and the opening of mouth or nose) until they emanate. Depending on the shape and length of the vocal tract, different frequencies may either be filtered or enhanced. The resulting frequency spectrum thus depends on the source signal and the filter function, typically resulting in different formants, i.e., peaks in the frequency spectrum (see Fitch & Hauser 1995, for review). The peak frequency is the frequency with the highest energy. This acoustic variable is calculated for each time segment within a call, and the mean peak frequency is then calculated over all time segments (see *Fig. 10.2a*).

Grunts

The sampling frequency was lowered from 44100 Hz to 5512.5 Hz to obtain a higher frequency resolution in the range of grunts, using Avisoft SASLab Pro Recorder 4.3 (R. Specht, Berlin / Germany). Duration was measured manually using the standard cursor function on spectrograms calculated by the same software (sampling frequency: 5512.5 Hz, FFT-length: 1024 points, Hamming window, overlap: 98.43 %, time resolution: 2.9 ms, frequency resolution: 10.8 Hz). The duration of each grunt was measured on the first harmonic. The binary spectrograms were saved and files exported into the bio-acoustic software LMA 2005 developed by K. Hammerschmidt (Schrader & Hammerschmidt 1997). The harmonic cursor tool was used to calculate mean fundamental frequency and mean peak frequency for each grunt (start and end thresholds: 10 %; cut-off frequency: set under the fundamental frequency and as far as possible above background noise, i.e., according to the call at 35 Hz, 42 Hz, 50 Hz, 58 Hz, and 70 Hz).The distributions of all values for each variable were checked afterwards and potential outliers corrected. When the fundamental frequency could not be detected or the calculation of the peak frequency was disturbed by extraneous noise, calls were replaced or excluded.

Clear Barks

The sampling frequency was again lowered, from 44100 Hz to 8000 Hz, to obtain a higher frequency resolution, using Avisoft. The binary spectrograms (sampling frequency: 8000 Hz, FFT-length: 1024 points, frequency resolution: 15.7 Hz, Hamming window, overlap: 96.87 %, time resolution: 4 ms) were exported from Avisoft to LMA 2005. The general macro of this software (cut-off frequency: 150 Hz; start and end thresholds: 5 %) was used to estimate the duration of the whole call ("wa" and "hoo" parts together). The harmonic cursor tool (cut off frequency: 150 Hz; start and end thresholds: 5 %) was used to calculate the mean fundamental frequency and the mean peak frequency of the "wa" part of the barks.

Statistical Analyses

Exact Mann-Whitney *U*-tests compared the mean number of individuals within 10 m around the focal animal between the two troops, as an estimate of group dispersal and also to compare grunt rates between troops and between sexes and bark rates between troops, because of the small sample size and the violation of assumptions for parametric tests (Mundry & Fischer 1998). A general linear model (GLM) with repeated measures was used to examine the variations of bark rates with the habitat type, defined as a within-subject factor. A linear mixed model with caller identity as a random factor and sex as a fixed factor compared the acoustic characteristics of grunts between males and females. A linear mixed model with caller identity as a random factor and habitat and context (and their interaction) as fixed factors tested the influence of habitat and context on the acoustic structure of grunts. For a further comparison of directed grunts given in infant handling context and undirected grunts in the non-social context of resting alone, we also used a linear mixed model with caller identity as a random factor and context as a fixed factor. We corrected p-values for multiple testing with a Step-up Hochberg correction (Westfall & Young 1993). SPSS 15.0 for Windows was used for all statistical analyses and the significance level was set at 0.05.

Comparison with Other Populations

Our data were interpreted against records for non-Nigerian baboons. (a) Uganda. A troop of olive baboons in Budongo Forest Reserve with 18 – 19 individuals (3 adult males, 7 adult females) ranged in moist semi-deciduous tropical forest and grassland and was studied for 2.5 months (Apr 07 – Jul 07; Rahn 2008). (b) Botswana. A troop of chacma baboons in the Moremi Wildlife Reserve in the Okavango delta with 75 – 84 individuals ranging in grassy floodplains and "islands" of woodland was studied from Feb 96 – Mar 97 (Rendall 2003), Jan 98 – Jun 99, and Mar 05 (Fischer *et al.* 2001a, Ey *et al.* 2007b).

Results

Behavioural Data

Group spread (i.e., number of individuals within 10 m of the focal animal) was similar in Gamgam (mean ± SD: 1.9 ± 0.3 individuals per scan, i.e., 9.2 ± 1.5 % of the group) and Kwano troop (1.9 ± 0.6 individuals per scan, i.e., 7.1 ± 1.9 % of the group; exact Mann-Whitney U-test: $n_1 = 6$, $n_2 = 13$, $U = 36$, $p = 0.831$).

Baboons spent about a fifth of their time in social activities (Gamgam: 22.1 % of scans; Kwano: 19.0 %). Among these, grooming and resting were most frequent, followed by infant handling (*Tab. 10.1a*). The remaining time was invested in non-social activities (Gamgam: 77.9 % of scans; Kwano: 81.0 %). Among these, feeding was most frequent, followed by resting and travelling (*Tab. 10.1b*).

Table 10.1 Proportion (%) of activity scans, and number (#) and proportion (%) of grunts in social and non-social contexts in olive baboon troops at Kwano and Gamgam in Nigeria

Social contexts		Approach	Infant handling	Rest	Groom	Embrace	Present	Mount	Aggression
Gamgam	% scans		2.5	49.0	47.0				1.5
	no.of directed grunts(1254)	491	467	161	63	66	6	0	0
	% directed grunts	39.2	37.2	12.8	5.0	5.3	0.5	0.0	0.0
Kwano	% scans		9.6	43.1	45.2				2.1
	no. of directed grunts (1141)	382	593	80	50	7	16	13	0
	% directed grunts	33.5	52.0	7.0	4.4	0.6	1.4	1.1	0.0
Non-social contexts		Forage	Rest alone	Travel					
Gamgam	% scans	52.6	24.5	22.9					
	no. of directed grunts (799)	178	506	115					
	% directed grunts	22.3	63.3	14.4					
	no. of undirected grunts (755)	215	430	110					
	% undirected grunts	28.5	56.9	14.6					
Kwano	% scans	59.3	21.9	18.8					
	no. of directed grunts (514)	194	218	102					
	% directed grunts	37.7	42.4	19.9					
	no.of undirected grunts(300)	141	138	21					
	% undirected grunts	47.0	46.0	7.0					

Grunts

Occurrence of Calls

Grunts were given by all age and sex classes (rarely by young infants), but only grunts from adults are considered here because continuous data were collected only on this age class. These vocalisations were either single calls or rapid series which at times were followed by grunts from other individuals.

In social contexts, grunts were always directed to another individual. Social activities elicited 44.7 % of 2808 grunts recorded in the Gamgam troop from 6 focal individuals and 58.4 % of 1955 grunts recorded in the Kwano troop from 13 focal individuals. Most grunts were emitted during friendly approaches (Gamgam: 39.2 %; Kwano: 33.5 %), and infant handling (Gamgam: 37.2 %; Kwano: 52.0 %). These contexts were either absent or underrepresented in the activity scans, in contrast to resting and grooming which were over-represented, but elicited fewer grunts (*Tab. 10.1a*). In non-social contexts, grunts clearly directed to a recipient and those apparently not addressed to any animal in the vicinity were separated. Most grunts – both directed and undirected – were uttered during feeding and resting. We recorded fewer grunts during travel (*Tab. 10.1b*).

Grunt Rate

Mean grunt rates were higher in Gamgam troop (17.4 ± 6.5 grunts / h) than in Kwano troop (11.7 ± 4.9 grunts / h), although this difference was not significant (exact Mann-Whitney *U*-test: $n_1 = 6$, $n_2 = 13$, $U = 18.0$, $p = 0.072$). In both troops, males (4 males: 15.5 ± 2.7 grunts / h) tended to have a slightly higher grunt rate than females (15 females: 12.9 ± 6.5 grunts / h), but the difference was again not significant (exact Mann-Whitney *U*-test: $n_1 = 4$, $n_2 = 15$, $U = 19.0$, $p = 0.307$). Because rates did not differ significantly between troops and sexes, data were pooled for habitat type. Again, we found no significant difference between forest (13.3 ± 8.9 grunts / h) and woodland-savannah (12.8 ± 9.6 grunts / h; linear mixed model: $F = 0.05$, $p = 0.833$).

Acoustic Characteristics

Habitat significantly affected mean duration, as grunts emitted in forest (males: 238 ± 50 ms; females: 181 ± 36 ms) were significantly longer than those in woodland-savannah (males: 230 ± 53 ms; females: 170 ± 29 ms; linear mixed model: $F = 8.48$, corrected $p = 0.034$). Fundamental and peak frequency, however, did not vary significantly between habitats. Social and non-social contexts did likewise not trigger differences, as grunts showed a similar duration, fundamental frequency, and peak

Table 10.2 Mean and standard deviation of the duration, mean fundamental frequency and mean peak frequency of grunts of adult olive baboons from Nigeria and Uganda and chacma baboons from Botswana

Baboon taxon	Olive			Chacma
Site	Nigeria	Nigeria	Uganda	Botswana
Individuals (n)	4 males	8 females	5 females	8 females
Calls (n)	112	218	75 (a)	606 (b)
Duration (ms)	234 ± 52	176 ± 33	194 ± 7	138 ± 39
Mean fundamental frequency (Hz)	44 ± 4	77 ± 14	92 ± 10	118 ± 22
Mean peak frequency (Hz)	212 ± 88	296 ± 103	280 ± 49	
Source	This study	This study	Rahn 2008	Rendall 2003

(a) Undirected grunts in non-social contexts
(b) Grunts during infant handling and before group movement

frequency. Interactions between context and habitat were also not significant after a Step-Up Hochberg correction for multiple testing in any of the three acoustic variables. However, differences did occur in some more specific contexts, e.g., directed grunts during infant handling were longer (189 ± 33 ms) than undirected grunts while resting alone (166 ± 29 ms; linear mixed model: $F = 20.46$, corrected $p < 0.001$). There were no significant differences in fundamental frequency or peak frequency, however. Sex of the caller had a significant effect on all acoustic variables. Males emitted longer grunts (linear mixed model: $F = 12.20$, corrected $p = 0.006$), with a lower fundamental frequency ($F = 34.77$, corrected $p < 0.001$) and a lower peak frequency ($F = 18.30$, corrected $p = 0.003$) than females (*Tab.* 10.2, first 2 columns).

Clear Barks

Occurrence of Calls

Most calls occurred as single barks (Gamgam: 69.4 %, Kwano: 65.9 %), separated by more than 5 min from any other bark of the same animal. The remaining calls were in bouts of 2 – 10 calls (except for one bout of 33 calls in Kwano troop after a mother had lost contact with her infant).

In the following analyses, we considered only barks from individually identified immature animals and adult females. Most clear barks were uttered when females lost contact with their infants or vice versa, but sometimes also by animals who were alone or in a separated subgroup. Usually, animals looked around as they vocalised. Some barks were uttered from perches on trees or rocks. Perched calls were more common in Kwano (immature individuals: 47.7 % of 44 barks, adult females: 44.8 % of 181 barks) than in Gamgam troop (immature individuals: 10.0 % of 10 barks, adult females: 35.3 % of 51 barks). Barks were uttered in non-social contexts, except for one call of an adult female, who sat near an infant and a juvenile when separated from the rest of the group (*Tab.* 10.3).

Table 10.3 Contexts of occurrences of clear barks recorded from adult females and immatures of olive baboons in Nigeria

	Troop	Individuals (*n*)	Barks (*n*)	Non-social Forage	Non-social Rest alone	Non-social Travel	Social Rest	Unknown context
Adult females	Gamgam	4	51	14	11	8	0	18
	Kwano	10	181	46	74	48	1	12
Immatures	Gamgam	5	10	0	5	3	0	2
	Kwano	9	44	0	18	22	0	4

Table 10.4 Acoustic features of clear barks of adult females. Duration was measured over whole calls ("wa" and "hoo" parts), but mean fundamental frequency and mean peak frequency were measured only for the "wa" part.

Baboon taxon	Olive	Chacma
Site	Nigeria	Botswana
Individuals (*n*)	11 females	22 females
Calls (*n*)	28	39
Duration (ms)	407 ± 101	373 ± 69
Mean fundamental frequency (Hz)	435 ± 58	471 ± 54
Mean peak frequency (Hz)	751 ± 175	936 ± 178
Source	This study	Fischer *et al.* 2001a, Ey *et al.* 2007b

Bark Rates

There was no significant difference in bark rate between forest and woodland-savannah in either troop (GLM with repeated measures: Gamgam: $n = 42$, $F = 0.051$, $p = 0.823$; Kwano: $n = 60$, $F = 1.769$, $p = 0.189$). We considered the troops separately because the mean bark rate in the Gamgam troop (1.03 ± 2.75 barks / h, or related to the number of individuals: 0.05 ± 0.13 barks / h / individual) was significantly lower than in the Kwano troop (2.93 ± 4.55 barks / h, or related to the number of individuals: 0.11 ± 0.16 barks / h / individual; exact Mann-Whitney *U*-test: $n_1 = 57$, $n_2 = 66$, $U = 969.0$, $p < 0.001$).

Acoustic Characteristics

The effect of the habitat on the acoustic structure could not be tested due to limited sample size, but general acoustic features of clear barks of adult females could be calculated for comparison with chacma baboons (*Tab. 10.4*).

Discussion

Our research focused on the degree of flexibility as a response to environmental conditions in usage and acoustic structure of contact vocalisations in olive baboons, while also presenting general features of usage and structure of these calls. Grunt rates were similar between the two study troops and between sexes. Contrary to our predictions of a higher call rate in forest than in woodland-savannah, however, grunt rates did not vary across habitat type, which might be explained by the fact that most grunts were given in social interactions at very close range. In contrast, baboons showed a notable flexibility in the duration of their grunts across habitats. Clear barks were given mostly as single calls by immature individuals and adult females in non-social contexts. Baboons of the Kwano troop uttered clear barks at a higher rate than those of the Gamgam troop, but neither troop varied its bark rate in relation to the habitat type. These findings are now discussed in light of studies in other baboon populations.

Grunts

Most importantly, in accordance with our predictions, grunts were longer in forest than in woodland-savannah. Such adjustment was also found in olive baboons from Budongo Forest Reserve in Uganda (Ey *et al.* 2009). This might improve call propagation since more reverberating surfaces are present in forest than in open habitat. Indeed, the amplitude of the reverberated sound waves might be added to the direct waves when both wave types are overlapping. This should allow sounds to carry further (Nemeth *et al.* 2006). Lengthening should also increase the likelihood of the calls being located and detected in such habitats of low visibility (Brumm *et al.* 2004). As a matter of fact, from the receiver perspective, spatial detection through the binaural system is more efficient on longer sounds (Macpherson & Middlebrooks 2000). In addition, for short signals like grunts, when the duration of the stimulus increases, the amplitude level of the signal required for detection decreases (Brown & Maloney 1986). Baboons might therefore have been selected to lengthen their grunts to counteract the larger loss of energy due to propagation through vegetation in forest compared to open habitat and to facilitate detection. Whether or not the increase in call duration is indeed perceptually salient is a matter of empirical research. The proximate mechanisms underlying these changes in call production currently remain unclear. The fundamental frequency and peak frequency, at least, did not vary systematically with the habitat. This is consistent with other studies suggesting that variation in acoustic structure is mainly related to amplitude variation and call duration (reviewed in Fischer 2008).

Contrary to our predictions, the grunt rate in Nigeria did not vary significantly in relation to habitat type. A comparison with the Ugandan troop reveals some striking differences. There, adult females grunted at a higher rate in forest than in an open habitat, in accordance with our predictions (EE, C. Rahn *et al.*, in prep.). This inter-population difference might be related to differences in the overall usage of grunts as mean individual grunt rate of adults in Uganda was 28.8 ± 11.7 grunts / h, approximately twice the grunt rate of the Nigerian troops. These higher grunt rates in Ugandan olive baboons were especially notable in non-social contexts where individuals are more scattered than in social contexts (Ey 2008).

Acoustic variables of grunts in the Nigerian troops were similar to those found in Uganda. A comparison with a more distantly related taxon, however, revealed some differences, as grunts of female adult chacma baboons had a shorter duration and a higher fundamental frequency (*Tab. 10.2*, *Fig. 10.3*). This is somewhat surprising since chacma baboons are on average slightly heavier than olive baboons (chacma: adult males: 28.8 ± 2.3 kg; adult females: 13.9 ± 1.0 kg – 16.0 ± 1.6 kg; Bulger & Hamilton 1987; olive baboons: adult males: 25.1 – 27.4 kg; adult females: 13.3 – 15.6 kg; Strum 1991, Smith & Jungers 1997). According to the mechanisms of sound production (Fitch & Hauser 1995), heavier or larger animals with larger lungs, thicker vocal folds and a longer vocal tract should utter calls with a longer

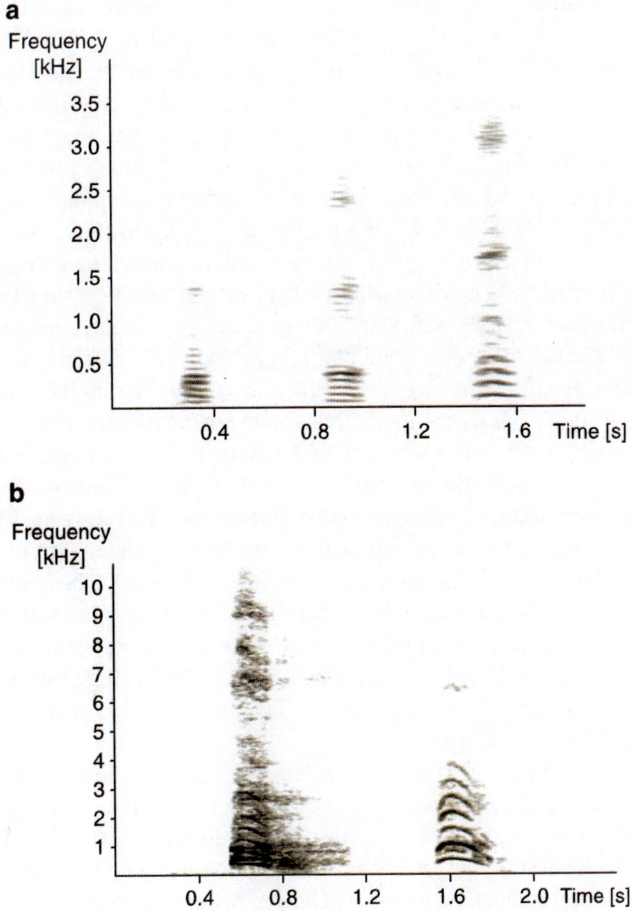

Fig. 10.3 Spectrograms of contact calls of adult females from different baboon populations and taxa. (**a**) Grunt (left: olive baboon from Nigeria; centre: olive baboon from Uganda; right: chacma baboon from Botswana). (**b**) Clear bark (left: olive baboon from Nigeria; right: chacma baboon from Botswana)

duration, a lower fundamental frequency and energy concentrated in lower frequencies (reviewed in Ey *et al.* 2007a; see also Pfefferle & Fischer 2006). This counter-intuitive result might be explained by more subtle anatomical differences. Olive baboons (*Fig. 10.1a*) seem to have a broader rib cage and a broader skull and neck than chacma baboons, which look more slender (*Fig. 10.1b*; EE, JF *pers. obs.*)., although detailed measures of skull size and neck and rib cage circumferences would be needed to corroborate this impression. Differences in duration, fundamental frequency and peak frequency between the sexes in olive baboons match the predictions from the mechanisms of sound production cited above, since lighter (and smaller) females have shorter grunts, with higher fundamental and peak frequencies than heavier (and larger) males.

Descriptions of vocal repertoire and contexts for chacma baboons (e.g., Owren *et al.* 1997, Rendall *et al.* 1999), hamadryas baboons (Pfefferle & Fischer 2006), Guinea baboons (Byrne 1981), and olive baboons (Hall & de Vore 1965, Rowell 1966, Ransom 1981) are similar to findings of the present study. Although social activities were rarer than non-social, many grunts were uttered in social contexts. While previous studies did not detail social and non-social situation, contexts of emission seem to be similar. Grunts are often given as a "group contact vocalisation" when the troop is spread out while foraging or resting in sleeping trees, or before a group movement (Ransom 1981, Rendall *et al.* 1999; C. Meise, C. Keller *pers. comm.*). In social interactions, grunts are uttered during infant handling and by individuals approaching others, especially mothers with infants (Ransom 1981, Rendall *et al.* 1999), in accordance with our findings. Here, grunts are used to mollify the recipient, as when a male or a dominant female approaches a female or a subordinate, respectively (Cheney *et al.* 1995, Silk *et al.* 1996, Palombit *et al.* 1999). The higher proportion of grunts during infant handling in the Kwano troop can be related to a higher number of youngsters (up to 3 infants present) compared to Gamgam troop (1 infant present). The proportion of grunts during resting and grooming was low in comparison to those during friendly approaches. This suggests that grunts are used mainly to engage in social interactions, while the maintenance of the interaction requires fewer grunts, as suggested by Gilmore (1983b). The context of infant handling appears to be particular as the handler may have to constantly reassure the mother that the infant is not mishandled. Grunts have also been associated with reconciliation (Silk *et al.* 1996, Cheney & Seyfarth 1997) and general enhancement of friendly interactions (Cheney *et al.* 1995, Palombit *et al.* 1999), although this was not investigated in this study.

The unexpected lack of difference in the acoustic structure between grunts in social and non-social contexts appeared to be due to variation in the delineation of contexts. When we compared the acoustic structure of grunts between more specific contexts, we found that grunts were longer during infant handling than resting alone. This partly replicates a study on chacma baboons in Namibia (Meise 2008), which also reported variations in other acoustic variables. In another group of chacma baboons in Botswana, grunts given during infant handling had a higher second formant frequency and a steeper spectral slope than grunts before a group movement (Owren *et al.* 1997). These differences appeared to be salient for the animals, independently from the situation (Rendall *et al.* 1999). Although these particular variables were not analysed here, the results suggest that there are at least some similarities between baboon taxa concerning the context-related differences in the acoustic structure of grunts.

Clear Barks

Bark rates were very low, especially in the Gamgam troop. Indeed, it was not rare to spend entire days with both troops without hearing any single clear bark (Gamgam: 32 of 57 observation days; Kwano: 13 of 66 observation days). A slightly higher group bark rate was found in the Ugandan troop of olive baboons (0.15 barks / h / individual), but it was likewise not uncommon to spend entire days without hearing any barks (C. Rahn *pers. comm.*). In contrast, bark rates of chacma baboons in Botswana were much higher. Adult females as well as juveniles and infants older than 6 months uttered on average 1.1 barks / h / individual and subadult and adult males 0.2 barks / h / individual (Cheney *et al.* 1996). These rates were calculated for the 2.5 h after the baboons left their sleeping site. A retrospective calculation over the same daytime period still revealed much lower rates for Nigeria (EE *unpubl.*). Guinea baboons in Senegal were found to have a high average bark rate (19.7 barks / h for a group of 150 – 200 individuals [calculated over 5-min intervals during various situations]; Byrne 1981) but individual rates (approximately 0.11 barks / h / individual) were similar to olive baboons in Nigeria. More data on group movement patterns and inter-individual distances would be needed to evaluate their contribution to the usage of long-distance contact calls.

The much lower bark rate in the Gamgam troop cannot be explained by a lower spread as individuals did not stay closer to one another than in the Kwano troop. Lower rates could therefore reflect reduced usefulness of vocal contact, since this troop ranged mostly in woodland-savannah where visual contact can be maintained over longer distances. In contrast, forest occupies most of the Kwano troop range. However, contrary to our expectations we did not find any variation in bark rate between forest and woodland-savannah. Thus, bark rates might be more dependent on the situation and the inner state of the caller (Cheney *et al.* 1996, Rendall *et al.* 2000). Perhaps, as the Gamgam troop was smaller than the Kwano troop, they were spread out less. Alternatively, Gamgam baboons might suppress barks as they frequently engage in crop-raiding, so as not to alert farmers and escape detection (Warren *et al.* this volume [Ch. 8]).

Clear barks are described for chacma baboons (e.g., Cheney *et al.* 1996, Fischer *et al.* 2001a, 2002), Guinea baboons (Byrne 1981), and olive baboons (Hall & de Vore 1965, Rowell 1966, Ransom 1981). Contrary to previous reports that most barks occur in bouts (within-bout rate in olive baboons: 1 – 20 barks / min; Ransom 1981), we found that they occurred mostly as single calls. As in chacma baboons in Botswana (Cheney *et al.* 1996), adult females and juveniles accounted for the majority of barks. These calls were uttered in the same contexts as described in the literature for both olive and chacma baboons, i.e., by animals separated from the rest of the troop (but not necessarily alone) or from particular individuals, such as a mother from her infant and vice versa (Ransom 1981, Cheney *et al.* 1996, Rendall *et al.* 2000, Fischer *et al.* 2001a). Baboons seem to call mostly according to their own spatial position in the group, which might also be linked to their rank (Cheney *et al.* 1996, Rendall *et al.* 2000).

When acoustic features of clear barks from adult females of Nigerian olive baboons are compared to those of adult female chacma baboons from Botswana (extended analyses of data from Fischer *et al.* 2001a, Ey *et al.* 2007b), we noticed that calls of olive baboons were longer (*Tab. 10.4*). This might be due to the fact that the "hoo" part was always present in olive but not always in those of chacma baboons (*Fig. 10.3*). Fundamental frequencies were similar. This does not corroborate the findings for grunts, but one has to consider that the fundamental frequency of barks is not the minimum fundamental frequency of the animal, which should theoretically be most strongly constrained by body size (Fitch & Hauser 1995). Finally, energy was concentrated in lower frequencies in calls of Nigerian olive baboons than in those of chacma baboons (*Tab. 10.4*). This difference might stem from slight anatomical differences, as hypothesised for grunts, and would require further investigation.

To summarise, olive baboons from Nigeria did not alter their call usage in relation to the habitat, while olive baboons from Uganda called more frequently when the visibility was poor. However, baboons from both populations increased the length of their calls in the forest, thus revealing some intra-individual plasticity in terms of their vocal production. Overall, call contexts were similar between baboon taxa from Nigeria, Uganda and Botswana, and the same call types were found. Future studies will need to integrate a more fine-grained description of contextual variation, group dispersal, as well as genetic differences and phylogenetic relatedness to further unravel the driving forces in the evolution of baboon vocal communication.

Acknowledgements We thank the Nigerian National Park Service, Volker Sommer and Caroline Ross for permissions to conduct research in Gashaka Gumti National Park / Nigeria, and to join the *Gashaka Primate Project*. We are also grateful to Halidu Ilyusun, Mustapha Isa, Abdulkadiri Adamu and Vincent Thuet for invaluable help with field work. Charlotte Rahn kindly provided data about her study conducted with olive baboons in Uganda for her diploma thesis. We also thank Kurt Hammerschmidt for helpful comments on the manuscript. Financial support was provided by the German Primate Centre, Göttingen, and the DAAD (German Academic Exchange Service).

Literature

Akinsoji, A. (1996). *Vegetation Types and Ethnobotanical Studies of Gashaka Gumti National Park, Nigeria*. Report in conjunction with Department of Biological Sciences, University of Lagos, Nigeria, to NCF & WWF-UK.

Altmann, J. (1980). *Baboon Mothers and Infants*. London: Harvard University Press.

Altmann, S. A. & Altmann, J. (1970). *Baboon Ecology – African Field Research*. Basel: Karger.
Bergman, T. J. & Beehner, J. C. (2004). Social system of a hybrid baboon group (*Papio anubis* x *P. hamadryas*). *International Journal of Primatology* 25: 1313–1330.
Brown, C. H. & Maloney, C. G. (1986). Temporal integration in two species of Old World monkeys: blue monkeys (*Cercopithecus mitis*) and grey-cheecked mangabeys (*Cercocebus albigena*). *Journal of the Acoustical Society of America* 79: 1058–1064.
Brumm, H., Voss, K., Köllmer, I. & Todt, D. (2004). Acoustic communication in noise: regulation of call characteristics in a New World monkey. *Journal of Experimental Biology* 207: 443–448.
Bulger, J. & Hamilton, W. J. (1987). Rank and density correlates of inclusive fitness measures in a natural Chacma baboon (*Papio ursinus*) troop. *International Journal of Primatology* 8: 635–650.
Byrne, R. W. (1981). Distance vocalizations of Guinea baboons (*Papio papio*) in Senegal: an analysis of function. *Behaviour* 78: 283–312.
Chapman, J. D. & Chapman, H. M. (2001). *The Forest Flora of Taraba and Adamawa States, Nigeria. An Ecological Account and Plant species Checklist*. New Zealand: University of Canterbury.
Cheney, D. L. & Seyfarth, R. M. (1990). *How Monkeys See the World*. Chicago: The University of Chicago Press.
Cheney, D. L. & Seyfarth, R. M. (1997). Reconciliatory grunts by dominant female baboons influence victims' behaviour. *Animal Behaviour* 54: 409–418.
Cheney, D. L., Seyfarth, R. M. & Palombit, R. A. (1996). The function and mechanisms underlying baboon "contact" barks. *Animal Behaviour* 52: 507–518.
Cheney, D. L., Seyfarth, R. M. & Silk, J. B. (1995). The role of grunts in reconciling opponents and facilitating interactions among adult female baboons. *Animal Behaviour* 50: 249–257.
Cheney, D. L., Seyfarth, R. M., Fischer, J., Beehner, J., Bergman, T., Johnson, S. E., Kitchen, D. M., Palombit, R. A., Rendall, D. & Silk, J. B. (2004). Factors affecting reproduction and mortality among baboons in the Okavango Delta, Botswana. *International Journal of Primatology* 25: 401–428.
Davies, J. G. & Cowlishaw, G. (1996). Baboon carnivory and raptor interspecific competition in the Namib Desert. *Journal of Arid Environments* 34: 247–249.
De Vore, I. & Hall, K. R. L. (1965). Baboon ecology. In de Vore, I. (ed.), *Primate Behaviour. Field Studies of Monkeys and Apes*. New York: Holt, Rinehart and Winston, pp. 20–25.
Dunbar, R. I. M. & Nathan, M. F. (1972). Social organization of the Guinea baboon, *Papio papio*. *Folia Primatologica* 17: 321–334.
Dunn, A. (1993). *The Large Mammals of Gashaka-Gumti National Park, Nigeria: Line Transect Surveys from Forest and Savannah*. Report to WWF-UK, NCF & Federal Ministry of Agriculture, Water Resources and Rural Development / Nigeria.
Egnor, S. E. R. & Hauser, M. D. (2006). Noise-induced vocal modulation in cotton-top tamarins (*Saguinus oedipus*). *American Journal of Primatology* 68: 1183–1190.
Ey, E. (2008). *Influences of Ecological Factors on Vocal Communication in Olive Baboons (Papio hamadryas anubis)*. PhD thesis, Georg-August Universität Göttingen.
Ey, E., Pfefferle, D. & Fischer, J. (2007a). Do age- and sex-related variations reliably reflect body size in non-human primate vocalizations? A review. *Primates* 48: 253–267.
Ey, E., Hammerschmidt, K., Seyfarth, R. M. & Fischer, J. (2007b). Age- and sex-related variations in clear calls of *Papio ursinus*. *International Journal of Primatology* 28: 947–960.
Ey, E. & Fischer, J. (2009). The 'acoustic adaptation hypothesis' A review of the evidence from birds, anurans and mammals. *Bioacoustics* 19: 21–48.
Ey, E., Rahn, C., Hammerschmidt, K. & Fischer, J. (2009). Wild female olive baboons adapt their grunt vocalizations to environmental conditions. *Ethology* 115: 493–503.
Fischer, J. (2002). Developmental modifications in the vocal behaviour of nonhuman primates. In Ghazanfar, A. (ed.), *Primate Audition: Behaviour and Neurobiology*. Boca Raton: CRC Press, pp. 109–125.
Fischer, J. (2008). Transmission of acquired information in nonhuman primates. In Byrne, J. H. (ed.), *Learning and Memory: A Comprehensive Reference*. Oxford: Elsevier, pp. 299–313.

Fischer, J., Hammerschmidt, K., Cheney, D. L. & Seyfarth, R. M. (2001a). Acoustic features of female chacma baboon barks. *Ethology* 107: 33–54.

Fischer, J., Hammerschmidt, K., Cheney, D. L. & Seyfarth, R. M. (2002). Acoustic features of male baboon loud calls: Influences of context, age, and individuality. *Journal of the Acoustical Society of America* 111: 1465–1474.

Fischer, J., Kitchen, D. M., Seyfarth, R. M. & Cheney, D. L. (2004). Baboon loud calls advertise male quality: acoustic features and their relation to rank, age, and exhaustion. *Behavioural Ecology and Sociobiology* 56: 140–148.

Fischer, J., Metz, M., Cheney, D. L. & Seyfarth, R. M. (2001b). Baboon responses to graded bark variants. *Animal Behaviour* 61: 925–931.

Fitch, W. T. & Hauser, M. D. (1995). Vocal production in nonhuman primates: acoustics, physiology, and functional constraints on "honest"advertisement. *American Journal of Primatology* 37: 191–219.

Gilmore, H. A. (1978). Message analysis of a baboon vocalization. *American Journal of Physical Anthropology* 48: 398–398.

Gilmore, H. A. (1979). Function of the female post-copulatory vocalization in olive baboons. *American Journal of Physical Anthropology* 50: 442–442.

Gilmore, H. A. (1983a). Approaching others: a functional analysis of the grunt vocalization among olive baboons. *American Journal of Primatology* 4: 343–343.

Gilmore, H. A. (1983b). Grunting and groaning: rethinking the baboon contact call. *American Journal of Physical Anthropology* 60: 197–197.

Hall, K. R. L. (1963). Variations in the ecology of the chacma baboon, *Papio ursinus. Symposium of the Zoological Society London* 10: 1–28.

Hall, K. R. L. & de Vore, I. (1965). Baboon social behavior. In de Vore, I. (ed.), *Primate Behaviour. Field Studies of Monkeys and Apes*. New York: Holt, Rinehart and Winston, pp. 53–110.

Hammerschmidt, K. & Fischer, J. (2008). Constraints in primate vocal production. In Griebel, U. & Oller, K. (eds.), *The Evolution of Communicative Creativity: From Fixed Signals to Contextual Flexibility*. Cambridge, MA: The MIT Press, pp. 93–119.

Hammerschmidt, K., Freudenstein, T. & Jürgens, U. (2001). Vocal development in squirrel monkeys. *Behaviour* 138: 1179–1204.

Harcourt, C. S. & Ellerton, N. G. (1995). *A Brief Look at the Primates of Gashaka Gumti National Park, Nigeria*. Report to NCF & WWF-UK.

Harding, R. S. O. (1976). Ranging patterns of a troop of baboons (*Papio anubis*) in Kenya. *Folia Primatologica* 25: 143–185.

Heffner, H. E. (1998). Auditory awareness. *Applied Animal Behaviour Science* 57: 259–268.

Heffner, R. S. (2004). Primate hearing from a mammalian perspective. *The Anatomical Record Part A* 281A: 1111–1122.

Henzi, P. & Barrett, L. (2003). Evolutionary ecology, sexual conflict, and behavioral differentiation among baboon populations. *Evolutionary Anthropology* 12: 217–230.

Hill, R. A., Barrett, L., Gaynor, D., Weingrill, T., Dixon, P., Payne, H. & Henzi, S. P. (2003). Day length, latitude and behavioural (in)flexibility in baboons (*Papio cynocephalus ursinus*). *Behavioural Ecology and Sociobiology* 53: 278–286.

Jamain, S., Radyushkin, K., Hammerschmidt, K., Granon, S., Boretius, S., Varoqueaux, F., Ramanantsoa, N., Gallego, J., Ronnenberg, A., Winter, D., Frahm, J., Fischer, J., Bourgeron, T., Ehrenreich, H. & Brose, N. (2008). Reduced social interaction and ultrasonic communication in a mouse model of monogenic heritable autism. *Proceedings of the National Academy of Science of the USA* 105: 1710–1715.

Janik, V. M. & Slater, P. J. B. (2000). The different role of social learning in vocal communication. *Animal Behaviour* 60: 1–11.

Jürgens, U. (2002). Neural pathways underlying vocal control. *Neuroscience and Behavioral reviews* 26: 235–258.

Kingdon, J. (1997). *The Kingdon Field Guide to African Mammals*. San Diego: Academic Press.

Kummer, H. (1968). *Social Organization of Hamadryas Baboons. A Field Study*. Basel: Karger.

Lemasson, A., Palombit, R. A. & Jubin, R. (2008). Friendships between males and lactating females in a free-ranging group of olive baboons (*Papio hamadryas anubis*): evidence from playback experiments. *Behavioural Ecology and Sociobiology* 62: 1027–1035.

Macpherson, E. A. & Middlebrooks, J. C. (2000). Localization of brief sounds: effects of level and background noise. *Journal of the Acoustical Society of America* 108: 1834–1849.

Maestripieri, D., Mayhew, J., Carlson, C. L., Hoffman, C. L. & Radtke, J. M. (2007). One-male harems and female social dynamics in Guinea baboons. *Folia Primatologica* 78: 56–68.

Marcus, G. F. & Fisher, S. E. (2003). FOXP2 in focus: what can genes tell us about speech and language? *Trends in Cognitive Sciences* 7: 257–262.

Meise, K. (2008). *Information Content of Female Chacma Baboon Grunts – The Signallers Potential to Utter Functionally Referential Grunts*. Diploma thesis, Universität Bielefeld.

Morton, E. S. (1975). Ecological sources of selection on avian sounds. *The American Naturalist* 109: 17–34.

Mundry, R. & Fischer, J. (1998). Use of statistical programs for nonparametric tests of small samples often leads to incorrect P-values: Examples from Animal Behaviour. *Animal Behaviour* 56: 256–259.

Nemeth, E., Dabelsteen, T., Pedersen, S. B. & Winkler, H. (2006). Rainforests as concert halls for birds: are reverberations improving sound transmission of long song elements? *Journal of the Acoustical Society of America* 119: 620–626.

O Connell, S. M. & Cowlishaw, G. (1994). Infanticide avoidance, sperm competition and mate choice: the function of copulation calls in female baboons. *Animal Behaviour* 48: 687–694.

Oda, R. (1996). Effects of contextual and social variables on contact call production in free ranging ringtailed lemurs (*Lemur catta*). *International Journal of Primatology* 17: 191–205.

Owren, M. J., Seyfarth, R. M. & Cheney, D. L. (1997). The acoustic features of vowel-like grunt calls in Chacma baboons (*Papio cynocephalus ursinus*): Implications for production processes and functions. *Journal of the Acoustical Society of America* 101: 2951–2963.

Palombit, R. A., Cheney, D. L. & Seyfarth, R. M. (1999). Male grunts as mediators of social interaction with females in wild Chacma baboons (*Papio cynocephalus ursinus*). *Behaviour* 136: 221–242.

Pfefferle, D. & Fischer, J. (2006). Sound and size: Identification of acoustic variables that reflect body size in Hamadryas baboons, *Papio hamadryas*. *Animal Behaviour* 72: 43–51.

Pierce, J. D. (1985). A review of attempts to condition operantly alloprimate vocalizations. *Primates* 26: 202-213.

Rahn, C. (2008). *Usage, Structure and Function of Olive Baboon Contact Calls. A Field Study in Budongo Forest, Uganda.* Diploma thesis, Freie Universität Berlin.

Ransom, T. W. (1981). *Beach Troop of the Gombe*. USA: Associated University Press Inc.

Rendall, D., Cheney, D. L. & Seyfarth, R. M. (2000). Proximate factors mediating "contact" calls in adult female baboons (*Papio cynocephalus ursinus*) and their infants. *Journal of Comparative Psychology* 114: 36–46.

Rendall, D., Owren, M. J., Weerts, E. & Hienz, R. D. (2004). Sex differences in the acoustic structure of vowel-like grunt vocalizations in baboons and their perceptual discrimination by baboon listeners. *Journal of the Acoustical Society of America* 115: 411–421.

Rendall, D., Seyfarth, R. M., Cheney, D. L. & Owren, M. J. (1999). The meaning and function of grunt variants in baboons. *Animal Behaviour* 57: 583–592.

Ron, T., Henzi, S. P. & Motro, U. (1996). Do female chacma baboons compete for a safe spatial position in a southern woodland habitat? *Behaviour* 133: 475–490.

Rowell, T. E. (1966). Forest living baboons in Uganda. *Journal of Zoology of London* 149: 344–364.

Sarmiento, E. (1997). Current problems with Papio taxanomies. *African Primates* 3: 48–52.

Schrader, L. & Hammerschmidt, K. (1997). Computer-aided analysis of acoustic parameters in animal vocalizations: a multi-parametric approach. *Bioacoustics* 7: 247–265.

Semple, S., McComb, K., Alberts, S. & Altmann, J. (2002). Information content of female copulation calls in yellow baboons. *American Journal of Primatology* 56: 43–56.

Seyfarth, R. M. & Cheney, D. L. (1997). Some general features of vocal development in nonhuman primates. In Snowdon, C. T. & Hausberger, M. (eds.), *Social Influences on Vocal Development*. Cambridge: Cambridge University Press, pp. 249–273.

Silk, J. B., Cheney, D. L. & Seyfarth, R. M. (1996). The form and function of post-conflict interactions between female baboons. *Animal Behaviour* 52: 259–268.

Smith, R. J. & Jungers, W. L. (1997). Body mass in comparative primatology. *Journal of Human Evolution* 32: 523–559.

Strum, S. C. (1987). *Almost Human – A Journey into the World of Baboons*. New York: Random House.

Strum, S. C. (1991). Weight and age in wild olive baboons. *American Journal of Primatology* 25: 219–237.

Sugiura, H. (2007). Effects of proximity and behavioural context on acoustic variation in the coo calls of Japanese macaques. *American Journal of Primatology* 69: 1–13.

Sutton, D., Larson, C., Taylor, E. M. & Lindeman, R. C. (1973). Vocalizations in rhesus monkeys: conditionability. *Brain Research* 52: 225–231.

Warren, Y. (2003). *Olive Baboons (Papio cynocephalus anubis): Behaviour, Ecology And Human Conflict In Gashaka Gumti National Park, Nigeria*. PhD thesis, Roehampton University.

Waser, P. M. & Waser, M. S. (1977). Experimental studies of primate vocalization: specializations for long-distance propagation. *Zeitschrift für Tierpsychologie* 43: 239–263.

Westfall, P. H. & Young, S. S. (1993). *Resampling-based Multiple Testing. Examples and Methods for P-value Adjustments*. New York: Wiley, J.

Wiley, R. H. & Richards, D. G. (1978). Physical constraints on acoustic communication in the atmosphere: implications for the evolution of animal vocalizations. *Behavioural Ecology and Sociobiology* 3: 69–94.

Winter, P., Handley, P., Ploog, D. & Schott, D. (1973). Ontogeny of squirrel monkey calls under normal conditions and under acoustic isolation. *Behaviour* 43: 230–239.

Zinner, D., Groeneveld, L. F., Keller, C. & Roos, C. (2009). Mitochondrial phylogeography of baboons (*Papio* spp.) – Indication for introgressive hybridization? *BMC Evolutionary Biology* 9: 83.

Chapter 11
Not Words but Meanings? Alarm Calling Behaviour in a Forest Guenon

Kate Arnold, Yvonne Pohlner, and Klaus Zuberbühler

Abstract Vervet monkey alarm call behaviour has long been the paradigmatic example of how primates use vocalisations to protect themselves from predators. In these primates, there is a close link between different types of vocalisations and different types of predators. More recent work has shown that there are alternative ways in which primates use vocalisations to cope with predation, suggesting that the one-call / one-predator type alarm calling system may not be universal among primates. Here, we describe studies on the alarm call system of putty-nosed monkeys at Gashaka Gumti National Park / Nigeria. We found that the adult males regularly produce two acoustically distinct alarm call types but neither is given exclusively to one predator. There are striking regularities in how calls are ordered into more complex combinations, one of which, the pyow-hack sequence, is tightly linked to subsequent behaviour and apparently functions to initiate group movements. The remainder are each associated with a range of contexts and so appear to provide information which differs from the alarm calls of other guenon species studied to date. Our research focuses on understanding whether the different call series might encode information at different levels, such as predator type, degree of threat or urgency, or the callers imminent behaviour. More broadly, our aim is to characterise the cognitive mechanisms underlying primate communication from which human language has evolved.

Keywords Referential communication • Anti-predator behaviour • Call combinations • Evolution of language • Alarm call

---

K. Zuberbühler (✉)
School of Psychology, University of St. Andrews, St. Andrews KY16 9JP, Scotland UK
e-mail: kz3@st-andrews.ac.uk

V. Sommer and C. Ross (eds.), *Primates of Gashaka*, Developments in Primatology: Progress and Prospects 35, DOI 10.1007/978-1-4419-7403-7_11,

Introduction

Alarm Calling in Primates

One persistent debate in the animal communication literature concerns the kinds of psychological processes that underlie and drive an animal's vocal response. One classic school of thought assumes that vocalisations are the product of different underlying affective states, such as various types of emotions or motivations (e.g., Lancaster 1975; Morton 1982). For example, the reason why rhesus macaques produce different types of acoustically distinct calls in agonistic interactions is because they are experiencing different levels of "fear" (Gouzoules *et al.* 1984).

The "affective state" account is often taken as the parsimonious level of explanation. One area in which it has enjoyed particular attention is in the alarm calling literature (Marler *et al.* 1992), especially in cases where animals produce acoustically distinct alarm call types in response to different levels of threat imposed by a predator, e.g., Californian ground squirrels (Owings & Virginia 1978, Leger *et al.* 1980, Owings & Leger 1980), Richardson's ground squirrels (Davis 1984, Warkentin *et al.* 2001), and white-browed scrubwrens (Leavesley & Magrath 2005). Urgency-dependent alarm calling appears to work well for species that rely on a single escape strategy, such as bolting into a burrow, to deal with all potential threats of predation. In these species, the level of threat is the only information required for an appropriate escape response (Macedonia & Evans 1993). Thus, in *response urgency* alarm call systems, one call type is given when a predator is in a position to attack, and another is given when the predator is far away and perhaps only requires monitoring. So far, response urgency alarm calling systems have not been identified in any primate species.

Another type of alarm call system has been documented in a growing number of species, many of them primates. These alarm-calling systems consist of two or more acoustically distinct alarm call types, each of which is given to different categories of predator types. The best-known case is the alarm calling system of vervet monkeys. These primates have several distinct alarm calls, each of which is closely associated with detection of their most important predators, which include leopards, eagles, and pythons (Struhsaker 1967). On hearing a leopard alarm call, for example, they respond by climbing into nearby trees, whereas eagle alarm calls result in monkeys diving down into dense undergrowth. In the vast majority of cases, they respond to these predator-specific alarm calls by taking evasive action appropriate to the hunting tactics of the predator in question, as if they had detected the predator first hand. They also respond appropriately to playbacks of alarm calls, in the absence of any visual or other forms of evidence of the presence of the predator (Seyfarth *et al.* 1980). Because different call types are usually given to different predator types, it has been argued that the vervet monkey calling system is referential. Different alarm call types act as symbols for different external objects, the *referents* (e.g., eagles, leopards, snakes), which are mediated by corresponding mental concepts, the *references* (Ogden & Richards 1923).

The notion of referentiality in animal communication has raised a number of additional issues. First, one important finding, especially in relation to previously described urgency-based alarm call systems, is that vervet and also Diana monkeys (*Cercopithecus diana*) produce their predator-specific alarm calls irrespective of the distance at which the predator is detected (Cheney & Seyfarth 1990, Zuberbühler 2000a). Second, another crucial distinction is that each alarm call type in vervet or Diana monkeys is tightly linked to a very narrow range of stimuli (the referents) thereby providing listeners with relatively specific information (Seyfarth & Cheney 1980, Zuberbühler *et al.* 1997). A third and largely unresolved problem, highlighted by the analogies invoked by the notion of referentiality, is whether or not animal signallers have an active desire to inform others about an object or ongoing event. The currently accepted position is that signallers do not produce calls deliberately to inform others about what they have just witnessed. Consequently, the term functionally referential has been adopted to emphasise the notion that animal calls given in response to particular objects or events do not necessarily have to be the product of a human-like urge to inform an ignorant audience. From the perspective of the recipient, the main feature of referential signals is that they effectively come to designate external objects or events, i.e., they become meaningful to listeners and thus obtain *semantic* status (Seyfarth *et al.* 1980, Cheney & Seyfarth 1990, Seyfarth & Cheney 1993, Zuberbühler *et al.* 1999, Seyfarth & Cheney 2003). For example, a leopard alarm call means "leopard" (Seyfarth *et al.* 1980; Cheney & Seyfarth 1992), just as the human word "leopard" refers to a leopard. Another interpretation is that signals *represent* objects or events, e.g., a leopard alarm call triggers a representation of "leopardness" in the mind of the signaller and receiver (Gouzoules *et al.* 1984).

Several primate alarm-calling systems have been termed "functionally referential", thereby acknowledging the fact that signallers are probably displaying an evolved behavioural propensity rather than a desire to inform an ignorant audience. Ring-tailed lemurs (Pereira & Macedonia 1990), Diana monkeys (Zuberbühler 2000a) and Campbell's monkeys (Zuberbühler 2001) all produce acoustically distinct alarm calls to ground predators and large raptors, irrespective of the level of threat, and playbacks of alarm calls cause recipients to respond as though the corresponding predator were present.

White-faced capuchins (Fichtel *et al.* 2005, Digweed *et al.* 2005), redfronted lemurs, white sifakas (Fichtel & Kappeler 2002), and Coquerel sifakas (Fichtel & van Schaik 2006) have alarm calling systems where one type of call is given to a relatively restricted range of potentially dangerous raptors. Another call type is given to a range of terrestrial predators, but the same calls are often also given to non-predatory events, such as inter-group encounters or to harmless animals. Although these calls are given to some well-defined range of external events, but not others (e.g., finding food, losing the group), it is often difficult to think of a unifying mental concept (i.e., the reference) that would include the different referents the monkeys' calls refer to.

Two points are relevant here. First, non-human primates have little voluntary control over the acoustic products that they are able to produce and thus are forced to operate with extremely restricted vocal repertoires (Riede *et al.* 2005). Field research has shown that there are a large number of external events that warrant loud call responses. It is possible that the observed heterogeneity of referents linked with one call type is the result of these constraints. It does not necessarily follow from these observations that monkeys do not possess discrete mental categories that clearly separate events such as inter-group encounters from spotting a leopard on the ground. Second, in human language the same utterances commonly relate to different and unrelated referents (e.g., drawers, chest), but this rarely generates difficulties for transmitting semantic content. In these cases, the meaning of the utterances is usually determined by the context or other pragmatic cues. Field observations suggest that context is equally important in helping animal receivers to disambiguate the meaning of calls given to a variety of external referents that do not share any apparent common semantic ground.

Animal Call Combinations

Comparative studies concerning the problem of what aspects of the language faculty are uniquely human and what are shared with non-human animals have been most productive when focused on the more peripheral components such as conceptual-intentional systems, number representation, and rule learning (Hauser *et al.* 2002). Evidence that non-human animals share with humans the capacity to communicate about specific objects or events is limited, both in terms of the number of species that have been shown to exhibit this capacity and the contexts in which such communication occurs (e.g., Seyfarth *et al.* 1980, Pereira & Macedonia 1990, Evans & Marler 1995, Zuberbühler *et al.* 1999, Manser 2001). As mentioned in the previous section, most primates have a very restricted repertoire of calls that are innate and largely unmodifiable (Hauser *et al.* 2002, Ey & Fischer this volume [Ch. 10], but see Lemasson *et al.* 2005).

In contrast, human language draws on a relatively large repertoire of discrete sound units that are combined to build more complex meaningful units. This vast combinatorial power, both at the level of morphology and syntax, is thought to set human communication most decisively apart from any animal communication system. In contrast, animal vocalisations are generally considered to be holistic (Bickerton 1990) and lack the features that mark particular relationships between one call and another. Examples of calls being combined, such that utterances mean something other than the sum of their parts, have proven elusive. Call combinations in which calls are concatenated according to simple ordering rules generally function as advertisements of identity, status, or quality (Kroodsma & Miller 1982, Payne & Payne 1985, Mitani & Marler 1989, Suzuki *et al.* 2006), although certain call combinations might subtly modify call meaning.

Examples of what has been termed *phonological syntax*, in which rule-governed combinations of calls generate meaningless compound signals, are fairly common, especially in birds (Marler 1977). In primates, there are a number of examples in which combinations appear to act as a contextual modifier for each other, i.e., where two calls are given in combination, the presence of one of the calls changes the "meaning" of the other. Call combinations in tamarin (Cleveland & Snowdon 1992) and capuchin monkeys (Robinson 1984) have been cited as examples. Here, the compound call is in a direct linear relationship to the meanings of the

Fig. 11.1 Male putty-nosed monkey *Cercopithecus nictitans martinii* (photo: KA)

constituent calls (e.g., alert calls and alarm calls combined equal high alert or low level alarm). In another example, Diana monkeys respond to the predator specific alarm calls of sympatric Campbell's monkeys with their own corresponding alarm calls, unless the Campbell's alarm call is modified by a preceding "boom", in which case Diana monkeys do not respond, although the meaning of the predator alarm call appears to be retained (Zuberbühler 2002). White-handed gibbons, finally, use different orderings of song units when singing as part of their regular morning routine or when singing in response to terrestrial predators (Clarke *et al.* 2006).

We review some recent progress concerning the question of whether primates are able to combine calls into more complex sequences that are meaningful to receivers. For this we summarise our empirical work (Pohlner 2005, Arnold & Zuberbühler 2006b, Arnold & Zuberbühler 2008, Arnold *et al.* 2008) on the alarm calling behaviour in male putty-nosed monkeys (*Cercopithecus nictitans martini*) living in Gashaka Gumti National Park / Nigeria (*Fig. 11.1*). First, we address the question of reference and referential signalling with regard to how these monkeys use their alarm signals in response to different types of external events. Second, we report how these monkeys combine elements of their call repertoire to build more complex call combinations and how these carry meaning independent of the meanings of the constituent calls. Our overall aim is to stimulate further work, both at the theoretical and empirical level, on this currently understudied topic in animal communication and, more generally, on comparative studies focused on understanding the evolutionary roots of human language.

Methods

Study Animals

Putty-nosed monkeys are a West African species of guenon. They are large, dark grey monkeys typically weighing 4 kg (females) to 8 kg (males) with a very conspicuous white nose. They live in large groups of up to 30 individuals comprising one adult male together with females and their offspring. They are widely distributed across central and west Africa from Nigeria to the Democratic Republic of Congo with a separate population extending from Ivory Coast to Liberia and Guinea (Oates 1996). They are primarily arboreal and frugivorous (Gautier-Hion *et al.* 1980, 1983, Gautier-Hion 1988, Mitani 1991) and spend a significant proportion of their time in polyspecific associations with other primate species (Gautier-Hion & Gautier 1974, Mitani 1991). Data on the social behaviour of this species are sparse.

Field experiments were conducted in Gashaka Gumti National Park / Nigeria (Mar 03 – May 05). The study area consisted of approximately 10 km^2 of primary rain forest in the Kwano region, near the Gashaka Primate Research Station (07°19′ N – 11°35′ E). Here putty-nosed monkeys live in one-male groups of typically up to about 20 individuals, with 6 – 9 adult females and their offspring (KA, *unpubl.*) sometimes forming poly-specific associations with mona monkeys (*C. mona*) and black-and-white colobus monkeys (*Colobus guereza*). Density has been estimated at 3 – 4 groups per km^2 (Dunn 1993). One study group has been habituated to human presence by KA and a number of field assistants since May 04. This group contained 9 – 13 adult and subadult females together with ca. 8 juveniles and offspring and was observed from Jan 05 – May 07.

Aims and Hypotheses

The males have a repertoire of three basic loud call types: "booms", "pyows", and "hacks" (Struhsaker 1970). Booms are very rarely heard and occur in a wide range of contexts (KA *pers. obs.*). Pyows and hacks are frequently heard and have previously been interpreted as functioning primarily as calls used for intra-group cohesion and the maintenance of inter-group spacing (Gautier & Gautier-Hion, 1977). There are also reports of pyows and hacks being used in a variety of contexts (e.g., falling trees, thunderclaps, aerial predators, the approach of humans), all of which have a disturbing effect on the caller and other group members (Struhsaker 1970). Putty-nosed monkeys (*C. nictitans stampflii*) in the Ivory Coast use pyows and hacks as alarm calls (Eckardt & Zuberbühler 2004), but the degree of context-specificity of the various call sequences has not yet been investigated.

Our initial aim was to investigate how male putty-nosed monkeys use their repertoire of calls in different predatory contexts. We hypothesised that if male putty-nosed monkeys use different calls to refer to particular types of predators, then we should find direct relations between different call types and predator types. If the different calls referred to different degrees of threat, regardless of predator types, then the distance of the predator at the time of detection should have the strongest effect on call type selection. In order to distinguish between these two hypotheses we carried out field experiments for which we simulated the presence of a predator by playing back recordings of the vocalisations of crowned eagles (*Stephanoaetus coronatus*) and leopards (*Panthera pardus*) which have been previously shown to be a reliable way of simulating predator presence in several primate species (Macedonia & Yount 1991, Zuberbühler *et al.* 1997, 1999b, Zuberbühler 2000a, 2001, Fichtel & Kappeler 2002, Eckardt & Zuberbühler 2004, Rainey *et al.* 2004, Fichtel & van Schaik 2006, Arnold & Zuberbühler 2006a).

Experimental simulations of predator presence are necessary because encounters with real predators are rarely witnessed, particularly in arboreal species living in densely forested environments, and good systematic observations are difficult to obtain. However, acoustic models might not provide subjects with sufficient evidence of the presence of the predator since they tend to be short in duration and continued presence cannot be confirmed by visual or other modes of detection. It is possible that these factors might not elicit the species-typical response to the corresponding real predator. To ensure our acoustic models were effective we also presented realistic, visual models of predators for comparison. Such models have the advantage of providing sustained visual stimuli and have been used successfully in other studies of alarm calling behaviour in primates (Macedonia & Polak 1989, Pereira & Macedonia 1991, Brown *et al.* 1992, Ramakrishnan & Coss 2000, Wich & Sterck 2003, Fichtel & van Schaik 2006).

Primates often respond with long series of alarm calls when they detect a predator, but some studies have analysed only the first few calls, mainly because of the assumption that important predator information should be transmitted early on (Zuberbühler 2000a, 2001, Eckardt & Zuberbühler 2004). Nevertheless, primates often produce exceedingly long series of alarm calls and it is often unclear what the function of this behaviour is (but see Zuberbühler *et al.* 1999b, Wich & de Vries 2006). Thus, an additional aim of our research was to understand why alarm calling involves multiple calls, what types of call combinations occur, and whether they can be linked with external events or contexts.

Male Loud Alarm Calls

Of the three male putty-nosed monkey loud call types (hacks, pyows, and booms) booms were never produced in response to the experimental stimuli. Hacks are low-frequency tonal calls that can be characterised as having an abrupt onset, a duration of 18 – 68 ms, and a major band of acoustic energy between 0.6 – 1.1 kHz. Pyows are a more variable higher-frequency tonal call characterised by descending frequency modulation from ~ 2.5 – 0.6 kHz. Pyow calls vary in length from 28 – 289 ms (*Fig. 11.2*). Both vocalisations are loud and conspicuous discrete call types that carry over considerable distances in a rain forest habitat and that can easily be distinguished by ear (see Arnold & Zuberbühler 2006a).

Calls are usually given in bouts and, although exceptions occur, these bouts can typically be classified into three main categories: (a) hack series – repetitions of hacks; (b) pyow series – repetitions of pyows; (c) transitional series – a hack series followed by a pyow series. A fourth combination of calls, the P-H sequence, is made up 1 – 4 pyows followed by 1 – 4 hacks and is a distinct unit that either occurs alone or is inserted within the other three series categories.

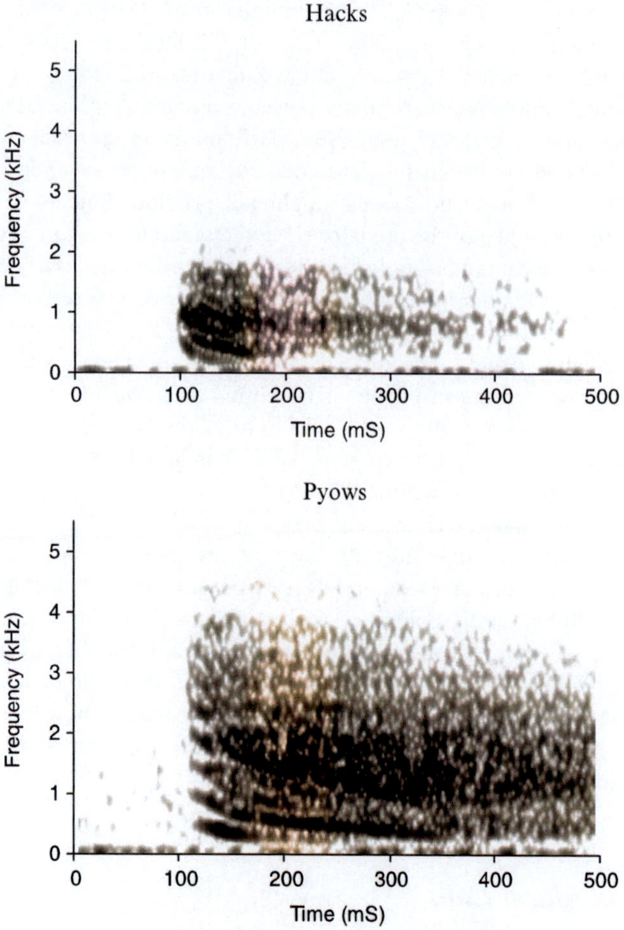

Fig. 11.2 Spectrographic illustration of representative exemplars of the two types of male putty-nosed loud alarm calls. Pyows are acoustically more variable than the hacks, but both call types are perceptually discrete (Arnold & Zuberbühler 2006a)

Alarm Calling Responses to Predators

Acoustic Predator Models

Different putty-nosed monkey groups within the study area were searched for and located by sight or their vocalisations. Once detected, KA and field assistant positioned themselves 25 – 100 m away and out of sight. All vocal behaviour was monitored for 30 min to ensure that the group was not aware of the presence of the experimenter, and that no other disturbance, indicated by alert or alarm calls, had occurred prior to an experiment. If this condition was met, a playback speaker connected to a portable CD player was positioned 0 – 2 m from the ground, in preparation for broadcasting the playback stimulus, whilst continuing to monitor the group. Recording of vocalisations began 5 min before exposure to the playback stimulus and continued until all anti-predator vocalisations had ceased. Playback stimuli were presented as natural series lasting 15 s and consisted of either leopard growls (n = 2: purchased from the Natural Sound Archive, London / UK) or crowned eagle shrieks (n = 2: recorded by KZ in the Taï National Park / Ivory Coast; eagle shrieks recorded at Taï are identical to those heard in the study area). Spectrograms depicting examples of these stimuli are shown in *Fig. 11.3*.

Vocal responses were recorded with a professional DAT Walkman (sampling rate 44.1 kHz) and a directional microphone. Trials were abandoned or discarded if the monkeys detected the experimenter or equipment at any time before the end of the trial. The following contextual information was also recorded: (1) distance between speaker and group male, later coded as "close" (< 50 m) or "far" (> 50 m), (2) local density of vegetation, coded as "dense" (thick undergrowth, upper canopy not visible) or "open" (little undergrowth, several tree crowns visible), (3) the general illumination, coded as "dark" (no shadows on the ground, sky heavily overcast or twilight) or "light" (shadows visible, sky slightly overcast or full sunlight). The location of the group was recorded using a GPS receiver. No group was tested with the same stimulus more than once thereby avoiding the possibility of dependencies in the data. Groups were therefore not tested if found within 1 km of previously a tested group. Calls produced spontaneously were also recorded *ad libitum* throughout the day, together with the context in which the calls was given where possible.

Visual Predator Models

A crowned eagle visual model was constructed by YP. For the visual leopard model, we used a commercially produced replica. Size, shape, posture, and colouration of both models matched those of real animals well (cf. *Fig. 11.3*).

In these experiments, YP systematically searched the study area for monkey groups with the help of a field assistant. Groups were located using either their vocalisations, or noises originating from individuals moving through the canopy. Locations were determined using a GPS. Once located, the most likely direction of group progression was estimated. The experimenter then circumnavigated the group at a distance far enough to avoid detection and positioned the predator model along the anticipated path in relatively open locations, either on the ground (< 1 m) or in a tree (range 1.0 – 13.0 m, average 4.0 m). The experiment simulated a natural situation since both predators are encountered on the ground and in trees (Shultz 2001, Jenny & Zuberbühler 2005, KA *pers. obs.*). The distance between the model and the subject was noted and coded as either < 10 m or ≥ 10 m. An experimental trial was usually terminated 45 min after detection of a stationary model, 20 min after detection of moving models, or when the group moved away. Again, the possibility of pseudo-replication was avoided by ensuring that no group within each area was tested with the same stimulus more than once.

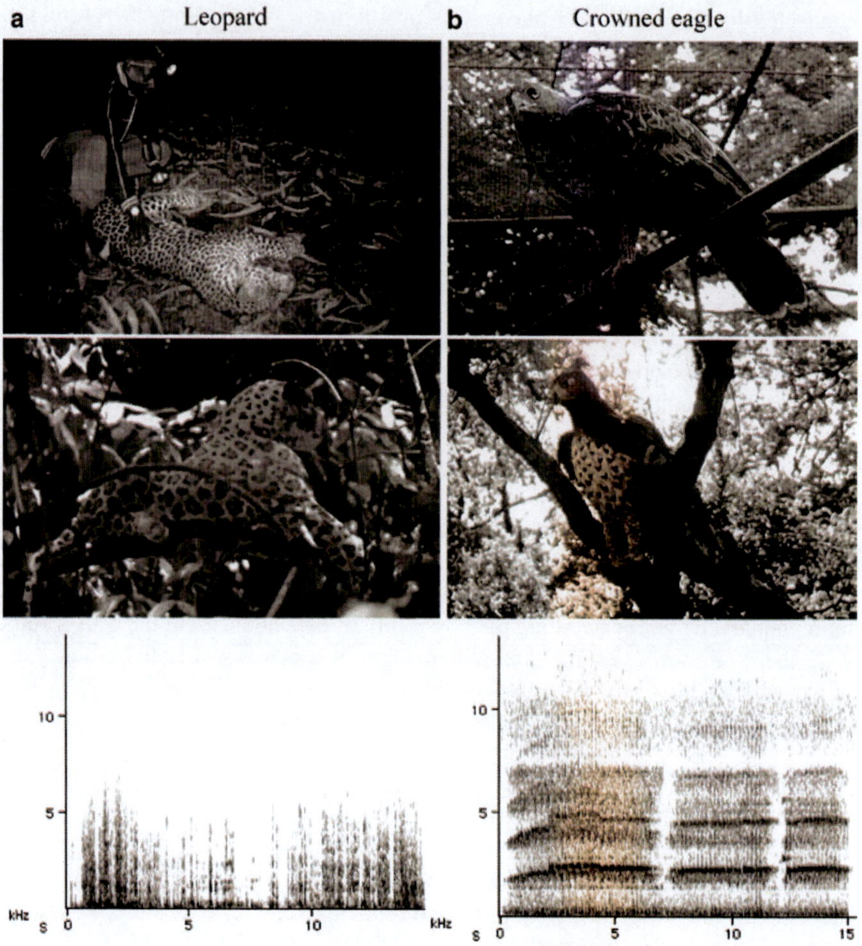

Fig. 11.3 Predators and visual and acoustic predator models used to simulate predation events. Visual predator models manufactured by YP and spectrographic representations of typical vocalisations by leopard and crowned eagles (photos: D. Jenny, KZ, YP, KA)

– Experiment 1. Responses to stationary predators

After positioning the predator model the experimenter moved away and, while hiding under a camouflage cover, started recording the approaching monkeys' vocal behaviour for at least 5 min, but often longer, before the first monkey detected the model. Vocalisations were tape-recorded with a cassette recorder and a directional microphone.

– Experiment 2. Responses to moving ground predators

Nigerian monkeys are confronted with different types of ground predators that vary in their hunting techniques and predation pressure, particularly large cats, chimpanzees, and human poachers. Although hunting pressure by humans is probably much lower at Gashaka Gumti compared to other parts of West Africa, illegal poaching does occur. To investigate whether putty-nosed monkeys respond differentially to different types of ground predators, we conducted the following experiment. A field assistant wore either (a) olive-brown clothing as typically used by local hunters or (b) a large piece of leopard-print fabric, which completely covered the body. While the experimenter began recording, using the equipment specified for experiment 1, the field assistant either walked quietly (poacher model) or moved with a hunched posture (leopard model) through the area occupied by the monkeys. All vocal responses produced from the time that the field assistant approached the group until he moved away after ca. 20 min from first detection. Trials were abandoned if the subjects detected the experimenter at any time.

The Pyow-Hack Sequence

Does the Pyow-Hack Sequence Elicit Group Movement?

Pilot observations during the early parts of this research indicated that when males produced a pyow-hack (P-H) sequence as part of their loud calling bouts, the group often subsequently travelled away from the area.

To investigate whether the P-H sequence possessed a communicative function, putty-nosed monkey groups were systematically searched for throughout the study area. Once detected, the experimenters positioned themselves approximately 50 m from the group and out of sight. Vocal behaviour was monitored for 30 min to ensure that the group was not aware of the presence of the experimenters, and that no other disturbance, indicated by alert or alarm calls, had occurred prior to an experiment. The playback equipment was then positioned 0 – 2 m from the ground, in preparation for broadcasting the playback stimulus, while continuing to monitor the group. Recording of all vocalisations began 5 min prior to the playback stimulus and continued until all anti-predator vocalisations ceased. Playback stimuli were presented as natural series and consisted of leopard growls in the first instance. This regularly elicited the P-H sequence from males. Twenty minutes later, a second stimulus was broadcast, consisting of a rapid series of hacks, recorded during an encounter with an unidentified large low flying eagle. This stimulus reliably elicited male vocal responses and enabled the experimenter to relocate the group. Initial and final group locations were recorded using a GPS. Vocalisations were recorded using as described in the section before. Trials were abandoned or discarded if the experimenter, field assistant or equipment were detected at any time before the end of the trial.

To better interpret our experimental findings we simultaneously collected observational data. A single habituated group of putty-nosed monkeys was followed for 2 months from 06:30–12:00 and again from 15:00 – 18:30 in 5-day blocks separated by at least 2 days over 2 months (total = 30 days). The position of the estimated centre of mass of the group was recorded using a GPS. The group location was recorded at 15-min intervals thereafter to determine the distance travelled. If the male produced loud calls, the group location was recorded by the field assistant. The time and sequence of calls was noted by KA. From that point on, the estimated centre of mass of the group was again recorded at 15-min intervals. Distance travelled was calculated for each uninterrupted 45-min block.

Are Pyow-Hack Sequences Causally Responsible for Eliciting Group Movement?

To investigate this hypothesis, we conducted experiments on a habituated group of putty-nosed monkeys. From an extensive library of recordings of the group male's calls, we selected 5 different examples of P-H sequences, pyow series, and hack series. A fourth set of stimuli consisted of 5 P-H sequences where the pyows and hacks used were recorded as part of call series, which did not contain P-H sequences and edited together. These stimuli were required to test whether it was the arrangement of pyows and hacks in the sequence that was responsible for eliciting movement, or whether there were any acoustic cues available in natural P-H sequences that might be responsible for listeners responding differently to the individual calls. Before each trial the location of the male was established and his behaviour monitored by a field assistant throughout the trial. When the male moved to the periphery of the group KA selected an adult female who was at least 50 m from the adult male thus ensuring that she did not have the opportunity to observe his behaviour. Using a walkie-talkie, a second field assistant was directed to a suitable location (*Fig. 11.4*), where he set up the playback equipment (see previous section).

Playback from this position ensured that the direction from which the female heard the stimulus was consistent with the known location of the male, and that other individuals were not able to detect the equipment and learn to associate it with the stimulus. One of the four stimulus types was broadcast and the behaviour of the

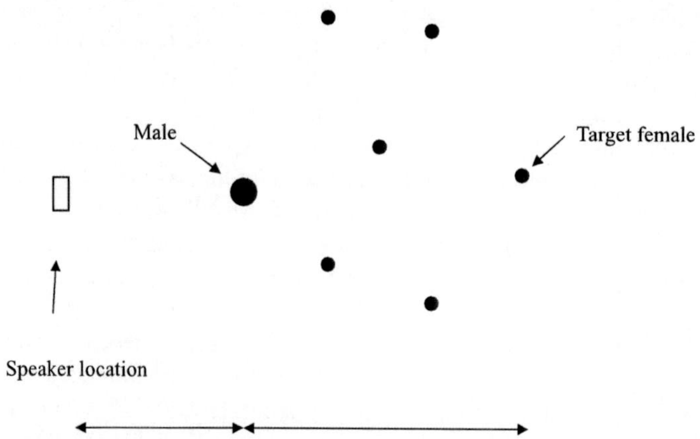

Fig. 11.4 Experimental set up and speaker location in relation to male and target females

target female monitored for 20 min. Her initial position was read off a GPS unit at the time of the broadcast and at the end of the 20 min monitoring period. Verbatim descriptions of relevant behaviours, such as visual scanning, the time at which any locomotion began, the direction of the locomotion, and any vocalisations produced by the target female were recorded. For each sequence type we conducted 10 trials (2×5 examples) in a randomised order resulting in a total of 40 trials. Trials were conducted at different times of the day and not more than once every 3 days. Trials were abandoned if the equipment setup was observed or if the group male responded with his own calls.

Statistical Analysis

For experiments investigating responses to the acoustic predator models, we conducted 49 trials in which leopard growls were presented and 34 trials with eagle shrieks. Adult males produced loud calls in 38.8 % (19 / 49) of leopard trials and in 58.8 % (20 / 34) of eagle trials. Three leopard trials were discarded where the male appeared to respond to non-experimental stimuli such as large branches snapping. Analyses were therefore conducted on the remaining 16 leopard trials and on 20 eagle trials. All trials within each predator category represent the vocal response of a different male thus avoiding problems of pseudo-replication. Probabilities associated with the results of Wilcoxon signed-ranks tests are exact. All tests were two tailed with a significance level set at 0.05.

We completed a total of 87 visual model trials (visual eagle, $n = 23$; visual leopard, $n = 29$; moving human, $n = 23$; moving leopard, $n = 12$). Trials were excluded if the male did not respond to the models (visual eagle, $n = 2$; visual leopard, $n = 4$; moving human, $n = 16$; moving leopard, $n = 0$). In these cases it was often impossible to be sure that the lack of a response was because the male had simply not seen the model. These numbers therefore cannot be assumed to be meaningful. We also excluded a number of trials because the target males called from a distance and never approached the site (visual eagle, $n = 13$; visual leopard, n = 8; moving human, $n = 1$; moving leopard, $n = 1$). This suggested that the male might have called in response to the females' alarm calls rather than because he had detected the predator model himself. The number of calls, calling duration, call rate and the proportion of hacks given in response to the models were compared using an exact Mann-Whitney *U*-test. Responses to each model type were analysed at two levels. First, the distributions of the three main call series types were compared. Given that P-H sequences can play a specific role in anti-predator strategies, we were interested in the proportion of responses in which these sequences occurred. Consequently we compared the distribution of all responses in which P-H sequences were given alone or inserted within the three major series types. We compared the distribution of response types using Fisher's exact test. All tests were two-tailed. Critical *p*-values were adjusted to $\alpha = 0.025$ due to multiple comparisons between visual and acoustic predator models. For comparisons between moving human and leopard models $\alpha = 0.05$. Multinomial logistic regression was used to identify the contextual variables that predict the category of response produced in acoustic and visual model experiments. Sequence categories with $n < 4$ are not included in the analysis.

Distance travelled following male responses to leopard growls in which P-H sequences either occurred or did not were compared using Kruskal-Wallis analysis of variance. Where the results of these tests were significant, we conducted exact Mann-Whitney *U post hoc* tests to identify where significant differences lay. The same tests were used in the accompanying observational study. Distance travelled and latency to travel following playbacks of pyows, hacks, and pyow-hack sequences we conducted Friedman tests. Where the results of these tests were significant, we conducted exact Wilcoxon signed ranks *post hoc* tests to determine where significant differences lay. All tests were two-tailed with α set at 0.05 except for *post hoc* comparisons. In these cases a Bonferroni correction was applied resulting in $\alpha = 0.025$ (0.05 / 2). All statistical analyses were conducted using SPSS 14.0.

Results

Alarm Calling

General Response Characteristics

Statistical comparisons of general response characteristics to each model type are given in *Tab. 11.1*. The latency to call in response to acoustic stimuli was shorter after hearing eagle stimuli compared to leopard stimuli (median latency to respond to eagle v leopard = 18.6 s vs. 53.1 s. Mann-Whitney U test: $U = 61$, $n_1 = 20$, $n_2 = 16$, $p = 0.001$). Measures of the latency to call in response to visual models were considered to be unreliable since the time at which the male first saw the models could not be accurately assessed. According to other measures, the responses of adult males to the visual models of the predators were similar in many respects to those produced to acoustic models. Male responses to the acoustic and visual models of crowned eagles did not differ in terms of the number of calls given and call series duration. Responses to the visual leopard model were comparable in these respects although responses to the acoustic leopard model were shorter. Calls were given at similar rates irrespective of model type. Males responded more strongly to acoustic eagle models than leopard models both in terms of the number of calls produced and in the duration of the call series although there was no difference when faced with visual models of these two predator types. The proportion of hacks given to each of the model types was dependent on the predator category, with more hacks being given to eagle models, but was unaffected by whether the model was acoustic or visual (cf. *Tab. 11.1*).

Table 11.1 Call series characteristics (median values) to visual and acoustic models of predators

	Visual models		Acoustic models		Comparisons of predator categories (a)		Comparisons of model types (a)	
Series characteristics	Eagle	Leopard	Eagle	Leopard	Visual	Acoustic	Eagle	Leopard
	$n = 8$	$n = 17$	$n = 20$	$n = 16$	E vs. L (b) n(E) = 8, n(L) = 17	E vs. L (b) n(E) = 20, n(L) = 16	V vs. A (b) n(V) = 8, n(A) = 20	V vs. A (b) n(V) = 17, n(A) = 16
Number of calls	9	10	13.5	4.5	$U = 55.5$ $p = 0.482$	$U = 64$ $p = 0.002$	U = 53.5 p = 0.185	U = 62 p = 0.007
Call series duration (s)	87.4	133.5	110.8	20.6	$U = 61$ $p = 0.711$	$U = 81$ $p = 0.011$	U = 73 p = 0.746	U = 58 p = 0.004
Call rate (calls / s)	0.12	0.14	0.17	0.24	$U = 52.5$ $p = 0.38$	$U = 128$ $p = 0.32$	U = 57 p = 0.254	U = 84 p = 0.062
Proportion of hacks	0.63	0.06	0.93	0.30	$U = 21.5$ $p = 0.005$	$U = 67$ $p = 0.002$	U = 71 p = 0.652	U = 102.5 p = 0.225

(a) Mann-Whitney U test for comparisons of call series within and between model types
(b) E = eagle, L = leopard, V = visual model, A = acoustic model

Analysis of Alarm Call Types

Males produced between 2 – 40 calls in response to eagle stimuli and hacks were the first calls given in response to eagle stimuli in 90.0 % of cases (18 / 20). When the call series as a whole was considered, pyows occurred, especially from call 5 onwards although hacks continued to predominate (median proportion of hacks = 0.93, pyows = 0.07; Wilcoxon signed ranks test: $z = -2.35$, $n = 20$, $p = 0.016$). In response to leopard stimuli, males produced between 2 – 20 calls. Pyows were the first calls given in response to leopard stimuli in 93.8 % of cases (15 / 16). Again over the whole call series, pyows predominated (median proportion of hacks = 0.21; pyows = 0.79. Wilcoxon signed ranks test: $z = -2.18$, $n = 16$, $p = 0.027$). However, hacks occurred quite often during the early part of the call series at positions and analysis of the first 5 calls in a series demonstrate that hacks were just as likely to occur as pyows (median proportion of hacks = 0.45, pyows = 0.55; Wilcoxon signed ranks test: $z = -1.55$, $n = 16$, $p = 0.131$).

Alarm Call Series Characteristics

Despite the strong biases in favour of one call type over the other in response to the two categories of predator stimuli, the frequency with which both call types occurred within a single call series was surprising (*Fig. 11.5*).

However, close examination of individual cases revealed distinct patterns in the alarm call responses. Calls were given as one of three basic series types: (a) pyow series, (b) hack series, and (c) transitional series. In addition, the P-H sequence was given alone or inserted at the beginning of, or within, one of these call series'. We categorised responses according to these series types. In 50.0 % of acoustic eagle model trials and 37.5 % of visual eagle model trials, the male responded with a hack series. Transitional series, accounted for a further 25.0 % of responses in acoustic eagle model trials and 37.5 % of visual eagle model trials. Pyow series were the least common response to eagle models accounting for only 5.0 % of responses to the acoustic eagle model and for 12.5 % of responses to the visual eagle model. Some alarm calling series had P-H sequences inserted within them. This occurred in 20.0 % (4 / 20) of acoustic eagle trials and in 12.5 % (1 / 8) of visual eagle trials. In acoustic eagle trials, P-H sequences were inserted into hack series on two occasions and once into transitional once into a pyow series. In visual eagle trials, one transitional series contained two P-H sequences.

In 31.3 % of acoustic leopard model trials and 47.1 % of visual leopard model trials, the male responded with a pure series of pyows. Transitional series were rare, occurring only once in response to the acoustic leopard model. Hack series were equally rare and occurred only once in response to the visual leopard model. In contrast with responses to the eagle models, responses to leopard models often contained P-H sequences. They were either inserted into pyow series, often at the beginning (25.0 % of acoustic leopard model trials and 41.1 % of visual leopard model trials) or given alone (30.0 % of acoustic leopard model trials and 5.9 % of visual leopard model trials).

EAGLE ACOUSTIC MODEL

No.	n	1	2	3	4	5	6	7	8	9	10	11
1	23	H	H	H	H	**P**	H	H	P	P	P	P
2	8	H	H	H	H	H	H	H	H			
3	29	H	H	H	H	H	**P**	H	H	H	**H**	H
4	13	H	H	H	H	H	**P**	P	**P**	P	P	P
5	38	H	H	H	H	H	H	H	H	H	H	H
6	33	H	H	**P**	H	P	H	H	**H**	H	H	H
7	12	H	H	H	H	H	H	H	H	H	H	H
8	12	P	P	P	P	P	P	P	P	P	P	P
9	40	H	H	H	H	H	H	H	H	H	H	H
10	13	H	H	H	H	H	H	H	H	H	H	H
11	20	H	H	H	H	H	H	H	H	H	H	H
12	5	H	H	H	H	H						
13	14	H	H	H	H	H	H	H	**P**	P	P	P
14	19	H	H	H	H	H	H	H	H	**P**	P	P
15	13	H	H	H	**P**	P	P	P	P	P	P	P
16	19	P	P	P	P	P	P	P	P	**P**	H	**P**
17	12	H	H	H	H	**P**	P	P	**P**	P	P	P
18	7	H	H	H	H	H	H	H				
19	2	H	H									
20	17	H	H	H	H	H	H	H	H	H	H	H

EAGLE VISUAL MODEL

n	1	2	3	4	5	6	7	8	9	10	11
11	H	H	H	H	**P**	P	P	P	P	P	P
33	P	P	P	P	**P**	P	P	P	P	P	P
6	H	H	H	H	**P**	P					
8	H	H	H	H	**P**	P	P	P			
3	H	H	H								
57	H	H	H	H	H	H	H	H	H	H	H
5	H	H	H	H	H						
10	H	H	H	H	**P**	H	**P**	P	**P**	H	

LEOPARD ACOUSTIC MODEL

No.	n	1	2	3	4	5	6	7	8	9	10	11
1	14	P	P	P	P	P	P	P	P	P	P	P
2	4	P	P	P	P							
3	20	P	P	H	H	H	P	P	P	P	P	P
4	9	P	P	H	**P**	P	P	P	P	P		
5	5	P	H	H	**P**	H						
6	11	P	H	H	P	P	P	P	P	**P**	H	**P**
7	2	P	H									
8	3	H	H	**P**								
9	15	P	H	H	H	**P**	P	P	P	P	P	P
10	5	P	P	P	P	P						
11	14	P	P	P	P	P	P	P	P	P	P	P
12	3	P	P	H								
13	4	P	P	H	H							
14	4	P	H	H	H							
15	4	P	H	H	H							
16	2	P	P									
17												

LEOPARD VISUAL MODEL

n	1	2	3	4	5	6	7	8	9	10	11
7	P	P	P	P	P	P	P				
10	P	P	P	P	P	P	P	P	P	P	
37	P	P	P	P	P	P	P	P	P	P	P
63	P	P	P	H	H	P	P	P	P	P	P
8	P	H	H	**P**	P	P	P	P			
10	P	H	H	H	H	**P**	P	P	P	P	
3	P	H	H								
40	P	P	P	P	P	P	P	P	P	P	P
49	H	H	H	H	H	H	H	H	H	H	H
9	P	P	P	H	**P**	H	H	**P**	P		
6	P	P	P	P	P	P					
18	P	P	P	P	P	P	P	P	P	P	P
21	P	P	H	H	**P**	P	P	P	P	P	P
5	P	H	H	P	P						
10	P	P	P	P	P	P	P	P	P	P	
16	P	P	P	P	P	P	P	**P**	H	**P**	P
9	P	P	P	P	P	P	P	P	P		

Fig. 11.5 Raw data of calling patterns of the first 11 calls given in response to visual and acoustic models of a crowned eagle and a leopard. Trials are depicted in chronological order. *n* = total number of alarm calls given. All calls following the last call depicted are of the same type as that indicated by the last call depicted. Different call series types are indicated by coloured boxes: dark grey = pyow sequence; white = hack sequence; P-H sequence = light grey; transitional series consist of a sequence of hacks followed by a sequence of pyows. Significantly long pauses (mean + 3 SD of pauses between all preceding calls) between sequences are indicated by marking the first call of the sequence in bold

Table 11.2 Proportion of call series types (%) given in response to each model type

	Acoustic models		Visual models	
	Eagle	Leopard	Eagle	Leopard
Series type	$n = 20$	$n = 16$	$n = 8$	$n = 17$
Hack series	60.0 (12)	0.0	37.5 (3)	5.9 (1)
Transitional series	30.0 (6)	6.3 (1)	50.0 (4)	0.0
Pyow series	10.0 (2)	56.3 (9)	12.5 (1)	88.2 (15)
P-H sequence given	20.0 (4)	62.5 (10)	12.5 (1)	47.1 (8)
P-H sequence given alone	0.0	37.5 (6)	0.0	5.9 (1)
P-H sequence inserted	20.0 (4)	25.0 (4)	12.5 (1)	41.2 (7)

Comparison Between Acoustic and Visual and Different Predator Models

The type of predator model did not affect the frequency with which male alarm calling series were given (*Tab. 11.2*, *Fig. 11.6*). Pyow series were usually given in response to leopard models whereas hack and transitional series were generally associated with eagle models irrespective of whether they were presented visually or acoustically. However, there was a non-significant bias in favour of hack series being given to acoustic eagle models as opposed to transitional responses whereas the converse was true for visual eagle models. This relatively low proportion of transitional responses to acoustic eagle models also resulted in the non-significant difference between the proportion of transitional series given to eagle and leopard acoustic models. P-H sequences were given far more often in response to leopard models than to eagle models although this distinction was less clear for visual models. This was because males often inserted the P-H sequence within a longer alarm call series when presented with the visual leopard model whereas there was a slight bias in favour of giving the P-H sequence but no further alarm calls to acoustic models.

Contextual Factors

Why though did male putty-nosed monkeys produce a range of calling patterns to both predator types? For example, why would males respond to an eagle model with a transitional as opposed to a hack series? A number of environmental variables were investigated which could have influenced the responses. For example, variation in the distance between speaker and caller should simulate differences in the level of perceived threat experienced by the listeners. Differences in luminosity and the openness of the vegetation might also have a bearing on the perceived level of threat since this should affect local visibility and therefore the monkeys ability to locate a predator visually. However, a multinomial logistic regression carried out on the acoustic model data revealed that only the predator category affected which of the three basic alarm call series types was produced (*Tab. 11.3*).

A similar analysis for visual predators also revealed that the predator category predicted the type of call series produced, whereas the position of the model, on the ground or in a tree, and the distance between the caller and model had no effect (*Tab. 11.4*).

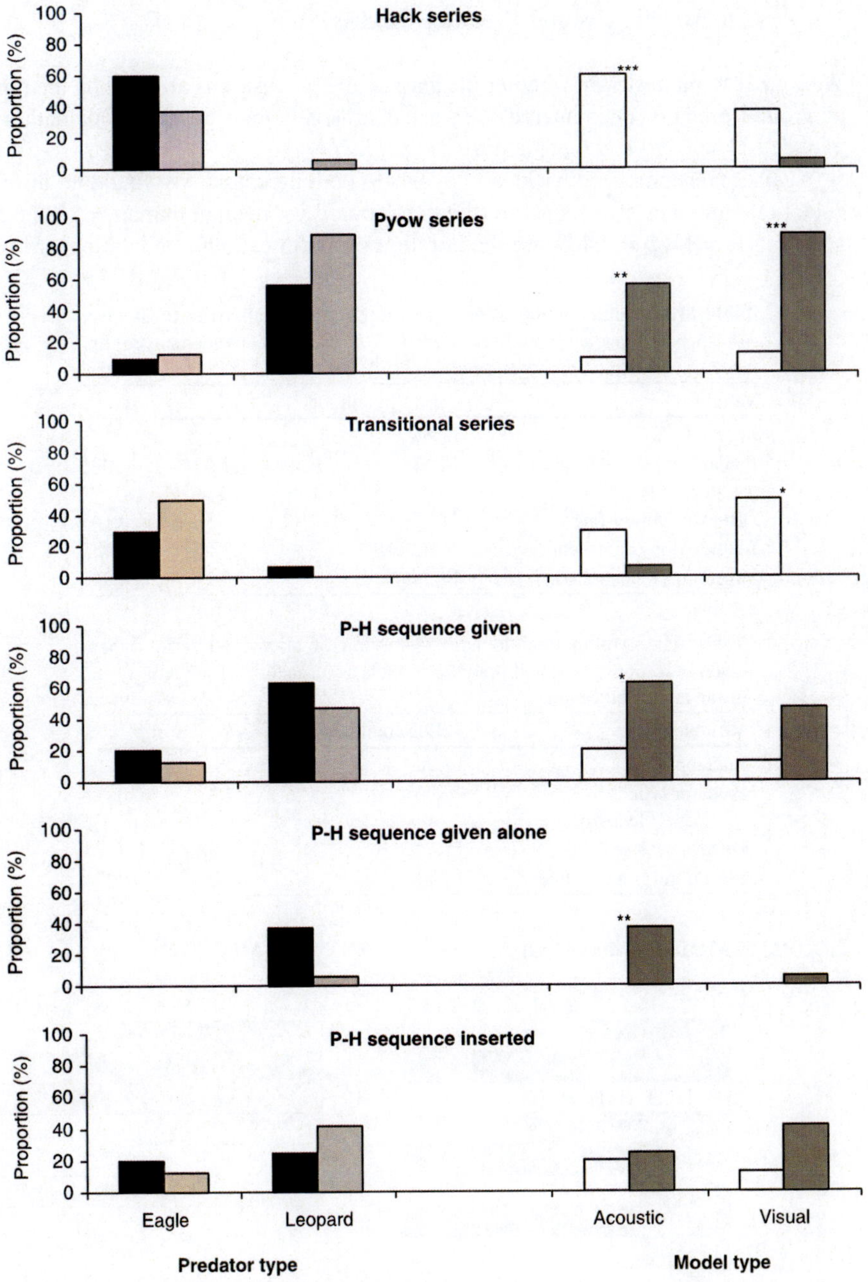

Fig. 11.6 The proportion of responses in which (**a**) hack series, (**b**) transitional series, (**c**) pyow series, (**d**) at least one P-H sequence is given to each stimulus type (eagle / leopard, acoustic / visual). The last category is further broken down to show the proportion of responses in which (**e**) a P-H sequence was given alone and, (**f**) a P-H sequence was inserted into a longer call series. Significant results of Mann-Whitney U tests are shown as, $*p < 0.025$, $**p < 0.005$, $***p < 0.001$

Response to Moving Ground Predator Models

We were also interested in whether the alarm call response was affected by the type of ground predator encountered, i.e., a leopard or human poacher. Our data set included 11 leopard and 6 eagle trials (*Fig. 11.7*).

Again, males produced hacks and pyows to both predators. We found no difference in the number of calls given (median leopard vs. median human = 5 vs. 8.5; $U = 25.5$, $n = 17$, $p = 0.469$), in the duration of alarm calling (median leopard vs.

Table 11.3 Multinomial logistic regression of alarm call series types given in response to acoustic predator models as a function of predator type, distance, illumination and vegetation

Variable	–2Log likelihood	χ^{24}	*p*
Intercept	17.88		
Predator type (eagle / leopard)	45.15	27.27	0.001
Distance (close / far)	21.77	3.88	0.143
Illumination (light / dark)	21.62	3.73	0.155
Vegetation (open / closed)	17.99	0.11	0.947

Table 11.4 Multinomial logistic regression of alarm call series types given in response to visual predator models as a function of predator type, distance, and elevation

Variable	–2Log likelihood	χ^{24}	*p*
Intercept	17.76		
Predator type (eagle / leopard)	29.06	14.3	0.001
Distance (close / far)	15.44	0.68	0.712
Elevation (ground / tree)	14.99	0.23	0.892

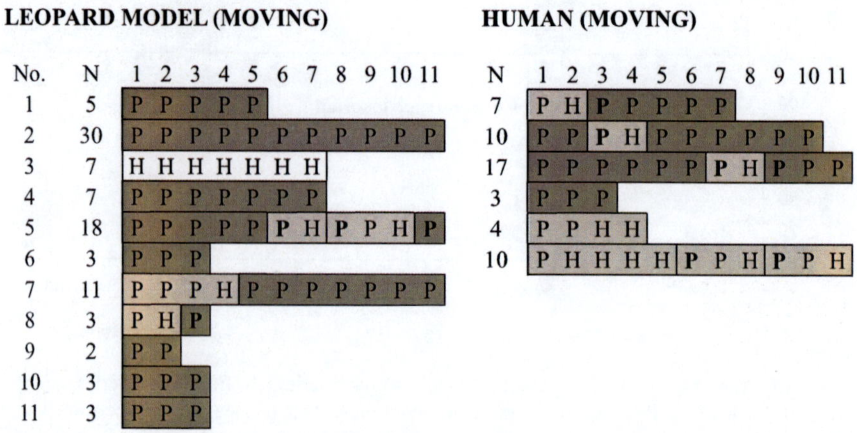

Fig. 11.7 Raw data of calling patterns of the first 11 calls given in response to approaching leopard and human predator models. See *Fig. 11.6* for explanation

median human = 25.6 s vs. 90.9 s; $U = 26$, $n = 17$, $p = 0.525$), in the call rate (median leopard vs. median human = 0.17 calls / s vs. 0.15 calls / s; $U = 30$, $n = 17$, $p = 0.808$), or in the proportion of hacks (median leopard vs. human = 0 vs. 0.19; $U = 16.5$, $n = 17$, $p = 0.085$).

In this experiment, the two model types elicited very similar calling patterns in males. Pyow sequences were the most common (proportion of responses given to leopard vs. human = 0.91 vs. 0.83, Fisher s exact $p = 1.000$). P-H sequences often preceded or were inserted with pyow series given to the human model but were less common in responses to the leopard model although this difference was not significant (proportion of responses given to leopard vs. human = 0.13 vs. 0.67, Fisher s exact $p = 0.162$). On one occasion, a P-H sequence and no other call structure was given to the human model but never to the leopard model. Overall, P-H sequences were given significantly more often to the human model than to the leopard model (proportion of responses in which P-H sequences were given to leopard vs. human = 0.13 vs. 0.83, Fisher s exact $p = 0.05$).

Loud Calling Under Natural Conditions

We were interested in the variety of contexts in which males produced different loud calling patterns as the stronger the association between call patterns and context, the more likely that call patterns might function referentially. We were able to record 173 call series together with the contexts in which they occurred (hack series, $n = 22$; transitional series $n = 24$; pyow series, $n = 127$). *Fig. 11.8* shows that there was a substantial degree of overlap in the contexts in which different call series types where given, many of them nonpredatory.

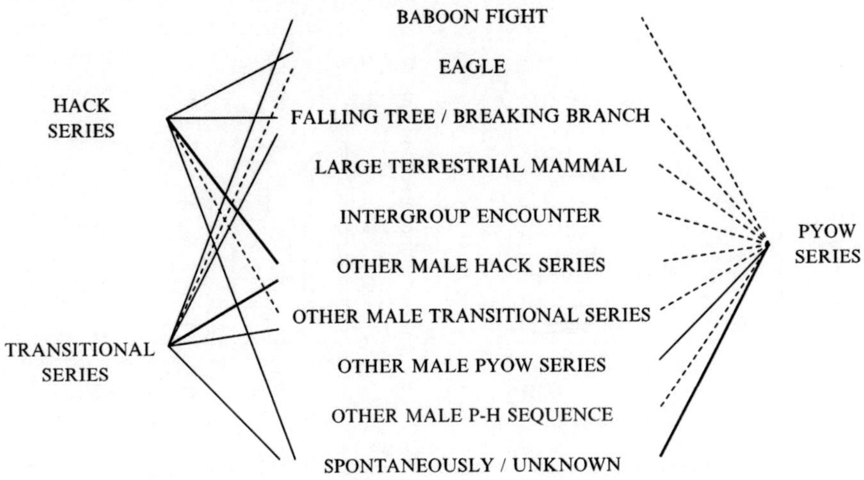

Fig. 11.8 The natural contexts in which hack, transitional and pyow call series occurred. The proportion of recorded call series of each type given in each context are indicated by; dashed line < 10 %, solid line 11 – 20 %, bold line > 30 %

The P-H Sequence

Does the P-H Sequence Elicit Group Movement?

Alarm calling series that included the P-H sequence resulted in groups travelling significantly further (median distance = 85 m) than when this sequence was not included (median distance = 17 m; $n_1 = 9$, $n_2 = 8$, $U = 2$, $p = 0.0003$; *Fig. 11.9a*). In the parallel observational study (*Fig. 11.9b*), the habituated group travelled significantly further in the 45 min following call series which included a P-H sequence (median distance = 110 m) than when the male produced another type of sequence (median distance = 30 m) or when he did not call at all (mean distance = 14 m; n_1 = 29 m, n_2 = 43 m, n_3 = 65 m, df = 2, $\chi^2 = 19.27$, $p = 0.00007$; post hoc tests, P-H vs. Other, $U = 325.5$, $p = 0.0005$, P-H vs. None, $U = 426.5$, $p = 0.0001$, Other vs. None, $U = 1236.5$, $p = 0.314$).

Are P-H Sequences Causally Responsible for Eliciting Group Movement?

Results showed that playback of hack series (often indicating eagle presence) inhibited movement in females (median = 1 m) and caused long latencies to move (median = 7.4 min). Playback of pyow series (indicating a range of disturbances, including leopard presence) elicited relatively rapid responses (median = 1.1 min), although the distance travelled towards the speaker was moderate (median = 4.0 m). Movement was often accompanied by scanning behaviour in the direction of the calls, as if trying to acquire additional information about the cause, a pattern also seen with natural pyows. P-H sequences resulted in significant travel towards the

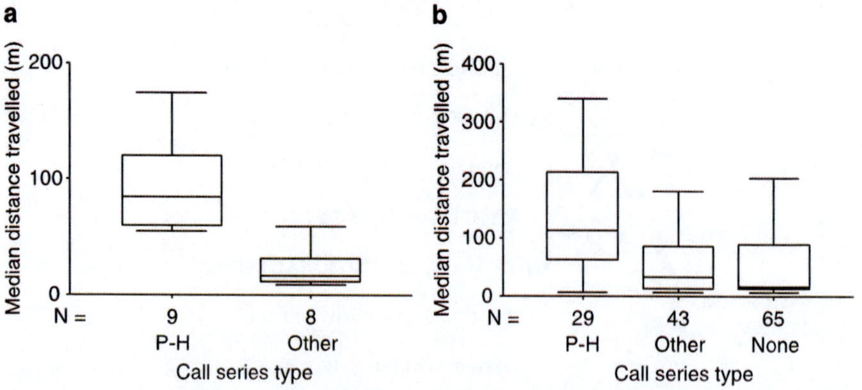

Fig. 11.9 Median distance travelled following call series, (**a**) produced in response to experimental stimuli, or (**b**) occurring naturally, according to whether call series included a P-H or not. P-H = call series including the P-H sequence. Other = any call series not including the P-H sequence. None = median distance travelled in the absence of preceding calls. Box plots indicate medians, inter-quartiles and ranges; outliers are indicated by open circles

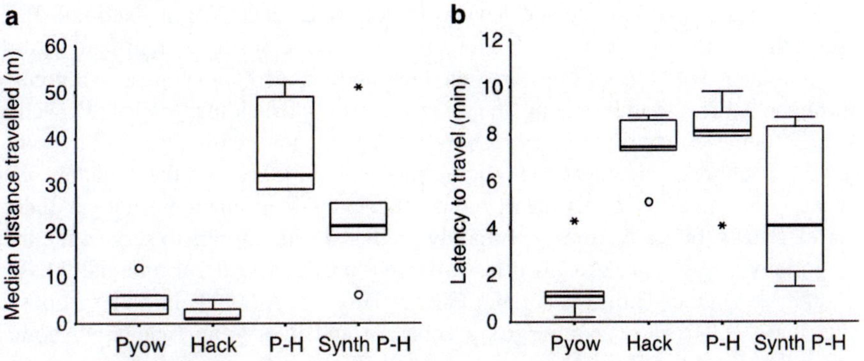

Fig. 11.10 (**a**) Median distance travelled and (**b**) median latency to travel after hearing playbacks of different call series by the groups male: "pyow" series, "hack" series, natural P-H sequences (P-H), and artificially composed P-H sequences (Synth P-H). Box plots indicate medians, interquartiles and ranges; outliers are indicated by open circles

calls (median = 32 m), although females rarely responded quickly (median = 8.1 min). Statistically, both travel distance and latencies differed significantly across the three conditions (distance, $n_{\mathrm{H}} = n_{\mathrm{P}} = n_{\mathrm{P\text{-}H}} = 5$, df = 2, $\chi^2 = 8.4$, $p = 0.008$, Fig. 11.10a; latency, $n_{\mathrm{H}} = n_{\mathrm{P}} = n_{\mathrm{P\text{-}H}} = 5$, Friedman test df = 2, $\chi^2 = 7.6$, $p = 0.024$, *Fig. 10.10b*).

At this point, it could be argued that there were subtle acoustic differences between pyows and hacks produced as part of P-H sequences compared to calls given in response to predators, and that listeners responded to these differences, rather than call combinations. Although acoustic analyses did not support this hypothesis (Arnold & Zuberbühler 2006a, 2008b) we carried out an experiment, this time using artificially composed P-H sequences edited from pyows and hacks originally given to predator stimuli.

We found no differences between artificial ($\mathrm{P\text{-}H_{synth}}$) and natural P-H sequences ($\mathrm{P\text{-}H_{nat}}$) in the distance travelled by females and their response latencies (distance, $\mathrm{P\text{-}H_{nat}}$ median = 32 m, $\mathrm{P\text{-}H_{synth}}$ median = 21 m: $n_1 = n_2 = 5$, $z = -1.483$, $p = 0.188$, *Fig. 11.10a*; $\mathrm{P\text{-}H_{nat}}$ median = 8.1 min, $\mathrm{P\text{-}H_{synth}}$ median = 4.1 min: $n_1 = n_2 = 5$, $z = -1.214$, $p = 0.313$. *Fig. 11.10b*). We concluded that the communicatively salient property of the signal was how calls were assembled, rather than any context-specific acoustic variation within pyows and hacks.

Discussion

Our work focuses on primate communication, cognition and the evolution of human language, the foundations of which are likely to be present in our closest cousins. Putty-nosed guenons are a good model for study as they are a typical arboreal Old World monkey that faces the kinds of ecological and selection pressures common to most primate species.

Putty-nosed monkeys in the Kwano region of Gashaka Gumti National Park spend the majority of their time feeding, resting and socialising with little to disturb them in their daily lives. Males produce their aptly named loud calls intermittently throughout the day advertising their presence to surrounding groups. However, when trouble does arise, males respond with these loud calls, warning even the most distant group members and other groups within the area. Of the three loud call types, booms are very rarely heard and consequently their function is not yet understood. Pyows, by far the most commonly produced call, appears to serve a number of purposes. Hacks are produced less often but nonetheless occur on a daily basis. Gautier & Gautier-Hion (1977) and Gautier-Hion *et al.* (1983) initially proposed that these calls promote intra-group cohesion and inter-group spacing. Another study demonstrated that male putty-nosed monkeys, like many other primate males, also use their repertoire of loud calls as alarm calls (Eckardt & Zuberbühler 2004). In our studies, we used experimental techniques to investigate what information these calls might carry.

We broadcast the vocalisations of two of their main predator types to different groups of monkeys to see if their alarm call responses differed depending on whether they heard the shrieks of a crowned eagle or the growls of a leopard. First, we found that males called for longer, produced more calls and with a shorter delay before calling in response to the eagle stimuli than to the leopard stimuli. They were also more likely to call in response to eagle stimuli than to leopard stimuli suggesting that aerial predators produce a stronger response. There is good reason why this might be so as eagles are likely to represent a greater threat. First, the density of leopards is quite low in the study area. Second, eagles can attack at all heights whereas leopards rely on ambush from the ground and are, therefore, likely to be a lesser threat to primarily arboreal primates. Third, mobbing by numerous group members tends to drive leopards and other terrestrial predators away (Zuberbühler *et al.* 1999b). The best defence against crowned eagles, on the other hand, is to hide in dense foliage since this restricts their access to smaller monkeys. The larger males may be relatively invulnerable to attack and can thus take an active role in driving predatory eagles away (KA, YP *pers. obs.*; Cordeiro 2003). It should be noted that responses to the visual eagle and leopard models were equally strong, probably because the stationary models, unlike acoustic stimuli, were located and did not react to detection. It is likely, therefore, that males adopted similar calling strategies that both informed the predators that they had been detected and promoted mobbing by other group members.

The Notion of Reference in Primate Vocalisations

Previous studies have proposed that the alarm calls of male cercopithecine monkeys serve as functionally referential signals (Seyfarth *et al.* 1980, Zuberbühler 2000a, c, 2001). In these studies, call production was tightly coupled with the perception of a particular class of objects or events, while calling was largely unaffected by other contextual factors such as the distance and presumed degree of threat imposed by the predator (Evans 1997). For recipients, these types of alarm calls come to effectively label predator types, potentially creating interesting parallels with, and sharing at least some of the properties of human words (e.g., Cheney & Seyfarth 1990). Do the alarm calls of male putty-nosed monkeys exhibit a similar type of referentiality? Our results show that we were unable to replicate the general findings reported from Diana or vervet monkeys, i.e., that one call type was consistently produced in response to leopard-related stimuli and another consistently to eagle-related stimuli, with little or no overlap. For example, in response to the visual eagle model, roughly one third of calls produced in an average call series were pyows. Similarly, an average call series given to the visual leopard model was made up almost entirely of pyows, whereas 30 % of all calls given to the acoustic leopard model were hacks. Crucially though, pyows and hacks also occur as part of P-H sequences which are often produced in non-predatory contexts (Arnold & Zuberbühler 2006b). Pyows and hacks, in other words, do not refer exclusively to leopards and eagles, but are part of a more flexible communication system that can convey information at the level of call series. An important additional point is that pyows and hacks given as part of P-H sequences are acoustically indistinguishable from those given as part of other sequences, either to predators (Arnold & Zuberbühler 2006b) or in other contexts. At the level of the individual call, male putty-nosed monkey alarm calls do not refer to different predator classes.

Do the Alarm Call Series of Male Putty-Nosed Monkeys Refer to Particular Events?

Male putty-nosed monkeys almost exclusively give calls in series and these call series often contain distinctive patterns that are easily recognised. "Pyow series" consist entirely of pyows, "hack series" consist entirely of hacks, and "transitional series" consist of a series of hacks followed by a switch to a series of pyows. P-H sequences can be produced alone or inserted at different positions within the other three main call series types. Other more randomly organised series do occur but are relatively rare (Arnold & Zuberbühler 2006a).

P-H sequences were given more often in response to leopard stimuli than eagle stimuli and responses consisting entirely of one or more P-H sequences were given only to leopard stimuli. As already mentioned, leopards that have been detected represent little threat to monkeys and so movement away from this predator, elicited by the P-H sequence, should be a safe option. However, movement through the canopy when eagles are present increases exposure and is likely to heighten the risk of attack. This call sequence appears to function referentially as a cue to group travel but not about predator categories since it is often produced in non-predatory contexts (Arnold & Zuberbühler 2006b). Instead, its differential use in the face of different types of predators is most likely to reflect the selection of sensible anti-predator responses in accordance with the hunting strategies of the predators concerned.

The distribution of pyow series, hack series and transitional series given in response to eagle and leopard stimuli indicates great potential for the encoding of predator category information. With P-H sequences removed from alarm call responses for the sake of simplicity and discarding responses consisting only of P-H sequences, a distinct pattern emerges. Hack series and transitional series, which begin with a series of hacks, were given almost exclusively to eagle stimuli (acoustic = 18 / 20; visual = 7 / 8) whereas pyow series were given equally exclusively to leopard stimuli (acoustic = 9 / 10; visual = 15 / 16). The very strong associations between call series types and predator categories should enable listeners to predict which of the two predator categories has been detected with a considerable degree of accuracy using call series information alone. Thus, the potential of these call series to encode referential information is very great.

The Contexts of Naturally Occurring Loud Call Series

When "asked" to respond to the presence of eagles and leopards by experimentally simulating their presence, males generally produced hack or transitional series to eagles and pyow series to leopards, sometimes including a P-H sequence as well. However, these same calling patterns series were elicited by a much wider range of natural (i.e., non-experimental) stimuli, such as non-predatory mammals and inanimate objects. Pyows were often given in situations in which it was impossible to detect any cause. The male could start calling, without any apparent reason, while resting or feeding after which he would resume his previous activity. It appeared that the male s attention was not directed to any particular location, nor did he become more vigilant. The apparently relaxed attitude of the male during calling in many "unknown" contexts suggested that these calls were not produced in response to any external event but were spontaneous. This behaviour contrasted sharply with calls in response to a detectable event. In these cases, the male ceased all other activities, became vigilant, and, when the object of attention was in view, oriented his body toward the disturbance and monitored it closely. If not in view, he would often move in the direction of noisy disturbances. Another context in which males often called was in response to hack, transitional and pyow series of neighbouring males.

These observations demonstrate that hack, pyow, and transitional series are given in a variety of contexts, and not exclusively to the presence of eagles or terrestrial mammals, respectively. Since it is difficult to think of a unifying reference for all these different calling occasions, it is unlikely that the different call series are linked to just one reference (i.e., mental concept), as proposed for Diana and vervet monkeys. Instead the same call series are regularly triggered by multiple sources, which suggests that a variety of mental concepts are involved to the same calling behaviour.

What Information Do the Loud Calls of Male Putty-nosed Monkeys Convey?

One strategy for successfully linking a small number of call series with a large variety of external events is to combine calling with discrete external stimuli available during the calling event. In this way, meaning arises by combining calls with contexts. For hacks, this is not a difficult problem since the range of contexts in which they are given is relatively small, such as hack or transitional series of other males, breaking branches or falling trees, eagles, noisy baboon fights and, on one rather exceptional occasion, to a duck flying at the forest edge (KA *pers. obs.*). In experimental contexts they are given to eagle stimuli and, exceptionally, to leopard stimuli. It was also possible to elicit hack series with loud stimuli that were unfamiliar to the monkeys, such as a rape alarm operated at close quarters (KA *pers. obs.*). It is possible that the unifying theme underlying these events is that the monkey found himself in a state of high arousal but this hypothesis would require independent verification through other variables, ideally some simple physiological measure such as heart rate. In sum, hacks and transitional series are all given in response to disturbances. They are linked to a variety of contexts with few intrinsic commonalities and appear to function exclusively as alarm calls.

Pyows, on the other hand, are given in a much wider range of contexts. Spontaneous calling has been mentioned, in which the calls appear to function simply to advertise the males presence to members of his own group and neighbouring groups, supporting the view that pyows function to regulate cohesion and spacing (Gautier & Gautier-Hion 1977). The next most common context was in response to the calls of other males. A proportion of these calls are likely to serve the same function as those given spontaneously, especially in response to the pyows of other males. Thus, at the proximate level, pyows appear to draw the attention of conspecifics to the calling male, whether they are members of his own group or an adjacent group. Pyows are also given, although much less frequently, to events that can trigger hacks and transitional series, such as falling trees and breaking branches and, with the lowest frequency, to noisy baboon fights. More specifically, pyow series are common during inter-group encounters and in response to large terrestrial mammals. In experimental contexts, pyow series are often given to leopard stimuli and, as part of transitional responses, to eagle stimuli. Again, pyow series given during inter-group encounters probably serve to advertise the males presence both to his own group and to the encountered group. The remaining contexts are all related to disturbances of some kind, suggesting that pyows can also function as alarm calls.

An Exceptional Guenon?

Why are putty-nosed monkeys different from other previously described guenon species, particularly Diana and vervet monkeys? Macedonia & Evans (1993) suggested that the evolution of referential alarm-calling systems is driven by the incompatibility of escape strategies in response to different predator types. This possibility is well illustrated by ground dwelling mammals that respond in a single plane, for example by bolting into burrows. In these species, alarm call behaviour often appears to reflect the level of threat experienced by the caller regardless of the predators biological class. Arboreal primates, on the other hand, operate in three-dimensional space and anti-predator responses are more complex, making referential alarm calling systems an adaptive trait. The actual escape strategies employed by arboreal monkeys are, in fact, not so clear-cut. Increased visual scanning is often the most common response in listeners, who often do not take any evasive action. In Diana monkeys, for example, detection of a leopard also results in group members descending to lower levels in the canopy to keep the predator in sight while eagle detection results in movement into dense vegetation (Zuberbühler *et al.* 1997, Eckardt & Zuberbühler 2004, KZ *pers. obs.*). This contrasts with the behaviour of mainly terrestrial savannah species, such as vervet monkeys (Struhsaker 1967) and ringtailed lemurs, which spend much of their time on the ground in riverine, drought-adapted scrub forests (Jolly 1966, Sussman 1974, 1977). For these species, escape strategies are more clear-cut, either climbing into or descending from trees or remaining on the ground and finding cover (Seyfarth *et al.* 1980, Pereira & Macedonia 1991).

Putty-nosed monkeys may not be the exception but the rule. Studies of arboreal white-faced capuchins (Fichtel *et al.* 2005, Digweed *et al.* 2005), redfronted lemurs, white sifakas (Fichtel & Kappeler 2002) and Coquerel sifakas (Fichtel & van Schaik 2006) all describe mixed alarm calling systems where aerial alarm calls appear to be referential in most cases but calls given to terrestrial predators are not.

The evolutionary origins of divergence thus remain something of a mystery. Putty-nosed monkeys live in habitats that are largely comparable to those of Diana or Campbells monkeys, suggesting that ecological differences are unlikely to account for differences in the evolution of referential calling. It should be pointed out, however, that analyses of Diana and Campbells monkeys alarm calling behaviour included only the first few alarm calls given, respectively (Zuberbühler *et al.* 1997, Zuberbühler 2000b). Important variation in alarm call production, particularly in the later parts of a series, might therefore have been missed. Diana monkey call sequences are often short and there is ample experimental evidence that recipients respond to these sequences as if the corresponding predator were present, suggesting that the first few calls sufficiently serve as carriers of referential information in this species. With putty-nosed monkeys, comparable experiments are currently ongoing and with the exception of P-H sequences (Arnold & Zuberbühler 2006b) it still remains unclear what the relevant meaningful units are in this species communication system. Investigations into other closely related primates, particularly blue monkeys (*Cercopithecus mitis*; Butynski *et al.* 2002) are likely to provide important additional information concerning the evolutionary origins of these different calling systems.

The use of the P-H sequence is a phenomenon that appears to set putty-nosed monkeys apart from other cercopithecines, indeed all other primates. Examples of apparently meaningless rule-governed combinations of calls are fairly common in animals (Marler 1977). By contrast, a limited number of primate studies have described call combinations in which one call type appears to act as a contextual modifier for another (Cleveland & Snowdon 1982, Robinson 1984, Zuberbühler 2002). In all of these cases, the difference in the meaning of the call combination compared with the meanings of the constituent calls is a matter of degree and not kind.

The P-H sequence is unlike any previously described call sequence since it is meaningful, and its meaning is functionally distinct since it does not appear to be related to any meaning conveyed by its constituent elements even in a modified form. It also differs from other examples of call sequences insofar that it is often produced within a call series, which is composed of the same call types, and therefore within the same context, and yet is distinguishable as a functional unit.

Intuitively, the type of call combinations produced by tamarins and capuchin monkeys seem more interesting than the more common, meaningless variety because of the semantic relationships that hold between the sequences and their elements, i.e., the meanings of the constituent calls contribute directly to the overall meaning of the call combination. For example, alarm and alert calls produced in combination are given in contexts described as intermediate between the contexts in which each call type is produced alone (Cleveland & Snowdon 1982). But, it is precisely this relationship that limits the range of things that can be communicated about since these combinations can only generate information that is very closely related to the information that the individual calls can convey. However, call combinations that are not constrained in this way can potentially generate as many meaningful expressions as there are combinations. The P-H sequence is a rare example of the realisation of this potential.

Acknowledgements We thank the Nigerian National Park Service for providing the research permits, the Nigerian Conservation Foundation (NCF) for help with logistics and the Chester Zoo Nigeria Biodiversity Programme for core support to the *Gashaka Primate Project*. We also thank Ali Tappare, Buba Bello, Isaac Timothy, and Halidu Ilyusu for help in the field and David Stothard for invaluable technical assistance. This research has been funded by research grants from the Wenner Gren Foundation, the European Science Foundation ("Origins of Man, Language, and Languages"), the British Academy, the Leverhulme Trust, and the EC Pathfinder FP6 ("Origins of Referential Communication").

Literature

Arnold, K. & Zuberbühler, K. (2006a). The alarm calling system of adult male putty-nosed monkeys, *Cercopithecus nictitans martini*. *Animal Behaviour* 72: 643–653.

Arnold, K. & Zuberbühler, K. (2006b). Semantic combinations in primate calls. *Nature* 441: 303.

Arnold, K. & Zuberbühler, K. (2008). Meaningful call combinations in a nonhuman primate. *Current Biology* 18 (5).

Arnold, K., Pohlner, Y. & Zuberbühler, K. (2008). A forest monkey's alarm call series to predator models. *Behavioral Ecology and Sociobiology* 62: 549–559.

Bickerton, D. (1990). *Language and Species*. Chicago: The University of Chicago Press.

Brown, M. M., Kreiter, N. A., Maple, J. T. & Sinnott, J. M. (1992). Silhouettes elicit alarm calls from captive vervet monkeys (*Cercopithecus aethiops*). *Journal of Comparative Psychology* 106: 350–359.

Butynski T. M., Chapman C. A., Chapman L. J. & Weary D. M. (1992). The use of male blue monkey pyow calls for long-term individual identification. *American Journal of Primatology* 28: 183–189.

Cheney, D. L. & Seyfarth, R. M. (1990). *How Monkeys See the World*. Chicago: The University of Chicago Press.

Cheney, D. L. & Seyfarth, R. M. (1992). Meaning, reference, and intentionality in the natural vocalizations of monkeys. In Nishida, T., McGrew, W. C., Marler, P., Pickford, M. & de Waal, F. (eds.), *Topics in Primatology, vol 1, Human Origins*. Tokyo: Tokyo University Press, pp. 315–330.

Clarke, E., Reichard, U. & Zuberbühler, K. (2006). The syntax and meaning of wild gibbon songs. *PLoS One* 1: e73.

Cleveland, J. & Snowdon, C. T. (1992). The complex vocal repertoire of the adult cotton- top tamarin (*Sanguinus oedipus oedipus*). *Zeitschrift für Tierpsychologie* 58: 231–270.

Cordeiro, N. J. (2003). Two unsuccessful attacks by crowned eagles (*Stephanoaetus coronatus*) on white-throated monkeys (*Cercopithecus mitis*). *African Journal of Ecology* 41: 190–191.

Davis, L. S. (1984). Alarm calling in richardsons ground-squirrels (*Spermophilus- richardsonii*). *Zeitschrift für Tierpsychologie* 66: 152–164.

Digweed, S. M., Fedigan, L. M. & Rendall, D. (2005). Variable specificity in the antipredator vocalizations and behavior of the white-faced capuchin, *Cebus capuchinus*. *Behaviour* 142: 997–1021.

Dunn, A. (1993). *The Large Mammals of Gashaka-Gumti National Park, Nigeria: Line Transect Surveys from Forest and Savannah*. Report to WWF-UK, NCF & Federal Ministry of Agriculture, Water Resources and Rural Development / Nigeria.

Eckardt, W. & Zuberbühler, K. (2004). Cooperation and competition in two forest monkeys. *Behavioral Ecology* 15: 400–411.

Evans, C. S. (1997). Referential signals. In Owings, D. H., Beacher, M. D. & Thompson, N. S. (eds.), *Perspectives in Ethology, Volume 12 Communication*. New York: Plenum Press, pp. 99–143.

Evans, C. S. & Marler, P. (1995). Language and animal communication: parallels and contrasts. In Roitblatt, H. (ed.), *Comparative Approaches to Cognitive Science*. Cambridge, MA: MIT Press, pp. 241–282.

Fichtel, C. & Kappeler, P. M. (2002). Anti-predator behavior of group-living Malagasy primates: mixed evidence for a referential alarm call system. *Behavioral Ecology and Sociobiology* 51: 262–275.

Fichtel, C. & van Schaik, C. P. (2006). Semantic differences in sifaka (*Propithecus verreauxi*) alarm calls: a reflection of genetic or cultural variants? *Ethology* 112: 839–849.

Fichtel, C., Perry, S. & Gros-Louis, J. (2005). Alarm calls of white faced capuchin monkeys: an acoustic analysis. *Animal Behaviour* 70: 165–176.

Gautier, J. P. & Gautier-Hion, A. (1977). Communication in old world monkeys. In Sebeok, T. A. (ed.), *How Animals Communicate*. Bloomington, Indiana: Indiana University Press, pp. 890–964.

Gautier-Hion, A. & Gautier, J. P. (1974). Les associations polyspecifiques de Cercopitheques du Plateau de M Passa (Gabon). *Folia Primatologica* 22: 134–177.

Gautier-Hion, A. (1988). The diet and dietary habits of forest guenons. In Gautier-Hion, A., Bourliere, F., Gautier, J. P. & Kingdon, J. (eds.), *A Primate Radiation: Evolutionary Biology of the African Guenons*. Cambridge: Cambridge University Press, pp. 257–283.

Gautier-Hion, A., Emmons, L. H. & Dubost, G. (1980). A comparison of the diets of three major groups of primary consumers of Gabon (primates, squirrels and ruminants). *Oecologia* 45: 182–189.

Gautier-Hion, A., Quiris, R. & Gautier, J. P. (1983). Monospecific vs. polyspecific life: a comparative study of foraging and antipredatory tactics in a community of Cercopithecus monkeys. *Behavioral Ecology and Sociobiology* 12: 325–335.

Gouzoules, S., Gouzoules, H. & Marler, P. (1984). Rhesus monkey (*Macacca mulatta*) screams: Representational signalling in the recruitment of agonistic aid. *Animal Behaviour* 32: 182–193.

Hauser, M. D., Chomsky, N. & Fitch, W. T. (2002). The faculty of language: What is it, who has it, and how did it evolve? *Science* 298: 1569–1579.

Jenny, D. & Zuberbühler, K. (2005). Hunting behaviour in West African forest leopards. *African Journal of Ecology* 43: 197–200.

Jolly, A. (1966). *Lemur Behavior*. Chicago: The University of Chicago Press.

Kroodsma, D. E. & Miller, E. H. (1982). *Acoustic Communication in Birds, Vol. 2*. New York: Academic Press.

Lancaster, J. B. (1975). *Primate Behavior and the Emergence of Human Culture*. New York: Holt Rinehart & Winston.

Leavesley, A. J. & R. D. Magrath (2005). Communicating about danger: urgency alarm calling in a bird. *Animal Behaviour* 70: 365-373.

Leger, D. W., Owings, D. H. & Gelfand, D. L. (1980). Single note vocalizations of California ground squirrels: graded signals and situation – specificity of predator and socially evoked calls. *Zeitschrift für Tierpsychologie* 52: 227–246.

Lemasson, A., Hausberger, M. & Zuberbühler, K. (2005). Socially meaningful vocal plasticity in adult Campbell s monkeys (*Cercopithecus campbelli*). *Journal of Comparative Psychology* 119: 220–229.

Macedonia, J. M. & Evans, C. S. (1993). Variation among mammalian alarm call systems and the problem of meaning in animal signals. *Ethology* 93: 177–197.

Macedonia, J. M. & Polak, J. F. (1989). Visual assessment of avian threat in semi-captive ringtailed lemurs (*Lemur catta*). *Behaviour* 111: 291–304.

Macedonia, J. M. & Yount, P. L. (1991). Auditory assessment of avian predator threat in semicaptive ringtailed lemurs (*Lemur catta*). Primates 32: 169–182.

Manser, M. B. (2001). The acoustic structure of suricates alarm calls varies with predator type and the level of response urgency. *Proceedings of the Royal Society of London Series B – Biological Sciences* 268: 2315–2324.

Marler, P. (1977). The structure of animal communication sounds. In Bullock, T. H. (ed.), *Recognition of Complex Acoustic Signals: Report of the Dahlem Workshop*. Berlin: Abakon Verlag, pp. 17–35.

Marler, P., Evans, C. S. & Hauser, M. D. (1992). Animal signals: motivational referential or both? In H. Papousek, H. Jürgens, U. & Papousek, M. (eds.), *Nonverbal Vocal Communication: Comparative and Developmental Approaches*. Cambridge: Cambridge University Press, pp. 66–86.

Mitani, J. C. & Marler, P. (1989). A phonological analysis of male gibbon singing behavior. *Behaviour* 106: 20–45.

Mitani, M. (1991). Niche overlap and polyspecific associations among sympatric cercopithecoids in the Campo Animal Reserve, Southwestern Cameroon. *Primates* 32: 137–151.

Morton, E. S. (1982). Grading, discreteness, and motivational structural rules. In Kroodsma, D. E. & Miller, E. H. (eds.), *Acoustic Communication in Birds: Volume 1*. New York: Academic Press, pp. 183–212.

Oates, J. F. (1996). *African Primates Status Survey and Conservation Action Plan*. IUCN / SSC Primates Specialist Group.

Ogden, C. K. & Richards, I. A. (1923). *The Meaning of Meaning: A Study of the Influence of Language upon Thought and of the Science of Symbolism*. New York: Harcourt, Brace & World, Inc.

Owings, D. H. & Leger, D. W. (1980). Chatter vocalisations of California ground squirrels: predator-social-role specificity. *Zeitschrift für Tierpsychologie* 54: 163–184.

Owings, D. H. & Virginia, R. A. (1978). Alarm calls of California ground squirrels (*Spermophilus beecheyi*). *Zeitschrift für Tierpsychologie* 46: 58–70.

Payne, K. & Payne, R. (1985). Large scale changes over 19 years in songs of humpback whales in Bermuda. *Zeitschrift für Tierpsychologie* 68: 89–114.

Pereira, M. E. & Macedonia, J. M. (1991). Ringtailed lemur anti-predator calls denote predator class, not response urgency. *Animal Behaviour* 41: 543–544.

Pohlner, Y. (2005). *Ruf-Muster arborealer Affen in Gashaka, Nigeria*. Diplomarbeit, Institut für Zoologie, Universitat Freiburg / Germany.

Rainey, H. J., Zuberbühler, K. & Slater, P. J. B. (2004). Hornbills can distinguish between primate alarm calls. *Proceedings of the Royal Society of London Series B – Biological Sciences* 271: 755–759.

Ramakrishnan, U. & Coss, R. G. (2000). Age differences in the response to adult and juvenile alarm calls by bonnet macaques (*Macaca radiata*). *Ethology* 106: 131–144.

Riede, T., Bronson, E., Hatzikirou, H. & Zuberbühler, K. (2005). Vocal production mechanisms in a non-human primate: morphological data and a model. *Journal of Human Evolution* 48: 85–96.

Robinson, J. G. (1984). Syntactic structures in the vocalizations of wedge-capped capuchin monkeys, *Cebus olivaceus*. *Behaviour* 90: 46–79.

Seyfarth, R. M. & Cheney, D. L. (1980). The ontogeny of vervet monkey alarm calling: a preliminary report. *Zeitschrift für Tierpsychologie* 54: 37–56.

Seyfarth, R. M. & Cheney, D. L. (1993). Meaning, reference and intentionality in the natural vocalizations of monkeys. In Roitblat, H. L., Herman, L. M. & Nachtigall, P. E. (eds.), *Language and Communication: Comparative Perspectives*. Hillsdale, New Jersey: Lawrence Earlbaum Associates, pp. 196–220.

Seyfarth, R. M. & Cheney, D. L. (2003). Signalers and receivers in animal communication. *Annual Review of Psychology* 54: 145–173.

Seyfarth, R. M., Cheney, D. L. & Marler, P. (1980). Monkey responses to three different alarm calls: Evidence of predator classification and semantic communication. *Science* 210: 801–803.

Shultz, S. (2001). Notes on interactions between monkeys and African crowned eagles in Taï National Park, Ivory Coast. *Folia Primatologica* 72: 248–250.

Struhsaker, T. T. (1967). Ecology of vervet monkeys (*Cercopithecus aethiops*) in The Masai-Amboseli Game Reserve, Kenya. *Ecology* 48: 891–904.

Struhsaker, T. T. (1970). Phylogenetic implications of some vocalisations of *Cercopithecus* monkeys. In Napier, J. R. & Napier, P. H. (eds.), *Old World Monkeys: Evolution, Systematics and Behaviour*. London: Academic Press, pp. 365–407.

Sussman, R. W. (1974). Ecological distinctions between two species of lemur. In Martin, R. D., Doyle, D. A. & Walker, C. (eds.), Prosimian Biology. London: Duckworth, pp. 75–108.

Sussman, R. W. (1977). Socialization, social structure and ecology of two sympatric species of lemur. In Chevalier-Skolnikoff, S. & Poirier, F. E. (eds.), Primate Bio-Social Development: Biological, Social, and Ecological Determinants. New York: Garland, pp. 515–528.

Suzuki, R., Buck, J. R. & Tyack, P. L. (2006). Information entropy in humpback whale songs. *Journal of the Acoustical Society of America* 119: 1849–1866.

Warkentin, K. J., Keelay, A. T. H. & Hare, J. F. (2001). Repetitive calls of juvenile Richardsons ground squirrels (*Spermophilus richardsonii*) communicate response urgency. *Canadian Journal of Zoology* 79: 569–573.

Wich, S. A. & de Vries, H. (2006). Male monkeys remember which group members have given alarm calls. *Proceedings of the Royal Society B-Biological Sciences* 273: 735–740.

Wich, S. A. & Sterck, E. H. M. (2003). Possible audience effect in Thomas langurs (Primates; *Presbytis thomasi*): an experimental study on male loud calls in response to a tiger model. *American Journal of Primatology* 60: 155–159.

Zuberbühler, K. (2000a). Referential labelling in Diana monkeys. *Animal Behaviour* 59: 917–927.

Zuberbühler, K. (2000b). Interspecific semantic communication in two forest monkeys. *Proceedings of the Royal Society of London Series B – Biological Sciences* 267: 713–718.

Zuberbühler, K. (2000c). Causal cognition in a nonhuman primate. *Cognition* 76: 195–207.

Zuberbühler, K. (2001). Predator-specific alarm calls in Campbell s guenons. *Behavioral Ecology and Sociobiology* 50: 414–422.

Zuberbühler, K. (2002). A syntactic rule in forest monkey communication. *Animal Behaviour* 63: 293–299.

Zuberbühler, K. Jenny, D. & Bshary, R. (1999b). The predator deterrence function of primate alarm calls. *Ethology* 105: 477–490.

Zuberbühler, K., Noë, R. & Seyfarth, R. M. (1997). Diana monkey long-distance calls: messages for conspecifics and predators. *Animal Behaviour* 53: 589–604.

Zuberbühler, K., Cheney, D. L. & Seyfarth, R. M. (1999a). Conceptual semantics in a nonhuman primate. *Journal of Comparative Psychology* 113: 33–42.

Chapter 12
Patriarchal Chimpanzees, Matriarchal Bonobos: Potential Ecological Causes of a *Pan* Dichotomy

Volker Sommer, Jan Bauer, Andrew Fowler, and Sylvia Ortmann

Abstract Chimpanzees and bonobos, despite being closely related hominoid primates, differ in female gregariousness and dominance style. Violent male aggression is not atypical in chimpanzee societies and is vented against both other males and females in intra- as well as inter-group conflicts; relationships amongst females are rather weak. Bonobo societies, on the other hand, are female-centred; reports about inter-group conflict are rare to absent but there are numerous reports of blood-drawing injuries inflicted upon males by coalitions of females.

This dichotomy is of potential interest for the understanding of social dynamics in contemporary human societies, too, given that *Pan* and *Homo* shared a last common ancestor 5 – 6 million years ago. For example, political agendas to achieve a greater equality of the sexes might have to work against our natural inclinations, if the last common ancestor exhibited the patriarchal tendencies found in chimpanzees. Vice versa, if the last common ancestor possessed the matriarchal tendencies of bonobos, then patriarchal tendencies in contemporary human societies could be understood as rather recent cultural developments that can be more easily undone by counter-measures, i.e., changes in socio-economic dynamics.

Such assertions are not unproblematic, given millions of years of evolution. Nevertheless, a reconstruction of the ancestral roots of the behavioural suites of *Homo* and *Pan* will have to rely on a causal understanding of the different species psychologies of chimpanzees versus bonobos. These should in some ways be related to ecology. Both species have a mixed diet dominated by fruit with a similar composition. To test if their diet differs in availability and quality, we collected data on habitat phenology and analysed nutritional content of food plants and non-food plants from a community of bonobos in Salonga National Park, Democratic Republic of Congo, and a community of chimpanzees living in Gashaka Gumti National Park / Nigeria.

V. Sommer (✉)
Department of Anthropology, University College London, London WC1E 6BT, UK
e-mail: v.sommer@ucl.ac.uk

V. Sommer and C. Ross (eds.), *Primates of Gashaka*, Developments in Primatology: Progress and Prospects 35, DOI 10.1007/978-1-4419-7403-7_12,

We found that chimpanzee diet is more diverse, whereas bonobos can rely on a few staple species for longer periods of time – which reflects the more seasonal climate at the chimpanzee site. Both species prefer fruit with elevated contents of water, sugar and fat, but chimpanzees have to cope with much higher levels of antifeedants such as tannins. Moreover, only bonobos have access to a herb with low levels of fibre but high protein. In addition, chimpanzees invest more time and energy in the removal of seeds from fruit and in digestion. The costs of acquisition of high quality food are thus higher in chimpanzees than in bonobos. The greater constraints in terms of food availability and quality are reflected in greater levels of female-female competition as evidenced by consistently lower levels of gregariousness in chimpanzees measured through the size of nest groups.

Thus, local ecologies can modify social behaviours. However, support for the hypothesis that ecological differences are at the heart of the dichotomy of sociality in *Pan* is not unequivocal because of considerable intra-specific variability throughout the geographical range of bonobos and, in particular, chimpanzees with respect to social processes and fluctuating parameters of flora, fauna and climate. Accordingly, future studies will have to explore the extent of this flexibility and if and how it covaries with local ecologies.

Keywords *Pan troglodytes* • *Pan paniscus* • Matriarchy • Food quality

Introduction

Hobbes, Rousseau, and Apes as Role Models

The closest living relatives of humans are members of the genus *Pan*: the chimpanzee (*Pan troglodytes*) and the bonobo (*Pan paniscus*) (*Fig. 12.1*). Research in the wild and in captivity has established that these two African apes not only share certain behavioural traits, but also differ in remarkable ways (for the following, see Susman 1984, Goodall 1986, Heltne & Marquardt 1989, Kano 1992, Wrangham *et al.* 1994, McGrew *et al.* 1996, Boesch & Boesch-Achermann 2000, Boesch *et al.* 2002, Reynolds 2005, Campbell *et al.* 2007).

The degree of these similarities and differences, apart from scientific interest, has entered public debates because of alleged implications for the rights and wrongs of modern human societies in areas such as gender equality and conflict management. We here aim to contribute to a better understanding of the potential causes of behavioural differences in bonobos and chimpanzees – information on which political and societal debates with a solid naturalistic framework should be based.

Bonobos typically occupy the moist evergreen lowland forests south of the Congo River, although more southern habitat may be a mosaic of primary forest, secondary growth, and grassland. Chimpanzees live north of the Congo, where they often coexist with gorillas. They occur in varied habitats, ranging from woodland-savannah

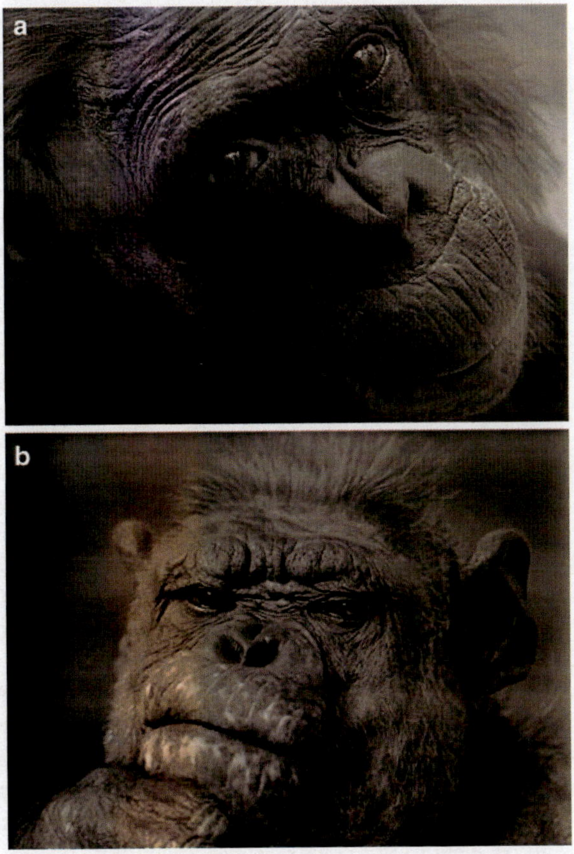

Fig. 12.1 The closest living relatives of humans are the two members of the genus *Pan*, bonobos (**a**) and chimpanzees (**b**) (photos: Jutta Hof)

and dry forests to moist evergreen forests, from sea level to altitudes several thousand metres high.

Despite varied biotopes, both species feed mostly and often almost exclusively on fruit from trees and climbers (average feeding time for chimpanzees 64 %, for bonobos 55 %). Leaves, flowers, tubers, and pith are minor diet components, complemented further by insects and vertebrates.

Both species form communities with multiple males and females that can number more than 100 members. These are typically split into smaller parties of various and often changing age-sex composition. Such a "fission-fusion society", with its varied party sizes for foraging, travel, and nesting, allows mitigation of resource competition. Females of both species tend to transfer from their natal groups, whereas males remain philopatric. Therefore, resident males are typically more closely related to each other than are females (Morin *et al.* 1994, Gerloff *et al.* 1999).

However, it is not these communalities that have created considerable attention in popular media, but the differences in behaviour – as these involve sexual behaviour, aggression, and male-female relationships that are of predictable interest to the general public (de Waal & Lanting 1997, Parker 2007, Sommer 2008). Nevertheless, the popular characterisation of bonobos and chimpanzees is often biased and imprecise. A typical stereotype is the night-and-day opposition "chimpanzees are from Mars, bonobos are from Venus". This suggests that chimpanzees are generally more aggressive, whereas bonobos are more peaceful and engage in constant sex. This simplification might well reflect a desire found in Western societies to identify a politically correct role model, which post-modern humans can emulate and which trumps the philosophy of Hobbes in favour of that of Rousseau. Thomas Hobbes (1588 – 1679) considered war as the natural state of human societies. Taming these aggressive tendencies would mean to stir humans through science and laws "against nature". Jean-Jaques Rousseau (1712 – 1778), on the other hand, believed that competition and animosity was unheard of in ancient human societies, and that our basically peaceful nature was corrupted only through the rise of technology and science. His mantra was therefore "back to nature". Various groups were romanticised, which allegedly incarnated this "good nature" – from the "noble savage" of the 19th century via hunter-gatherers and Pacific Islanders to chimpanzees – but all fell from grace, once their peace-loving attitudes were debunked as myths (for philosophical ramifications, see Vogel 1999, Sommer 2000).

So, what *do* we know about bonobos and chimpanzees? It is indeed correct that chimpanzee males form strategic alliances that may stalk, attack, and kill males of neighbouring communities and commit infanticide, and that adult males invariably dominate females and physically attack them in competition over resources. Such patterns have not been observed in wild bonobo communities – although much less is known about wild populations of *P. paniscus* than about those of *P. troglodytes*. Still, it is unlikely that future studies might reveal patterns of male-committed aggression in bonobos, which will come close to the frequency reported in some chimpanzee societies. Another clear-cut difference is the frequency, mode, and relative ease with which female bonobos interact. Unrelated female mammals tend to avoid each other, as their reproduction is limited by access to resources, and proximity to other females will increase the degree of resource competition (Trivers 1972). Nevertheless, grooming interactions between bonobo females are common, as are bouts of homosexual behaviour, in particular the so-called genito-genital rubbing (Fruth & Hohmann 2006).

In spite of this, bonobo societies are a far cry from the peace-loving sex-sedated hippie-like utopia typically portrayed in the popular press. Competition for resources still exists, but the balance of power over their control is tipped in favour of females. The crucial tool that enables females to dominate males is their ability to form coalitions – despite the fact, that they are, on average, as unrelated to each other as chimpanzee females (Gerloff *et al.* 1999). Such female coalitions are known to attack males, and they regularly inflict blood-drawing injuries including the loss of appendices such as fingers and toes (Parish 1996). Thus, alliances enable bonobo females to control access to food and how it is shared (Vervaecke *et al.* 2000, Fruth & Hohmann 2006).

Some debate exists as to how "real" these differences between bonobos and chimpanzees are, or if they are just gradual with the bonobo pattern well in the range of behaviours displayed by chimpanzees across Africa (Stanford 1998). But, it seems as if a majority of researchers maintain that the differences are profound and reflect quite distinct and deeply engrained species psychologies (see reviews in Wrangham 1986, Wrangham *et al.* 1996, Boesch *et al.* 2002).

Bonobo communities thus earn the label "matriarchal" (Greek, *archein*, to rule), given that a matriarchy characterises a gynocentric societal structure in which females have power and / or which is centred around females. In contrast, chimpanzee societies fulfil the criteria of "patriarchal", i.e., rule by males. Good evidence exists for human societies that are female-centred or matrilineal. However, there is widespread agreement that truly matriarchal societies did not exist in human prehistory – at least not as matriarchy in the sense of a reversed patriarchy. Of course, such statements are likely tainted by or embroiled in political controversy (e.g., Goldberg 1993, Eller 2000). Still, although female dominance is not uncommon in other mammals including primates (Hemelrijk *et al.* 2008), it is of more than scientific interest that such structure characterises also one of our closest relatives.

What might have caused the dichotomy between bonobos and chimpanzees? One can hypothesise that parties with female chimpanzees are forced to fragment because of severe intra-sexual competition for resources. Conversely, the friendly social relationships of female bonobos should reflect reduced levels of competition. These have been linked in the past to factors such as (a) an abundance of terrestrial herbs (Badrian & Malenky 1984), (b) high quality of terrestrial herbs (Wrangham *et al.* 1991), or (c) large food patch size (White & Wrangham 1988). Data for distinct populations of bonobos and chimpanzees provided some support for these hypotheses, but more factors are clearly involved (Chapman *et al.* 1994). In any case, there is little disagreement that local ecologies have, over evolutionary time, "led to differences in species psychology" (Wrangham *et al.* 1996: 45).

Basics of Nutritional Ecology

The two *Pan* species provide a fruitful model for the interplay between the distribution and quality of food resources and how this influences modes and levels of competition and thus social relations. Abundance and distribution of food resources are key elements of evolutionary theories trying to explain inter-specific variation of social systems (Wrangham 1979), particularly with respect to primates who exhibit an exceptional diversity (Kummer 1971, Wrangham 1980), including a considerable flexibility of how to manage conflicts (van Schaik 1989).

Food selection and foraging efficiency is affected by parameters such as nutritional quality and distribution and abundance of sources (Carlo *et al.* 2003, Saracco *et al.* 2004). One major working hypothesis is the assumption that individuals will select sources with high nutrient and energy levels (Emlen 1966), while avoiding chemical components that impede digestion, so-called anti-feedants (Alm *et al.* 2000,

Clauss *et al.* 2003). Plants that are typical for young, regenerating forests contain relatively low amounts of secondary compounds as defenses against predators, parasites, and diseases, or for interspecies competition (Harbourne 1993), while vegetation that grows under difficult conditions contains higher levels (Mueller-Harvey & McAllen 1992).

Still, nutritional requirements will vary both between and within species (related to age, sex, and life-history stage) and thus lead to varied preferences for different nutrients and abilities to tolerate anti-feedants. Therefore, feeding strategies incorporate factors such as digestive physiology (Witmer & van Soest 1998, Schaefer *et al.* 2003) and modes of food preparation (Corlett & Lucas 1990, Lambert 2006), in particular how seed is handled (Witmer 1998). Intake rates are high when seeds are simply ingested, but large seed loads decrease the density of nutrients and energy per unit of ingesta. Removal of seeds, on the other hand, increases handling time and decreases intake rate but maximises the proportions of desirable ingesta. Clues about feeding strategies can thus already be derived from macroscopic analyses of faecal samples, whereas faecal particle size can provide information on digestive strategies including masticatory effort before the food is swallowed (Campos-Acreiz *et al.* 2004, Clauss & Lechner-Doll 2001, Clauss *et al.* 2002). Larger particles have a relatively smaller surface area that decreases the accessibility of nutrients. Particle size also affects retention, i.e., gut passage time during which food is exposed to enzymes and microbes, as well as the turnover of food and, consequently, foraging behaviour.

To date, only few studies of primate nutritional ecology integrate measures of resource abundance, quality, and digestive physiology (e.g., howler monkeys: Milton 1978, Milton & McBee 1983; baboons: Altmann 1998; orang-utans: Leighton 1993). The feeding ecology of *Pan* species has been studied in a range of habitats but typically focused on the general type of food ingested and its temporal and spatial distribution and abundance, whereas information on nutritional ecology is rather limited (bonobos, e.g.: Lomako / Democratic Republic of Congo, Malenky 1990, Malenky & Stiles 1991; chimpanzees, e.g.: Ipassa / Gabon, Hladik 1977; Bossou / Guinea, Takemoto 2003; Kanyawara / Uganda, Wrangham *et al.* 1991; Budongo / Uganda, Reynolds *et al.* 1998; Gombe / Tanzania, Wrangham & Waterman 1983; Mahale / Tanzania, Matsumoto-Oda & Hayashi 1999). The picture emerging from this array of approaches is rather complex, and standardised methodologies are needed to improve comparability of results.

Pan food choice in general seems to be guided by a preference for macronutrients and / or avoidance of anti-feedants (bonobos: Malenky 1990; chimpanzees: Wrangham & Waterman 1983, Reynolds *et al.* 1998, Takemoto 2003). Probably because of such discriminatory abilities, the overall nutritional quality of chimpanzee diet, compared to that of sympatric monkeys, is higher than predicted by body mass (Conklin-Brittain *et al.* 1998), and only chimpanzees maintained high levels of macro-nutrients and relatively low levels of anti-feedants year-round (Wrangham *et al.* 1998). Nevertheless, some of these results are inconclusive as corresponding information on the nutritional content of non-food plant items were absent.

Comparing Chimpanzees and Bonobos

Previous studies on feeding and nutritional ecology are often not suitable for straightforward comparisons of chimpanzees and bonobos, because of different methodologies. We therefore ensured standardised methods of both data collection and analyses for a comparison of a chimpanzee community in Nigeria with a community of bonobos in the Democratic Republic of Congo (Hohmann *et al.* 2006, 2007, Bauer 2006, Fowler 2006) (*Fig. 12.2*).

Here, we synthesise and extend findings about a particularly important ecological factor – the availability and chemical quality of the major food of both *Pan* species: fruit. For this, we (a) identified annual variation of plant food production in both habitats; (b) compared fruits eaten compared to fruits not eaten for functionally different chemical components (nutrients, fibre, anti-feedants); (c) reconstructed food choice and digestive strategies from faecal samples; and (d) related the pattern of plant food availability and consumption to annual changes in gregariousness.

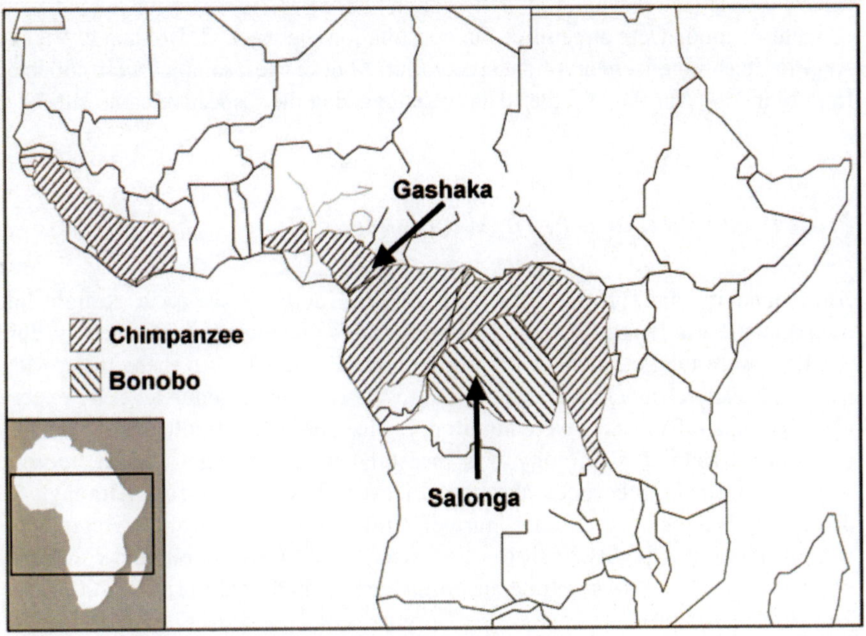

Fig. 12.2 Area of distribution of chimpanzees and bonobos across Africa, with study sites indicated at Gashaka Gumti National Park / Nigeria and Salonga National Park / Democratic Republic of Congo. Note that latitudinal and longitudinal distribution of bonobos is entirely overlapped by that of chimpanzees

Our research thus aims to understand better the constraints of plant food acquisition on large frugivores such as bonobos and chimpanzees, and the social consequences derived from these limitations – in particular the differences in gregariousness and dominance style.

Methods

Study Sites

The chimpanzee study site was inside Gashaka Gumti National Park / Nigeria at Kwano (elevation 583 m; 07°19′ N – 11°35′ E), 11 km from the village of Gashaka (Sommer *et al.* 2004). The bonobo study site was inside Salonga National Park / Democratic Republic of Congo at Lui Kotal (elevation 320 – 413 m; 02°45.610′ S – 20°22.723′ E; Hohmann & Fruth 2003a). The sites belong to different biomes. Salonga is situated in the southern part of the Cuvette Central (Hohmann & Fruth 2003a). The terrain is flat and covered by a closed canopy forest occasionally interspersed by small, circular savannah patches. Gashaka belongs to the Guinea savannah belt, the terrain is undulating and encompasses savannah-woodland and gallery forests (see Adanu *et al.* this volume [Ch. 3]).

Field-methods were streamlined in consultations between G. Hohmann, VS and AF prior to commencement of data recording. At both sites, samples were collected from Mar 02 – Apr 03. Weather data were logged at the respective camp sites.

Fruit Production and Fruit Abundance

Plant diversity and productivity was assessed at both sites via straight-line transects of 8 km length. About 1000 trees with a diameter at breast height (dbh) > 10 cm were tagged, along with woody climbers attached to these trees. Most trees and many climbers were identified to the species or genus level by experienced botanists. Transects were monitored twice per month at intervals of roughly two weeks. Habitat phenology was measured by recording for each tree and woody climber the presence / absence of leaves, flowers, and fruit. Although the distinction between "ripe" and "unripe" fruit is somewhat arbitrary, comments were made on the quality of fruit, using size, colour, taste, smell, and conditions of seeds. Crop size was estimated on an exponential scale (1, 10, 100, 1000, 10000 fruits / plant).

Monthly food abundance indices (Am) were calculated as $Am = \sum (Dk \times Bk \times Pkm)$, where Dk refers to the density of species k in the transect area; Bk is the average basal area of this species; and Pkm is the monthly proportion of species k of all fruit- bearing trees (Anderson *et al.* 2002). Since trees growing along phenology trails were selected only by size criteria, phenology samples consisted of both food plants and plants not consumed by the apes.

Terrestrial herbs (mostly Marantaceae and Zingiberaceae) were monitored within several plots along the line transects.

Collection of Plant Food for Chemical Analyses

Efforts were made to collect fruit samples from all taxa of trees and woody climbers on the line transects as well as from selected terrestrial herbs (Marantaceae, Zingiberaceae), whether they were eaten by the apes (food) or not (non-food). Plants known to be eaten but not found on line transects or plots were also collected, preferentially from individual trees or plots that had been visited by the apes.

For transport back to camp, samples were stored in paper bags or leaves. Bags with soft fruit were put into an open plastic container to avoid damage. Plastic bags were not used, as this might have changed the consistency of the plants. Intact fruits were placed on grid paper and photographed, and descriptive notes made to illustrate their appearance. Weight and size (vertical and horizontal circumference) was recorded for at least 10 specimens per sample. The fruit was then cut, and the parts that were eaten separated from the rest, such as seeds and hard shells. If this information was lacking, samples were created of parts likely to be ingested, i.e., the soft meso- and exocarp. The inner section of stems (pith) were conserved from terrestrial herbs.

At least 5 g of dry weight were needed to determine nutrients and about 20 g for anti-feedants. Thus, at least 25 g of dry matter were required from each species. This corresponded to 50 – 250 g of fresh material, depending on estimated water content (low content < 50 %, as, e.g., in hazel nuts; medium content, ca. 50 %, as, e.g., in bananas and apples; high content, ca. 90 %, as, e.g., in oranges and pineapples). The capacity of sample tubes was 10 ml, which equalled about 10 g of dry material. Therefore, 4 tubes were to be filled with plant material of rather dry consistency, 6 for those of medium and 25 for those with a high water content. Tubes were labelled with a permanent marker as well as scratch marks on the lid. Tubes were stored in liquid nitrogen to prevent changes in the chemical content by mold and other biochemically active sources.

Samples for Salonga included 47 food species (of which 72 % were identified to at least the family level) and 39 non-food species (of which 49 % where identified). Samples for Gashaka included 53 food species (with 81 % identified) and 46 non-food species (with 57 % identified).

Phytochemistry of Plant Food

For shipment from Africa to laboratories in Germany, tubes were removed from the nitrogen containers, packed in styrofoam and dry ice and either flown via air cargo or carried as hand-luggage. Samples were analysed for nutrients (Leibniz Institute for Zoo and Wildlife Research; working group of SO) and anti-feedants (Institut für Tierernährung in den Tropen und Subtropen, Universität Stuttgart-Hohenheim; working group of Klaus Becker).

Macro-nutrients and Energy

Samples were freeze-dried and ground. Dry matter content was determined by drying a portion at 105 °C overnight. Nitrogen was determined via Dumas-Combustion. Total nitrogen (N) provides an estimate of crude protein (protein level = N × 6.25). Crude fat was assessed by Soxhlett petroleum ether extraction, starch and mono- / disaccharides enzymatically, and energy content by bomb calorimetry. Detergent Fibre Analysis was performed after van Soest (1994) and provided a rapid stepwise procedure for determining soluble cellular components as well as the insoluble cell wall matrix and its major subcomponents: hemicellulose, cellulose, and lignin. Cell contents and soluble components were estimated by boiling the sample in neutral detergent solution. The residue Neutral Detergent Fibre (NDF) contains hemicellulose, cellulose, and lignin. Hemicellulose was extracted by boiling the sample in acid detergent solution. The residue Acid Detergent Fibre (ADF) contains only cellulose and lignin. The last step extracts cellulose by acid hydrolysis and burns the sample to ash at 550 °C. The residue Acid Detergent Lignin (ADL) contains only lignin. Hemicellulose and cellulose contents are calculated by weighing and subtracting residues with: hemicellulose (NDF – ADF) and cellulose (ADF – ADL).

Anti-feedants

For extraction of simple phenolics and tannins, 100 mg of plant material was dried and finely ground (< 0.18 mm particle size). The sample was mixed with 5 ml of aqueous acetone (70 % v / v) and sonicated (i.e., exposed to energy produced by sound waves) for 20 min at room temperature. The mixture was then centrifuged for 10 min at 3000 g at 4 °C. The supernatant was collected in a fresh tube and used for the following steps as "sample phenolic extract".

Total phenolics were determined according to Makkar *et al.* (1993). Suitable quantities of the aliquots (determined by trial and error so that the absorbance value below was within the range of the calibration curve) of the phenolic extract were placed in a test tube and made up to 500 μl with distilled water. 250 μl of the Folin Ciocalteau reagent followed by 1.25 ml of the sodium carbonate solution was added and the mixture stirred on a vortex machine. Absorbance of the solution was recorded at 725 nm after a 40-min incubation period in the dark. The total amount of phenols was then calculated as tannic acid equivalent from a calibration curve prepared with tannic acid. Results were expressed as g / 100 g tannic acid equivalent on a dry matter basis.

Estimation of total tannins from total phenolic extracts: 100 mg of polyvinyl polypyrrolidone (PVPP) was weighed into a 100 × 12-mm test tube. 1.0 ml distilled water was followed by 1.0 ml of the sample phenolic extract added to the test tube

(100 mg PVPP is sufficient to bind 2 mg of total phenols; if total phenolic content is more than 10% on a dry matter basis, the extract has to be diluted). The contents of the tube were shaken on a vortex machine and kept at 4 °C for 15 min, after which it was shaken again and centrifuged (3000 g for 10 min) to collect the supernatant, which only contains simple phenolics other than tannins (the tannins are precipitated with the PVPP). The phenolic content of the supernatant was measured as described above. The content of the non-tannin phenolics was expressed on a dry matter basis. The tannin content (%) was then calculated as follows: total phenolics – non-tannin phenolics = tannins. The result is then expressed as tannic acid equivalent on a dry matter basis.

Determination of condensed tannins followed Porter *et al.* (1986). Phenolic extract (0.50 ml) was diluted with 70 % acetone and pipetted into a 100-mm × 16-mm glass test tube. To this, 3.0 ml of the butanol-HCI reagent was added and 0.1 ml of ferric reagent. The sample was heated on a block set at 95 °C for 60 min. After cooling, the absorbance of the mixture was recorded at 550 nm. Condensed tannins (% in dry matter) as leucocyanidin equivalent are calculated using the formula: (A 550 nm × 78.26) / (% dry matter).

Structural Analyses of Faecal Samples

Direct observations of feeding were rare, as the apes were not fully habituated to human observers – although numerous food plants could be identified from actual feeding sites. Therefore, assessments of ape diet were largely based on macroscopic inspection of dung samples (n = 147 at Salonga, n = 104 at Gashaka) – although this introduces some bias in favour of food with undigestible components, which remain visible in the faeces. The study aimed to collect about 40 samples of fresh faeces each month from nest sites or other places and to spread sample collection evenly across the month. In fact, the number of sampled nest groups / month ranged from 1 – 8. Samples were weighed and a rough macroscopic inspection carried out before they were stored in liquid nitrogen.

Faecal samples were processed in Berlin by JB. For this, dry weight was taken before items larger than > 5 mm (such as seeds and large fibres) were removed. Length, height, and width were recorded for at least 10 specimens of each species of seeds found in faecal samples. From these, seed volume was calculated, assuming that most seeds are roughly ellipsoid. Since faecal samples from the same nest group were likely to contain the same type of seeds, we used average values of 10 seeds from a sample for inter-species comparison.

Particle size of faeces reflects how much food is broken down during ingestion (chewing) and gut passage and is measured through a modulus of fineness (MoF; Poppi *et al.* 1980). Accordingly, dry material < 5 mm was transferred in a solution of 30 % H_2O_2 and mixed. After 30 min, KOH was added and mixing continued for another 60 min. The wet matrix was then washed through a layer of 6 sieves (RETSCH VS 1000, HAAN) of decreasing mesh width (4, 2, 1, 0.5, 0.25, 0.125 mm) for 10 min and at 50 Hz. Particles > 0.125 mm retained on different layers were stored in separate containers ("large matrix"), dried at 100 °C for 12 h and weighed. Their combined weight was subtracted from that of the dry matrix before washing to calculate the weight of material lost by wet sieving ("small matrix", < 0.125 mm).

Particle sizes are expressed as % of all retained particles. For this, the proportion of particles retained in the finest sieve is multiplied by 1, and the figure added to the multiplication of the proportion of particles retained on the second-finest sieve by 2. This iteration is repeated up to the proportion of particles retained on the sieve with the largest pore size, which is multiplied by 6. The resulting sum is divided by 100. A MoF of 1, accordingly, would indicate that all of the retained particles were on the finest sieve. Note, that values for MoF are comparable only if sieves of same sizes have been used. Comparisons between chimpanzees and bonobos are based on one (average) value per taxon.

Party Size as a Measure of Sociality

Communities of chimpanzees and bonobos split into foraging parties during the early hours of the day and tend to merge into somewhat larger parties at dusk when each weaned individual constructs a night nest in a tree (Fruth 1995, Fowler 2006). Nest group sizes can thus serve as an indirect measure for seasonal fluctuations in gregariousness during daytime hours. Many nest sites were detected by vocalisations, and because large parties are more likely to vocalise than small ones, the results are probably biased towards larger groups. However, assuming that the detectibility of the distance calls of both species is similar (Hohmann & Fruth 1995), it is unlikely that this bias alone would produce consistent differences between the two species.

Chimpanzees and bonobos nest in various parts of their home-range. Assignment of a cluster of nests to the same group was based on: (a) direct observation of nest construction; (b) close spatial aggregation; (c) freshness of nest materials (i.e., green and fresh leaves and twigs); (d) fresh faeces and / or urine underneath the nest. Most nest counts were made immediately after the site had been vacated. The forest surrounding a given nest site was checked for the presence of fresh nests within an area of 50 m in each direction.

Results

Environmental Features

The bonobo habitat at Salonga is flat closed canopy forest interspersed by small savannah patches. The chimpanzee terrain at Gashaka is undulating and a mosaic of primary rain forest, gallery forest, and savannah-woodland.

The climate at the bonobo site (rainfall 2304 mm, monthly range 45 – 375 mm; average minimum monthly temperature 21 °C, range 19 – 23 °C; average maximum monthly temperature 27 °C, range 26 – 29 °C) is less seasonal and considerably cooler than at the chimpanzee site (rainfall: 1864 mm, monthly range 0 – 446 mm; average minimum monthly temperature 21 °C, range 14 – 24 °C; average maximum monthly temperature 33 °C, range 30 – 38 °C) with a pronounced dry season including 3 – 4 months without rains (Dec – Mar; *Fig. 12.3*).

The random composition of the forest was reconstructed from the floristic composition of the about 1000 trees tagged along the 8 km of straight line transect.

– At Salonga, 959 trees and 1874 vines were tagged over the 8 km. For this, a transect width of 1.3 m was needed (total area: 1 ha), corresponding to a density of 1 tree / 104 m^2.

– At Gashaka, 985 trees and 794 vines (including common taxa such as *Landolphia*) were tagged over 8 km, which required a width of 2.0 m (total area: 1.6 ha). This corresponds to a density of 1 tree / 160 m^2. Overall, 78 % of the transect was classified as forest and 22 % as savannah-woodland.

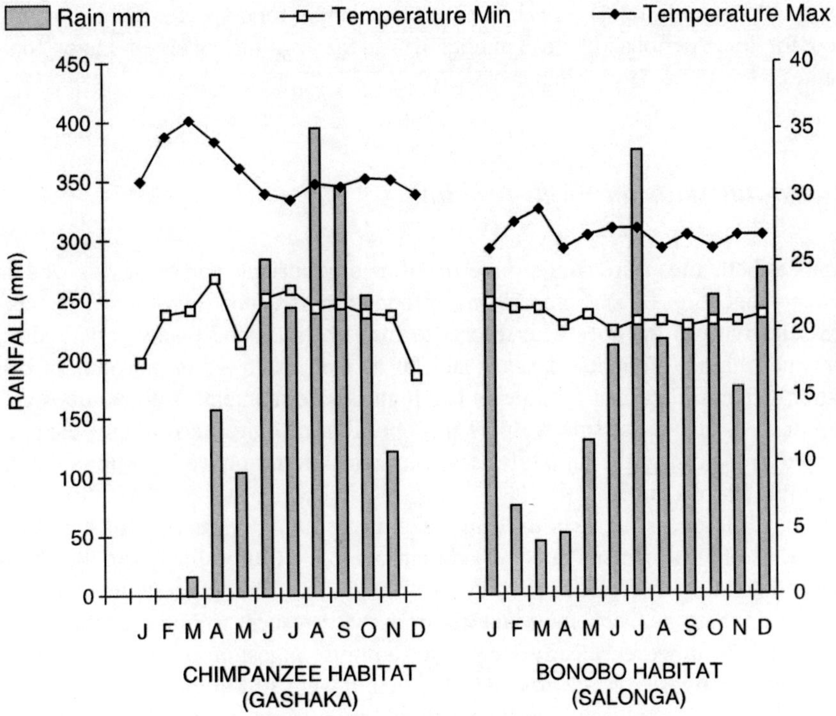

Fig. 12.3 Climatic profiles of Salonga and Gashaka during the study period

Although tree density was higher at Salonga, overall distribution of size categories was similar at both sites (*Fig. 12.4a*). Nevertheless, Salonga hosts more thin trees < 10 cm dbh and more large trees > 50 cm dbh. However, thin trees are far more common than thick trees, and the basal area of trees at Salonga was therefore significantly smaller (mean 218 cm^2) than at Gashaka (mean 283 cm^2; Mann-Whitney *U*-test, $Z = -3.831$, $p < 0.001$). Climbers were more abundant at Salonga (mean 1.8 / tree, range 0 – 6) than at Gashaka (mean = 0.8 / tree, range 0 – 4).

During each study month, Gashaka had a higher proportion of fruit bearing trees and climbers than Salonga (*Fig. 12.4b*). There was also substantial monthly variation of fruit abundance at Gashaka where the peak of fruit production coincided with the end of the wet season – whereas no discernible annual pattern existed at Salonga. This corresponds to the less seasonal climate in Salonga.

Diet Composition

Fresh faeces were screened for identifiable remains such as seeds, fibres, and fragments of fruit skin, indicating that relatively few species accounted for the majority of remains at any given time. The number of fruit species found in faeces was 22 for bonobo samples (mean 5.5 / month, range 3 – 10) and 25 for chimpanzees (mean 4.5 / month, range 1 – 7). For bonobos, single food species often provided food for long periods (up to 6 months) whereas fruiting cycles of major food species were shorter for chimpanzees (*Tab. 12.1*).

Macro-nutrients and Anti-feedants

Plants at both sites were similar in terms of macro nutrients and energy as well as fibre content (*Fig. 12.5*). Comparisons of food with non-food items revealed a certain selectivity of the apes with respect to fruit quality. Food plants at both sites, contained higher amounts of sugar and fat as well as lower proportions of dry matter, which means that food items had higher water content. Both bonobos and chimpanzees also chose fruit with lower proportions of fibre. Accordingly, the apes seemed to maximise the input of macro-nutrients and minimise ingestion of non-digestible fibrous parts.

Clear differences between the *Pan* taxa emerged with respect to anti-feedants. The levels of anti-feedants in non-food items were 2 – 3 times higher in the chimpanzee habitat, but levels for food items were similar. This suggests bonobos did not discriminate between food and non-food with respect to anti-feedant levels. Chimpanzees, however, clearly seemed to minimise ingestion of phenol and total tannin, while ingested levels of condensed tannin were still high.

Food plants at both sites contain various proportions of fibre versus protein (*Fig. 12.6*). However, only bonobos – and not chimpanzees – have access to one

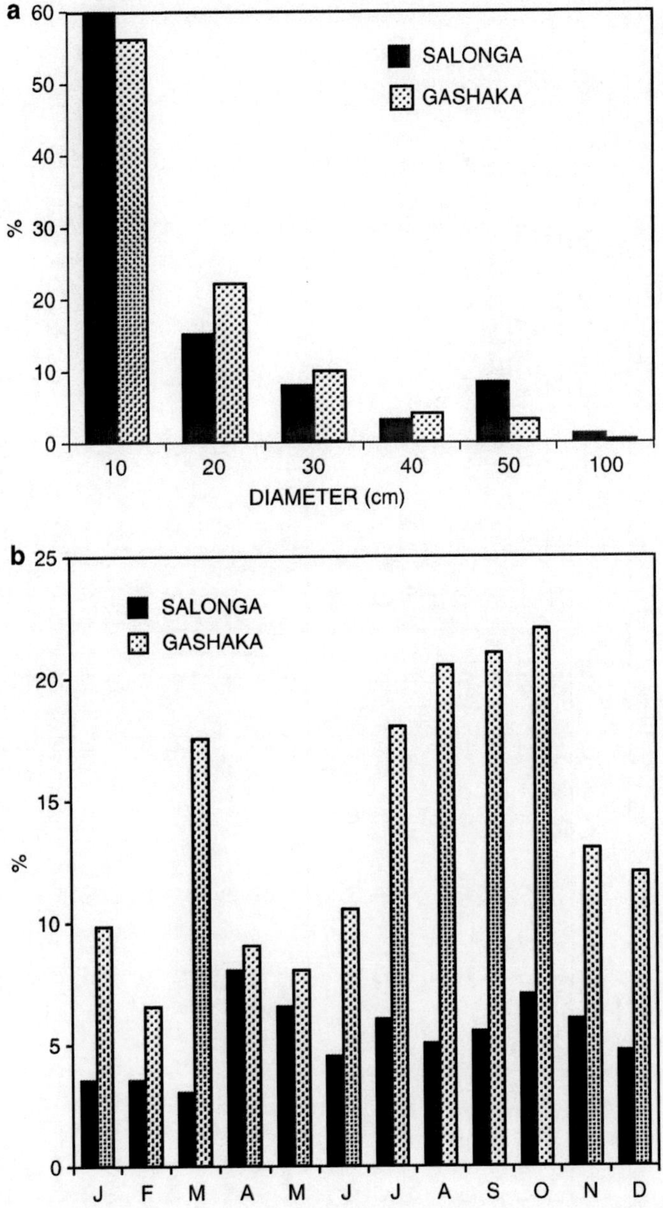

Fig. 12.4 Basic forest phenology at the study sites of bonobos at Salonga and chimpanzees at Gashaka. (**a**) Percentage of transect trees of different size classes ranging from 10 cm to > 100 cm diameter at breast height. (**b**) Proportion of transect plants fruiting each month

Table 12.1 Dominant fruit obtained from visual inspection of fresh faecal samples from bonobos and chimpanzees. Only species that appeared in at least 50 % of the monthly faecal sample collection

		Jan	Feb	Mar	Apr	May	Jun	Jul	Aug	Sep	Oct	Nov	Dec	Months with > 50% (a)
Bonobo (Salonga)	Faecal samples (*n*)	63	29	46	15	24	43	36	45	35	37	24	36	
	Dialium	0	0	0	67	96	100	94	100	86	14	4	0	6
	Ficus	2	59	46	0	0	0	14	0	0	57	63	31	3
	Grewia	75	100	100	0	0	0	0	0	0	0	0	25	3
	Polyathia	2	0	0	0	0	0	0	0	9	89	100	44	2
	Meliaceae	0	0	0	87	0	0	0	0	0	0	0	0	1
	Cissus	0	0	0	0	0	0	3	64	49	30	0	0	1
	Drypetes	0	0	0	53	0	0	0	0	0	0	0	0	1
	No ID	0	0	0	100	0	0	0	0	0	0	0	0	1
Chimpanzee (Gashaka)	Faecal samples (n)	27	3	0	33	27	2	23	19	11	2	2	0	
	Anthiaris	56	100		0	0	0	0	0	0	0	0		2
	Landolphia	11	0		27	96	100	17	0	0	0	0		2
	Lasianthus	0	0		0	0	0	0	0	0	50	100		2
	Leea	19	0		0	0	0	0	11	9	50	100		2
	Malacantha	4	0		0	0	0	0	0	36	50	100		2
	Afromomum	0	0		0	0	0	52	37	0	0	0		1
	Bridelia	0	0		0	0	0	0	0	81	0	0		1
	Canarium	0	0		0	0	0	43	63	27	0	0		1
	Cola	0	0		0	4	100	9	0	0	0	0		1
	No ID	0	0		0	0	0	0	0	55	0	0		1

(a) Number of months with certain food species appearing in at least 50 % of all faecal samples

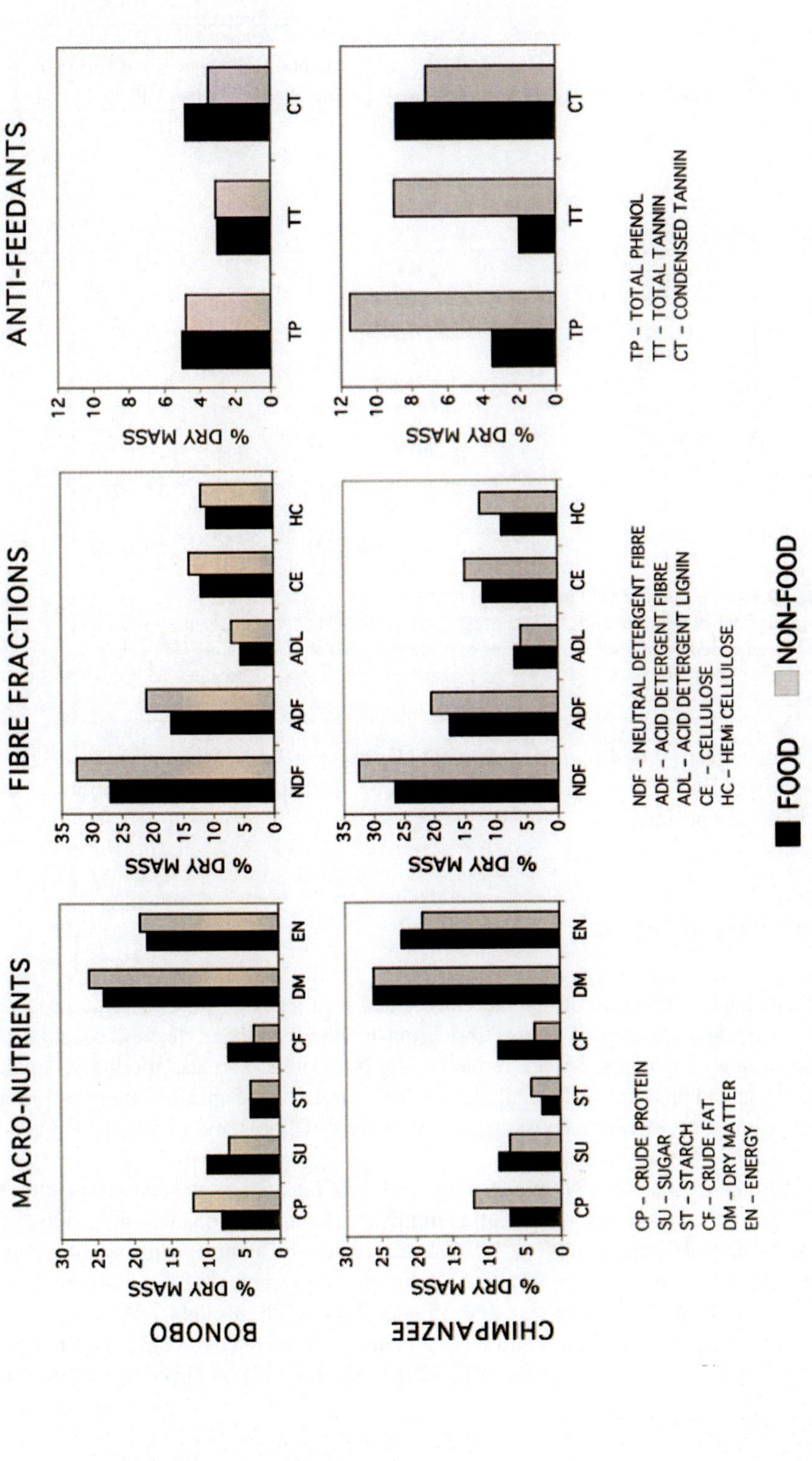

Fig. 12.5 Phytochemical characteristics (macro-nutrients, fibre fractions, anti-feedants) of food plants and non-food plants at the chimpanzee site of Gashaka and the bonobo site at Salonga. The unit for energy is kj/g DM

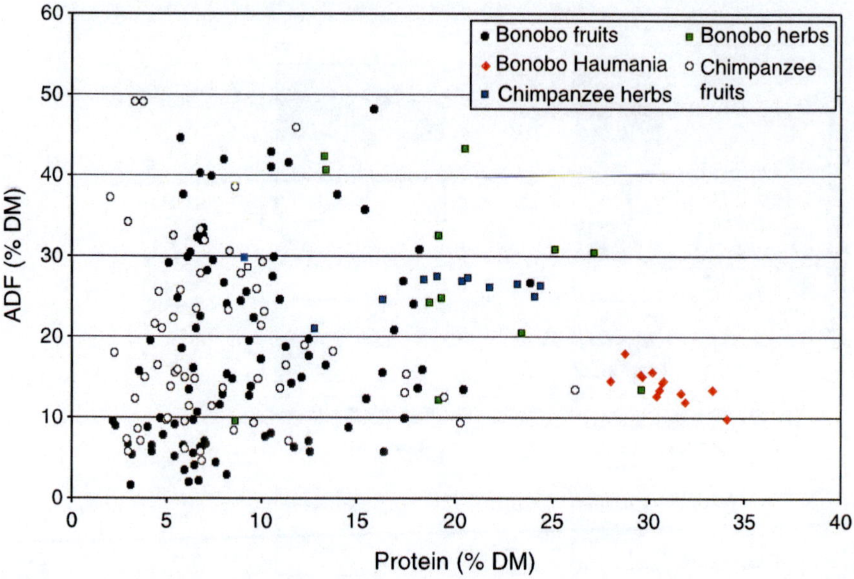

Fig. 12.6 Proportion of acid detergent fibre (% dry mass) as a function of protein (% dry mass) in food plants of bonobos and chimpanzees. Food quality is better when values for protein are high and those for ADF are low. Note cluster of outliers by herbs of the genus *Haumania* that is only available to bonobos

type of terrestrial herbaceous vegetation (*Haumania liebrechtsiana*) that is exceptionally rich in protein with, at the same time, particularly low concentrations of acid detergent fibre.

Structure of Ingesta

Macroscopic inspection of dry faeces revealed a variety of items, including seeds, mesocarp and exocarp of fruits, fibre from monocotyledones, leaves, coagulated latex, bone fragments, insect remains, shells of invertebrates, molluscs, small stones and an unstructured, small and amorphic fraction ("Schluff"). Items > 5 mm were almost exclusively seeds, apart from pellets of fibres and leaves, which constituted < 1 % of dry mass.

A bonobo bolus ($n = 81$) weighed on average 47 ± 35 g, with seeds constituting 72.3 %. The average dry weight of a chimpanzee bolus ($n = 63$) was 45 ± 32 g, of which only half were seeds (Fig. 12.7). Seeds in bonobo samples were significantly larger (639 ± 458 mm^3; $n = 73$) than in chimpanzee samples (431 ± 238 mm^3, $n = 60$; exact Mann-Whitney U-test, $U = 1596.0$, $Z = -2.686$, $p = 0.007$).

Moreover, particles in bonobo faeces tended to be coarser (modulus of fineness 4.02 ± 0.82) than in chimpanzees (3.72 ± 0.83; Fig. 12.8). This was not due to

Fig. 12.7 Seed load in faecal samples of bonobos (n = 145) and chimpanzees (n = 103)

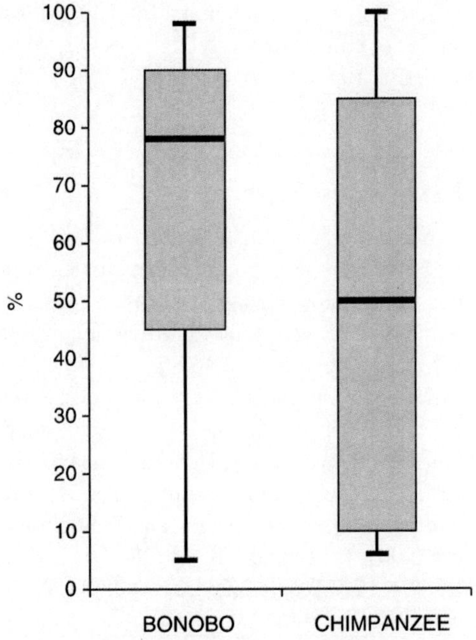

Fig. 12.8 Modulus of fineness (MoF) of faecal particle sizes in bonobos and chimpanzees

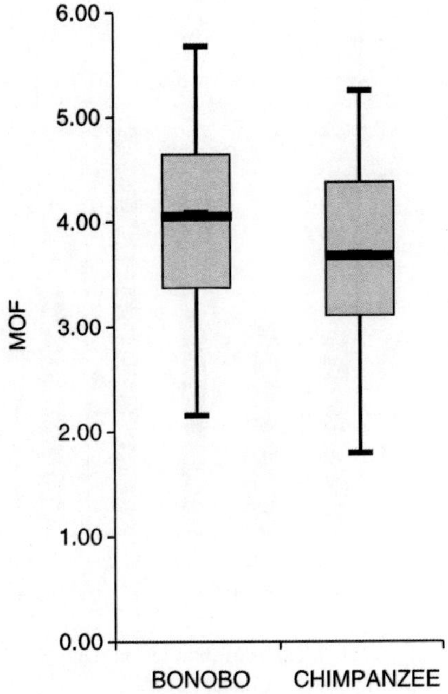

different fibre contents in the food, as these were virtually identical (% for bonobos vs. % for chimpanzees: NDF 27 vs. 27, ADF 17 & vs. 18, ADL 5 vs. 7, CEL 12 vs. 12, HEC 10 vs. 9; see also *Fig. 12.5*).

Nest-Group Sizes

At both sites, nest-group size fluctuated from month to month but groups of bonobos were almost always larger than those of chimpanzees (*Fig. 12.9*). The median monthly size was thus significantly different between the two species (Mann-Whitney *U*-test, $Z = -2.457$, $p = 0.014$).

To test relationships between food abundance and foraging party size, we used the median of monthly nest group size as a proxy and related it to the monthly fruit index at the two sites. A General Linear Model analysis with party size as dependent variables, and fruit index and species as independent variables revealed a significant interaction between species and fruit abundance (species: df = 1, 18, $F = 16.212$, $p = 0.001$, fruit index: df = 1, 18, $F = 2.518$, $p = 0.13$, species × fruit index: df = 1, 18, $F = 10.502$, $p = 0.005$, $R^2 = 0.57$). This meant that the effect of the fruit index was different at the two sites.

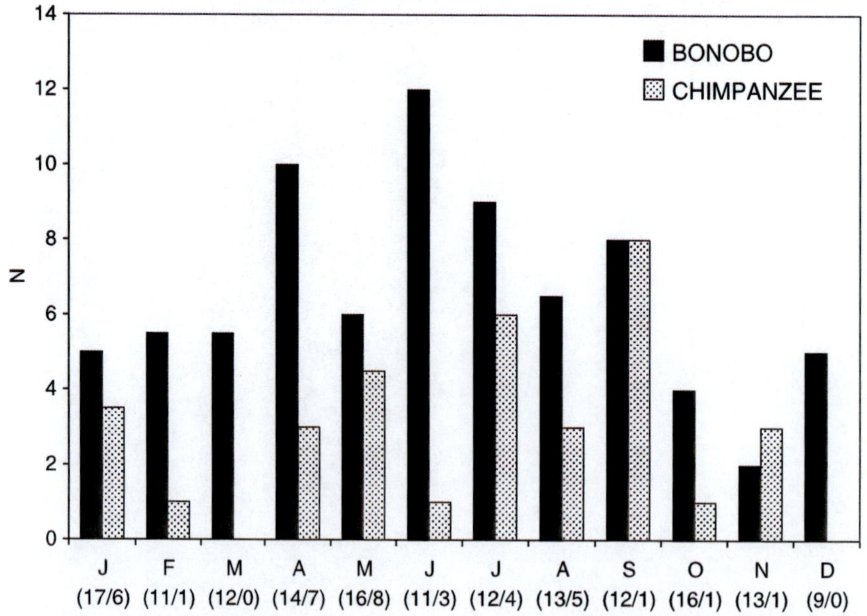

Fig. 12.9 Median of monthly nest group sizes of bonobos and chimpanzees. Numbers indicate monthly sample of nest groups

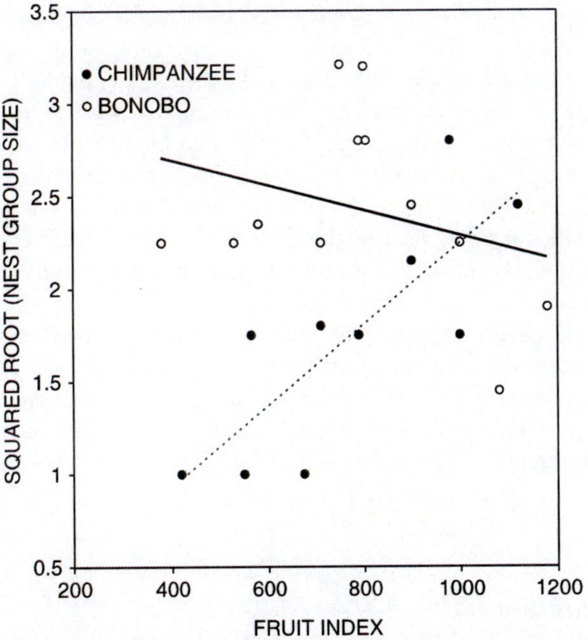

Fig. 12.10 Linear regression between monthly fruit availability and transformed count data (squared-root) on minimum nest group size in bonobos at Salonga (open circles, solid line) and chimpanzees at Gashaka (filled circles, dotted line)

Moreover, a significant and strictly positive relationship was found between nest group size and fruit production for Gashaka chimpanzees (df = 1, 8, $F = 14.834$, standardised beta = 0.806, $p = 0.005$), as opposed to Salonga bonobos (df = 1, 10, $F = 1.223$, standardised beta = –0.33, $p = 0.295$; Fig.12.10).

Discussion

Fruits are the main diet of chimpanzees and bonobos. We quantify fruit availability and quality for two populations of these apes, and relate similarities and differences to observed levels of gregariousness.

Climate and Habitat Phenology

The bonobo habitat at Salonga is cooler by several degrees, has 500 mm more rain per annum and lacks the pronounced dry season of the chimpanzee habitat at Gashaka (cf. *Fig. 12.3*). These conditions profoundly influence vegetation cover, in that they foster a closed canopy forest at the bonobo site and a more open biome with considerable proportions of woodland at the chimpanzee habitat. Not least because of the mosaic nature of the landscape at Gashaka, plant diversity exceeds that of Salonga. Gashaka is situated in a transition zone between two biogeographical regions, and therefore boosts an exceptionally high biodiversity (Fjeldsa & Lovett 1997, Oates *et al.* 2004). Moreover, because of the rugged terrain at Gashaka Gumti National Park (300 – 2419 m above sea level), microhabitats are more diverse than at Salonga National Park. Finally, the flora at Gashaka is characterised by plant forms that can withstand the pronounced dry season with no rains for several consecutive months.

These broad environmental conditions are reflected in the habitat phenology. Trees in the chimpanzee habitat grow further apart and have fewer vines than those in the bonobo habitat. Trees and vines in the chimpanzee habitat produce fruit more often, although their availability is more seasonal, and fewer species serve as a staple food throughout a given month. The chimpanzee diet is hence more diverse, overall.

Food Availability

Throughout the year, fruit production by trees and climbers was higher at the chimpanzee site. The proportion of fruit-bearing trees and climbers at Salonga was generally very low (cf. *Fig. 12.4b*), perhaps because of the high proportion of small transect trees, which may include many immature and as yet non-reproductive plants (cf. *Fig. 12.4a*). Nevertheless, the low fruit scores for Salonga are within the reported range from other sites, e.g., for chimpanzees at Kanyawara / Uganda, where just 3 % of trees bear fruit per month (Chapman *et al.* 1994).

The highly seasonal forest productivity in the chimpanzee habitat is due to climatic shifts at the onset and end of the dry season, whereas the low dynamics in the bonobo habitat reflect comparatively modest fluctuations in temperature and rain (cf. *Fig. 12.3*). As a knock-on effect, a striking difference was found for food diversity: A single species could produce fruit for up to half a year at Salonga, and bonobos indeed relied on a relatively small number of fruit species for extensive periods (cf. *Tab. 12.1*). This is probably caused by a low degree of synchrony in fruit production of individual specimen of the same species (Poulin *et al.* 1999). Pronounced shifts between wet and dry seasons as in the chimpanzee habitat trigger highly synchronised flowering and fruiting processes (Leigh & Windsor 1982). This, in turn, results in brief periods of fruit availability for a given species – such as those found at Gashaka.

The asynchronous pattern at Salonga may also be responsible for the surprising finding of only marginal fruit production in climbers – despite the fact that trees harbour more vines than those at Gashaka. Thus, a one-year study is probably not sufficient to encompass periods when bonobos, too, rely heavily on fruit from vines (Berkhoudt *et al.* 2005).

Longer-term data may lead to revisions of the rather clear-cut differences found between the chimpanzee and bonobo habitats, as substantial inter-annual variation in fruit production is not atypical for wild trees (Poulin *et al.* 1999). For example, at Lopé / Gabon, 98 % of sample trees did not produce fruit during some years, whereas high synchrony within and between species lead to peaks of fruit availability during other years (Tutin & Fernandez 1993).

Food Quality

Both bonobos and chimpanzees appeared to be choosy and selective. Compared to non-consumed fruits, those eaten had higher levels of water, sugar, and crude fat, and lower levels of fibre components (cf. *Fig. 12.5*). The combination of relatively high levels of macro-nutrients and low levels of fibre in food items as compared to non-food matches predictions of optimal diet theory and is in line with findings from other studies (Rogers *et al.* 1990, Reynolds *et al.* 1998). Nevertheless, it should be kept in mind that samples came only from fruit with physical properties of "ripeness", although bonobos and chimpanzees consume substantial amounts of unripe fruit (Reynolds *et al.* 1998). Therefore, the chemical composition of measured fruit is likely biased towards above-average quality.

Some species differences were also obvious, for example, fruit in the chimpanzee habitat contained higher overall levels of anti-feedants (cf. *Fig. 12.5*). High intake levels of tannins create energetic costs for detoxification and reduce access to macro-nutrients such as protein. This, in turn, requires higher food consumption to compensate the loss. However, ingestion volume is constrained by gut capacity and passage time. A generally tannin-rich diet may therefore require more selective feeding and frequent shifts between food patches.

Tannin levels vary with soil, water, and other environmental factors (Mueller-Harvey & McAllan 1992). For example, young, regenerating forests contain relatively low amounts of plant secondary compounds (Harbourne 1993), while vegetation growing under difficult conditions contains higher levels of defensive substances (Mueller-Harvey & McAllen 1992). Tannin levels and activity also increase with ambient temperature (Makkar & Becker 1998). As a consequence, tannins might bind more protein at Gashaka than in the relatively cooler Salonga habitat.

An earlier study at the bonobo site of Lomako found very low levels of tannins in fruit, which led to the assumption that avoidance of tannins guides food selection (Malenky 1990). However, values for Salonga are ten times higher. Moreover, tannin-rich food is also eaten by chimpanzees at Gombe / Tanzania (Wrangham & Waterman 1983), Kanyawara / Uganda (Wrangham *et al.* 1998), Budongo / Uganda (Reynolds *et al.* 1998) and to a certain degree also at Bossou / Guinea (Takemoto 2003). Likewise, experimentally fed chimpanzees did not avoid tannin-enriched foods (Remis 2002). Thus, it is unlikely that food selection in general aims to discriminate against tannins.

Nutrients in wild plants are regularly associated with components that reduce palatability and / or digestibility (Dearing & Schall 1992). Functional links between chemical composition and food choice have been found for insects, birds, and herbivorous mammals (Belovsky 1978, Lotz & Nicolson 1996, Raubenheimer & Simpson 1997). However, the same type of macro-nutrient may be of different importance for a given species or population because digestive strategies differ (Milton 1981, Witmer & van Soest 1998). For example, experiments with herbivores demonstrated that food choice was triggered by tannin content rather than concentrations of macro-nutrients (Alm *et al.* 2000, Clauss 2003).

In any case, the detoxification constraints hypothesis (Freeland & Janzen 1974) predicts that animals should adopt strategies to reduce the negative impact of undesirable food components. Compared to bonobos, Gashaka chimpanzees can thus be expected to (a) have shorter feeding bouts, (b) visit more food patches per day, (c) not exploit food patches of the same species in close succession, so as to avoid ingesting too many of the same toxic substances, (d) have longer retention times, as gut passage time facilitates detoxification (Lambert 1998), (e) reduce tannin activity by eating soils that are rich in clay (Krishnamani & Mahaney 2000). Future research will be needed to test if these predictions are met.

Food Processing

In any case, the relatively low quality of chimpanzee food is associated with more careful processing. Accordingly, chimpanzees swallow fewer and smaller seeds than bonobos – probably because they remove more of these indigestible items beforehand (cf. *Fig. 12.7*). Chimpanzees also seem to chew and / or absorb their food better, as their faeces have smaller particles, compared to bonobos (cf. *Fig. 12.8*). This difference is not due to larger fibres in bonobo fruit, as the fractions are virtually identical at Salonga and Gashaka (cf. *Fig. 12.5*).

Feeding Ecology and Gregariousness

For chimpanzees, then, food was of generally reduced quality, was more difficult to locate due to considerable seasonal fluctuations in availability and diversity, and required more sophisticated processing. These constraints are most likely reflected in lower levels of sociality compared to bonobos. Throughout the year, gregariousness was lower in chimpanzees, as indicated by consistently smaller sizes of nest groups (cf. *Fig. 12.9*). Moreover, chimpanzee nest group size decreased when monthly fruit availability decreased (cf. *Fig. 12.10*). This is expected, reduced food availability increases competition at food patches and thus forces the animals to disperse. However, no such correlation existed for bonobos who are therefore clearly less constrained by fruit availability. This relatively greater flexibility is perhaps related to a lower diversity of food species and easier availability of staple-food species (cf. *Tab. 12.1*), which reduces the costs of searching and food processing. In addition, shortfalls can probably also be compensated through consumption of terrestrial herbaceous vegetation. Not only are herbs more common at the moister Salonga habitat (G. Hohmann pers. comm.), but they do also include the genus *Haumania*, which favourably combines high protein value with low fibre content (cf. *Fig. 12.6*).

Nest groups can serve as a proxy for sociality during the previous day, as their sizes, while slightly larger, correlate with those of foraging parties (Fruth 1995, Sommer *et al.* 2004). The finding of larger bonobo nest groups compared to chimpanzees support previous studies (Fruth & Hohmann 1996). Interestingly, chimpanzee party size does not always vary with food abundance, as has been found for East African populations (Hashimoto *et al.* 2004), suggesting a strong link between abundance and gregariousness only for times when resources are scarce. Of course, factors other than food availability and quality will also influence the degree of sociality – for example predation and hunting pressure (Miller 2002) as well as prevalence of pathogens (Nunn & Altizer 2006). Party size in bonobos (Hohmann & Fruth 2003b) and chimpanzees (Anderson *et al.* 2002, Mitani *et al.* 2002) increases also with the number of estrous females as these attract more males to their presence. This is also true for Gashaka chimpanzees (Sommer *et al.* 2004), independent from any link between food and nest group size.

Ecology and Species Psychology

Bonobos at Salonga exhibit greater social cohesion than chimpanzees at Gashaka – probably due to more favourable ecological conditions. The chimpanzees cope with a more seasonal habitat, lower fruit quality and absence of highly nutritious herbs. As a result, opportunity costs for gregariousness increase.

These shortcomings seem to force Gashaka chimpanzees to supplement their plant food through animal matter. They frequently use stick tools to obtain honey from colonies of stingless-bees and honey-bees and also exploit army ant colonies as ant remains are found in about half of all chimpanzee faecal samples (Fowler & Sommer 2007, Schöning *et al.* 2007, Fowler *et al.* this volume [Ch. 13]). Honey provides sugar and energy, although the nutritional benefits of insect imagos are unclear (McGrew 2004, Deblauwe & Janssens 2007). In any case, Gashaka chimpanzees engage in the highest frequency of myrmecophagy so far measured for any *Pan* population. Ants are therefore no fall-back food, but may reflect the perennially marginal nature of the Gashaka habitat. So far, tool use and insect harvesting has not been recorded for Salonga, despite apparently high densities of suitable insect colonies (McGrew *et al.* 2007). On the other hand, Salonga bonobos have recently been found to consume meat with frequencies similar to some chimpanzee populations (Hohmann & Fruth 2008). Future analyses of nutritional ecology at the two sites will need to incorporate more information about dietary components other than fruit, in particular herbs and animals matter.

Thus, on a first glance, our data seem to support the assumption that ecological differences are at the heart of the dichotomy of sociality in *Pan*. However, such conclusion would be premature, given a remarkable degree of intra-specific variability in both the fine-tuning of social processes (Boesch *et al.* 2002) and varying floristic, faunistic, and climatic parameters throughout the geographical range of *Pan*. Plant consumption is likely to reflect geographic variation of vegetation cover rather than preferences for certain taxa (Rodman 2002). In other words: apes have to make do with what they find, both in terms of species and quality. This requires a considerable degree of flexibility. Thus, neither do Gashaka chimpanzees represent chimpanzees as a whole nor might Salonga bonobos be representative for bonobos – although habitats probably vary less throughout the bonobo range. Future studies will need to explore the extent of flexibility and if, and how, it covaries with local ecologies.

In any case, we can expect more correspondence, if sample populations live, for example, in a relatively rich habitat. It would be interesting to know if, e.g., near-absent levels of tool use and extractive foraging at the chimpanzee site of Budongo / Uganda (Reynolds 2005) are in fact caused by better and more abundant resources. This would support a gradient of subsistence technology of "rich habitat = low frequencies" / "poor habitat = high frequencies".

The more fundamental question is, of course, whether local ecologies also alter patterns of *social* behaviour. There is some evidence that, for example, female relationships are more relaxed in chimpanzee communities found in the rainforest of Taï / Ivory Coast (Wittig & Boesch 2003). However, the basic species psychology seems to be resilient against environmental fluctuations, as illustrated through captive studies. For example, groups of chimpanzees and bonobos are kept in virtually identical enclosures under the same feeding regime at a zoological garden in Stuttgart / Germany. Yet, social relationships of the former remain male-centred in that males dominate females, whereas the latter are female-centred in that females dominate males (Parish 1994). Similar patterns are found the world over, wherever the two *Pan* species are kept or studied.

As a consequence, food provisioning with low or even absent levels of food competition cannot change the basic dominance regime of either *Pan* species. Different selection processes over the last two million years or so must have hardwired the psychology of *Pan* populations, depending on which bank of the Congo River they lived.

Nevertheless, it is important to note that the Congo – while serving as a geographical boundary between the two species – flows in such a way north- and southwards that the latitudinal as well as longitudinal distribution of bonobos overlaps entirely with that of chimpanzees (cf. *Fig. 12.2*). Accordingly, "the range of climates and habitats experienced by bonobos is merely a subset of those experienced by chimpanzees" (Wrangham *et al.* 1996: 46). Hence, it is necessary to identify a factor, which is not constant in areas of overlap of geographic coordinates.

An obvious candidate is the absence of gorillas south of the Congo. Gorillas are known to often consume large quantities of terrestrial herbaceous vegetation (reviews in Harcourt & Stewart 2007, Robbins 2007). One can easily imagine a scenario in which gorillas regularly competed with, or even outcompeted, ancestral populations of chimpanzees for access to herbs (Wrangham & Peterson 1996). Such inter-species conflict skewed chimpanzee feeding ecology even further towards frugivory. This in turn might have re-enforced the default psychology of female-female resource competition, which was then maintained even when chimpanzees began to colonise habitats where gorillas were absent. In turn, more relaxed relationships developed at the Congo south bank where habitat richness was not "diluted" by competition with another ape. Here, bonobo females ultimately learned to form coalitions – which they maintained and managed through same-sex sexual contacts (Fruth & Hohmann 2006). Coalitions enabled bonobo females to physically dominate males and thus control resource access in their favour.

A first test of this model would not require exposing chimpanzees to a resource-saturated habitat, but to confront bonobos with an impoverished habitat to see if female coalitions can still be maintained.

Lessons for a Politically Correct Agenda?

Much more information is available for chimpanzees, compared to bonobos, who were not only much later recognised as a species, but remain, to date, much less studied (de Waal & Lanting 1997). This fosters a tendency to see bonobo-traits as derived and those of chimpanzees as more resembling the state of our shared common ancestor – an assumption that is, a priori, clearly wrong. It is simply not known whether the patriarchal or the matriarchal pattern represents the primitive state – and with this ancestral conditions for the *Homo* line, too. More than two-thirds of contemporary human cultures follow a patrilocal pattern of residency after marriage (review in Vogel & Sommer 1994). The bride leaves her kin-group to move in with the groom, allowing the groom's family to effectively control her sexuality and reproduction. Coercion of wives through husbands is common in such societies, including the use of force. More freedom and a greater equality of the sexes exists for women in hunter-gatherer societies, if only for the fact that control of female behaviour is more difficult to exert if women are on foraging trips (review in Parish & de Waal 2000).

It may therefore well be that strictly patriarchal structures were enforced only once agriculture developed, and with it stratified societies. Richer males where then not only able to attract multiple females and provision them, but also to effectively sever the support network these females had with their kin-groups.

Simone de Beauvoir ([1949] 1973), icon of the early feminist movement, lamented that emancipation was difficult to achieve as long as woman did not cooperate but continued to live dispersed, attached to individual males. This pattern simply mimics a general mammalian trend, where coalitions between females who have left their natal group cannot easily develop, as these would require unrelated individuals to form alliances. Difficult indeed, but not impossible – as the bonobos exemplify.

Modern Western societies denounce patriarchal structures and try to design policies, which promote greater equality between the sexes and peaceful resolution of conflict. Such policies will in some way have to counteract natural tendencies, if the ancestral behavioural suite of early humans resembled the patriarchal pattern of chimpanzees. Conversely, if our ancestral state resembled the matriarchal pattern of bonobos, then we can expect such policies to be more easily effective. Given our ignorance about the state of the last common ancestor, we have to remain agnostic whether Hobbe's "against nature" or Rousseau's "back to nature" would make a more fitting motto for realistic politics.

In any case, humans possess not only an extraordinary flexibility with respect to subsistence technology – and this was the recipe that allowed *Homo* to settle in most diverse terrains across the globe. Humans also display a much greater flexibility with respect to social systems, including polygyny, monogamy and polyandry (Vogel & Sommer 1994). The development of a certain pattern depends on the given ecological and, accordingly, economic framework – which again supports the general idea that differential access to resources is at the heart of the bonobo–chimpanzee dichotomy. Clearly, the recognition of a causal link between ecology and female gregariousness can provide policymakers with fruitful insights – for example, that power asymmetries change with resource availability. But, as the example of bonobo societies teaches, where females often forcefully suppress males: However successful equal opportunity policies might be, they are not likely to lead to a world without conflict. The challenge will be to mitigate conflicts in the most peaceful way.

Acknowledgements Permission to conduct field work was granted for Salonga National Park / DRC, by the Institut Congolaise pour la Conservation de la Nature, and for Gashaka Gumti National Park / Nigeria by the National Park Service. – Funding came from The Leakey Foundation for Anthropological Research, the Max Planck Gesellschaft, Volkswagen-Stiftung, Leventis Ltd. & ProNatura International / Lagos, Primate Conservation Inc., and Chester Zoo / North of England Zoological Society. – Data collection and processing of samples was made possible by field assistants and volunteers (Salonga: Verena Brauer, Julia Gessner, Alex Gregoriev, Etienne Mabonzo, Esperance Miezi, Musuyu Muganza, Valentin Omasombo, Veronica Vecellio, Leilani Zeumer; Gashaka: Claire Bracebridge, Julia Greenslade, Asako Saegusa, Judith Bovensiepen, Jennifer Rogan, Maurice Melle Ekane, Ymke Warren, Umaru Buba, Yakubu Wakirwa, Hammaunde Guruza, Salamu Waziri, Buba Bello, Bobbo Buba, Ali Tappare; Taiwo Oviasuyi, Institute of Tropical Agriculture, Ibadan). – Heidrun Barleben coordinated food quality analyses at Berlin. – Klaus Becker (Stuttgart-Hohenheim) lent his expertise for nutritional analyses. – Gottfried Hohmann inspired the study.

Literature

Alm, U., Birgersson, B. & Leimar, O. (2000). The effect of food quality and relative abundance on food choice in fallow deer. *Animal Behaviour* 64: 439–445.

Altmann, S. A. (1998). *Foraging for Survival*. Chicago: The University of Chicago Press.

Anderson, D. P., Nordheim, E. V., Boesch, C. & Moermond, T. C. (2002). Factors influencing fission-fusion grouping in chimpanzees in the Taï National Park, Cote d'Ivoire. In Boesch, C.,

Hohmann, G. & Marchant, L. F. (eds.), *Behavioural Diversity in Chimpanzees and Bonobos*. Cambridge: Cambridge University Press, pp. 90–101.

Badrian, N. L. & Malenky, R. K. (1984). Feeding ecology of *Pan paniscus* in the Lomako Forest, Zaire. In Susman, L. S. (ed.), *The Pygmy Chimpanzee: Evolutionary Biology and Behavior*. New York: Plenum Press, pp. 275–299.

Bauer, J. (2006). *Vergleichende Faecesuntersuchungen bei* Pan paniscus, Pan troglodytes vellerosus *und* Pan troglodytes verus. Diplomarbeit, Humboldt-Universität Berlin / Germany.

Belovsky, G. E. (1978). Diet optimization in a generalist herbivore: the moose. *Theoretical Population Biology* 14: 105–134.

Berkhoudt, K., Fruth, B. & Garber, P. A. (2005). *Food Patch Choice of Bonobos* (Pan paniscus*) in Lui Kotal, the Democratic Republic of Congo*. Poster presented at the American Association of Physical Anthropologists, April 6–9, 2005.

Boesch, C. & Boesch-Achermann, H. (2000). *The Chimpanzees of the Taï Forest. Behavioural Ecology and Evolution*. Oxford: Oxford University Press.

Boesch, C., Hohmann, G. & Marchant, L. F. (eds.) (2002). *Behavioural Diversity in Chimpanzees and Bonobos*. Cambridge: Cambridge University Press.

Campbell, C. J., Fuentes, A., Mackinnon, K. C., Panger, M. & Bearder, S. K. (Hg.) (2007). *Primates in Perspective*. Oxford: Oxford University Press.

Campos-Arceiz, A., Takatsuki, S. & Lhagvasuren, B. (2004). Food overlap between Mongolian gazelles and livestock in Omnogobi, southern Mongolia. *Ecological Research* 19: 455–460.

Carlo, T. A., Collazo, J. A. & Groom, M. J. (2003). Avian fruit preferences across a Puerto Rican forest landscape: pattern consistency and implications for seed removal. *Oecologia* 134: 119–31.

Chapman, C. A., White, F. J. & Wrangham, R. W. (1994). Party size in chimpanzees and bonobos: a reevaluation of theory based on two similarly forested sites. In Wrangham, R. W., McGrew, W. C., de Waal, F. B. M. & Heltne, P. G. (eds.), *Chimpanzee Cultures*. Cambridge: Harvard University Press, pp. 41–57.

Clauss, M. & Lechner-Doll, M. (2001). Differences in selective reticulo-ruminal particle retention as a key factor in ruminant diversification. *Oecologia* 129: 321–27.

Clauss, M. (2003). Tannins in the nutrition of wild animals: a review. In Fidgett, A., Clauss, M., Gansloser, U., Hatt, J. M. & Nijboer, J. (eds.), *Zoo Animal Nutrition, Volume 2*. Fürth: Filander Verlag, pp. 53–89.

Clauss, M., Lechner-Doll, M. & Streich, W. J. (2002). Faecal particle size distribution in captive wild ruminants: an approach to the browser / grazer from the other end. *Oecologia* 131: 343–49.

Clauss, M., Lason, K., Gehrke, J., Lechner-Doll, M., Fickel, J., Grune, T. & Streich, W. J. (2003). Captive roe dee (*Capreolus capreolus*) select for low amounts of tannin acid but not quebracho: fluctuation of preferences and potential benefits. *Comparative Biochemistry and Physiology B* 136: 369–382.

Conklin-Brittain, N. L., Wrangham, R. W. & Hunt, K. D. (1998). Dietary response of chimpanzees and Cercopithecines to seasonal variation in fruit abundance. II. Macro-nutrients. *International Journal of Primatology* 19: 971–998.

Corlett, R. T. & Lucas, P. W. (1990). Alternative seed-handling strategies in primates: seed-spitting by long-tailed macaques (*Macaca fascicularis*). *Oecologia* 82: 166–171.

De Waal, F. B. M. & Lanting, F. 1997. *Bonobo: The Forgotten Ape*. Berkeley, Los Angeles: University of California Press.

Dearing, M. D. & Schall, J. J. (1992). Testing models of diet assembly by the generalist herbivore lizard, *Cnemidophorus murinus*. *Ecology* 73: 845–867.

de Beauvoir, S. [1949] (1973). *The Second Sex*. New York: Vintage Books.

Deblauwe, I. & Janssens, G. P. J. (2007). New insights in insect prey choice by chimpanzees and gorillas in southeast Cameroon: the role of nutritional value. *American Journal of Physical Anthropology* 135: 42–55.

Eller, C. (2000). *The Myth of Matriarchal Prehistory. Why an Invented Past Won't Give Women a Future*. Boston: Beacon Press.

Emlen, J. M. (1966). The role of time and energy in food preferences. *The American Naturalist* 100: 611–617.

Fjeldsa, J. & Lovett, J. C. (1997). Geographical patterns of old and young species in African forest biota: the significance of specific montane areas as evolutionary centres. *Biodiversity and Conservation* 6: 325–346.

Fowler, A. (2006). *Socio-ecology of Nigerian Chimpanzees at Gashaka.* PhD thesis, University of London.

Fowler, A & Sommer, V. (2007). Subsistence technology in Nigerian chimpanzees. *International Journal of Primatology* 28: 997–1023.

Freeland, W. J. & Janzen, D. H. (1974). Strategies in herbivory by mammals: the role of plant secondary compounds. *The American Naturalist* 108: 269–289.

Fruth, B. (1995). *Nests and Nest Groups in Wild Bonobos (Pan paniscus): Ecological and Behavioural Correlates.* Aachen: Shaker.

Fruth, B. & Hohmann, G. (1996). Nest building behaviour in the great apes: the great leap forward? In McGrew, W. C., Marchant, L. F. & Nishida, T. (eds.), *Great Ape Societies.* New York: Cambridge University Press, pp. 225–240.

Fruth, B. & Hohmann, G. (2006). Social grease for females? Same-sex genital contacts in wild bonobos. In Sommer, V. & Vasey, P. (eds.), *Homosexual Behaviour in Animals: Evolutionary Perspectives.* Cambridge: Cambridge University Press, pp. 294–315.

Gerloff, U., Hartung, B., Fruth, B., Hohmann, G. & Tautz, D. (1999). Intracommunity relationships, dispersal pattern and paternity success in a wild living community of bonobos (*Pan paniscus*) determined from DNA analysis of faecal samples. *Proceedings of the Royal Society of London, B* 266: 1189–1195.

Goldberg, S. (1993). *Why Men Rule.* Chicago, Illinois: Open Court Publishing Company.

Goodall, J. (1986). *The Chimpanzees of Gombe: Patterns of Behaviour.* Cambridge, MA: Harvard University Press.

Harbourne, J. B. (1993). *Introduction to Ecological Biochemistry.* San Diego: Academic Press.

Harcourt, A. H. & Stewart, K. J. (2007). *Gorilla Society: Conflict, Compromise, and Cooperation Between the Sexes.* Chicago: The University of Chicago Press.

Hashimoto, C., Suzuki, S., Takenoshita, Y., Yamagiwa, J., Basabose, A. K. & Furuichi, T. (2004). How fruit abundance affects the chimpanzee party size: a comparison between four study sites. *Primates* 44: 77–81.

Heltne, P. G. & Marquardt, L. A. (eds.) (1989). *Understanding Chimpanzees.* Cambridge, MA: Harvard University Press.

Hemelrijk, C. K., Wantia, J. & Isler, K. (2008). Female dominance over males in primates: Self-organisation and sexual dimorphism. *PLoS ONE* 3: e2678.

Hladik, C. M. (1977). A comparative study of the feeding strategies of two sympatric species of leaf monkeys: *Presbytis senex* and *Presbytis entellus.* In Clutton-Brock, T. H. (ed.), *Primate Ecology: Studies of Feeding and Ranging Behaviour in Lemurs, Monkeys and Apes.* London: Academic Press, pp. 323–353.

Hohmann, G. & Fruth, B. (1995). Loud calls in great apes: sex differences and social correlates. In Zimmermann, E., Newman, J. D. & Jürgens, U. (eds.), *Current Topics in Primate Vocal Communication.* New York: Plenum Press, pp. 161–184.

Hohmann, G. & Fruth, B. (2003a). Lui Kotal – a new site for field research on bonobos in the Salonga National Park. *Pan African News* 10: 25–27.

Hohmann, G. & Fruth, B. (2003b). Intra- and inter-sexual aggression by bonobos in the context of mating. *Behaviour* 140: 1389–1413.

Hohmann, G., Sommer, V. & Ortmann, S. (2007). *What's on the menue? Variation in nutritional ecology of bonobos and chimpanzees.* Abstract. (Italy Conference).

Hohmann, G., Fowler, A., Sommer, V. & Ortmann, S. (2006). Frugivory and gregariousness of Salonga bonobos and Gashaka chimpanzees: the abundance and nutritional quality of fruit. In Hohmann, G. (ed.), *Feeding Ecology of Apes and Other Primates.* Cambridge: Cambridge University Press, pp. 123–159.

Hohmann, G. & Fruth, B. (2008). New records on prey capture and meat eating by bonobos at Lui Kotale, Salonga National Park, Democratic Republic of Congo. *Folia Primatologica* 79:103–110.

Kano, T. (1992). *The Last Ape: Pygmy Chimpanzee Behavior and Ecology.* Stanford, CA: Stanford University Press.

Krishnamani, R. & Mahaney, W. C. (2000). Geophagy among primates: adaptive significance and ecological consequences. *Animal Behaviour* 59: 899–915.

Kummer, H. (1971). *Primate Societies*. Chicago: Aldine.

Lambert, J. E. (1998). Primate digestion: interactions among anatomy, physiology, and feeding ecology. *Evolutionary Anthropology* 7: 8–20.

Lambert, J. E. (2006). Primate nutritional ecology: feeding biology and diet at ecological and evolutionary scales. In Campbell, C., Fuentes, A., MacKinnon, K. C., Panger, M. & Bearder, S. (eds.), *Primates in Perspective*. New York, Oxford: Oxford University Press, pp. 482–495.

Leigh, E. G. & Windsor, D. M. (1982). Forest production and regulation of primary consumers on Barro Colorado Island. In Stanley Rand, A., Windsor, D. M. & Leigh, E. G. (eds.), *The Ecology of a Tropical Forest: Seasonal Rhythms and Long-term Changes*. Washington: Smithsonian Institution Press, pp. 111–122.

Leighton, M. (1993). Modeling dietary selectivity by Bornean orangutans: evidence for integration of multiple criteria in fruit selection. *International Journal of Primatology* 14: 257–313.

Lotz, C. N. & Nicolson, S. W. (1996). Sugar preference of a nectarivorous passerine bird, the lesser double-collared sunbird (*Nectarinia chalybea*). *Functional Ecology* 10: 360–365.

Makkar, H. P. S. & Becker, K. (1998). Do tannins in leaves of trees and shrubs from African and Himalayan regions differ in level and activity? *Agroforestry Systems* 40: 59–68.

Makkar, H. P. S., Bluemmel, M., Borowy, N. K. & Becker, K. (1993). Gravimetric determination of tannins and their correlations with chemical and protein precipitation methods. *Journal of the Science of Food and Agriculture* 61: 161–164.

Malenky, R. (1990). *Ecological Factors Affecting Food Choice and Social Organization in Pan paniscus*. PhD thesis, State University of New York.

Malenky, R. K. & Stiles, E. W. (1991). Distribution of terrestrial herbaceous vegetation and its consumption by *Pan paniscus* in the Lomako Forest, Zaire. *American Journal of Primatology* 23: 153–169.

Matsumoto-Oda, A. & Hayashi, Y. (1999). Nutritional aspects of fruit choice by chimpanzees. *Folia Primatologica* 70: 154–162.

McGrew, W. C. (2004). *The Cultured Chimpanzee: Reflections on Cultural Primatology*. Cambridge: Cambridge University Press.

McGrew, W. C., Marchant, L. F. & Nishida, T. (eds.) (1996). *Great Ape Societies*. Cambridge: Cambridge University Press.

McGrew, W. C., Marchant, L. F., Beuerlein, M. M., Vrancken, D., Fruth, B. & Hohmann, G. (2007). Prospects for bonobo insectivory: Lui Kotal, Democratic Republic of Congo. *International Journal of Primatology* 28: 1237–1252.

Miller, L. (ed.) (2002). *Eat or be Eaten: Predator Sensitive Foraging Among Primates*. Cambridge University Press, Cambridge.

Milton, K. (1978). Behavioral adaptations to leaf-eating by the mantled howler monkey *(Alouatta palliata)*. In Montgomery, G. G. (ed.), *Ecology of Arboreal Folivores*. Washington, D. C.: Smithsonian Press, pp. 535–549.

Milton, K. (1981). Food choice and digestive strategies of two sympatric primate species. *The American Naturalist* 117: 476–495.

Milton, K. & McBee, R. H. (1983). Rates of fermentative digestion in the howler monkey (*Alouatta palliate*) (Primates: Cebidae). *Comparative Biochemistry and Physiology* 7: 4–29.

Mitani, J. C., Watts, D. P. & Lwanga, J. S. (2002). Ecological and social correlates of chimpanzee party size and composition. In Boesch C., Hohmann, G. & Marchant L. F. (eds.), *Behavioural Diversity in Chimpanzees and Bonobos*. Cambridge: Cambridge University Press, pp. 102–111.

Morin, P. A., Moore, J. J., Chakraborty, R., Jin, L., Goodall, J. & Woodruff, D. (1994). Kin selection, social structure, gene flow, and the evolution of chimpanzees. *Science* 265: 1193–1201.

Mueller-Harvey, I. & McAllan, A. B. (1992). Tannins, their biochemistry and nutritional properties. *Advances in Plant Cell Biochemistry and Biotechnology* 1: 151–217.

Nunn, C. L. & Altizer, S. (2006). *Infectious Diseases in Primates: Behavior, Ecology and Evolution*. (Oxford Series in Ecology & Evolution). Oxford: Oxford University Press.

Oates, J. F., Bergl, R. A. & Linder, J. M. (2004). *Africa's Gulf of Guinea Forests: Biodiversity Patterns and Conservation Priorities*. (Advances in Applied Biodiversity Science 6). Washington, DC: Conservation International.

Parish, A. R. (1994). Sex and food control in the "uncommon chimpanzee": How bonobo females overcome a phylogenetic legacy of male dominance. *Ethology and Sociobiology* 15: 157–179.

Parish, A. R. (1996). Female relationships in bonobos (*Pan paniscus*): evidence for bonding, cooperation, and female dominance in a male-philopatric species. *Human Nature* 7: 61–96.

Parish, A. R. & de Waal, F. (2000). The other "closest living relative". How bonobos (*Pan paniscus*) challenge traditional assumptions about females, dominance, intra- and intersexual interactions, and hominid evolution. *Annals of the New York Academy of Sciences* 907: 97–113.

Parker, I. (2007). Swingers. Bonobos are celebrated as peace-loving, matriarchal, and sexually liberated. Are they? *The New Yorker* July 30: 48–61.

Poppi, D., Norton, B., Minson, D. & Hendricksen, R. (1980). The validity of the critical size theory for particles leaving the rumen. *Journal of Agricultural Science Camb.* 94: 275–280.

Porter, L. J., Hrstich, L. N. & Chan, B. G. (1986). The conversion of procyanadins and prodelphinidins to cyanidin and delphinidin. *Phytochemistry* 25: 223–230.

Poulin, B., Wright, S. J., Lefebvrev, G. & Calderon, O. (1999). Interspecific synchrony and asynchrony in the fruiting phenologies of congeneric bird-dispersed plants in Panama. *Journal of Tropical Ecology* 15: 213–227.

Raubenheimer, D. & Simpson, S. J. (1997). Integrative models of nutrient balancing: application to insects and vertebrates. *Nutritional Research Reviews* 10: 151–179.

Remis, M. J. (2002). Food preferences among captive western gorillas (*Gorilla gorilla gorilla*) and chimpanzees (*Pan troglodytes*). *International Journal of Primatology* 23: 231–249.

Reynolds, V. (2005). *The Chimpanzees of the Budongo Forest*. Oxford: Oxford University Press.

Reynolds, V., Plumptre, A. J., Greenham, J. & Harborne, J. (1998). Condensed tannins and sugars in the diet of chimpanzees (*Pan troglodytes schweinfurthii*) in the Budongo Forest, Uganda. *Oecologia* 115: 331–336.

Robbins, M. M. (2007). Gorillas: Diversity in ecology and behavior. In Campbell, C. J., Fuentes, A., MacKinnon, K. C., Panger, M., Bearder, S. K. (eds.), *Primates Perspective*. New York: Oxford University Press, pp. 305–317.

Rodman, P. S. (2002). Plants of the apes: is there a hominoid model for the origins of the hominoid diet? In Ungar, P. S. & Teaford, M. F. (eds.), *Human Diet: Its Origin and Function*. Westpoint: Bergin & Harvey, pp. 77–109.

Rogers, M. E., Maisels, F., Williamson, E. A., Fernandez, M. & Tutin, C. E. G. (1990). Gorilla diet in the Lopé Reserve, Gabon: a nutritional analysis. *Oecologia* 84: 326–339.

Saracco, J. F., Collazo, J. A. & Groom, M. J. (2004). How frugivores track resources? Insight from spatial analyses of bird foraging in a tropical forest. *Oecologia* 139: 235–45.

Schaefer, H. M., Schmidt, V. & Baierlein, F. (2003). Discrimination abilities for nutrients: which difference matters for choosy birds and why? *Animal Behaviour* 65: 531–541.

Schöning, C., Ellis, D., Fowler, A. & Sommer, V. (2007). Army ant prey availability and consumption by chimpanzees at Gashaka (Nigeria). *Journal of Zoology* 271: 125–133.

Sommer, V. (2000). The holy wars about infanticide. Which side are you on? And why? In van Schaik, C. & Janson, C. (eds.), *Infanticide by Males and its Implications*. Cambridge: Cambridge University Press, pp. 9–26.

Sommer, V. (2008). Welcher Affe steckt in uns? *Bild der Wissenschaft* 4: 18–25.

Sommer, V., Adanu, J., Faucher, I. & Fowler, A. (2004). The Nigerian Chimpanzee (*Pan troglodytes vellerosus*) at Gashaka: Two years of habituation efforts. *Folia Primatologica* 75: 295–316.

Stanford, C. (1998). The social behavior of chimpanzees and bonobos: Empirical evidence and shifting assumptions. *Current Anthropology* 39: 399–420.

Susman, R. L. (Hg.) (1984). *The Pygmy Chimpanzee: Evolutionary Biology and Behavior*. New York: Plenum.

Takemoto, H. (2003). Phytochemical determination for leaf food choice by wild chimpanzees in Guinea, Bossou. *Journal of Chemical Ecology* 29: 2551–2573.

Trivers, R. L. (1972). Parental investment and sexual selection. In Campbell, B. (Hg.) (ed.), Sexual Selection and The Descent of Man, 1871–1971. Illinois, Chicago : Aldine Press, pp. 136–179.

Tutin, C. E. G. & Fernandez, M. (1993). Relationships between minimum temperature and fruit production in some tropical forest trees in Gabon. *Journal of Tropical Ecology* 9: 241–248.

van Schaik, C. P. (1989). The ecology of social relationships amongst female primates. In Standen, V. & Foley, R. A. (eds.), *Comparative Socioecology*. Oxford: Blackwell Scientific Publications. 195–218.

van Soest, P. J. (1994). *Nutritional Ecology of the Ruminant*. Ithaca: Cornell University Press.

Vervaecke, H., de Vries, H. & van Elsacker, L. (2000). Function and distribution of coalitions in captive bonobos (*Pan paniscus*). *Primates* 41: 249–265.

Vogel, C. (1999). *Anthropologische Spuren. Zur Natur des Menschen*. Ed. & introd. by Sommer, V.; Edition Universitas. Stuttgart: Hirzel.

Vogel, C. & Sommer, V. (1994). Drum prüfe, wer sich ewig bindet. In Schiefenhövel, W., Vogel, C., Vollmer, G. (eds.), *Vom Affen zum Halbgott*. Stuttgart: Trias, pp. 13–42.

White, F. J. & Wrangham, R. W. (1988). Food competition and patch size in the chimpanzee species *Pan paniscus* and *Pan troglodytes*. *Behaviour* 105: 148–164.

Witmer, M. C. (1998). Do seeds hinder digestive processing of fruit pulp? implications for plant / frugivore mutualisms. *The Auk* 115: 319–326.

Witmer, M. C. & van Soest, P. J. (1998). Contrasting digestive strategies of fruit-eating birds. *Functional Ecology* 12: 728–741.

Wittig, R. & Boesch, C. (2003). The choice of post-conflict interactions in wild chimpanzees (*Pan troglodytes*). *Behaviour* 140: 1527–1559.

Wrangham, R. W. (1979). On the evolution of ape social systems. *Social Science Information* 18: 335–368.

Wrangham, R. W. (1980). An ecological model of female-bonded primate groups. *Behaviour* 75: 262–300.

Wrangham, R. W. (1986). Ecology and social evolution in two species of chimpanzees. In Rubenstein, D. I. & Wrangham, R. W. (eds.), *Ecology and Social Evolution: Birds and Mammals*. Princeton: Princeton University Press, pp. 352–378.

Wrangham, R. & Peterson, D. (1996). *Demonic Males. Apes and the Origins of Human Violence*. London: Bloomsbury.

Wrangham, R. W. & Waterman, P. G. (1983). Condensed tannins in fruits eaten by chimpanzees. *Biotropica* 15: 217-222.

Wrangham, R. W., Conklin-Brittain, N. L. & Hunt, K. D. (1998). Dietary response of chimpanzees and Cercopithecines to seasonal variation in fruit abundance. I. Anti-feedants. *International Journal of Primatology* 19: 949–970.

Wrangham, R. W., Chapman, C. A., Clark-Arcadi, A. P. & Isabirye-Basuta, G. (1996). Social ecology of Kanyawara chimpanzees: implications for understanding the costs of great ape groups. In McGrew, W. C., Marchant, L. F. & Nishida, T. (eds.), *Great Ape Societies*. New York: Cambridge University Press, pp. 45–57.

Wrangham, R. W., Conklin, N. L., Chapman, C. A. & Hunt, K. D. (1991). The significance of fibrous foods for Kibale Forest chimpanzees. *Philosophical Transactions of the Royal Society of London, B* 334: 171–178.

Wrangham, R. W., McGrew, W. C., de Waal, F. B. M. & Heltne, P. G. (eds.) (1994). *Chimpanzee Cultures*. Cambridge, MA: Harvard University Press.

Chapter 13
Panthropology of the Fourth Chimpanzee: A Contribution to Cultural Primatology

Andrew Fowler, Alejandra Pascual-Garrido, Umaru Buba, Sandra Tranquilli, Callistus Akosim, Caspar Schöning, and Volker Sommer

Abstract A trademark of *Homo sapiens* is the enormous variation in behavioural patterns across populations. Insight into the development of human cultures can be aided by studies of *Pan* communities across Africa, which display unique combinations of social behaviour and elementary technology. Only cross-population comparisons can reveal whether this diversity reflects differential genetics, environmental constraints, or arbitrary cultural patterns. However, the recently recognised and most endangered subspecies *Pan troglodytes vellerosus* remains completely unstudied in this respect. We report on the Nigerian chimpanzees at Gashaka. At this site, diet composition is highly varied and the apes have to cope with high concentrations of anti-feedant defenses of plants against consumption. It is not surprising therefore, that Gashaka chimpanzees use a varied tool-kit for extractive foraging. For example, they harvest insects throughout the year, employing digging sticks and probes to obtain honey from nests of stingless bees and honey bees, dipping wands to prey on army ants and fishing rods to eat arboreal ants. Tools appeared to be custom-made with a considerable degree of standardisation and preferential use of distal ends. Many of these expressions of subsistence technology seem to be environmentally constrained. Most notably, the absence of termite eating could reflect a low abundance of mounds. Other traits may represent arbitrary cultural variation. For example, two types of hard-shelled nuts found in the habitat are not opened with tools, unlike what is observed elsewhere in West Africa. The prevalence of elementary technology may indicate that the material culture of Gashaka chimpanzees is most closely related to core cultural tendencies of Central African populations of these apes.

Keywords: Tool use • Insectivory • Cultural Primatology • Chimpanzee

A. Fowler (✉)
Max-Planck-Institute for Evolutionary Anthropology, Leipzig 04103, Germany
e-mail: andrew_fowler@eva.mpg.de

V. Sommer and C. Ross (eds.), *Primates of Gashaka*, Developments in Primatology: Progress and Prospects 35, DOI 10.1007/978-1-4419-7403-7_13,

Introduction

Behavioural patterns between human populations may vary tremendously. People even from neighbouring villages can differ in the way they speak, how they greet each other, what they consider acceptable conduct and what counts as offensive, how and what they like to eat, and which implements they employ. There is simply a certain way of doing things, which defines identity as well as the boundaries between in-group and out-group (McGrew 2004, Miller 2005).

Such variation is a trademark of *Homo sapiens* and constitutes the basis of our "cultural diversity". There is no consensus about what constitutes "culture", but social anthropologists tend to engage in a "humanist" stance, thus reserving the label "culture" exclusively for our own species. Biological anthropologists, on the other hand, tend to be "universalists", assuming an evolutionary continuum of traits that constitute culture (McGrew 1992, 2004).

The biological paradigm asserts that studies of non-human primates, our closest living relatives, are particularly suitable to reveal how the capacity for intra-specific diversity (Lott 1984) might have evolved – including cultural variety in humans. The assumption is supported by investigations of species that display considerable ecological and social flexibility. For example, habitats where Indian langur monkeys can thrive range from semi-arid conditions at the fringe of deserts, to Himalayan mountains, urban settings and moist evergreen forests. These primates are likewise socially flexible; in some places, they form reproductive units of the multi-male-multi-female type, but strict one-male-multi-female units in other places (review in Sommer 1996).

The potential mechanisms that cause behavioural diversity include (a) genetic make-up, (b) environmental influences, and (c) social traditions independent from ecological constraints. For example, many adult humans are unable to drink milk without adverse affect. They do not have the genotype to produce the enzyme lactase beyond infancy, a trait that evolved rather recently among certain pastoralist populations (Ingram *et al.* 2009). The consumption of milk or its absence is therefore largely genetically determined. An example of a behaviour constrained or influenced by the environment is the way people eat rice – as varied customs reflect to a large degree the consistency of this food. Thus, chop-sticks are the implement of choice when rice is sticky, whereas forks are more feasible when lose long-corn rice is consumed, whereas rice that is cooked into a mush is often eaten by hand. However, there are also behaviours that lack moulding by plausible genetic or environmental causes. For example, people in different countries tend to greet each other in different ways, by bowing (Thailand), shaking hands (Germany), kissing on one cheek or both (France), or by moving the right hand towards the heart (Nigeria), etc. These rather arbitrary customs thus represent pure "cultural variants".

An inclusive definition would define "culture" as "socially transmitted behaviour" (McGrew 2004). This definition would not include traits that are genetically determined; of course, whether or not adults drink milk would still contribute towards intra-specific variability, but the patterning does not count as a "cultural trait" in the above specified sense, because milk-drinking will not be learned if a person lacks lactase persistence. On the other hand, it is important to note that traits influenced or determined by the ecology – such as rice-eating – *can* be socially transmitted. These customs, together with arbitrary variants, would thus constitute the *cultural profile* of a population.

This scheme can be readily applied to non-human animals, too. Let us again consider the case of Indian langur monkeys. Discernible genetic differences between populations do exist, and they correspond, for example, with the way a tail is carried – if in a graceful loop across the back, or simply letting gravity take its course (Roonwal 1980). This is not unexpected, given that this taxon ranges across the Indian subcontinent, with outposts as distant as Sri Lanka, Pakistan, Nepal, and China. However, the two basic types of social system – harems or polygynandrous groups – are not tied to any genetic marker. Instead, environmental conditions strongly encourage one or the other type. If food is found throughout the year, then females de-synchronise their menstrual cycles, allowing a strong male to guard the one or two mates that are simultaneously fertile. This polygynous system breaks down, however, if food availability has seasonal bottlenecks. Under these conditions, females can only conceive during certain months, and their fertility will be synchronised. It is impossible in such a situation for even the most powerful male to guard all simultaneously fertile females. Consequently, a multi-male-multi-female system will develop. But langurs also display behaviours that obviously reflect arbitrary cultural variants. For example, at the site of Jodhpur in Rajasthan, adult monkeys will almost never huddle during rests – except perhaps for the coldest winter mornings. Jodhpur langurs will keep a small distance, even in the thickest crowd. However, in the nearby town of Jaipur, langurs huddle throughout much of the day and night, rain or shine, hot or cold, male with male, female with female, and male with female. These populations thus seem to adhere to different social conventions (Sommer 1996) – very much of the type "this is how things are done here" (McGrew 2004).

An astounding degree of behavioural diversity is exhibited by the chimpanzee (*Pan troglodytes*) and the bonobo (*P. paniscus*) (Boesch *et al.* 2002). Research across Africa has revealed for each study community a unique combination of the presence or absence of traits related to social customs, communication, territorial aggression, war-like raiding, hunting strategies, tool-kits, food-processing and consumption, and ingestion of plant matter for self-medication (Wrangham *et al.* 1994, McGrew *et al.* 1996, Hohmann & Fruth 2003, McGrew 1992, 2004; see also Sommer *et al.* this volume [Ch. 12]). This degree of plasticity in behavioural patterns is perhaps not surprising given that *Pan* and *Homo* shared a common ancestor until about 5 – 7 million years ago.

A classic study has compiled the behavioural patterns at 9 long-term chimpanzee research sites. Behaviours for which ecological explanations seemed plausible were carefully discerned from a couple of dozen traits customary or habitual among some groups but absent in others (Whiten *et al.* 1999, 2001). A well-known example is nut-cracking with stone or wooden hammers against an anvil. This technique is restricted to West Africa, despite an abundance of nuts and potential tools elsewhere (McGrew 1992). The practice is in all likelihood neither genetically determined nor a reflection of particular environmental conditions because some communities exhibit the behaviour, while others within a closely related population, separated only by the banks of a river but exposed to virtually identical environmental conditions, do not (Boesch *et al.* 1994).

Many primatologists will therefore not hesitate to label the diversity of chimpanzee behaviour as an expression of different "cultures" (e.g., McGrew 1992, de Waal 1999, Sommer 2003). This argument is based on such criteria as: the behaviour is *learned* and not "instinctive", *socially* acquired, *normative* and not random nor idiosyncratic in its execution, and a *collective* characteristic of a group (McGrew 2004).

The genus *Pan* includes the bonobo – restricted to the forested basin south of the Congo River – and the chimpanzee with subspecies in Central Africa (*troglodytes*), West Africa (*verus*) and East Africa (*schweinfurthii)*. A fourth subspecies – *vellerosus* from eastern Nigeria and eastern Cameroon – has recently been recognised (for a proposal to use the name *P. t. ellioti*, see Oates *et al.* 2008). While genetically the most distinct (Gonder *et al.* 1997, Gagneux *et al.* 2001), they are also the most endangered (Kormos *et al.* 2003, Hughes 2003, Hughes *et al.* this volume [Ch. 13]). Long-term studies of the socioecology of *P. t. vellerosus* are restricted to Gashaka Gumti National Park in Nigeria (Sommer *et al.* 2004).

In depth-research of a primate population can be likened to the cultural anthropological practice of *ethnography* i.e., a descriptive account of behavioural patterns observed within this particular population. The next step is *ethnology*, the analysis across populations, which aims to detect pattern and causes for them. Major tools for cross-cultural comparisons of human populations are the HRAF ("Human Relations Area Files"). The dynamic development of the paradigm of "cultural primatology" thus comes with the explicit aim of creating CRAF – "Chimpanzee Relations Area Files" (McGrew 2004).

Our contribution attempts to at least partly fill in the noticeable gap in chimpanzee research with respect to the "fourth chimpanzee" – the subspecies *vellerosus*. For this, we summarise findings that have been accumulated over nearly a decade of research in Nigeria (Sommer *et al.* 2004, Fowler 2006, Fowler & Sommer 2007, Fowler *et al.* 2007, Schöning *et al.* 2007, Sommer 2008). Nevertheless, our study is limited because we are still rarely able to directly observe the chimpanzees. We therefore know nothing about potential cultural variants in social behaviour, such as different techniques of grooming, courtship, or playing. Instead, most of our evidence is indirect, i.e., based on traces, objects and tools left behind by the apes while they make use of *elementary technology* – in particular implements they use to harvest social insects and their products (*Fig. 13.1*).

The "knowledgeable use of [...] physical objects as a means to achieve an end" (McGrew 2004: 103) is an expression of chimpanzee *material culture* (McGrew 1992) that can be studied even when the users are absent, or with unhabituated apes (McGrew *et al.* 2003). The situation is thus somewhat similar to the challenges

Fig. 13.1 Field assistant Hammaunde Guruza with a bundled up collection of ateliers of stick tools left behind by the chimpanzees of Gashaka-Kwano (photo: VS)

faced by archaeologists or palaeoanthropologists who cannot watch their study subjects and reconstruct their likely actions based on inference.

Only through investigations of "as many groups of chimpanzees in as many parts of Africa as possible" (Goodall 1994: 397) can universal behavioural patterns be discerned from variants and whether these differences reflect genetics, environment, or arbitrary customs. A call for cross-population comparison is echoed by many primatologists (e.g., Wrangham *et al.* 1994, Whiten *et al.* 2001, McGrew 2004). The Nigerian chimpanzee is clearly a missing piece of the jigsaw.

Studies in a West African mosaic chimpanzee habitat will ultimately broaden our knowledge about conditions encountered by Mio-Pliocene hominids (Hunt & McGrew 2002). A better understanding of "the fourth chimpanzee" will therefore facilitate our comprehension of that other creature, fittingly nick-named *The Third Chimpanzee* (Diamond 1992).

Materials and Methods

Much of the material culture employed by chimpanzees of the Gashaka-Kwano study community in Nigeria relates to extractive foraging of social insects and their products, although there are notable exceptions. A general introduction to chimpanzee socioecology is therefore followed by a description of characteristics of potential insect prey, before we introduce the study site and data collection procedures.

Basic Chimpanzee Socioecology

Chimpanzees (for general review, see Goodall 1986, Heltne & Marquardt 1989, Wrangham *et al.* 1994, Boesch & Boesch-Achermann 2000, Reynolds 2005) are African apes. Populations survive in at least 18 nations from Tanzania and Uganda in the east to Mali and Senegal in the west. They inhabit a variety of biotopes such as evergreen and semi-deciduous rain forests, open woodland-savannah, gallery forests, and mosaic habitats that may include plantations and grassland.

Chimpanzees live in "communities" or "unit-groups" of 10 – 140 members which range over 5 – 38 km^2 in forests, and 25 – 560 km^2 in open habitats. Patchy distribution of food causes communities to forage in small parties of 6 members on average (range 3 – 10). Different members may join these parties ("fusion") or split from them ("fission"). Parties communicate through vocalising and drumming, utilising buttress roots. Males are philopatric whereas females tend to leave their natal community upon sexual maturity. Males are generally closely related and cooperate to defend their range against neighbouring communities with whom they may engage in violent conflicts ("lethal raiding").

Each night – and often also during the day – every group member (except dependent offspring) builds a new nest ("sleeping platform") from leafy twigs, typically in trees. The architecture of nests, the location, and size of nest groups and whether or not ground nests occur may be influenced by the risk of predation by leopards, lions, and humans. Occasionally, conspecifics may also prey upon chimpanzees.

Chimpanzees feed on ripe fruit 56 – 71 % of foraging time, on leaves 18 – 21 %, and 11 – 23 % on other plant-parts, in particular terrestrial herbs. Faunivory constitutes 0.1 – 4 %, comprising at least 25 vertebrate species, which may be hunted cooperatively (80 % colobus monkeys 20 % mammals such as duikers, bush pigs, baboons, and rodents) as well as social insects and their products (see below).

Wild chimpanzees manufacture and / or use a variety of tools from materials such as bark, leaves, sticks and rocks, as sponges, wipes, probes, hooks, drills, missiles, hammers, and toys, to extract resources (water, insect prey, honey, seeds) and in social contacts (sexual invitations, conflicts, play).

Many local populations of chimpanzee have disappeared over the last few decades or are in danger of extinction due to hunting, deforestation and other forms of human encroachment (Ammann *et al.* 2003).

Characteristics of Chimpanzee Insect Prey

Chimpanzees have been observed to eat insects such as caterpillars, larvae, and imagos of beetles or fig wasps. These are ingested occasionally or incidentally while foraging on leaves and fruit (Goodall 1986, Reynolds 2005). However, at several sites, chimpanzees systematically exploit the colonies of eusocial insects such as ants, termites, and bees (review in McGrew 1992; see below), often using tools manufactured from plant parts.

Bees

Various chimpanzee populations prey upon brood and stored honey of bees (e.g., Yamagiwa *et al.* 1988, Sanz *et al.* 2004), including tribes of both Apini (honey bees) and Meliponini (stingless bees).

Honey bees (*Apis mellifera*) occur at low densities of about 3.3 colonies per km^2 in African equatorial forests (Roubik 1991). A hive may contain 20000 or more bees, including queen, drones (males) and workers (sterile females). The queen can produce more than 1000 eggs per day (Hart & Ratnieks 2002) and suppress worker reproduction by secreting a pheromone (Gould & Gould 1995). Workers feed the queen and maintain the nest. Honey bees prefer tree holes with a small entrance, well protected from predators, and attack intruders with mandibles and painful stings. Some bees patrol the nest periphery, alerting the colony of danger through pheromones (Fletcher 1978). Combs, produced by wax from worker glands, contain hexagonal cells to hold honey, pollen, eggs and pupae that, when filled, are sealed with a cap. Pollen collected from flowers is the main supply of protein and vitamins for the hive (Winston 1987). Body-movements ("dance") communicate to nest mates the existence, distance and location of food patches (von Frisch 1967).

Other honey-storing bees include the stingless Meliponini, restricted to the tropics. These have not been as intensively studied as honey bees (Nieh 2003), but include the small *Trigona* (3 – 4 mm) and the larger *Melipona* (6 – 8 mm; Anzenberger 1977, Tutin & Fernandez 1992, Tutin *et al.* 1995). Their perennial colonies include a queen plus rarely more than 500 workers (sterile females) and males (drones). The small nests are typically built in tree cavities. The bees have access through a cerume tube made of wax and resin, which often sticks out a few centimetres from the entrance. Some colonies nest in the ground, with tunnels more than a meter deep. Meliponini construct hexagonal cells for their brood, which are placed in combs. Meliponini also feed on nectar and pollen of flowers (Hofstede 2005), but they store food in different cells or storage pots, in contrast to Apini bees (Hart & Ratnieks 2002). Stingless bees, as their name indicates, cannot sting, but defend their nests by swarming, buzzing, and biting.

Termites

Chimpanzees at various sites – but not all – feed on different types of termites, often using pliable tools (Goodall 1963, 1986, McGrew *et al.* 1979, McGrew 1992, Sanz *et al.* 2004). In particular, they harvest brood, soldiers, and workers of the genus *Macrotermes*, the most ubiquitous taxon across sub-Saharan Africa (Howse 1970), which also produces the largest imagos of any termite (nickname: big mac).

Macrotermes such as *M. bellicosus* construct nests from subsoil material and saliva, often recognizable above ground as they extend into characteristic mounds (Stoops 1964) – although mound morphology may differ, according to species and environment (Gumnior & Thiemeyer 2003). Above-ground construction activity is highest during the wet season as wet soil is moulded more easily. A constant microclimate inside the nest is regulated by airflow through chimneys. This facilitates growth of a particular fungus, *Termitomyces* sp., which the termites cultivate via collected plant litter. The litter is pre-digested by the fungus, to then become the major termite food (Bourliere 1983). Termites retreat deeper underground during the dry season, to mitigate higher temperatures, sealing the outer walls with clay, sand, and faeces (McGrew *et al.* 1979).

Workers – small blind, wingless and sexually immature – dig the nest, control its temperature, locate food, and provision other castes. Soldiers have much larger heads and bodies and defend the colony from attackers. The reproductive forms are an egg-lying queen, up to 6 cm long, and a king who provides sperm (Wilson 1971). Winged reproductive forms (alates) swarm early in the wet season.

Ants

Ants are an important food source for many African mammals, including pangolins, aardvarks, gorillas and chimpanzees (Redford 1987, Watts 1989, Yamagiwa *et al.* 1991, Gotwald 1995, Kingdon 1997, Ganas & Robbins 2004).

Weaver ants, *Oecophylla longinoda*, do not occur at Gashaka but are the species most commonly eaten by chimpanzees elsewhere (e.g., at Bossou / Guinea; Sugiyama 1995). They are named after the way they construct arboreal nests, using silk produced by their larvae to bind living leaves together. These cocoons buffer against fluctuations of humidity and temperature that would be harmful to the larvae. Each colony has a single queen but usually multiple nests. The ants are stingless but can inflict painful bites to invaders and spray formic acid (Peeters & Andersen 1989, Hölldobler & Wilson 1990).

Crematogaster spp. form small colonies in tree branches or in hollow dead stems of certain plants (Nishida 1973, Tutin *et al.* 1983). They are very small and reddish brown or black. Their heart-shaped abdomen can rise over their heads when threatened – earning them the nick-name "acrobat ant" – and they also use a chemical defense against attackers. These ants are eaten by chimpanzees at, e.g., Lopé / Gabon (Tutin *et al.* 1995). *Crematogaster* does occur at Gashaka, but seems to be rare.

The arboreal *Camponotus* spp. represent the most widely distributed ant genus. Some species nest in the soil, although others form small colonies in hollow trees or cavities (Nishida 1973). They respond to predators with rapid escape; some taxa react by biting. Chimpanzees consume them, e.g., at Mahale / Tanzania (Nishida 1973, Nishida & Hiraiwa 1982).

The stinging ants *Pachycondyla analis* (formerly *Megaponera foetens*; Bolton 1994), are specialised predators of fungus-growing termites. Workers enter termite galleries and, in characteristic single file, carry immobilised prey back to their nests, often recognizable by a small mound of earth around the entrance (Longhurst *et al.* 1978). Chimpanzees prey on them at, e.g., Mt. Assirik / Senegal (McGrew *et al.* 1988).

Dorylus (subgenus *Anomma)* ants are a favourite chimpanzee prey (e.g., at Gombe / Tanzania; Goodall 1963). These "army ants" differ from all other listed social insects by their enormous colony sizes of up to 12 million. The total fresh mass (including brood and workers) of an average colony with 6.5 million workers is about 40 kg (Leroux 1982, Kronauer 2008).

Not all army ant species are available as prey for chimpanzees (Schöning *et al.* 2005). Species with a "subterranean" life-style hunt in the soil and their nests are not indicated by obvious cues on the surface. These species have never been reported to be consumed by chimpanzees. Species with an "intermediate" life-style hunt in the leaf litter but do not climb vegetation when foraging, and inhabit less conspicuous nests. These include taxa such as *D. gribodoi* (Schöning *et al.* 2008a), which is consumed by chimpanzees at Bossou / Guinea (Humle & Matsuzawa 2002) and Taï / Ivory Coast (Boesch & Boesch 1990). Species with an "epigaeic" life-style hunt above ground, organising conspicuous swarm raids on the surface and up in the vegetation; any animal able to move will flee the approach of these fierce predators. This is why the term "driver ant" (often applied to all members of the genus) should be used *only* for the epigaeic species. These include species such as *D.* (*A.*) *molestus*, *D.* (*A.*) *nigricans*, *D.* (*A.*) *sjoestedti*, and *D.* (*A.*) *wilverthi* (Raignier & van Boven 1955; Gotwald 1974; Leroux 1982). *Dorylus* (*A.*) *nigricans* is preyed on by chimpanzees at Bossou / Guinea (Humle & Matsuzawa 2002) and Taï / Ivory Coast (Boesch & Boesch 1990).

Army ants, by definition, always hunt as a group, by both day and night, as they are completely blind. Communal hunts have the advantage that creatures much bigger than the ants themselves can be captured. Swarm raids of epigaeic ants cover the forest floor like a thick black carpet. The swarm structure, easily 10 m wide, sweeps through the forest like a drag net. Even larger creatures, such as rats, snakes, and frogs, hasten to escape. Raiding ants also climb bushes and trees, simply letting themselves drop to the ground when these higher places hold no food for them. Countless workers that fall simultaneously from vegetation produce the sound of light rain. Every day, tens of thousands of animals – spiders, stick insects, slugs, caterpillars, crickets, beetles, grasshoppers, woodlice, other ants, and earthworms – fall victim to a driver ant colony (Schöning 2005).

Epigaeic army ants build underground nests, for which they may use pre-existing holes such as burrows of small mammals, with depths of up to 1.7 m underground (Leroux 1982). The highly specialised queen resides here. *Dorylus* queens, unlike other ant queens, never embark on a mating flight, but store the sperm of various males who, as winged forms, previously emerged from other colonies and manage to locate and visit a foreign queen in her nest. *Dorylus* queens are thus permanently wingless, and so lack the enlarged thorax that in most insects houses the flight muscles. These queens are the largest ants in existence. They may produce an astounding 3 – 4 million eggs a month, i.e., one egg every 1.5 s!

Dorylus (*Anomma*) nests are well defended. Many of the largest workers will rush to the surface if their nests are disturbed, often forming a ball-like plug at the entrance. *D.* (*Anomma*) ants – contrary to the related Ecitoninae of South America – lack a functional sting. However, their falcate (sickle-shaped) mandibles can inflict painful bites and easily pierce human skin.

Chimpanzees will harvest army ants with so-called dipping wands that help them to avoid getting bitten. The chimpanzees at Taï / Ivory Coast represent a notable exception as they often open nests and take out workers and brood directly with their hands (Boesch & Boesch 1990).

Army ants differ from other social insect prey targeted by chimpanzees because colonies move to new nest sites, once food in the vicinity of the current nest is depleted (Wilson 1958, Schöning *et al.* 2005). The alternate name "safari ants" alludis to these movements. Some species migrate in synchrony with their brood development, e.g., when young workers have just emerged or young larvae hatched (Schneirla 1971). This, however, does not apply to *Dorylus* species. Migrations occur on average every 17 days (*D. molestus* in Kenya; Schöning *et al.* 2005b), but may be as frequent as every 8 days or so (*D. nigricans* in Ivory Coast; Leroux 1982). Migratory trails can be recognised because brood is carried to the new location. Workers will guard these trails particularly well when and where the queen is making the hazardous journey to the new home.

The Gashaka Study Site

Gashaka Gumti National Park (GGNP) lies in southern Taraba State in eastern Nigeria on the border with Cameroon (06° 55′ – 08° 13′ N and 11° 13′ N – 12° 11′ E). GGNP is, at about 6700 km^2, Nigeria's largest national park (for the following, see Dunn 1999, Chapman & Chapman 2001 Chapman *et al.* 2004, Adanu *et al.* this volume [Ch. 3], Sommer *et al.* this volume [Ch. 13]).

Our project maintains a field station inside the park at Kwano (583 m asl; 07°19′ N – 11°35′ E), location of an abandoned settlement, 11 aerial km from the nearest village of Gashaka. An ancient footpath connects Gashaka with Kwano, leading to the highlands and on to Cameroon (traffic approx. 1 person / h).

Terrain in the southern Gashaka sector is rugged with altitudes from about 300 – 2400 m, including the northern outcrops of the Cameroonian Highland chain. Abundant rivers flow continuously, even throughout the distinct dry season. Pronounced annual wet and dry seasons are associated with corresponding fluctuations in temperature and humidity. Heavy downpours from mid-April to mid-November are followed by a 5 month period with very little or no rainfall at all. This dry period coincides with the Harmattan, a dry dusty wind that blows from the Sahara. Weather data collected at Kwano from 2001 – 2008 document the following average percentages of rainy days / month: J 0, F 4, M 11, A 53, M 71, J 69, J 64, A 69, S 71, O 58, N 11, D 0. The overall average of 40 % rainy days / month correponds to 146 days with rain per annum. The wettest day (Jul 08) saw 125 mm of rain. The yearly average rainfall was 1973 mm (2001, 1683; 2002, 2056; 2003, no data; 2004, 2337; 2005, 1945; 2006, 2279; 2007, 1786; 2008, 1726). Mean monthly humidity at sunset fluctuated between 59 % (Feb) and 87 % (Sep). The mean minimum temperature was 20.9 °C, the coolest recorded temperature 14 °C (Jan, Dec), the mean maximum 31.9 °C, and the hottest day on record 43 °C (Mar 01).

The Gashaka area is located in the sub-Saharan Guinea zone, representing a mosaic of habitats. Montane forests and open grassland are found outside the Kwano study area, which is characterised by savannah-woodland, lowland, and gallery forest.

The region has experienced anthropogenic influences for centuries. Most notable is the deliberate yearly burning of grass (Dec – Feb), which has probably turned considerable parts of previously semi-deciduous forest into grassy woodland (Louppe *et al.* 1995). Large scale logging does not occur, but some timber and non-timber forest products are extracted. Sporadically, Fulani pastoralists graze cattle.

The wider study area harbours a great diversity of wildlife, with several monkey species (olive baboon, tantalus monkey, mona monkey, putty-nosed monkey, black-and-white colobus) as well as a population of *P. t. vellerosus.* Other large mammals include carnivores (African civet, golden cat, leopard), ungulates (red river hog, giant forest hog, African buffalo, bushbuck, red-flanked duiker, yellow-backed duiker, waterbuck), rodents (crested porcupine), and aardvark. All hunting is prohibited in GGNP although ungulates such as buffalo, duiker, and pigs are sometimes poached. Primates, with rare exceptions, are not hunted in the Gashaka-Kwano region, not least because of Islamic religious taboos (but see Adanu *et al.* this volume [Ch. 3]).

Gashaka-Kwano Chimpanzee Study Community

Our studies concentrate on a chimpanzee group in the surroundings of our field station, called the Gashaka-Kwano community (Sommer *et al.* 2004). Their range encompasses at least 26 km^2, and we estimate that the community has about 35 members. This would correspond to a density of 1.3 / km^2. The community might have immediate neighbours only in the south-west of their range where the forest is continuous. A hostile interaction between males was observed here during 2002 by AF.

We calculated party sizes from 95 sightings of chimpanzees during 2000 – 2001, and from > 700 counts of sleeping sites (Fowler 2006). Accordingly, day-parties averaged 3.7 (range 1 – 17). The average of the monthly mean party sizes was greater (4.1), and, not unexpectedly, nest-group size was even larger (mean 5.7, range 1 – 23). The overall socionomic sex-ratio (AM / AF = 0.90) was similar to other sites. Party size tended to increase when estrous females were present, and encounters lasted significantly longer, in line with many other reports (e.g., Anderson *et al.* 2002, Matsumoto-Oda 2002, Mitani *et al.* 2002, Wallis 2002). It is typically hypothesised that party sizes increase because males seek sexually receptive females or because such females seek copulations (Wallis 2002, Wrangham 2002). A tendency for cycle onset during the dry season is reported from Budongo / Uganda and Gombe / Tanzania, probably as a result of feeding conditions, which stimulate hormonal activities (Wallis 2002). The Gashaka data support this, since 18 % of dry season parties but only 4 % of wet season parties contained females sporting ano-genital swellings.

It is known that the smaller the community, the larger the relative party size (i.e., mean party size / community size × 100; Boesch & Boesch-Achermann 2000). If we assume a size of 35 for the Gashaka-Kwano community, party sizes of 3.7, 4.1, and 5.7 translate into relative sizes of 11 %, 12 % resp. 16 %. These values are well within the observed cross-population range and, for example, close to Budongo / Uganda (12 %) or Taï / Ivory Coast (13 %). Thus, our data seem to confirm that smaller communities are socially less fluid, indicating that not only food competition influences fission-fusion, but perhaps also predation pressure (Boesch & Boesch-Achermann 2000), which could be considerable at Gashaka.

Vegetation and climate at Gashaka is similar to mixed-vegetation Tanzanian sites (Gombe: 1775 mm rain during 152 days, mean max temp. 28 °C, mean min. temp. 19 °C ; Mahale: 1836 mm rain during 141 rainy days, mean max. temp. 27 °C, mean min. temp. 19 °C). The climate thus occupies a middle position between very dry (Assirik / Senegal: 954 mm rain) and wet sites (Boussou / Guinea: 2230 mm; cf. Hunt & Mc Grew 2002). The wider Gashaka area has a larger set of potential predators (leopard, lion, hyena, wild dog) compared to most other chimpanzee study sites. The area therefore represents the West African equivalent of a chimpanzee site similar to the hypothesised forest-woodland-habitat in which early humans lived (Hunt & Mc Grew 2002).

Candidate Behavioural Patterns Reflecting Subsistence Technology

We analysed the dataset accumulated for the Gashaka chimpanzees for evidence of elementary technology. This included tools manufactured and used by the chimpanzees and then left behind, i.e., *artefacts* ("the end-product of modification of an object to fulfil a useful purpose"; McGrew 2004: 104, citing the definition given by Oswalt 1976) as well as *naturefacts* ("a natural form, used without prior modification"; ibid.).

It could be argued that the construction of sleeping platforms from branches attached to trees or the swallowing of leaves for assumed medicinal purposes (Koutsioni & Sommer this volume [Ch. 5]) likewise constitutes elementary technology. However, we do not address these patterns here.

Instead, we scrutinised the 65 behavioural patterns described in a benchmark-paper on "Charting cultural variation in chimpanzees" (Whiten *et al.* 2001) to seek evidence as to whether our field data established reasonable details about the presence of a certain technology trait, or if we had good enough reasons to assume their absence at our field site. The resulting list of candidate patterns (*Tab. 13.1*) includes subsistence technology related to insectivory and herbivory (*sensu* McGrew 2004). The particular activity associated with these categories is not always synonymous with a name provided in Whiten *et al.* (2001: *Tab. 1*) since any given pattern may encompass elements of several related patterns (e.g., "dig" is similar to "perforate" or "expel / stir"). A brief description of the behaviour therefore concludes the definition of candidate pattern provided in our tabulation.

Scrutinising the data produced 6 candidate behavioural patterns, which encompass 18 related patterns. We are thus able to make a comparison with about one quarter of the original list of Whiten *et al.* (2001) of potential cultural variants of behavioural patterns in wild chimpanzees.

Table 13.1 Candidate behavioural patterns involving material culture in Nigerian chimpanzees at Gashaka

Subsistence category	Definition of behavioural pattern (a)	Definition may encompass elements of the following related behavioural pattern (b)	Artefact (A) / Naturefact (N)	Description (1)
Insectivory	Dig	Dig [14], Perforate [13], Open-and-probe [22], Lever open [44], Expel/stir [45]	A: Digging stick	Stout stick used as lever or spade to enlarge bee nest entrance
	Bee-probe	Brush-stick [15], Fluid-dip [41], Bee probe [42], Expel/stir [45]	A: Probing stick	Stick used to probe bee nest entrance and / or obtain honey
	Ant-dip	Ant-dip-wipe [39], Ant-dip [40]	A: Dipping stick	Slender wand used to harvest army ants
	Ant-fish	Ant-fish [38] probes	A: Fishing rod	Fine probe used to extract arboreal (wood-boring) ants from tunnels
	Termite-fish	Termite-fish using leaf midrib [36] or non-leaf materials [37]	A: Probing leaf mid-rib or other probe	Leaf mid-rib or other probing instrument used to extract termites from tunnels
Herbivory	Nut-hammer	Nut-hammer: wood or stone hammer on wood or stone anvil or hard ground [29-33]	A / N: Hammer on anvil	Stone or wood used to crack nuts

(a) Modified after Whiten *et al.* 2001, McGrew 2004
(b) Descriptors as named and numbered in Whiten *et al.* 2001: 1

Data Collection

Field work on the chimpanzees of Gashaka-Kwano was conducted by researchers, field assistants, volunteers and students of the *Gashaka Primate Project* over 8 years (Jan 00 – Dec 08) on various socioecological topics (habitat phenology, nest-building behaviour, party-size variation, plant food ecology, insectivory, tool use; cf. Sommer *et al.* 2004, Fowler 2006, Fowler & Sommer 2007, Fowler *et al.* 2007, Schöning *et al.* 2007).

Tools discarded by the chimpanzees at usage sites ("ateliers"; *Fig. 13.2*) were gathered *ad libitum* from Apr 01 – Dec 05, mostly by AF, who also measured them (n = 313 tools, n = 41 ateliers). The results were published by Fowler & Sommer

Fig. 13.2 Tools used by chimpanzees to harvest honey of stingless bees. (**a**) Entrance to nest at base of tree indicated by index finger; 4 stick tools are visible; note fray at tool end (photo: VS); (**b**) tool left inserted in the trunk of a tree (photo: APG)

(2007). Here, we replicate this analysis with an expanded sample that now includes 290 ateliers and 1116 tools. From Jan 06 onwards, several hundred nest sites of social insects were marked and revisited at least once a month (number of colonies: 55 *Melipona* stingless bees, 59 *Trigona* stingless bees, 89 honey bees, 148 army ants). Tools were mostly recovered from these marked nests and measured by ST (Feb – May 06), APG (Feb 07 – Jun 07, Nov 07 – May 08), and UB (Jan 06 – Dec 08).

Moreover, a total of 381 chimpanzee faecal samples found under night nests were collected to assess insect prey taxonomy and frequency of myrmecophagy. These include 254 samples collected from Mar 01 – Dec 04, mostly by AF, and 127 samples from Jan – Apr 05, collected by Darren Ellis under guidance of AF, and examined by Ellis in greater detail. Army ant availability in the Gashaka habitat was determined from 76 samples, with about 10 large workers each, obtained from trails or nest sites, plus samples of winged army ant males ("sausage flies") collected at night at the field station's light sources (Ellis 2005).

All samples of imagos plus ant remains found in faecal samples were later identified by CS. Other samples of termites, ants and stingless bees gathered opportunistically were identified by Judith Korb (Regensburg), David W. Roubik (Washington), and Barry Bolton (Isle of Wight), respectively.

Results

Field work around the project station at Gashaka-Kwano produced ample evidence that wild Nigerian chimpanzees employ a varied tool-kit in elementary technology. At the same time, some forms of subsistence technology reported from other chimpanzee study sites seem to be absent. We report on these first.

Hard-shelled Nuts: Absence of Cracking Technology

West-African populations of chimpanzees employ hammers and anvils of stone or wood to crack hard-shelled nuts (review in Boesch & Boesch-Achermann 2000). Hard-shelled nuts at Gashaka are restricted to those of oil-palms (*Elaeis guineensis*) and sweet detar (*Detarium microcarpum*), a sought-after product sold in local markets. However, we found no evidence for percussive activity of chimpanzees.

Both species of nut-bearing trees occur at a low density, as records from a permanent 8-km straight line transect suggest, where *Detarium* does not appear at all and *Elaeis* only 4 times. However, oil-palms occur also in clumped distribution at sites of abandoned farms and human settlements. Humans, when passing through the chimpanzee home-range, crack both types of nuts with hammers and anvils of stone. Sympatric olive baboons open nuts of *Detarium* and *Elaeis* with their teeth and eat the seeds (Warren 2003). Judging from feeding remains, other animals – e.g., wild pigs – also bite open both types of nuts, but it is not known if chimpanzees are amongst them. Chimpanzees eat at least the fruit of *Detarium*. The outer skin and flesh is normally removed, exposing the fibrous husk, which covers the nut shell. Freshly discarded husks were found at chimpanzee feeding sites.

Several specimens of nut-bearing trees were visited by AF and VS to assess the abundance of hammers and anvils that chimpanzees could reasonably use for nut-hammering (for weight and size criteria, see McGrew *et al.* 1997, Humle & Matsuzawa 2004). Within a 5 m radius of the nut-producing tree, we tried to find 5 stones or pieces of wood which seemed suitable as hammers, and a further 5 as anvils. The latter could also be embedded in the ground. Stones or pieces of wood that fulfilled certain dimensions (hammers: weight between > 100 g and 2.5 kg, hammering surface at least 5 × 10 cm; anvils: embedded in the ground or weight > 400 g – 10 kg and with a surface area of at least 7 cm × 10 cm) were pounded against a rock to test whether they would crack or fragment. The selection of stones was repeated until 5 suitable hammers and anvils were identified, or until no more fitting material was available within a 5 m radius.

Suitable hammers and anvils can indeed be found in the vicinity of nut-bearing trees, particularly if they grow at the banks of small, dry riverbeds. However, these materials were not common. Within a 5 m radius around 3 *Detarium* and 3 *Elaeis* trees, we measured the abundance of stone hammers (SH), wooden hammers (WH), stone anvils (SA), and wooden anvils (WA). Many stones of suitable size were brittle and fragmented on impact. Of 48 potential hammer stones, 67 % broke upon usage, and of 17 potential anvil stones 18 % broke. The procedure yielded the following numbers for useable tools at each tree:

- *Detarium 1*: SH 5, WH 0; SA 5, WA 0;
- *Detarium 2*: SH 0, WH 0; SA 0, WA 0;
- *Detarium 3*: SH 1, WH 0; SA 0, WA 0;
- *Elaeis 1*: SH 3, WH 0; SA 3, WA 0 but trunk portions suitable;
- *Elaeis 2*: SH 2, WH 0; SA 1, WA 0 but trunk portions suitable;
- *Elaeis 3*: SH 5, WH 0; SA 5, WA 0 but trunk portions suitable.

Thus, by our definition, nut-cracking would have been possible under two-thirds (4 / 6) of nut-bearing trees.

A brief survey was conducted in another habitat of *P. t. vellerosus*, 400 km SW of Gashaka in Korup National Park / Cameroon (05 – 09 Feb 01, VS). The forest harbours trees of *Poga oleosa* (Pierre) (Anisophylleacea), which produce the very hard-shelled poga or inui nuts. At least some suitable stones for cracking were present. Local humans passing through the chimpanzee range crack these nuts with stone hammers and anvils of stone or wood. However, evidence for nut-cracking by chimpanzees was likewise absent.

Ignored Insect Prey: Termites

Wild chimpanzees at various African sites use thin probes as instruments to extract termites from tunnels. However, this behaviour is again conspicuously absent at Gashaka.

Mounds indicating nests of the fungus-farming genus *Macrotermes* exist in the Gashaka area, particularly in savannah-woodland, often with dimensions of 2 m height and a base diameter of 2 m. However, we never recorded discarded tools on or near termite mounds, and 381 faecal samples covering each month of the year were likewise completely devoid of termite remains.

The abundance of *Macrotermes* mounds at Gashaka was measured by walking the 8 km straight line transect, which covers both forests (78 %) and woodland-savannah areas (22 %). Each mound of at least 20 cm height within a 20 m wide strip along the transect was documented (total survey area = 16 ha). However, only 2 small mounds were recorded in a forested section. Nevertheless, in savannah-woodland off transect, inhabited mounds are regularly found in clumped distribution with, for example, a density of 6 / 50 m^2, a maximum height of 1.8 m and a base diameter of 2.4 – 2.9 m. Mounds are also present at the forest edge.

It is noteworthy that field workers, using plant probes in efforts to imitate chimpanzee harvesting techniques, were unable to extract a single termite from a mound. Moreover, even when mounds were partly broken and the termites directly provoked with the plant part, not a single one, including the largest workers, would attach itself to any blade or stem – despite the fact that soldiers bite readily into human fingers, inflicting blood-drawing injuries.

Chimpanzee behavioural patterns associated with subsistence technology such as *nut-hammer* and *termite-fish* were thus seemingly absent from Gashaka. However, there was evidence for other patterns such as *dig*, *bee-probe*, *ant-fish* and *ant-dip*.

Insect Prey: Stingless Bees, Honey Bees

Meliponini stingless bees at Gashaka include species such as *Hypotrigona gribodoi* (Magretti) of the smaller taxon *Trigona*, and *Meliponula erythra* (Schletterer) of the larger taxon *Melipona*.

Chimpanzees attacked only colonies of *Melipona*, but not those of *Trigona* (contrary to a statement in Fowler & Sommer 2007). Relatively short and sturdy sticks, typically soiled and with frayed ends (*Fig. 13.3a*) were found at entrances to the subterranean nests of *Melipona*, at times together with dead insects. Chimpanzees enlarged the nest entrances with these sticks. They could then extract honey and / or insects, usually by employing smaller bee probes (see below). *Melipona* nests in tree trunks were likewise exploited with tools (cf. *Fig. 13.2*). Similar tool use was documented during a Feb 02 survey to the small montane forest fragment at Ngel Nyaki, about 70 aerial km from Kwano (see Adanu *et al.* this volume [Ch. 3]). Here, about a dozen chimpanzees survive (Beck & Chapman 2008), and recently disturbed earth and sticks with frayed ends were found around a nest entrance. Local guides were familiar with digging tools, and also reported ant dipping as well as probing for the honey of both honey bees and stingless bees (see below).

At Gashaka, relatively long sticks, slender or sturdy (*Fig. 13.3b*), were left behind at sites where chimpanzees obtained honey of either the African honey bee (*Apis mellifera)* or stingless bees. The type of insect was again discernible by the

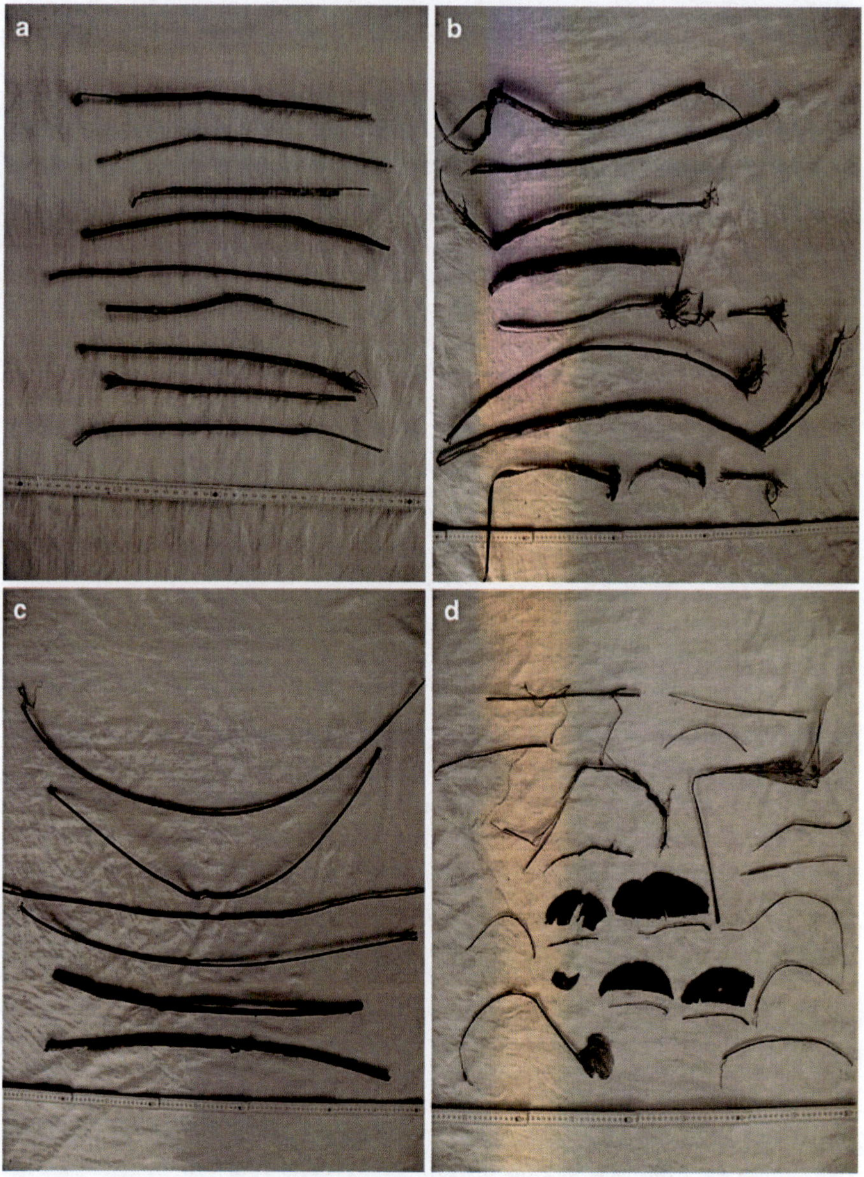

Fig. 13.3 Tool-kit of Gashaka chimpanzees. (**a**) Stingless bee digging sticks; (**b**) stingless bee probing sticks; (**c**) army ant dipping wands; (**d**) arboreal ant fishing rods including leaf mid-ribs (photos: VS)

presence of dead imagos. Sticks with the bark stripped and exhibiting traces or the odour of honey, with frayed ends indicating manipulation, biting or chewing, were regularly found beneath trees, sometimes with honey-comb fragments and "wads" of chewed honey-comb and bees.

Insect Prey: Ants Other Than Army Ants

Species other than army ants (see below) were found in 3.9 % (5 / 127) of the more closely examined chimpanzee faecal samples. Remains of the termite-hunting *Pachycondyla analis* were present in 3 samples. Two of these contained pupal cases, so that these ants were probably consumed intentionally from nests. Two other samples contained remains of the genus *Camponotus*, which may represent ground-dwelling or arboreal species. It is unclear if they were eaten accidentally or deliberately targeted. These two species are obviously rather unimportant as food sources compared to army ant prey.

Some direct observations revealed how chimpanzees use thin elastic twigs, grass, pieces of stripped bark and the mid-ribs of large leaves (*Fig. 13.3d*) to "fish" for arboreal ants such as *Camponotus chrysurus* (Gerst.).

Episode 1 (19 Apr 01, 67 min of direct observations by AF): 1 adult male chimpanzee, 3 adult females and an infant are sitting on the ground and fishing for ants from inside the trunk of a large tree, first selecting the twigs, often stripping them of leaves and then placing them in the nest hole for 5 – 10 s. The tools are removed and ants picked off with the lips.

Episode 2 (11 Jun 01, observers: Hammaunde Guruza, Yakubu Wakirwa): 3 adult chimpanzees are seen at the base of a tree. They flee at the arrival of the observers. The stripped mid-ribs of 4 leaves are found discarded on the ground. Ants are detected on the tree's surface.

Episode 3 (23 Feb 05, 14 min of direct observation by Hammaunde Guruza, Klaus Meister, AF): 1 adult female chimpanzee makes and uses stick tools to probe into the nest of wood-boring ants in a newly fallen tree. She holds the end of a stick in her mouth and strips leaves off in a sweeping motion, using a foot. She leans down from above the nest and probes, withdrawing and eating ants from the stick, hand and tree trunk. Nests of both ants and stingless bees are discovered in the tree.

Insect Prey: Army Ants

The vast majority (93.4 %) of 76 samples of army ant workers obtained from nests and trails belonged to a typical representative of the "epigaeic" life-style, *Dorylus* (*Anomma*) *rubellus* (Schöning *et al.* 2008a; identified as *D. rufescens* by Schöning *et al.* 2007 because the relevant type material had not been examined at the time). Species with an "intermediate" life-style were represented by 2 samples from *D. gribodoi* (Emery) and 1 sample from *D. (A.) kohli* (Wasmann) whereas the 2 remaining samples belonged to "subterranean" species in the subgenera *D.* (*Dorylus*) and *D.* (*Typhlopone*). Males collected at field station lights were either the epigaeic *D. rufescens* or *D. gribodoi*, with remaining males belonging to subgenera other than *D.* (*Anomma*). Therefore 3 army ant species – 1 with an "epigaeic" and 2 with an "intermediate" life-style- seem to be available as prey for chimpanzees at Gashaka-Kwano.

Long thin sticks with leaves and side-twigs removed, often with frayed or bitten-off ends (*Fig. 13.3c*), were frequently discovered at disturbed nests of army ants. These wands were obviously used to ferry ants to the mouth at a safe distance from the nest, thereby reducing the painful bites of the larger workers, which swarm out to defend the nest entrance. On only a single occasion was a dipping wand found at an ant trail, 15 m away from a nest, perhaps used to dip ants directly from the trail (field assistant Bobbo Buba, pers. obs.).

Dorylus remains were found in 42.3 % (161 / 381) of all chimpanzee faecal samples. *D. rubellus* was the only species present. The number of *Dorylus* heads in the 59 samples examined more thoroughly ranged from 3 to 4636 (median 181). The relative head number ranged from 0.2 heads / g dry mass to 120.6 heads / g dry mass (median 5.44 heads / g dry mass). The proportion of samples with *Dorylus* fragments did not vary with season (dry season: 115 samples without army ants vs. 74 with army ants; rainy season: 105 without vs. 87 with; χ^2-test, $\chi^2 = 2.87$, df = 1, $p = 0.09$)

Characteristics of Insect-harvesting Tools

We recovered a total of 1116 tools from 290 tool sites (*Tab. 13.2*).

Most common were ateliers of *army ant dipping wands* (35.9 %), followed by ateliers of *stingless bee probing sticks* (35.5 %), whereas ateliers with *stingless bee digging sticks* (22.8 %), *honey bee probing sticks* (4.1 %) and *arboreal ant fishing rods* (1.7 %) were much less common (*Fig. 13.4*). Half of all ateliers held only 1 – 2 tools, three quarters up to 4 tools, and just 5 % of all ateliers contained 10 or more tools (*Fig. 13.5*). Half of all tools were recovered from ateliers with 1 – 5 tools. Ateliers with arboreal ant fishing rods contained the greatest average number of tools ($n = 9.8$), followed by ateliers with stingless bee probing sticks (4.9). The average number of tools for other ateliers was lower (army ant dipping wands 3.4, stingless bee digging sticks 2.7, honey bee probing sticks 2.3). Stingless bee probing sticks made up almost half (44.8 %) of all recovered tools, and army ant dipping wands about a third (32.3 %). Stingless bee digging sticks were less common (16.0 %), and only a fraction of tools were arboreal ant fishing rods (4.4 %) or honey bee probing sticks (2.5 %). About a third (37.4 %) of all tools were new, i.e., less than a day old. Another 22.2 % were fresh, i.e., a few days old. Just 12.2 % had been used about a week prior to discovery, but 28.1 % were older than a week (*Fig. 13.6*). Tools were recovered throughout the year (*Fig. 13.7*) but fewer were found during the heavy rains (May – Nov).

Tools differed in their dimensions (cf. *Tab. 13.2*). Wands for ant-dipping were by far the longest (84 cm) and relatively thin (diameter at mid-point 6 mm), whereas stingless bee digging sticks were shorter (38 cm) and thicker (9 mm). Arboreal ant fishing rods were the shortest (19 cm) and thinnest (1 mm). Unlike most other tools, they were often not woody. A sub-sample of 38 recovered rods consisted of stripped twigs (40 %), mid-ribs of large leaves (24 %), grass-blades or vines (18 %), twigs with bark scraped off (13 %), twig (3 %) and bark fibre (3 %).

Table 13.2 Dimensions of tools Gashaka chimpanzees use for insectivory. Proximal = end of a tool which was closest to stem before its removal from mother plant; distal = the other (terminal) end

Tool		Stingless bee digging stick	Stingless bee probing stick	Honey bee probing stick	Army ant dipping wand	Arboreal ant fishing rod	Average	Total
Ateliers	*n*	66	103	12	104	5		290
Tools	*n*	179	500	28	360	49		1116
Tools per atelier	Mean	2.7	4.9	2.3	3.4	9.8	4.6	
	Min	1	1	1	1	4		
	Max	9	48	7	16	20		
	SD	2.0	6.4	2.2	2.5	7.0		
Tool length (cm)	*n*	179	499	28	359	49		
	Mean	38.2	33.1	44.1	83.8	18.8	43.6	
	Median	36	31	46	83	16	43	
	SD	18.2	17.1	16.1	25.2	7.7		
	Min	3.7	5.2	13.1	17.4	5.0		
	Max	113.4	111.8	74.3	165.5	44.4		
Tool diameter at mid-point of proximal end (mm)	*n*	179.0	497.0	28.0	359.0	49.0		1112
	Mean	8.5	6.7	6.7	6.2	1.2	5.9	
	Median	8.0	6.0	6.0	6.0	1.0	5	
	SD	4.5	2.9	2.5	1.8	0.8		
	Min	3.0	2.0	2.0	2.0	0.4		
	Max	60.0	50.0	13.0	18.0	6.0		
Proximal end: diameter (mm)	*n*	179.0	497.0	28.0	359.0	49.0		1112
	Mean	8.9	6.7	7.7	7.0	1.4	6.3	
	Median	8.0	6.0	7.0	7.0	1.2	6	
	SD	4.8	2.9	3.0	1.9	0.8		
	Min	4.0	2.0	2.5	3.0	0.5		
	Max	65.0	50.0	14.0	13.4	5.6		

(continued)

Table 13.2 (continued)

Tool		Stingless bee digging stick	Stingless bee probing stick	Honey bee probing stick	Army ant dipping wand	Arboreal ant fishing rod	Average	Total
Tool diameter at mid-point of distal end (mm)	*n*	159	284	12	246	11		712
	Mean	8.2	6.6	6.0	5.1	1.2	5.4	
	Median	8	6	6	5	1	5	
	SD	4.7	2.1	1.8	1.4	0.4		
	Min	3	2	4	2	1		
	Max	60	14	11	10	02		
Distal end: diameter (mm)	*n*	179	499	28	359	49		1114
Proximal end: length of fray (mm)	Mean	7.7	6.2	6.1	4.9	1.1	5.2	
	Median	7	6	5	5	1	5	
	SD	4.6	3.0	2.4	1.6	1.0		
	Min	2	0.0	3	2	0.2		
	Max	60	50	11.5	14	5.4		
	n	90	248	9	166	11		524
	Mean	22.3	20.6	21.1	14.8	14.4	18.6	
	Median	15	15	20	9	4	13	
	SD	37.7	30.8	16.1	18.7	21.3		
	Min	1	2	1	1	1		
	Max	310	342	50	150	61		
Distal end: length of fray (mm)	*n*	116	342	14	210	6		688
	Mean	9.8	18.9	15.4	10.0	4.8	11.8	
	Median	6	12	15	7	6	9	
	SD	11.8	21.0	9.5	12.9	3.2		
	Min	2	1	3	1	0		
	Max	87	160	33	95	8.3		

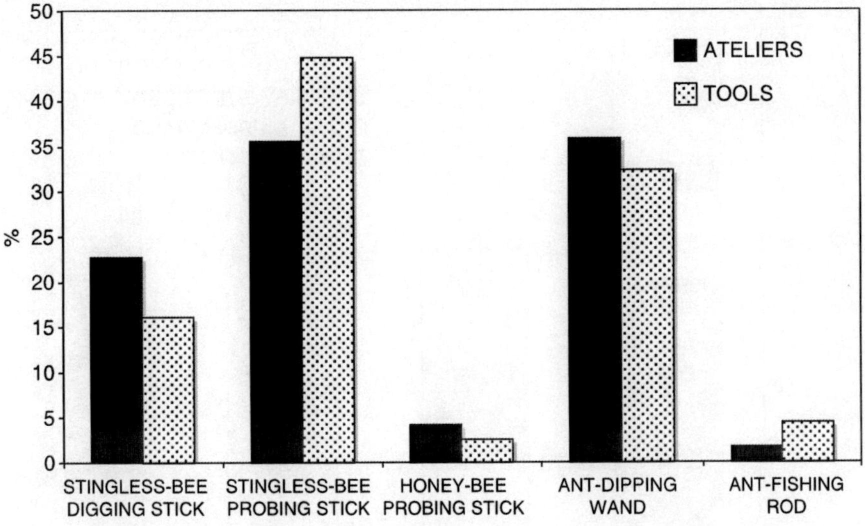

Fig. 13.4 Proportion of ateliers and tools of different categories recovered at the study site

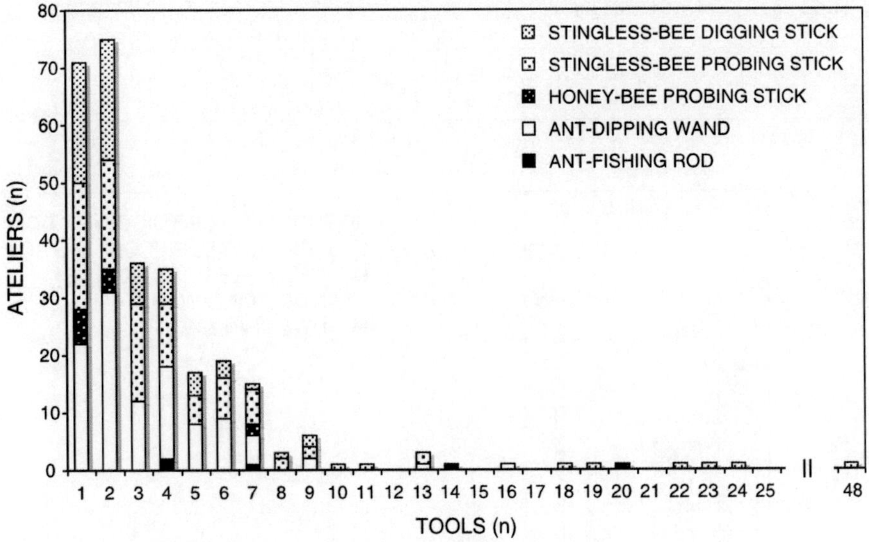

Fig. 13.5 Number of tools per atelier

Protrusions such as small twigs or leaves were attached to only 12 % of recovered tools, indicating that many projecting parts must have been removed before use. The bark of 82.2 % of all tools was at least partially stripped. Less than half of the bark was removed in 27.1 % of the tools, 6.7 % had 50 – 75 % removed, and 13.4 % were stripped of more than 75 % of the bark. Tools to obtain honey were stripped to a greater degree than army ant dipping wands or arboreal ant fishing rods.

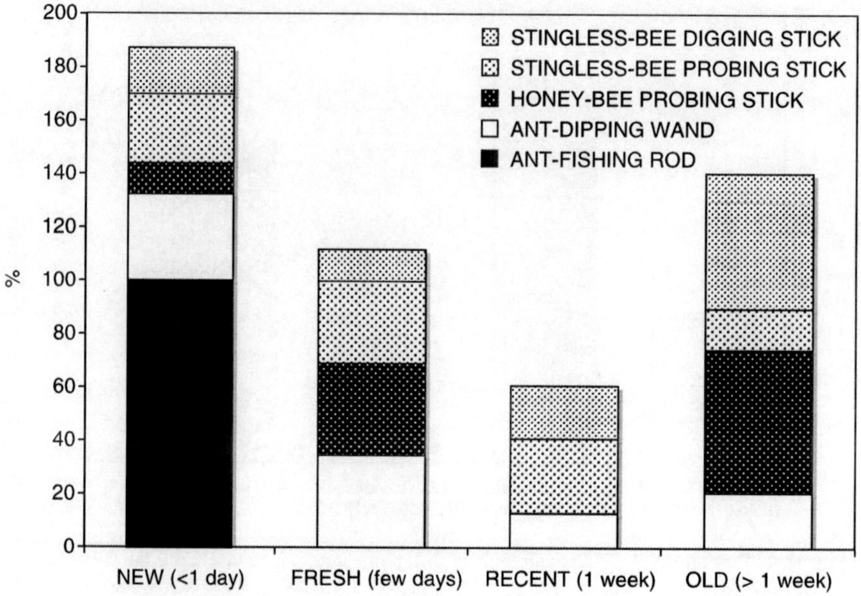

Fig. 13.6 Age of recovered tools

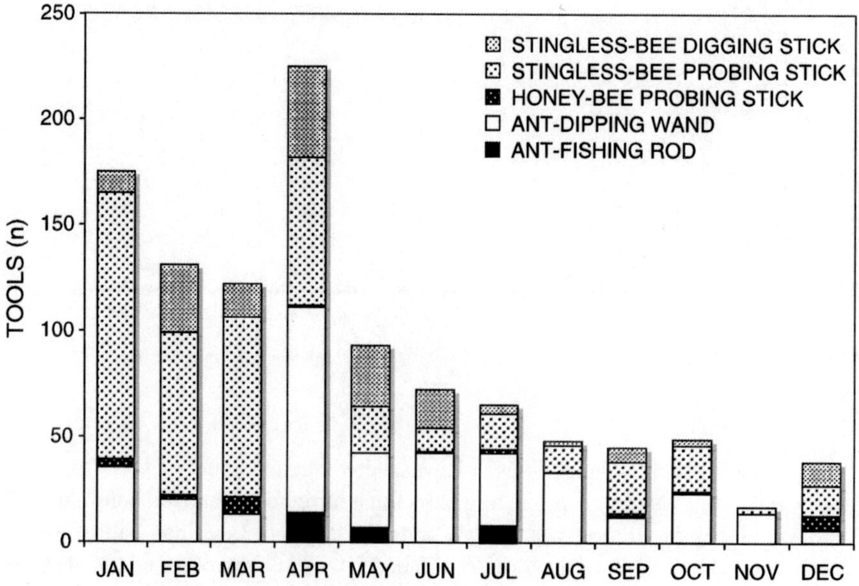

Fig. 13.7 Tool ateliers recovered during different months of the year

Some noticeable differences existed between a tool's proximal end (the end closer to the stem of the originating plant before the tool was removed) and the distal end (the end nearer to the terminal end of the twig, stem or leave). For example, average diameter of tools decreased by about one-fifth (18.8 %) from the proximal end (6.4 mm) towards the distal end (5.2 mm). This decrease held true for all types of tools (stingless bee digging stick proximal end vs. distal end 8.9 vs. 7.7 mm; stingless bee probing stick 7.2 vs. 6.2 mm, honey bee probing stick 7.7 vs. 6.1 mm; ant-dipping stick 7.0 vs. 4.9 mm; arboreal ant fishing rods 1.4 mm vs. 1.2 mm).

It is obviously easier to insert the thinner end of an object into an opening. Indeed, judging from the condition of tools, chimpanzees more often preferred to manipulate (poke, probe, stir) with the distal end of twigs (92.3 %; *Fig. 13.8*), whereas the proximal end was used in only about one sixth of the cases (16.6 %; total > 100 %, because both ends were in one-tenth of tools). The preference for distal ends held true for all types of tools. The chimpanzees thus exerted pressure preferentially with the thinner ends. This positioning is also reflected in the fact that distal ends were more frequently compacted (45.3 %) than proximal ends (14.2 %), and that distal ends were more often stripped of bark (75.3 %) than proximal ends (67.3 %).

Distal ends were also slightly more often frayed (52.5 %) than proximal ends (45.5 %). Frays at proximal tool ends (cf. *Tab. 13.2*) were one-third longer (18.6 mm) than those at distal ends (11.8 mm). This held true for all tool types.

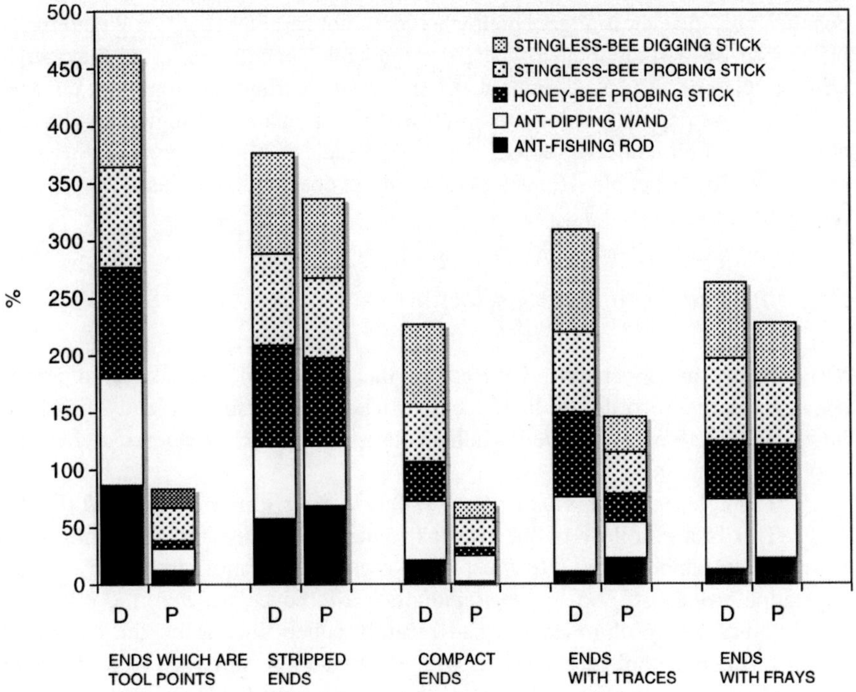

Fig. 13.8 Differences between distal (D) and proximal (P) ends of tools

Similarly, traces were found much more often on distal ends (46.3 %) than on proximal ends (20.3 %). Three quarters of identifiable traces were soil (76.5 %), whereas honey (13.4 %), remnants of cerume tubes (6.7 %) and bees-wax (3.2 %) was much rarer (*Tab. 13.3*). A single type of trace was found on 94.7 % of all tools, whereas 2 traces (such as soil with cerume or honey, or cerume with honey) was restricted to 5.1 %; three traces (e.g., soil, bees-wax, cerume) was even rarer (0.2 %). Stingless bee digging sticks had most often traces (62.6 %) and arboreal ant fishing rods least often (18.4 %). The type of trace reflected the technique or purpose for which the tool was employed. Traces on arboreal ant fishing rods were always soil – probably because grass-blades employed as ant-fishing tools were pulled directly from the earth. Similarly, 93.8 % of all army ant dipping wands exhibited traces of soil, and so did 90.0 % of all stingless bee digging sticks. Honey or cerume residue is not found at all or rarely on the latter three tool types as they are not employed to extract these resources. Instead, bees-wax, cerume tube residue and honey traces were found on many stingless bee and honey bee probing sticks.

Discussion

Our report on the material culture of Nigerian chimpanzees broadens our understanding of chimpanzee behavioural diversity. Moreover, the data allow us to draw conclusions, however limited, about whether or not certain behavioural patterns are influenced by particular genotypes, environmental conditions, or if they reflect arbitrary patterns. Traits considered as pure cultural variants are regularly observed in at least one population, but are absent without plausible ecological explanation at other sites (Whiten *et al.* 2001: 1493). Nevertheless, it is notoriously difficult to exclude ecological explanations and to weed out observational bias.

Potential Biases in Data Collection

We probably under-report the variety of the tool kit of Gashaka chimpanzees because positive records and their accuracy tend to increase with study length and how well animals are habituated to human observers (Boesch & Boesch-Achermann 2000, McGrew 2004).

For example, *more than half of all recovered tools were new or fresh* (cf. *Fig. 13.5*). This is a corollary of the fact that older sites had deteriorated more than fresher sites, which were more conspicuous, with freshly moved earth or unwilted vegetation broken off and / or scattered about. Moreover, we were more likely to survey areas where chimpanzees had recently called, increasing the chances of encountering new tool sites. Ateliers with arboreal ant fishing rods were invariably new and recovered only after direct observations of ant-fishing. Given that many ant-fishing tools were fragile leaf mid-ribs and grass blades, it is unlikely that tools

Table 13.3 Traces found on tools discarded by Gashaka chimpanzees

Traces	Tools total (*n*)	Stingless bee digging stick	Stingless bee probing stick	Honey bee probing stick	Army ant dipping wand	Arboreal ant fishing rod	Total %
Unidentified (*n*)	108	4	83	2	17	2	
Identified (*n*)	464	220	491	15	325	14	
Total (*n*)	572	224	574	17	342	16	
Tools with traces (%)		62.6	57.4	30.4	47.5	18.4	
Soil (%)	355	90.0	43.6	40.0	93.8	100.0	76.5
Honey (%)	62	4.1	24.6	33.3	0.0	0.0	13.4
Cerume (%)	31	7.3	23.6	0.0	0.0	0.0	6.7
Bees-wax (%)	15	0.0	1.4	13.3	0.0	0.0	3.2
Insects (%)	1	0.5	0.0	0.0	0.6	0.0	0.2
Faeces (%)	0	0.0	0.0	0.0	0.3	0.0	0.1

would have been found after the chimpanzees had left the site. Nevertheless, in our previous analyses (Fowler & Sommer 2007), old tools constituted just a fraction of about 5 %, whereas they represent 28.1 % in our new sample. The increased likelihood of recovering old tools is due to the systematic re-visits of marked colonies of social insects, which lead to the increased retrieval even of tools that were abandoned more than a week earlier. The relative paucity of recovered tools for honey bee probing and arboreal ant fishing is probably also influenced by the fact that these are arboreal activities, which leave fewer traces for researchers who move on the ground compared to the more terrestrial ant dipping and stingless bee probing.

Similarly, studies at other sites have found a seasonality in tool use (Goodall 1986; McGrew *et al.* 1979). However, our records of the *annual distribution of recovered tool* (cf. *Fig. 13.7*) was not controlled for time spent in the field as many artifacts were collected opportunistically (cf. McGrew *et al.* 2005). The relative rarity of finds during the wet season might therefore be misleading, as chimpanzee tracking is less easy during this period and because ateliers quickly disintegrate in heavy rains.

Moreover, we have few direct observations of extractive foraging and may therefore have *lumped tools with different functions into a single category*. The varying length within a certain category of tools might in fact reflect specific situations. For example, an enormous number of army ants might swarm out from larger nests attacked by the chimpanzees, in which case they might retreat further from the nest entrance and use longer dipping wands, whereas smaller nests could be exploited with shorter sticks. Similarly, Tutin *et al.* (1995) compared tools used to obtain honey from *Meliponula* and *Trigona*, finding that the former were longer, with a greater range of lengths. However, this is an unlikely scenario for Gashaka, given that chimpanzees did not attack colonies of the smaller *Trigona* species. Finally, tools may form a "set", in which two or more types are used sequentially (Brewer & McGrew 1990, Sanz & Morgan 2007) but limited direct observations may prevent us from noticing that. We are currently looking into the exact dynamics of how stingless bee nests are exploited, as entrances are often enlarged with digging sticks before the honey is scooped out by smaller probes (APG, UB, in prep.).

Lack of Behavioural Variation ("Universals")

Certain behavioural patterns, such as "buttress-beat", in which chimpanzees drum on a tree base occur at all study sites and thus lack obvious behavioural variation. However, they may still represent "cultural" traits: "Just that if chimpanzees do acquire them by social learning, they do so in all communities studied" (Whiten *et al.* 2001: 1496). In any case, none of the subsistence technology patterns present or absent at Gashaka are a "putative chimpanzee universal", as they may or may not occur at certain study sites (see below, *Fig. 13.9*). To assume that cultural variation

BEHAVIOUR	DEFINITIONS OF WHITEN et al. (2001)	WEST AFRICA			WEST-CENTRAL		EAST AFRICA				
		As	Bs	Tai	Ga	Lo	Ma	Mk	Go	Kib	Bd
Dig	Lever open (stick used to enlarge entrance)	U	--	H	x	C	--	--	C	--	--
Bee-probe	Bee-probe (disable bees, flick with probe)	U	--	C	x	U	--	+	--	--	--
	Brush-stick (probing stick with brush end)	U	--	--	x	--	--	--	--	--	--
Ant-dip	Ant-dip (dip stick on ants to harvest)	U	C	C	x	--	--	--	C	--	--
	Ant-dip single (one handed dip stick on ants)	--	C	C	U	--	--	--	+	--	--
	Ant-dip-wipe (manually wipe ants off wand)	H	+	--	x	--	--	--	C	--	--
Ant-fish	Ant-fish (probe used to extract ants)	H	+	--	x	C	C	C	+	--	--
Termite-fish	Termite-fish using leaf midrib	H	+	e	e?	e?	--	C	--	e	e?
	Termite-fish using non-leaf materials	H	--	e	e?	e?	--	C	C	e	e?
Nut-hammer	Nut hammer, stone hammer on stone anvil	e	C	C	--	--	--	--	--	e?	e
	Nut-hammer, stone hammer on wood anvil	e	+	C	--	--	e	e	--	e?	e
	Nut-hammer, wood hammer on stone anvil	e	--	C	--	--	--	--	--	e?	e
	Nut-hammer, wood hammer on wood anvil	e	--	C	--	--	e	e	--	e?	e
	Nut-hammer, other (e.g. on ground)	e	--	H	--	--	--	--	--	e?	e

Fig. 13.9 Presence or absence of behavioural pattern associated with subsistence technology of Nigerian chimpanzees compared to 9 long-term study sites across Africa (Whiten *et al.* 2001). Study sites: West Africa – As, Assirik / Senegal; Bs, Bossou / Guinea; Tai / Ivory Coast; West-Central Africa – Lo, Lopé / Gabon; Ga, Gashaka / Nigeria; East Africa – Ma, Mahale M community / Tanzania; Mk, Mahale K community / Tanzania; Go, Gombe, Tanzania; Kib, Kibale Kanyawara community / Uganda; Bd, Budongo / Uganda. Occurrence codes of behavioural pattern (after Whiten *et al.* 2001, except code "X"): C = customary (occurs in all or most able-bodied individuals), H = habitual (occurs repeatedly in several individuals), + = present (occurs but not customary or habitual), X = occurs (individual frequencies unknown; can be C, H or P) , – – = absent(absent with no ecological explanation), e = absence ecological (explained by local environmental or ecological constraint), U = unknown (not recorded, but inadequate observations)

is the cause of this diversity requires us to exclude genetic differences as well as environmental constraints.

Genetic Explanations for Behavioural Variation

It seems reasonable to reject genetic explanations if the variation occurs between gene-exchanging neighbouring communities – such as when the Tanzanian Mahale K group fishes for termites, and the Mahale M group does not (Whiten *et al.* 2001). Most behavioural studies of *P. t. vellerosus* are restricted to Gashaka, with only anecdotal comparative data for neighbouring communities or populations such as Ngel Nyaki (where bee-probing does likewise occur) and Korup (where nut-hammering seems likewise absent). On the other hand, both captive and wild studies of chimpanzees from various subspecies and populations demonstrate that they learn easily to manipulate with objects, rendering tool use as "poor candidates for merely instinctual variations" (Whiten *et al.* 2001: 1511). The same is, of course, true for humans, as distinct local genomic variants of people from, say, Thai-, Swazi- or Iceland do not prevent them from learning to use a corkscrew or a bicycle.

Environmental Explanations for Behavioural Variation

Subsistence behaviours are more easily influenced by ecological factors than, for example, social or hygienic behaviours. Thus, in some places, chimpanzees will attain a "grooming handclasp" in an A-frame position, i.e., the partners involved will each hold one hand up and grasp each other, while they manipulate each other's fur with the other hand. Elsewhere, this behaviour is never seen. This is as clear-cut an arbitrary cultural variant as one can possibly imagine, as it is hard to think of environmental factors that encourage or discourage such a body position (McGrew 2004).

To exclude environmental influences on tool use is more difficult, as many variables are involved (McGrew & Tutin 1978). For example, the likelihood with which *one or both ends of tools* are used varies between sites. Gashaka chimpanzees employ the distal end in 92 % of all cases, the proximal end in 17 %; consequently, one tenth of the tools are used at both ends. This is a different situation from Gombe, where usually both ends are used, and more similar to Assirik, where almost always only one end is used (McGrew 1992). A potential environmental explanation for frequent two-ended use could be that tools are made from vines, which are roughly uniformly cylindrical throughout their length and thus suitable for use from both ends. In reality, vines are employed as often in Gombe and Assirik; therefore, the usage seems to follow an arbitrary pattern (McGrew 1992). However, preference of distal ends at Gashaka seems to have a plausible environmental explanation, as most tools are made from twigs. These become thinner towards the distal end (cf. *Tab. 13.2)* and are thus simply better suited as a tool point.

Certain features of *harvesting techniques of army ants* (review in Whiten *et al.* 1999) are straightforward corollaries of prey characteristics. In the typical two-handed "pull-through" technique, the chimpanzee dips a wand into the ant nest, waits for defenders to crawl up, then withdraws the tool while sweeping off the insects with the other hand; the jumbled mass is then rapidly ingested. In the method of one-handed "direct-mouthing", the wand is also dipped with one hand into the nest, but the chimpanzee then uses the same hand to sweep the tool directly between the lips or teeth, or nibbles the insects directly off the stick. Schöning *et al.* (2008b) compared patterns of army ant biology and presence or absence of chimpanzee harvesting techniques across 14 sites in eastern, central, and western Africa. They found that epigaeically foraging species are harvested with longer tools and usually by the 'pull-through' technique, compared to taxa that forage in leaf-litter, which are harvested with shorter tools and by the method of 'direct-mouthing'. These differences are probably due to the fact that workers of epigaeic species have longer legs – presumably as an adaptation to above ground foraging and pursuit of prey, and also larger mandibles. Thus, workers of epigaeic species can run faster and inflict more painful bites (Schöning *et al.* 2008b) – which explains the use of longer tools and an ingestion technique that immobilises the prey faster.

At several sites, chimpanzee feeding techniques for ants have not been directly observed. It is thus not known if they use the direct-mouthing or pull-through technique. Based on the principle that form reflects function, McGrew *et al.* (2005) predicted that chimpanzees at Fongoli / Senegal employ the two-handed pull-through technique, given that tools recovered at this site are relatively long. The even greater average length of dipping wands at Gashaka suggests the same.

However, prey species characteristics do not explain several differences in army ant-eating between Bossou (Guinea) and Taï (Ivory Coast), where the same suite of prey species is available and consumed. For example, both one-handed and two-handed harvesting is practiced at Bossou / Guinea. At Taï, only direct-mouthing is seen, and only at Taï do the chimpanzees open nests by hand. This variation in the predator-prey relationship may thus be "solely sociocultural" (Schöning *et al.* 2008b: 48) – as discussed below.

A potential insect prey not attacked at Gashaka are colonies of *Trigona*, the genus of smaller stingless bees. These colonise the crevices of often dead tree trunks. Tunnels are typically only a few mm wide, and often snake through the substrate. Chimpanzees seem to be unable to break into the cavities, even with help of tools. The apes likewise fail to penetrate some tree-nests of the larger *Melipona*, if the wood is hard and the tunnels crooked; certain ground nests are likewise too difficult to access. Failed attempts are indicated by discarded digging sticks, while smaller, honey-gathering bee-probes are missing. We are currently comparing the physical traits of stingless bee nest sites that were successfully attacked, with those of failed attacks as well as nests that were ignored. This will help us to better understand the dynamics of environmental *constraints for exploitation of stingless bee colonies* (APG, UB, in prep.).

Another reflection of environmental constraints concerns the *absence of termite eating* at Gashaka. Termites are a favourite prey of chimpanzees across Africa, although they are not eaten everywhere (McGrew 1992: 155ff). At Gashaka, we found no evidence that chimpanzees use tools to harvest termites. Termite fishing tools are typically made from rather perishable material, such as grass and leaf mid-ribs (e.g., Goodall 1986; McGrew *et al.* 1979), which renders post-hoc discoveries difficult. Nevertheless, we never identified termite heads in faeces although chitin is indigestible for apes. This clearly contrasts with other sites, where termites show up in as many as 27 % of faeces (review in McGrew 1992). Various environmental factors could explain the absence of termite-eating. First, the mound abundance might be low – only 2 were located in a random 16-ha plot. Still, mounds exist, sometimes clumped in woodland-savannah, and large mounds also occur at forest edges and in the forest itself. Second, suitably flexible and resilient vegetation for tools might not be easily available, because the yearly burning of the savannah-woodland favours coarse grasses in the vicinity of mounds. However, termite fishing is common in Assirik where intense dry-season burning is likewise practised (Hunt & McGrew 2002). Thirdly, termites did not bite into probes of various materials with which we provoked them – despite the fact that soldiers will immediately bite into skin. Thus, on balance we tentatively conclude that the absence of termite-fishing reflects environmental constraints.

Cultural Variation

On the other hand, it is entirely possible that Gashaka chimpanzees do not consume termites, because "it is not something that is done here" (McGrew 2004) – similarly to the situation at Mahale / Tanzania, where termites are consumed in one community, but not the neighbouring (Whiten *et al.* 2001). There are corresponding findings for army ant consumption, as species available at the 5 sites where chimpanzees do not feed on them are all eaten elsewhere; for example, in Uganda, *Dorylus wilverthi* is consumed at Kalinzu, but ignored at Kibale and Budongo. Geographic differences with respect to absence and presence of consumption are not explained by parameters such as chimpanzee subspecies identity, consumption frequency of other insect prey, consumption frequency of meat, or habitat type including army ant colony density (Schöning *et al.* 2008b). These patterns are perhaps related to what would be called a "*food taboo*" in human societies. For example, humans in the Gashaka area will not consider eating dogs, while this is perfectly acceptable for the same or similar ethnic groups just 1 ½ days walk away, across the border in Cameroon (I. Faucher pers. comm.). The non-consumption of a perfectly edible food-item would thus serve as some sort of group-identifying trait.

In any case, some behavioural patterns of Gashaka chimpanzees seem to be shaped by arbitrary cultural variations. For example, many raw materials for tools are, similar to other sites, altered through "reduction", i.e., stripping of leaves, breaking off of twigs, peeling off of bark, and clipping of ends. Four out of five

Gashaka tools were at least partially stripped. A similar proportion of *peeling* (86 %) is reported from Assirik, whereas tools at Gombe are virtually never peeled (McGrew *et al.* 1979, cit. in McGrew 1992). It is hard to imagine that intrinsic characteristics of vegetation differ so much between West, Central and East Africa as to predetermine the degree to which implements are stripped of bark. Thus, varying proportions could well reflect cultural variation. In a human context, we would perhaps label this as an "ornamental" custom.

A second potential example concerns the lack of evidence for *nut-hammering* at Gashaka, although large hard-shelled and edible nuts as well as suitable hammers and anvils exist. A species similar to Gashaka's *D. microcarpum* – *D. senegalense* – is cracked by chimpanzees with hammers at Tiwai / Sierra Leone (Whitesides 1985; cit. in McGrew *et al.* 1997), and Taï / Ivory Coast (Boesch & Boesch-Achermann 2000). Oil-palm nuts are cracked with tools at Kanton / Liberia (Kortlandt & Holzhaus 1987), Nimba / Guinea (Yamakoshi & Matsuzawa 1993) as well as Bossou / Guinea (Sugiyama & Koman 1979) where no other hard-shelled nut suitable for cracking occurs (Matsuzawa 1994).

The nearby population of *P. t. vellerosus* in Korup / Cameroon seems to likewise ignore hard nuts, despite the at least occasional presence of suitable hammer stones. Again, environmental constraints might be fine-tuned so as to remain unrecognised. For example, suitable nuts might be so rare at Gashaka that it is uneconomical to exploit them – particularly as suitable hammers and anvils are hard to come by. Nevertheless, on balance the scenarios at Gashaka (and Korup) seem to be similar to Lopé / Gabon. There, nut-producing trees – including *D. microcarpum* – are also present at low density, but none of the nuts is cracked with any of the readily available tools (McGrew *et al.* 1997). Thus, the absence of nut-hammering is likely to reflect cultural variation.

These findings seemed to support the hypothesis that nut-hammering is indeed restricted to populations west of the N'Zo-Sassandra River in Ivory Coast (Boesch *et al.* 1994). Interestingly, this long-held assumption had to be revised when it was discovered that *P. t. vellerosus* chimpanzees in the Ebo forest, Cameroon, more than 1700 km east of the previously proposed riverine "information barrier", crack *Coula edulis* nuts with stone hammers, while sitting in trees and using thick branches as anvils (Morgan & Abwe 2006). The observation does not necessarily challenge the "cultural variant" explanation, but questions the existing model of the cultural diffusion of nut-cracking behaviour (Boesch *et al.* 1994; McGrew *et al.* 1997). Instead, nut-cracking would have been invented multiple times, or it went extinct in the region between the N'Zo-Sassandra River and the Ebo forest.

Finally, cultural variants could be reflected by the relation between *finer tools and tools with brush-ends*. Frayed ends are an altered form of digging and probing sticks, and, if > 30 mm, were termed "brush-sticks" (Sugiyama 1985). At Gashaka, 46.9 % of proximal tool ends had frays, compared to 61.6 % of distal ends. Frays at proximal ends were considerably longer than those of distal ends (cf. *Tab. 13.2*). This is probably because distal ends were more often used as tool-points, thus shortening the frays as they become compacted, are slivered or bitten off. In addition, proximal ends might fray inadvertently when removed from the stem, and these frays would then be a by-product of tool manufacture (Takemoto *et al.* 2005). Lacking sufficient direct observations, we do not know if frayed ends are also deliberately produced by pulling the tool sideways through partially closed teeth (Sanz *et al.* 2004) or if they are simply by-products of wear through repeated use (McGrew & Collins 1985).

While frays reduce the ease of insertion, they increase the working surface so that greater quantities of honey or insects can be extracted. Brush ends are also used at Lopé to harvest honey of both stingless bees and honey bees. Longer frays may allow for more honey to be gathered. Experiments suggest that tools with frayed ends can capture up to 6 times more honey than compacted ends (Tutin *et al.* 1995).

However, important differences exist between Gashaka and other sites with respect to brush-ends. Gashaka chimpanzees use them *only* to obtain honey whereas those at Lopé use them also to harvest arboreal ants (*Camponotus brutus;* Tutin *et al.* 1995). At other Central African sites, brush ends also catch termites (Cameroon, Equatorial Guinea, Central African Republic, Republic of Congo; references in Tutin *et al.* 1995, Sanz *et al.* 2004). Gashaka chimpanzees do not harvest termites with tools, and they fish arboreal ants with grass and leaf mid-ribs, a technique so far only reported from Mahale where *Camponotus vividus* and *C. brutus* are preyed upon (Nishida 1973, Nishida & Hiraiwa 1982).

Given the propensity for chimpanzees to adapt to different situations it seems reasonable to expect that a tool which is used in one context could be applied to another. Fishing for wood boring arboreal ants or for termites from their mounds are intrinsically similar activities. The insects are invisible in both cases and the probing tools require pliability and strength. However, neither at Mahale (Nishida 1973) nor at Gashaka does the technology of fine ant-fishing tools double for termite-fishing. Similarly, Gashaka chimpanzees also do not re-contextualise the brush-end tools to harvest arboreal ants, as Lopé chimpanzees do. Future research should investigate whether species such as *Camponotus chrysurus* at Gashaka would be amenable to this method of extraction. Currently, plausible ecological explanations are absent, and the trait therefore seems to be an arbitrary cultural variant.

"Core Cultural Tendencies" of Behavioural Variation?

Whiten *et al.* (2001) pondered the question of whether chimpanzee cultures – similar to those of humans – reflect a "central cultural core or theme". One suggestion is that some populations, like those at Taï / Ivory Coast, might show a greater *inclination towards technology*, whereas others, such as those at Budongo / Uganda, might be more non-technological, given a much smaller tool-kit (Whiten *et al.* 2001).

Still, environmental constraints of tool use are hard to rule out, as there could be fine-tuned, but crucial differences in climate, flora, fauna as well as the availability and quality of food. If a forest is rich in fruit, like at Budongo, chimpanzees might simply not need tools for extractive foraging. A similar argument can be made for bonobos in the "fruit bowl" of the central Congo cuvette, which are also not technologically inclined (Hohmann & Fruth 2003). Conversely, Gashaka chimpanzees employ a rather extensive tool-kit, probably to make the most of a low quality habitat saturated with anti-feedants such as tannins and phenols (Hohmann *et al.* 2006, Sommer *et al.* this volume [Ch. 12]).

Faunivory seems an obvious way to supplement a meagre diet. There is currently very little evidence for meat-eating at Gashaka, although apes that are not fully habituated tend to cease hunting when observers approach (Goodall 1986, Boesch & Boesch-Achermann 2000). In any case, mammalian meat makes up less than 5 % of chimpanzee diet (Stanford 1998), suggesting that the benefits of carnivory may lie in the gain of specific micro-nutrients (Tennie *et al.* 2009) or as a commodity to exchange for sex with females.

Entomophagy, in terms of macro-nutrients and calories, might be more important for chimpanzees than meat-eating (McGrew 1992: 154), although the nutritional details on insectivory are still quite confusing (Deblauwe & Janssens 2007). In any case, a cross-site review indicates that, while only few chimpanzee faecal samples contain traces of vertebrate prey (1 – 6 %), many more contain ants (22 – 24 %), termites (2 – 27 %) or bees (1 – 23 %; McGrew 1992). Gashaka has the highest proportion of faecal samples with ant remains of any chimpanzee study site (42 %). Myrmecophagy would be a daily activity, if we assume that chimpanzees defecate only twice a day. In any case, with such high consumption frequencies, ants cannot be considered to be a fall-back food. At least at Gashaka, ants would classify as a *staple* food. If and how this somehow compensates for the lack of termite-eating is not yet known.

Honey, although of high-quality, is less often consumed and would thus constitute a *treat* (McGrew 1992). Raiding bee nests not only provides energy through honey, but honey-combs also yield fat and protein from larvae, pupae, pollen, and imagos (McGrew 1992). Some authors doubt procurement of protein as a major function of ant-eating, assuming instead that specific nutrients (essential amino acids or vitamins of the B-group) might be important (Hladik 1977, Nishida & Hiraiwa 1982). Non-nutritional interpretations have also been proposed, such that the formic acid of ants is a spicy snack (Nishida & Hiraiwa 1982). Interestingly, in South-east Asia, weaver ants are an integral component of curry powder (Bodenheimer 1951, cit. in Nishida & Hiraiwa 1982). Termites might be a different matter, since winged reproductive forms are particularly rich in fat and protein (Nishida & Hiraiwa 1982). Nevertheless, the absence of termite eating at Gashaka is puzzling, as it is hard to reconcile with the assumption that food constraints drive tool use.

The Gashaka data are confusing with respect to the second potential cultural tendency Whiten *et al.* (2001) proposed, i.e., use of *brush tools* in Central Africa to harvest ants and termites (e.g., south-west Cameroon: Sugiyama 1985; Congo: Fay & Carroll 1992; Sanz *et al.* 2004) versus finer tools elsewhere. The Gashaka chimpanzees seem to defy either classification, since brush-tools harvest only honey, whereas fine tools such as leaf mid-ribs harvest ants. A geographic dichotomy between brush-tools and finer tools is therefore not supported by our data.

The significance of leaf mid-ribs should be investigated further. First, the material is perishable. This rules out a significant time-lag between manufacture and use, pointing clearly towards intentional production (*sensu* Beck 1980). Second, chimpanzees have to visualise that a tool is embedded in the leaf structure, rather than manufacturing a tool from a more obvious source, such as a grass blade. Whether it can be argued that this complexity makes it a less likely candidate for independent development and a more likely candidate for social transmission is an appealing question.

We compared the presence or absence of behavioural patterns associated with material culture of Gashaka chimpanzees with those for 9 long-term study sites (*Fig. 13.9*). Intuitively, one would assume that Nigerian chimpanzees are culturally more related to Central African populations than to those in West or East Africa. Indeed, the Gashaka profile has similarities with Lopé. On the other hand, Gashaka could share even more traits with Gombe. Thus, more data, particularly from Central Africa, are needed, to reconstruct potential regional cultural clusters.

We cannot currently speculate about *transmission mechanisms*, e.g., whether patterns develop through imitation, emulation or more individualistic processes such as social or local enhancement (cf. e.g., Whiten & Ham 1992, Byrne 1995). Sensible conclusions would require more direct, close-quarter observations than presently available for Gashaka – and it will be particularly difficult to be present if and when a new behaviour is invented. Nevertheless, we were able to partly document the emergence of a new behavioural pattern for another primate taxon at Gashaka (H. Guruza, *pers. comm.*). Oranges grow on trees in abandoned farms near the Kwano field station as well as around Gashaka village. However, neither baboons, guenons, chimpanzees or tantalus monkeys were ever known to eat the ripe fruits – probably because the acidic juice sprays into the eyes once the orange is bitten into. From 2005 onwards, tantalus monkeys near Gashaka started to increasingly raid the village's orange trees, once they had learned to close their eyes while initially biting through the peel.

Cultural Primatology and Conservation

Whiten *et al.* (2003) coined the fitting term "cultural panthropology" for an emerging research area at the interface of social sciences (with subdisciplines such as social anthropology, material culture, psychology, ethno-botany, archaeology) and biological sciences (with subdisciplines such as palaeoanthropology, primatology, physiology, genetics and evolutionary medicine).

However, "cultural panthropology" has become another "urgent anthropology", given that many wild populations of apes are threatened by extinction due to habitat destruction, disease and the trade in bush-meat (Sommer & Ammann 1998, Ammann *et al.* 2003, Caldecott & Miles 2005). Correspondingly, not only genetic diversity is lost, but also cultural diversity – similar to how globalisation has destroyed many traditional human ways of life (McGrew 2004). It is already clear that "we can never know the true extent of cultural diversity in chimpanzees because so many communities, along with their cultures, are already gone" (Goodall 1994: 397).

The situation is particularly dire for the Nigerian chimpanzee. Our survey work in north-eastern Nigeria suggests that *P. t. vellerosus* has a realistic chance of survival only in the Gashaka area since adjoining regions are already devoid of wildlife. Population viability models based on intake rates of ape sanctuaries in Nigeria and Cameroon predict extinction in as little as 2 decades (Adanu *et al.* this volume [Ch. 3], Hughes *et al.* this volume [Ch. 14]).

Studies like ours therefore not only shed light on the pathways of human evolution. They are also meant to shine a spotlight on a little known ape population, as such publicity may help to preserve their biological and cultural identity. Because, "all of the above will be academic [...] if the chimpanzee becomes extinct" (McGrew 2004: 194).

Acknowledgements The Nigeria National Park Service kindly granted a research permit to the *Gashaka Primate Project* (GPP). Gashaka Gumti National Park and the Nigerian Conservation Foundation (NCF) provided logistical support. Jeremiah Adanu, Judith Bovensiepen, Aylin McNamara, Jennifer Rogan, Asako Saegusa, and Yakubu Wakirwa contributed to the data collection. For identification of insects or plants, we are indebted to J. D. Chapman (Canterbury), Britta Kunz (Würzburg), Judith Korb (Regensburg), David W. Roubik (Washington), Barry Bolton (Isle of Wight). The work would have been impossible without local field assistants, in particular Hammaunde Guruza, Felix Vitalis, Bobbo Buba, Ali Tappare, Maigari Ahmadu, Buba Bello, Zacchariah Dollar, Isaac Timothy, Buba Hammaselbe, Sam Yusufu. The field work was supported by The Leakey Foundation (USA) and generous grants from the North of England Zoological Society / Chester Zoo Nigeria Biodiversity Programme.

Literature

Ammann, K., Rose, A. L., Mittermeier, R. A., Langrand, O., Ampadu-Agyei, O. & Butynski, T. M. (2003). *Consuming Nature. A Photo Essay on African Rain Forest Exploitation*. Palos Verdes, CA: Altisima.

Anderson, D. P., Nordheim, E. V., Boesch, C. & Moermond, T. C. (2002). Factors influencing fission-fusion grouping in chimpanzees in the Taï National Park, Cote d'Ivoire. In Boesch, C., Hohmann, G. & Marchant, L. F. (eds.), *Behavioural Diversity in Chimpanzees and Bonobos*. Cambridge: Cambridge University Press, pp. 90–101.

Anzenberger, G. (1977). Ethological study of African carpenter bees of the genus *Xylocopa* (Hymenoptera, Anthophoridae). *Zeitschrift für Tierpsychology* 44: 337–374.

Beck, B. (1980). *Animal Tool Behavior*. New York: Garland STMP Press.

Beck, J. & Chapman, H. (2008). A population estimate of the endangered chimpanzee *Pan troglodytes vellerosus* in a Nigerian montane forest: implications for conservation. *Oryx* 42: 448–451.

Bodenheimer, F. S. (1951). *Insects as Human Food*. The Hague: Dr W Junk.

Boesch, C. & Boesch, H. (1990). Tool use and tool making in wild chimpanzees. *Folia Primatologica* 54: 86–99.

Boesch, C. & Boesch-Achermann, H. (2000). *The Chimpanzees of the Taï Forest. Behavioural Ecology and Evolution*. Oxford: Oxford University Press.

Boesch, C., Hohmann, G. & Marchant, L. F. (eds.) (2002). *Behavioural Diversity in Chimpanzees and Bonobos*. Cambridge: Cambridge University Press.

Boesch, C., Marchesi, P., Marchesi, N., Fruth, B. & Joulian, F. (1994). Is nut cracking in wild chimpanzees a cultural behaviour? *Journal of Human Evolution* 26: 325–338.

Bolton, B. (1994). *A New General Catalogue of the Ants of the World.* Cambridge: Harvard University Press.

Bourliere, F. (ed.) (1983). *Tropical Savannas: Ecosystems of the World.* Oxford: Elsevier.

Brewer, S. M. & McGrew, W. C. (1990). Chimpanzee use of a tool-set to get honey. *Folia Primatologica* 54: 100–104.

Byrne, R. W. (1995). *The Thinking Ape. Evolutionary Origins of Intelligence.* Oxford: Oxford University Press.

Caldecott, J. & Miles, L. (ed.) (2005). *World Atlas of Great Apes and Their Conservation.* Los Angeles: University of California Press.

Chapman, H. M., Olson, S. M. & Trumm, D. (2004). An assessment of changes in the montane forests of Taraba State, Nigeria, over the past 30 years. *Oryx* 38: 1–9.

Chapman, J. D. & Chapman, H. M. (2001). *The Forest Flora of Taraba and Adamawa States, Nigeria. An Ecological Account and Plant Species Checklist.* University of Canterbury / NZ: WWF / DFID.

de Waal, F. B. M. (1999). Cultural primatology comes of age. *Nature* 399: 364–365.

Deblauwe, I. & Janssens, G. P. J. (2007). New insights in insect prey choice by chimpanzees and gorillas in southeast Cameroon: the role of nutritional value. *American Journal of Physical Anthropology* 135: 42–55.

Diamond, J. (1992). *The Third Chimpanzee.* New York: Harper Perennial.

Dunn, A. (1999). *Gashaka Gumti National Park: a Guidebook.* Lagos: Gashaka Gumti National Park, NCF / WWF-UK.

Ellis, D. (2005). *Seasonal Variation in Availability and Consumption of Army Ants by Nigerian Chimpanzees.* MSc in Human Evolution and Behaviour thesis, University College London.

Fay, J. M. & Carroll, R. W. (1992). Chimpanzee tool use for honey and termite extraction in central Africa. *American Journal of Primatology* 34: 309–317.

Fletcher, D. J. C. (1978). The African bee, *Apis mellifera adansonii*, in Africa. *Annual Review of Entomology* 23: 151–171.

Fowler, A. (2006). *Socio-ecology of Nigerian Chimpanzees at Gashaka.* PhD thesis, University of London.

Fowler, A. & Sommer, V. (2007). Subsistence technology in Nigerian chimpanzees. *International Journal of Primatology* 28: 997–1023.

Fowler, A., Koutsioni, Y. & Sommer, V. (2007b). Leaf-swallowing in Nigerian chimpanzees: Assumed evidence for self-medication. *Primates* 48: 73–76.

Gagneux, P., Gonder, M. K., Goldberg, T. L. & Morin, P. A. (2001). Gene flow in wild chimpanzee populations: what genetic data tell us about chimpanzee movement over space and time. *Philosophical Transactions of the Royal Society B Biological Sciences* 356: 889–897.

Ganas, J. & Robbins, M. M. (2004). Intrapopulation differences in ant eating in the mountain gorillas of Bwindi Impenetrable National Park, Uganda. *Primates* 45: 275–278.

Gonder, M. K., Oates, J. F., Disotell, T. R., Forstner, M. R. J., Morales, J. C. & Melnick, D. J. (1997). A new West African chimpanzee subspecies? *Nature* 338: 337.

Goodall, J. (1963). Feeding behaviour of free-living chimpanzees: A preliminary report. *Symposium, Zoological Society of London* 10: 39–47.

Goodall, J. (1986). *The Chimpanzees of Gombe: Patterns of Behaviour.* Cambridge, MA: Harvard University Press.

Goodall, J. (1994). Conservation and the future of chimpanzee and bonobo research in Africa. In Wrangham, R. W., McGrew, W. C., de Waal, F. B. M. & Heltne, P. G. (eds.), *Chimpanzee Cultures.* Cambridge, MA: Harvard University Press, pp. 397–404.

Gotwald, W. H. Jr. (1974). Foraging behavior of *Anomma* driver ants in Ghana cocoa farms. *Bulletin de l'Institut Fondamental d'Afrique Noire, Serie A* 36: 705–713.

Gotwald, W. H. Jr. (1995) *Army Ants – the Biology of Social Predation.* Ithaca & London: Cornell University Press.

Gould, J. L. & Gould, C. G. (1995). *The Honey Bee.* (Scientific American Library). New York: W. H. Freeman.

Gumnior, M. & Thiemeyer, H. (2003). Termitaria-Areale in der nordnigerianischen Savanne. Der Einfluß zweier Termitenspezies auf Genese und Stoffhaushalt von Sandböden. *Geo-Ökologie* 24: 321–336.
Hart, A. G. & Ratnieks, F. L. W. (2002). Task-partitioned nectar transfer in stingless bees: work organization in a phylogenetic context. *Ecological Entomology* 27: 163–168.
Heltne, P. G. & Marquardt, L. A. (eds.) (1989). *Understanding Chimpanzees*. Cambridge, MA: Harvard University Press.
Hladik, C. M. (1977). A comparative study of the feeding strategies of two sympatric species of leaf monkeys: *Presbytis senex* and *Presbytis entellus*. In Clutton-Brock, T. H. (ed.), *Primate Ecology: Studies of Feeding and Ranging Behaviour in Lemurs, Monkeys and Apes*. London: Academic Press, pp. 323–353.
Hofstede, F. E. (2005). *Foraging Task Specialization and Foraging Labour Allocation in Stingless Bees*. PhD thesis, Utrecht University.
Hohmann, G. & Fruth, B. (2003). Culture in bonobos? Between-species and within-species variation in behavior. *Current Anthropology* 44: 563–571.
Hohmann, G., Fowler, A., Sommer, V. & Ortmann, S. (2006). Frugivory and gregariousness of Salonga bonobos and Gashaka chimpanzees: the abundance and nutritional quality of fruit. In Hohmann, G. (ed.), *Feeding Ecology of Apes and Other Primates*. Cambridge: Cambridge University Press, pp. 123–159.
Hölldobler, B. & Wilson, E. O. (1990). *The Ants*. Cambridge, MA: Harvard University Press.
Howse, P. E. (1970). *Termites: A Study in Social Behaviour*. London: Hutchinson.
Hughes, N. (2003). *Great Ape Sanctuaries in Nigeria and Cameroon: a Window into the Bushmeat-Trade*. MSc in Human Evolution and Behaviour thesis, University College London.
Humle, T. & Matsuzawa, T. (2002). Ant-dipping among the chimpanzees of Bossou, Guinea, and some comparisons with other sites. *American Journal of Primatology* 58: 133–148.
Humle, T. & Matsuzawa, T. (2004). Oil palm use by adjacent communities of chimpanzees at Bossou and Nimba mountains, West Africa. *International Journal of Primatology* 25: 551–581.
Hunt, K. D. & McGrew, W. C. (2002). Chimpanzees in the dry areas of Assirik, Senegal and Semliki Wildlife Reserve, Uganda. In Boesch, C., Hohmann, G. & Marchant, L. F. (eds.), *Behavioural Diversity In Chimpanzees and Bonobos*. Cambridge: Cambridge University Press, pp. 35–51.
Ingram, C. J., Mulcare, C. A., Itan, Y., Thomas, M. G. & Swallow, D. M. (2009). Lactose digestion and the evolutionary genetics of lactase persistence. *Human Genetics* 124: 579–591.
Kingdon, J. (1997). *The Kingdon Field Guide to African Mammals*. San Diego: Academic Press.
Kormos, R., Boesch, C., Bakarr, M. I. & Butynski, T. M. (eds.) (2003). *West African Chimpanzees. Status Survey and Conservation Action Plan*. Gland & Cambridge UK: IUCN / SSC Primate Specialist Group.
Kortlandt, A. & Holzhaus, E. (1987). New data on the use of stone tools by chimpanzees in Guinea and Liberia. *Primates* 28: 473–496.
Kronauer, D. J. C. (2008). Recent advances in army ant biology (Hymenoptera: Formicidae). *Myrmecological News* 12: 51–65.
Leroux, J. M. (1982). Ecologie des populations de dorylines *Anomma nigricans* dans la Région de Lamto (Côte d'Ivoire). *Publications du Laboratoire de Zoologie, No. 22*. Ecole Normale Supérieure, Paris.
Longhurst, C., Johnson R. A. & Wood, T. G. (1978). Predation by *Megaponera foetens (Fabr.)* (Hymenoptera: Formicidae) on termites in the Nigerian southern Guinea savanna. *Oecologica* 1: 101–107.
Lott, D. F. (1984). Intraspecific variation in the social systems of wild vertebrates. *Behaviour* 88: 266–325.
Louppe, D., Oattara, N. & Coulibaly, A. (1995). The effects of brush fires on vegetation: the Audrville fire plots after 60 years. *Commonwealth Forestry Review* 74: 288–292.
Matsumoto-Oda, A. (2002). Social relationships between cycling females and adult males in Mahale chimpanzees. In Boesch, C., Hohmann, G. & Marchant, L. F. (eds.), *Behavioural Diversity in Chimpanzees and Bonobos*, Cambridge: Cambridge University Press, pp. 168–180.

Matsuzawa, T. (1994). Field experiments on the use of stone tools by chimpanzees in the wild. In Wrangham, R. W., McGrew, W. C., deWaal, F. B. M. & Heltne, P. G. (eds.), *Chimpanzee Cultures*. Cambridge, MA: Harvard University Press, pp. 351–370.

McGrew, W. C. (1992). *Chimpanzee Material Culture. Implications for Human Evolution*. Cambridge: Cambridge University Press.

McGrew, W. C. (2004). *The Cultured Chimpanzee: Reflections on Cultural Primatology*. Cambridge: Cambridge University Press.

McGrew, W. C. & Collins, D. A. (1985). Tool-use by wild chimpanzees (*Pan troglodytes*) to obtain termites (*Macrotermes herus*) in the Mahale mountains, Tanzania. *American Journal of Primatology* 9: 47–62.

McGrew, W. C. & Tutin, C. E. G. (1978). Evidence for a social custom in wild chimpanzees? *Man* 13: 234–251.

McGrew, W. C., Baldwin, P. J. & Tutin, C. E. G. (1988). Diet of wild chimpanzees (*Pan troglodytes verus*) at Mt Assirik, Senegal: I. Composition. *American Journal of Primatology* 16: 213–226.

McGrew, W. C., Marchant, L. F. & Nishida, T. (eds.) (1996). *Great Ape Societies*. Cambridge: Cambridge University Press.

McGrew, W. V., Pruetz, J. D. & Fulton, S. J. (2005). Chimpanzees use tools to harvest social insects at Fongoli, Senegal. *Folia Primatologica* 76: 222–226.

McGrew, W. C., Tutin, C. E. G. & Baldwin, P. J. (1979). Chimpanzees, tools and termites: cross-cultural comparisons of Senegal, Tanzania, and Rio Muni. *Man* 14: 185–214.

McGrew, W. C., Ham, R. M., White, L. T. J., Tutin, C. E. G. & Fernandez, M. (1997). Why don't chimpanzees in Gabon crack nuts? *International Journal of Primatology* 18: 353–374.

McGrew, W. C., Baldwin, P. J., Marchant, L. F., Pruetz, J. D., Scott, S. E. & Tutin, C. E. G. (2003). Ethnoarchaeology and elementary technology of unhabituated wild chimpanzees at Assirik, Senegal, West Africa. *Palaeoanthropology* 1: 1–20.

Miller, B. D. (2005). *Cultural Anthropology*. Boston, New York etc. : Pearson.

Mitani, J. C., Watts, D. P. & Lwanga, J. S. (2002). Ecological and social correlates of chimpanzee party size and composition. In Boesch, C., Hohmann, G. & Marchant, L. F. (eds.), *Behavioural Diversity in Chimpanzees and Bonobos*. Cambridge: Cambridge University Press, pp. 102–111.

Morgan, B. J & Abwe, E. E. (2006). Chimpanzees use stone hammers in Cameroon. *Current Biology* 16: R632–R633.

Nieh, J. C. (2003). Recruitment communication in stingless bees (Hymenoptera, Apidae, Meliponini). *Apidologie* 35: 159–182.

Nishida, T. (1973). The ant-gathering behaviour by the use of tools among wild chimpanzees of the Mahale mountains. *Journal of Human Evolution* 2: 357–370.

Nishida, T. & Hiraiwa, M. (1982). Natural history of a tool-using behaviour by wild chimpanzees in feeding on wood-boring ants. *Journal of Human Evolution* 11: 73–99.

Oates, J., Groves, C. & Jenkins, P. D. (2008). The type locality of *Pan troglodytes vellerosus* (Gray, 1862),and implications for the nomenclature of West African chimpanzees. *Primates* 50: 78-80.

Oswalt, W. H. (1976). *An Anthropological Analysis of Food-Getting Technology*. Canada: John Wiley & Sons.

Peeters, C. & Andersen, N. A. (1989). Cooperation between dealate queens during colony foundation in the green tree ant, *Oecophilla smaragdina*. *Psyche* 96: 39–44.

Raignier, A. & van Boven, J. K. A. (1955). Etude taxonomique, biologique et biométrique des Dorylus du sous-genre *Anomma* (Hymenoptera: Formicidae). *Annales du Musée Royal du Congo Belge, n. s. 4° (Sciences Zoologiques)* 2: 1–359.

Redford, K. H. (1987). Ants and termites as food: patterns of mammalian myrmecophagy. In Genoways, H. H. (ed.), *Current Mammalogy*. New York: Plenum Press, pp. 349–399.

Reynolds, V. (2005). *The Chimpanzees of the Budongo Forest*. Oxford: Oxford University Press.

Roonwal, M. L. (1980). Quantification of tail loop characters for clinal variations in the langur, *Presbytis entellus*, in peninsular India and Ceylon (Primates). *Zoologischer Anzeiger* 204: 177–185.

Roubik, D. W. (1991). Aspects of Africanized honey bee ecology in tropical America. In Spicak, M., Fletcher, D. J. C., & Breed, M. D. (eds.), *The "African" Honey Bee*. Colorado: Boulder Westview Press, pp. 259–281.

Sanz, C. M. & Morgan, D. B. (2007). Chimpanzee tool technology in the Goualougo Triangle, Republic of Congo. *Journal of Human Evolution* 52: 420–433.

Sanz, C. M., Morgan, D. B. & Gulick, S. (2004). New insights into chimpanzees, tools, and termites from the Congo Basin. *The American Naturalist* 164: 567–681.

Schneirla, T.C. (1971). *Army Ants. A Study in Social Organization*. San Francisco: W. H. Freeman.

Schöning, C. (2005). The driver ants of Mount Kenya. SWARA Apr-Jun: 32–37.

Schöning, C., Kinuthia, W. & Franks, N. R. (2005). Evolution of allometries in the worker caste of Dorylus army ants. *Oikos* 110: 231–240.

Schöning, C., Ellis, D., Fowler, A. & Sommer, V. (2007). Army ant prey availability and consumption by chimpanzees at Gashaka (Nigeria). *Journal of Zoology* 271: 125–133.

Schöning, C., Gotwald, W. H. Jr., Kronauer, D. J. C. & Vilhemsen, L. (2008a). Taxonomy of the African army ant *Dorylus gribodoi Emery*, 1892 (*Hymenoptera, Formicidae*). New insights from DNA sequence data and morphology. *Zootaxa* 1749: 39–52.

Schöning, C., Humle, T., Möbius, Y. & McGrew, W. C. (2008b). The nature of culture: Technological variation in chimpanzee predation on army ants revisited. *Journal of Human Evolution* 55: 48–59.

Sommer, V. (1996). *Heilige Egoisten. Die Soziobiologie indischer Tempelaffen*. Munich: Beck, C. H.

Sommer, V. (2003). Geistlose Affen oder äffische Geistwesen? Eine Exkursion durch die mentale Welt unserer Mitprimaten. In Becker, A., Mehr, C., Nau, H. H., Reuter, G. & Stegmüller, D. (eds.), *Gene, Meme und Gehirne. Geist und Gesellschaft als Natur*, Frankfurt / M. : Suhrkamp, pp. 112–136.

Sommer, V. (2008). *Schimpansenland. Wildes Leben in Afrika*. Munich: C. H. Beck.

Sommer, V. & Ammann, K. (1998). *Die Großen Menschenaffen: Orang-utan, Gorilla, Schimpanse, Bonobo*. Munich: BLV.

Sommer, V., Adanu, J., Faucher, I. & Fowler, A. (2004). The Nigerian Chimpanzee (*Pan troglodytes vellerosus*) at Gashaka: Two years of habituation efforts. *Folia Primatologica* 75: 295–316.

Stanford, C. (1998). The social behavior of chimpanzees and bonobos: Empirical evidence and shifting assumptions. *Current Anthropology* 39: 399–420.

Stoops, G. (1964). Application of some pedological methods to the analysis of termite mounds. In Bouillon, A. (ed.), *Etudes sur les Termites Africains*, Kinshasa: Universite' de Leopoldville, pp. 379–398.

Sugiyama, Y. (1985). The brush-stick of chimpanzees found in south-west Cameroon and their cultural characteristics. *Primates* 26: 361–374.

Sugiyama, Y. (1995). Tool-use for catching ants by chimpanzees at Bossou and Mount Nimba, West Africa. *Primates* 36 :193–205.

Sugiyama, Y. & Koman, J. (1979). Social structure and dynamics of wild chimpanzees at Bossou, Guinea. *Primates* 20: 513–524.

Tennie, C., Gilby, I. C. & Mundry, R. (2009). The meat-scrap hypothesis: small quantities of meat may promote cooperative hunting in wild chimpanzees (*Pan troglodytes*). *Behavioral Ecology and Sociobiology* 63: 421–431.

Tutin, C. E. G. & Fernandez, M. (1992). Insect-eating by sympatric lowland gorillas (*Gorilla g. gorilla*) and chimpanzees (*Pan t. troglodytes*) in the Lopé reserve, Gabon. *American Journal of Primatology* 28: 29–40.

Tutin, C. E. G., McGrew, W. C. & Baldwin, P. J. (1983) Social organization of savanna-dwelling chimpanzees, *Pan troglodytes verus*, at Mt. Assirik, Senegal. *Primates* 24: 154–173.

Tutin, C. E. G., Ham, R. & Wrogemann, D. (1995). Tool use by chimpanzees (*Pan t. troglodytes*) in the Lope Reserve, Gabon. *Primates* 36: 181–192.

von Frisch, K. (1967). *The Dance Language and Orientation of Bees*. Cambridge / MA: Harvard University Press.

Wallis, J. (2002). Seasonal aspects of reproduction and sexual behavior in two chimpanzee populations: a comparison of Gombe (Tanzania) and Budongo (Uganda). In Boesch, C., Hohmann,

G. & Marchant, L. F. (eds.), *Behavioural Diversity in Chimpanzees and Bonobos*, Cambridge: Cambridge University Press, pp. 181–191.

Warren, Y. (2003). *Olive Baboons (Papio cynocephalus anubis): Behaviour, Ecology And Human Conflict In Gashaka Gumti National Park*, Nigeria. PhD thesis, Roehampton University.

Watts, D. P. (1989). Ant eating behaviour by mountain gorillas. *Primates* 30: 121–125.

Whiten, A. & Ham, R. (1992). On the nature and evolution of imitation in the animal kingdom: reappraisal of a century of research. *Advances in the Study of Behavior* 21: 239–283.

Whiten, A., Horner, V. & Marshall-Pescini, S. (2003). Cultural panthropology. *Evolutionary Anthropology* 12: 92–105.

Whiten A., Goodall, J., McGrew, W. C., Nishida, T., Reynolds, V., Sugiyama, Y., Tutin, C. E. G, Wrangham, R. W. & Boesch, C. (1999). Cultures in chimpanzees. *Nature* 399: 682–685.

Whiten, A., Goodall, J., McGrew, W. C., Nishida, T., Reynolds, V., Sugiyama, Y., Tutin, C. E. G., Wrangham, R. W. & Boesch, C. (2001). Charting cultural variation in chimpanzees. *Behaviour* 138: 1481–1516.

Whitesides, G. H. (1985). Nut cracking by wild chimpanzees in Sierra Leone, West Africa. *Primates* 26: 91–94.

Wilson, E. O. (1958). The beginnings of nomadic and group predatory behaviour in the ponerine ants. *Evolution* 12: 24–36.

Wilson, E. O. (1971). *The Insects Societies*. Cambridge, MA: Belknap Press of Harvard University.

Winston, L. M. (1987). *The Biology of the Honey Bee*. Cambridge, MA: Harvard University Press.

Wrangham, R (2002). The costs of sexual attraction: is there a trade-off in female Pan between sex appeal and receiving coercion? In Boesch, C., Hohmann, G., Marchant, L. F. (eds.), *Behavioural Diversity in Chimpanzees and Bonobos*. Cambridge: Cambridge University Press, pp. 204–215.

Wrangham, R. W., McGrew, W. C., de Waal, F. B. M. & Heltne, P. G. (eds.) (1994). *Chimpanzee Cultures*. Cambridge, MA: Harvard University Press.

Yamagiwa, J., Mwanza, N., Yumoto, T. & Maruhashi, T. (1991). Ant eating by eastern lowland gorillas. *Primates* 32: 247–253.

Yamagiwa, J., Yumoto, T., Ndunda, M. & Maruhashi, T. (1988). Evidence of tool-use by chimpanzees (*Pan troglodytes schweinfurthii*) for digging out a bee-nest in the Kahuzi-Biega National Park, Zaire. *Primates* 29: 405–411.

Yamakoshi, G. & Matsuzawa, T. (1993). Preliminary surveys of the chimpanzees in the Nimba Reserve, Côte d'Ivoire. *Primate Research* 9: 13–17.

Chapter 14
Will the Nigeria-Cameroon Chimpanzee Go Extinct? Models Derived from Intake Rates of Ape Sanctuaries

Nicola Hughes, Norm Rosen, Neil Gretsky, and Volker Sommer

Abstract All great ape taxa are considered endangered due to a combination of habitat loss, hunting, disease outbreaks and human population increase. The bush-meat trade is thought to be the biggest threat to the survival of chimpanzees, bonobos and gorillas. However, assessments of the reduction of wild populations are notoriously unreliable since they are mostly based on indirect evidence such as brief surveys of markets and interviews with hunters. We use a direct approach and measure annual loss from the wild through intake rates of sanctuaries in Africa which currently hold close to 1000 apes. From these, we calculate loss to the wild populations by relating arrivals into sanctuaries to the proportions of infants in wild groups, hunting strategies, and the likelihood that captured babies make it to a sanctuary. We focus on the most endangered chimpanzee subspecies, *Pan troglodytes vellerosus*, found in eastern Nigeria / western Cameroon. Our calculations, based on intake rates since 1986 into four sanctuaries in Cameroon and Nigeria, suggest that current rates of hunting are 2 – 13 times higher than sustainable rates. Moreover, VORTEX – a population viability analysis tool – predicts that this chimpanzee subspecies will go extinct in as little as 20 years. The dramatic results emphasise the need for immediate conservation measures. Ape sanctuaries are, at times, considered to be a waste of resources, which could be better diverted to habitat protection. However, it is extremely unlikely that hunting pressure can be significantly reduced, and many local ape populations are already effectively extinct. Sanctuaries will therefore play an increasingly important role as ape conservation tools, if only for the fact that they may, in a couple of decades, harbour more apes than survive in the wild. We propose to apply our method of measuring extinction risk to other taxa of apes across Africa and Asia.

Keywords *Pan troglodytes vellerosus* • Bush-meat trade • Ape sanctuaries • Extinction • Vortex

V. Sommer (✉)
Department of Anthropology, University College London, London WC1E 6BT, UK
e-mail: v.sommer@ucl.ac.uk

V. Sommer and C. Ross (eds.), *Primates of Gashaka*, Developments in Primatology: Progress and Prospects 35, DOI 10.1007/978-1-4419-7403-7_14,

Introduction

Meat from wild animals is a traditional source of protein for humans across tropical Africa (Wilkie 2001, Fa *et al.* 2003). However, the trade in bush-meat is no longer a localised subsistence practice (de Merode *et al.* 2004) but a commercialised and highly profitable business. Much of the demand comes from expanding urban centres, facilitated by the opening up of previously inaccessible areas by logging companies (Rose *et al.* 2003, Peterson 2003). The use of guns has had a devastating impact on wildlife and on larger bodied animals in particular, as heavier animals are preferentially targeted to get more profit per cartridge (Fa *et al.* 2000, Wilkie & Godoy 2001).

The most commonly hunted species are ungulates and rodents. Primates constitute up to 40 % of the species typically found at bush-meat markets with great apes making up 1 – 2.5 % of the meat on offer (Bowen-Jones 1998, Bowen-Jones & Pendry 1999, Wilkie & Carpenter 1999, Fa *et al.* 2000). Even low level hunting seriously threatens ape populations due to their slow reproductive rates. In many areas, they have already been driven to local extinction due to the combined effects of habitat destruction and the bush-meat trade (Sommer & Ammann 1998, Peterson 2003, Rose *et al.* 2003, Caldecott & Miles 2005; but see Oates 2006).

In addition, market surveys are likely to underestimate ape hunting rates (Bowen-Jones & Pendry 1999). The large size of their carcasses hinders transport to central bush-meat markets, and the illegality of hunting great apes means that their meat will often not be freely on display. Moreover, chopping and smoking of bush-meat prior to sale makes identification rather difficult (Bowen-Jones & Pendry 1999). Preference for primate meat may also mean that apes are sold immediately from villages (Chapman & Peres 2001) or that they bypass markets as they are killed for specific buyers (Rose 1998), including African ex-patriots who continue to fuel demand overseas. Chimpanzee carcasses, for example, have been found in an African food shop in Brussels (*The Sunday Times* 1999). It is therefore difficult to relate data generated from bush-meat markets to the actual extinction risk faced by African great apes.

Another "window" into the bush-meat trade is potentially offered by arrival rates of apes into sanctuaries across Africa, particularly infants (Rosen *et al.* 2001, 2002). Because infants are so small, hunters stand to make more profit by selling them alive. It is sometimes stated that the demand for chimpanzees and gorillas as pets may intensify the targeting of nursing mothers with infants (Teleki 1989, Peres 1991, McManus 2005). Much more commonly, infant apes are seen as by-products or bonuses of meat hunting (Mittermeier & Cheney 1987, Ammann & Pearce 1995). Karl Ammann, who spent years investigating the bush-meat trade, has never come across a case where the demand for a live infant prompted the slaughter of an ape (Peterson 2003).

Currently, sanctuaries across Africa hold close to 1000 gorillas, bonobos, and chimpanzees (Rosen et al, in prep). Our paper uses sanctuary intake rates to estimate the impact of the bush-meat trade on the chimpanzee subspecies, *P. t. vellerosus* (*Fig. 14.1*). We concentrate on this taxon, because its restricted geographical range in eastern Nigeria and western Cameroon, between the rivers Niger / Benue and Sanaga, allows us to generate relatively reliable parameters for modelling purposes.

Fig. 14.1 Chimpanzee orphans in Africa. (**a**) This infant was removed from the wild on behalf of Ibadan Zoo / Nigeria from near the border of Cameroon. (photo: VS, 2000). (**b**) Carer Marie-Claire with some of the infants held in the Sanaga-Young Chimpanzee Rescue Centre / Cameroon (photo: NH, 2003)

P. t. vellerosus is considered to be the most endangered subspecies of chimpanzee (Kormos *et al.* 2003, Caldecott & Miles 2005). Cameroon represents the "front-line of logging and the unsustainable trade in bushmeat" (Ammann 2001: 76), and the situation is equally as desperate in Nigeria where habitat destruction is progressing at a rate of 2.6 % per year compared to a sub-Saharan average of 0.8 % (World Bank 2002).

Our analysis focuses on whether the hunting pressure on wild *P. t. vellerosus* is sustainable, and we model the risk of extinction for this chimpanzee subspecies. Finally, we emphasise the increasingly important role of sanctuaries as conservation tools.

Methods

Chimpanzee Subspecies

All great apes are classified as endangered or critically endangered by the current IUCN Red List of Threatened Species (Baillie *et al.* 2004, IUCN 2007). They also appear in Appendix 1 of the Convention on International Trade in Endangered

Species of Wild Flora and Fauna (CITES), which means that it is illegal to hunt, capture or move them across international boundaries (Bowen-Jones 1998).

Chimpanzees (*Pan troglodytes*), while under threat, are the most widely distributed of the great apes. They are thought to be extinct in Benin, Burkina Faso and Togo, but can still be found in 21 other African countries (Inskipp 2005). Chimpanzees inhabit savannah-woodland, mosaic grassland forest and tropical moist forests (Teleki 1989). They congregate in patrilocal multi-male, multi-female communities of 10 – 140 individuals. Chimpanzees are largely frugivorous, but also eat leaves, insects and hunt for larger vertebrate prey (Boesch & Boesch-Achermann 2000). Food distribution and quality, particularly of fruit, influences how chimpanzee communities aggregate and split into smaller parties (fusion-fission society). Studies across Africa reveal that each community has a unique combination of social customs, tool-kits, communication, territorial aggression, war-like raiding, hunting strategies, and plant consumption for food and self-medication (Whiten *et al.* 1999, McGrew 2004, Fowler & Sommer 2007).

The geographical range of 2340000 km^2 *(Fig. 14.2)* is believed to harbour between 152000 – 255000 wild chimpanzees (Butynski 2001). Current numbers are likely to be much lower, as, for example, the chimpanzee population of Ivory Coast, according to night nest counts, declined by 90 % over the last two decades (Campbell *et al.* 2008). All subspecies are classified by IUCN as at least endangered. They occur in East Africa (*P. t. schweinfurthii*), Central Africa (*P. t. troglodytes*) and West Africa (*P. t. verus*). An additional subspecies, *P. t. vellerosus*, has only recently been recognised (Gonder *et al.* 1997). This subspecies is either called the "Nigeria chimpanzee" or "Nigeria-Cameroon chimpanzee" (Inskipp 2005). (Note: The scientific name might need revision to *Pan troglodytes ellioti* [Matschie, 1914]; Oates *et al.* 2008). It is genetically the most distinct subspecies, and various members of the Section on Great Apes, Primate Specialist Group, IUCN, advocate that its status should be revised to "critically endangered" (Sommer *pers. comm.*).

P. t. vellerosus: Distribution, Population Size, and Life-history Parameters

The *geographical range* of *P. t. vellerosus* has a maximum east – west extension of 600 km (Mbam et Djerem National Park (NP) to Cross River NP) and a maximum south – north extension of 600 km (Ebo to Gashaka Gumti NP) (Caldecott & Miles 2005). This amounts to a range of about 152000 km^2 across eastern Nigeria and western Cameroon (Butynski 2001). However, it is commonly assumed that the actual area of occupancy is only 5 – 25 % of the range (Butynski 2001) or, in this instance, 7600 – 38000 km^2.

We assume that *P. t. vellerosus* is restricted to the area between the river Sanaga in the south-east and the rivers Benue / Niger in the west (cf. *Fig. 14.2*). The majority live in National Parks (NP) or forest reserves (FR). Major occurrences in

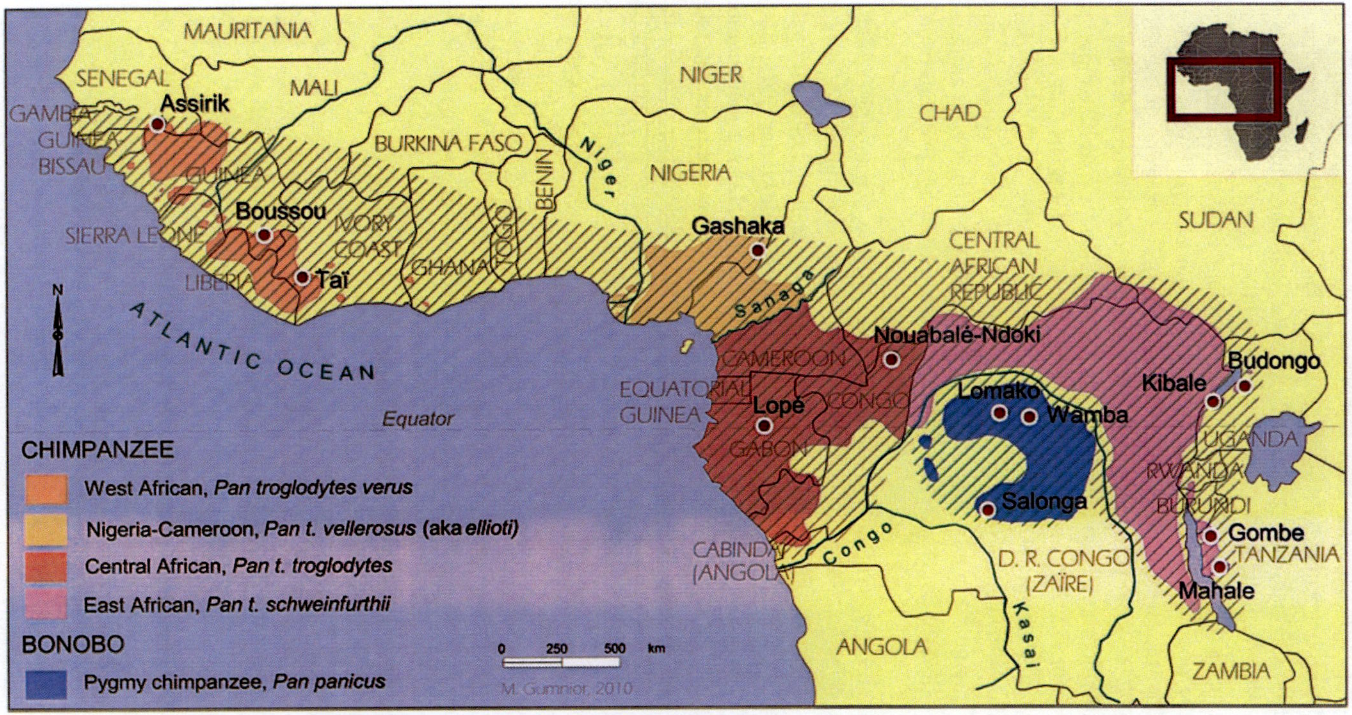

Fig. 14.2 Approximate geographical distribution of members of the genus *Pan* (chimpanzees, bonobos) (map: VS, Maren gumnior)

Cameroon include: Takamanda FR, Mone FR, Mbulu Hills Community Forest, Ebo forest, Korup NP, Banyang Mbo Wildlife Sanctuary, Rumpi Hills FR, Bakossi Mountains, Mount Kupé, Mount Manengouba. The largest groups are thought to occur at Korup, Takamanda and Ebo-Ndokbou (Ngalla *et al.* 2005). Major occurrences in Nigeria include the Gashaka Gumti NP and its vicinity and the Cross River NP and surroundings, including Afi Mountain Wildlife Sanctuary. Remnant chimpanzee populations can be found in the south-eastern forests of the wider Niger Delta area (west of the Benue / Niger divide), for example, in Ise FR, Omo FR, Owo FR, Okomu FR and Okomu NP. The subspecies affinity of these populations is still uncertain. Depending on analytical technique, they group either with *P. t. vellerosus* or *P. t. verus* of the Upper Guinea region (McManus 2005).

Population estimates for *P. t. vellerosus* are not reliable. A 2005 workshop in Brazzaville estimated that there are 3000 – 5000 individuals surviving in Cameroon (with one attempt for a more narrow estimate putting their number at around 3380) and 2000 – 3000 in Nigeria (with a narrower estimate of 3050) (Kormos *et al.* 2003; Ngalla *et al.* 2005, McManus 2005).

Nevertheless, several of the estimates listed in the Great Ape World Atlas (Caldecott & Miles 2005) that make up these country-wide figures are almost certainly too optimistic. For example, Gashaka Gumti NP is thought to host the largest population in Nigeria with "up to 1500 chimpanzees" (McManus 2005: 380), but surveys by the *Gashaka Primate Project* (2001, 2003, 2004) put their numbers at only 1000 (see Adanu *et al.* this volume [Ch. 3]). It is also stated that chimpanzees "occur in neighbouring areas, including the Mambilla Plateau (specifically in Ngel Nyaki Forest Reserve and Leinde Fadali and Akwaizantar forests) and the Donga River valley" (McManus 2005: 380). However, the forest fragment of Ngel Nyaki is home to only a dozen or so chimpanzees (Beck & Chapman 2008); the same is likely true for Leinde Fadali (which is also part of Gashaka Gumti NP), whereas the Akwaizantar forest has recently been destroyed (see Chapman *et al.* 2004) and no chimpanzees remain. Finally, the Donga valley is under extreme hunting pressure (Adanu *et al.* this volume [Ch. 3]). For the Okwango Division of Cross River NP, a number of 400 chimpanzees is estimated while the southern Oban Division is considered "important for chimpanzees" (McManus 2005: 380). Very few chimpanzees survive in the wider Niger delta area, whether or not they belong to *P. t. vellerosus* or *P. t. verus*.

Revised numerical estimates thus suggest a mere 1500 chimpanzees survive in Nigeria. It is close to impossible that the remaining habitats in Nigeria will harbour another 1500 of these apes. This renders the upper population estimate of 3000 chimpanzees for Nigeria unrealistic compared to the lower estimate of 2000.

Similar caveats apply to Cameroon. Unabated hunting in Korup NP – thought to contain one of the "largest populations" of *P. t. vellerosus* (Ngalla *et al.* 2005: 307) – means that only a remnant population survives here (Sommer *pers. obs.*). The vast majority of remaining chimpanzees will thus be found in the east-west axis of Mount Kupé – Bakossi Mountains – Banyang Mbo – Takamanda (Ngalla *et al.* 2005). However, to assume that these areas will contain close to 5000 chimpanzees is unrealistic. Even the lower estimate of 3000 seems too high.

To pre-empt concerns of being overtly pessimistic when modelling population extinction rates, we work with the upper population level of 8000 chimpanzees across Cameroon and Nigeria. All models are also run for a population of 5000 *P. t. vellerosus* (even though this number still likely overestimates the remaining population by a factor of 2).

Data for chimpanzee *population density* are also problematic. For example, density estimates for sites of *P. t. troglodytes* in Cameroon vary by a factor of 12, from as low as 0.1 / km^2 to as high as 1.2 / km^2 (Ngalla *et al.* 2005: 307). Comparable estimates for *P. t. vellerosus* across populations do not exist. However, several assessments with similarly considerable variation have been generated for a single population in Gashaka Gumti National Park, where estimates diverge by a factor of 7 (Dunn 1993: 1.1 chimpanzees / km^2; Hogarth 1997: 0.75; Foster 1998: 0.24 – 1.12; Adanu 1998: 0.2 – 2.1; Lameed 2002: 2.02; Sommer *et al.* 2004: 1.3). Such discrepancies reflect different techniques of nest-counting (straight line vs. non-random transects) and surveyed habitats (lowland forest, mountainous forest, woodland). Density measures also depend on assessments of suitable habitat within the park. This is a difficult task, given the mosaic nature of plant cover, the existence of lacunae and discontinuities in distribution, and the lack of updated vegetation maps based on remote sensing. Thus, one has to remain critical about reported density values. For our calculations, we use 0.42 / km^2 by averaging values for *P. t. troglodytes* and *P. t. verus* from numerous sites listed in Matthews & Matthews (2004).

Carrying capacity for *P. t. vellerosus* is calculated as 15960 individuals. We arrive at this number by applying the density of 0.42 animals / km^2 to 25 % of the total area of occupancy (one fourth of 152000 km^2 total range = 38000 km^2). The estimated lower population limit of 5000 will thus correspond to an area of occupancy of 7.8 % (11856 km^2) whereas the estimated upper population limit of 8000 corresponds to 12.5 % (19000 km^2).

Life-history parameters for *P. t. vellerosus* are not known. Instead, we use averaged data for other subspecies populations monitored at long-term study sites (*Tab. 14.1*; *Tab. 14.2*).

Ape Sanctuaries in Cameroon and Nigeria

The geographical range of *P. t. vellerosus* renders it likely that infant orphans will end up only in sanctuaries within Nigeria and Cameroon. Between April and June 2003, NH spent about 2 weeks at each of the 4 ape sanctuaries in these two nations (*Fig. 14.3*). Information about each ape that had passed through the facilities between 1986 and June 2003 was gathered from written records and interviews with sanctuary personnel (Hughes 2003).

- SYCRC: The Sanaga-Yong Chimpanzee Rescue Centre was founded in 1999 by Sheri Speede, an American veterinarian, near Belabo in the eastern Province of Cameroon. It initially aimed to be a refuge for adult chimpanzees held captive

Table 14.1 Life-history parameters of wild chimpanzees at long-term study sites (an independent compilation by Stumpf 2006 arrives at virtually the same values)

Site	Age at first birth (years)	Birth interval (months) (a)	Birth rate / female / year	Age at last birth (years)	Male reproductive maturity (years)	Secondary sex-ratio (M:F)	Source
Taï / Ivory Coast	13.7	69.5		43		1 : 1.3	Boesch & Boesch-Achermann 2000
Bossou / Guinea	13	61.2	0.197				Sugiyama 1994
Gombe / Tanzania	13.3	62.4	0.2	37	13 – 15 (b)	1 : 1.5	Teleki *et al.* 1976, Goodall 1986, Wallis 1997
Mahale / Tanzania	13.2	67.2	0.205	39		1 : 1.1	Nishida *et al.* 1990, 2003
Mean	13.3	65.1	0.201	40	13 – 15	1 : 1.3 (c)	

(a) Omitting instances where infants younger than 3 years died as this shortens intervals(b) Late adolescence
(c) Not significantly different from 1 : 1

Table 14.2 Yearly mortality rates of wild chimpanzees. Calculations based on Hill *et al.* (2001)

Age (years)	Females	Males
0	0.18 ± 0.05	0.22 ± 0.08
1	0.12 ± 0.07	0.17 ± 0.10
2	0.09 ± 0.08	0.09 ± 0.06
3	0.05 ± 0.06	0.05 ± 0.07
4	0.04 ± 0.06	0.07 ± 0.07
5	0.10 ± 0.18	0.06 ± 0.08
6	0.04 ± 0.06	0.10 ± 0.15
7	0.03 ± 0.03	0.02 ± 0.06
8	0.04 ± 0.04	0.02 ± 0.02
9	0.06 ± 0.09	0.09 ± 0.18
10	0.04 ± 0.04	0.05 ± 0.04
11	0.00 ± 0.00	0.00 ± 0.00
12	0.02 ± 0.02	0.05 ± 0.13
13	—	0.12 ± 0.12
14	—	0.11 ± 0.18
Mature mean (a)	0.09 ± 0.12	0.11 ± 0.19

(a) Females from 13 years, males from 15 years. Female average calculated up to 46 years, discounting two females listed with 53 and 55 years by Hill *et al.* (2001) as this would overtly bias the mean upwards

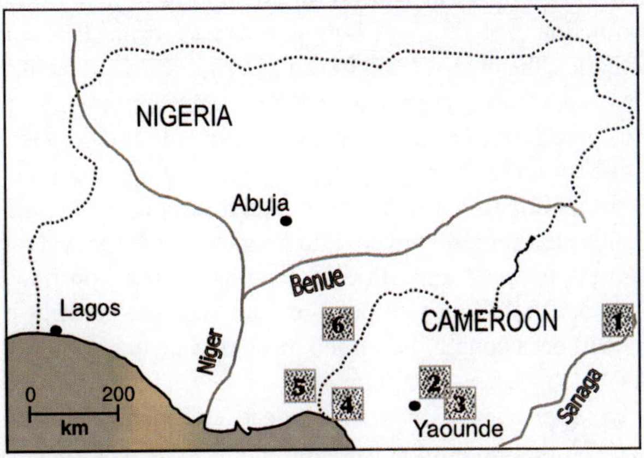

Fig. 14.3 Ape sanctuaries visited. 1: Sanaga-Young Chimpanzee Rescue Centre; 2: CWAF (Cameroon Wildlife Aid Fund) Mvog-Betsi Zoo Yaoundé; 3: CWAF Mefou National Park; 4: Limbe Wildlife Centre; 5: DRBC (Drill Rehabilitation and Breeding Centre), Calabar; 6: DRBC, Afi Mountains

in Cameroon for many years but its forested enclosures soon also became a shelter for bush-meat orphans.

- CWAF: Mvog-Betsi Zoo in Yaoundé was founded in the 1950s and the Cameroon Wildlife Aid Fund (CWAF) became involved in the modernisation of the zoo and the care of its primates in 1997. In 2001 CWAF also constructed a second facility just outside Cameroon's capital in Mefou National Park to create a more natural environment for some of the zoo's primates. The facilities are jointly run by CWAF and the Cameroonian Ministry of Environment and Forests (MINEF).

- LWC: Limbe Wildlife Centre, situated at the foot of Mount Cameroon in the south west Province, has been the site of a zoo since the 1950s. In 1993, Pandrillus, an NGO founded by Liza Gadsby and Peter Jenkins, became involved with the zoo and started its transformation into a primate sanctuary. LWC is jointly run by MINEF and Pandrillus.
- DRBC: Pandrillus, since 1991, also maintains the Drill Rehabilitation and Breeding Centre in Nigeria. DRBC breeds drills with the aim of releasing them back into the wild but the sanctuary also cares for chimpanzees at its two sites within Cross River State: a quarantine and holding facility in Calabar and a forested enclosure at the Afi Mountain Sanctuary.

How Many Dead Wild Chimpanzees Does One Sanctuary Orphan Represent?

A critical question we need to address in attempting to analyse hunting sustainability and extinction probability is how to estimate, from the number of orphans found in sanctuaries, the number of hunting deaths in wild chimpanzee populations. Several factors have to be taken into account when employing sanctuary intake rates of apes as corollaries of hunting pressure exerted on wild populations: (a) not all hunts capture an infant; (b) to capture an infant, other group members have to be killed; (c) not all infants survive capture and transport; (d) the majority of survivors will be held elsewhere; few make it to a sanctuary; (e) arrival rates also depend on the willingness of local authorities to confiscate pets, on how well known a sanctuary is, its accessibility, its reputation and how pro-actively it solicits donations; (f) apes will occasionally be turned away by sanctuaries because of health or legal reasons (A. Olivecrona *pers. comm.*).

How can we approximate how many dead chimpanzees a sanctuary orphan might represent? A first point of consideration has to be that infants cannot be captured other than by maiming or killing their mothers (Teleki 1989). Therefore, each orphan represents at least 2 chimpanzees (the infant and its mother) removed from the wild. However, this figure is still an underestimate, given that hunters do not, or very rarely, target just mothers to obtain infants. Instead, they will try to locate the approximate position of a party of chimpanzees by listening to vocalisations that are made while the apes construct their night nests. The hunters will then move in on the nesting trees before dawn and wait until there is sufficient light before indiscriminately firing at any ape trying to escape (Sommer 2008; see also Fowler 2006). Infants do not construct their own nests, but sleep with their mothers. However, it is normally impossible for hunters to pinpoint nests in which mothers rest with their infants. In any case, nursing females are not separately targeted but shot at random – as the hunters' main aim is not to obtain live infants, but meat. Even if hunters wanted (and managed) to target nursing mothers, additional deaths would result because of the protective behaviour of other group members, especially the adult males (Teleki 1989).

Data from Sierra Leone (Teleki 1980) suggest that each sanctuary infant represents 5 chimpanzees removed from the wild group – the infant, who is taken alive, and 4 others who are killed (its mother and 3 additional group members). We support this inference through calculations based on the number of infants in chimpanzee communities with individually known members. Chimpanzees are classified as "infants" up to the age of 4 – 5 years. Behavioural criteria commonly used to define this age-class are suckling, dorsal riding and constant association with their mother (Goodall 1986: 81, Boesch & Boesch-Achermann 2000: 18). The proportion of infants in well-studied communities across Africa averages 23.6 %, with some considerable variation between and within study sites (minimum 11.2 %, maximum 38.1 %; *Tab. 14.3)*.

Let us assume that 23.6 % represents a single infant individual. This means that the proportion of non-infants present in an average chimpanzee community is 76.4 %., and that (76.4 / 23.6 =) 3.2 individuals would have to be killed on average, to obtain that single infant by chance. This number could be as low as 1.6 individuals for the maximum reported proportion of infants of 38.1 % (which corresponds to a proportion of non-infants of 61.9 %). Or, the number of individuals killed could be as high as 7.9 for the minimum reported infant proportion of 11.2 % (as this would correspond to a proportion of non-infants of 88.8 %). Accordingly, if we add the surviving infant to the number of killed members of the chimpanzee community, we can say that an average of 4.2 (range 2.6 – 8.9) individuals are removed from the wild whenever a single infant is obtained alive as a by-product of a hunt.

This figure is still an underestimate, as nursing mothers and their infants will often travel on their own because infant carrying makes it difficult to travel with others

Table 14.3 Proportion of infants in wild chimpanzee communities

Site	Census years (*n*)	Mean % infants	min %	max %	Source
Budongo-Sonso	8	21.1	15.1	29.2	Reynolds 2005: 30
Kibale-Ngogo	1	17.4			Ghiglieri 1984: 14
Kibale-Kanyawara	8	24	16.7	26.1	Ghiglieri 1984: 14; Wrangham 2002: 205
Gombe-Kasakela	19	21	13.3	31	Goodall 1986: 82
Mahale-M	9	19.5	11.2	24	Nishida *et al.* 1990: 72; Nishida *et al.* 2003
Taï	15	28.4	24.2	34.5	Boesch & Boesch-Achermann 2000: 21
Bossou	6	33.8	31.6	38.1	Sugiyama 1984: 392
Overall	66	23.6	11.2	38.1	

Based on a total of 66 census years with 3661 individual counts, including 842 infants, of which 351 were identified as males and 387 as females.

and because lactating females are under pressure to reduce scramble competition (Williams *et al.* 2002, for Gombe / Tanzania; see also Wrangham *et al.* 1996). The proportion of (day-)parties made up of nursing mothers and infants averages 25.7 % for various long-term study sites (Assirik / Senegal 18 %, Bossou / Guinea 49 %, Taï / Ivory Coast 18 %, Gombe 24 %, Mahale 13 %; reviewed in Boesch 2000: 106, Boesch & Boesch-Achermann 2000: 93) and a population of *P. t. vellerosus* at Gashaka / Nigeria with a proportion of 32 % (Sommer *et al.* 2004).

Mother–offspring parties do not only often forage alone, they also tend to sleep alone, not least because they incur additional travel costs by teaming up with larger nest groups in the evening just to go their own way again in the morning. The larger nest-groups, however, are the explicit targets of hunters, as they are normally vocal or noisy and will therefore be much more likely located than mother–infant pairs. As infants are underrepresented in larger nest groups, this will once more increase the proportion of other wild chimpanzees which need to be killed, statistically, before a single infant is caught alive during a hunt by chance.

The proportion of lone mother–offspring parties (26 %) is very similar to the proportion of infants in a community (24 %). This is to be expected, and the finding therefore indicates the general reliability of the cross-population calculations.

Because infants depend on their mothers, we can thus safely assume that about half of all community members travel apart from larger parties during the daytime (the proportion is likely even greater because juvenile offspring often accompany mother-infant parties). A conservative estimate would assume that mother-offspring parties nest alone, but join larger nest-groups every other evening. Vice versa, they would be absent from every second larger nest group that is targeted by hunters. We would thus have to add at least 25 % to the 3.2 wild chimpanzees killed for each captured infant. This proportion of 0.8 would lead to a revised figure of 4.0 chimpanzees killed for each infant obtained alive, or, when counting in the infant itself, a total of 5.0 chimpanzees lost to the wild population for each orphan that initially survives the slaughter. Thus, one orphan can indeed safely be assumed to represent 5 individuals removed from the wild.

The Jane Goodall Institute (Peterson & Goodall 1993) maintains an even higher number, estimating that one sanctuary chimpanzee equates to 10 chimpanzees removed from the wild – the infant and 9 others who are killed, i.e., its mother and 8 additional group members. While this figure is not substantiated by actual data, we consider it a reasonable assumption and therefore make this higher estimate an independent part of our calculations.

It is also crucial to estimate how many orphans actually end up in a sanctuary. Teleki (1989) maintains that only one in five infants make it to their final destination. This assumption seems reasonable. Firstly, ape infants might be outright killed along with other group members, either accidentally or, in the case of older and thus heavier infants, to obtain their meat. They may also die as a result of being wounded, or perish when transported in adverse conditions, when they are incorrectly or inadequately fed or due to stress from capture and the loss of their mother (Peterson 2003). Moreover, the vast majority of orphans who are kept as pets by hunters, local dignitaries, expats, etc. will not reach a sanctuary (Peterson 2003).

Dale Peterson, an American author, travelled extensively through Central Africa in fact-finding expeditions for his books on primate socioecology and conservation. Together with Karl Ammann, the Swiss-born wildlife photographer and conservation activist, Peterson compiled narratives about ape orphans of which a few illustrate the fate of those who don't make it to a sanctuary:

- A woman in the Congo had kept a baby chimpanzee that was "too small for meat" but it had died the night before. She had tossed the carcass away into the grass on the other side of the road" (Peterson 2003: 32).
- The owner of a gorilla baby in Cameroon had kept it "at night in a suitcase", while it was "tied to a post" during the day – until it died. So it was thrown away into the underbrush (Peterson 2003: 47).
- A hunter in Cameroon brought a smoked female chimpanzee and her baby back to his house. "The baby had been injured in the jaw" by a pellet. The hunter "would have kept it alive for sale or to keep it in his house, but it was too badly injured and he slaughtered it with his machete" (Peterson 2003: 49).

Concrete figures are hard to come by, but the view that all bush-meat orphans arrive at a sanctuary is clearly untenable. Most probably, only a fraction of all infants removed from the wild will ever show up in sanctuary records. Thus, the assumption that one orphan in a sanctuary represents five infants taken from the wild is more likely an underestimate than an exaggeration.

When Does Hunting Become Unsustainable?

Hunting is not sustainable if current rates of "harvesting" – a euphemistic term borrowed from agricultural economics (Peterson 2003), as "harvesting" typically equates to "killing" – are higher than population growth rates. Various equations have been developed to calculate how many individuals can be removed from the wild without driving the population to extinction. Ideally, one would want to compare the *actual* rate of increase in a population (r) to the rates of hunting. However, detailed life-history parameters (e.g., age of first and last reproduction, infant mortality) are often unknown, as are stochastic events that limit growth (e.g., a catastrophic food-shortage because of an outbreak of a fruit-pest; off-take by natural predators). Therefore, conservationists tend to use the theoretical maximum growth rate r under optimal conditions.

It is also important to note that sustainability indices based upon r can only predict unsustainable harvest rates. They do not suggest that anything under this rate is necessarily sustainable. Thus, "not unsustainable" is not the same as "sustainable". In particular, it is entirely possible that, even though the hunting rate is smaller than the growth rate given by a particular model, the population is still unsustainable. Our simple criterion, that hunting is not sustainable if rates of harvesting are higher than population growth rates, is thus a sufficient but not necessary condition for unsustainability.

We review the merits and shortcomings of various calculations (for this, see also Milner-Gulland & Akçakaya 2001) and later use the level of loss represented by orphans to estimate if harvest rates are unsustainable.

A classic approach by Lamont Cole (1954) specifies a discrete-time equation that provides a model of population consequences for a species subpopulation. We will use the following notation in which we refer to discrete time increments as years:

α = female age at first birth
ω = female age at last birth
b_x = fecundity at age x (birth rate of female offspring)
l_x = survivorship at age x

The equation describing population growth can be expressed as:

$$\sum_{x=\alpha}^{\omega} l_x\, b_x\, e^{-rx} = 1 \qquad (14.1)$$

Crucial to this model is the quantity r as the intrinsic rate of natural increase of a given species. This is viewed as a fixed characteristic, which solely depends on the physiological (intrinsic) potential of the females of a given species to produce offspring. It describes the maximum continuous-time potential rate of increase under ideal conditions, i.e., when a population is "not limited by food, space, resource competition, or predation" (Robinson & Redford 1991: 417). The formula also assumes that there is no mortality (i.e., el_x = 1 for all x) and that fecundity is constant throughout the reproductive years (i.e., $b_x = b$ for all $\alpha \le x \le \omega$). Substituting these assumptions into (1) gives Cole's well-known equation:

$$1 = e^{-r} + b\, e^{-r\alpha} - b\, e^{-r(\omega+1)} \qquad (14.2)$$

If we are able to observe or infer values for the parameters α, ω, and b, then a solution for r may be found relatively easily by iterative numerical methods. Nevertheless, the Cole model assumes a "garden-of-Eden'" scenario that overstates real growth and is thus not anywhere close to actual rates of increase under real-world conditions.

Several authors suggest modifications to these equations to more closely approximate *actual* growth rates. For these alternative approaches, we first have to fix some terminology and notation for a species population:

r = intrinsic rate of natural increase in continuous time
R = maximum annual discrete-time rate of increase ($R = e^r$)
P = annual production
P_M = maximum possible annual production
K = population density at carrying capacity
D = population density, actual or estimated
N = population size
f = factor accounting for pre-reproductive mortality, depending on life span
b = female fecundity (assumed age-independent)
s = percentage of individuals surviving to average age of reproduction

We term the production in year n to be the population increase in the time interval $(n, n + 1]$; thus, the production in year n is $P(n) = N(n + 1) - N(n)$.

Robinson & Redford (1991) proposed a commonly cited alternative to Cole's equations. They suggested that maximum production occurs at 60 % of the carrying capacity density of the unhunted population. They also modified the intrinsic maximum discrete rate R by an ad hoc factor f intended to account for preproductive as well as adult mortality in the absence of harvesting. The values used by Robinson & Redford (1991) were 0.6, 0.4, and 0.2 for short-lived, medium-lived, and long-lived species. Apes are long-lived and slow-reproducing; therefore, they estimate for these species, only 20 % of maximum production may be harvested. Replacing R by the resulting effective rate of growth $1 + f(R - 1)$ yields:

$$P = 0.6K(R-1)f \tag{14.3}$$

Another modification was proposed by Slade *et al.* (1998). They first assumed a fixed fecundity rate b; second, specified a value 1_a of preproductive survival; and third, assumed a constant probability of adult survival p. These assumptions yield an equation more general than Cole's:

$$1 = p\,e^{-r} + 1_a\,b\,e^{-ar} - 1_a\,b\,p^{w-a+1}\,e^{-(w+1)r} \tag{14.4}$$

Solving this equation for r gives a value of growth rate that is more indicative of the actual population than is the intrinsic rate of natural increase. Moreover, empirically derived estimates of preproductive and adult survival rates can be easily used in the calculations as can the multipliers of Robinson & Redford (1991) to adjust for an effective growth rate.

Modelling Risk of Extinction

Models based on r overestimate the predicted real growth. Thus, as explained above, they can be used only to infer that a situation is *un*sustainable while inferences about sustainability are questionable. For these, rather than using analytic or numerical methods, we have to rely on simulations instead.

Milner-Gulland & Akçakaya (2001) modelled sustainability using RAMAS Metapop (Akçakaya 1998). We employ an intellectual successor to this program, the population viability analysis tool VORTEX (Lacy *et al.* 2003), which was designed to predict extinction risk to endangered species. VORTEX has, for example, been applied to *P. t. schweinfurthii* (Edroma *et al.* 1997) and *Gorilla gorilla beringei* (Werikhe *et al.* 1998) in Uganda. The program simulates the "effects of deterministic forces as well as demographic, environmental and genetic stochastic events on wild populations" (Edroma *et al.* 1997: 59). VORTEX performs population viability analysis by discrete event probabilistic simulation (Miller & Lacey 1997). The program can use empirical age-based survivorship data and introduces the use of statistical estimates of parameters with specification of uncertainty.

The input parameters that we used to simulate extinction risk of *P. t. vellerosus* in VORTEX can be summarised as follows:

- Each simulation was repeated 100 times for each scenario (sample paths resulting from different iterations were different as VORTEX uses a random number generator to provide stochasticity for the events in the population life cycle based on the parameters entered)
- Simulations were run over 100 years to investigate the likelihood of extinction in this timeframe.
- Extinction was defined as only one sex remaining.
- The simulations modelled only one population.
- Inbreeding was set to "No" because the vast majority of females migrate out of the natal group and chimpanzees seem to actively avoid breeding with relatives (Morin 1993).
- Environmental variation was deemed to be concordant with variation in survival and reproduction, i.e., "good years for reproduction are also good years for survival" (Miller & Lacey 1997: 32).
- Breeding system was specified as polygynous.
- Reproductive parameters (age of first breeding for males, age at first and last birth) were based on averages for long-term study sites (cf. *Tab. 14.1*); secondary sex-ratio was set to 50 : 50; litter size to 1.
- Reproduction was classified as density dependent (the proportion of reproducing females / year was set to 25 % at normal population levels and to 20 % at carrying capacity (Edroma *et al.* 1997)).
- Yearly mortality rates were taken from averages for long-term study sites (cf. *Tab. 14.2*).
- The starting population size was set to lower (5000) and upper population estimates (8000). VORTEX generates a stable age grade distribution from these initial population inputs.
- Carrying capacity was set to 15960, i.e., the estimated number of chimpanzees found in 25 % of the total area of occupancy.
- Harvesting rates were based on extrapolations of how many chimpanzees removed from the wild were represented by sanctuary arrivals.
- No supplementation (the purposeful addition of individuals to the population) was included.
- Habitat degradation was also omitted in these simulations as the intent was to demonstrate the effects of bush-meat hunting alone.

Results

Sanctuary Intake Rates

The 4 primate sanctuaries in Nigeria and Cameroon yielded records for 161 chimpanzees (SYCRC = 23, CWAF = 57, LWC = 54, DRBC = 27). Combined intake rates show a clear increase since 1986 (*Fig. 14.4*).

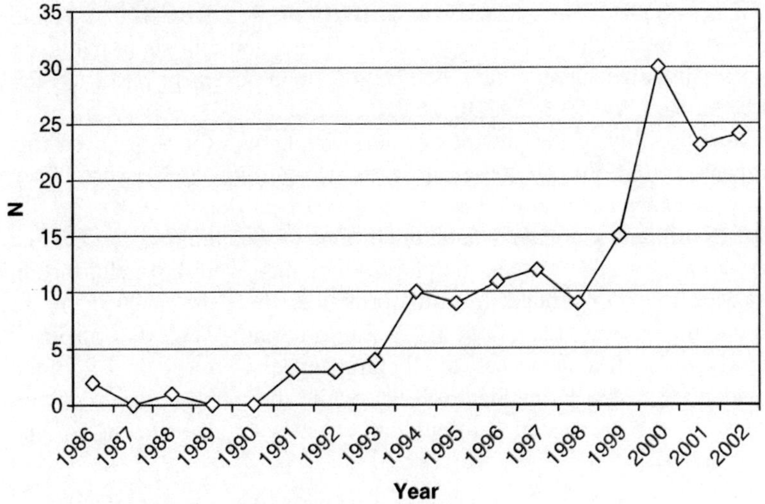

Fig. 14.4 Intake of chimpanzees into Nigerian and Cameroonian sanctuaries

The sex ratio of arrivals was relatively even, with 53 % males and 47 % females. At least 80.1 % of the arrivals were younger than 4 years of age and thus removed from the wild as dependent, nursing infants (23.6 % of arrivals < 1 year; 31.3 % 1 – 2 years; 25.5 % 3 – 4 years).

Subspecies could be allocated to about 70 – 80 % of chimpanzees at each sanctuary through DNA data and classification based on the area in which the chimpanzee was found. Accordingly, 50 % were *P. t. troglodytes*, 47 % *P. t. vellerosus*, and 3 % *P. t. verus*. The 40 chimpanzees that could not be attributed to a subspecies were, for modelling purposes, randomly and proportionally assigned (21 = *P. t. troglodytes*, 18 = *P. t. vellerosus* and 1 *P. t. verus*).

The total number of *P. t. vellerosus* used in the modelling analysis is 75 (1991 = 2, 1992 = 2, 1993 = 3, 1994 = 4, 1995 = 4, 1996 = 8, 1997 = 9, 1998 = 3, 1999 = 7, 2000 = 12, 2001 = 8, 2002 = 6, 2003 = 7) with an annual average intake of 6 (SD 3) per year.

Lives of Orphans Before Arrival

About half (50.5 %) of all chimpanzees were donated (SYRC 52 %, CWAF 42 %, LWC 54 %, DRBC 55 %). The proportion of confiscated apes was only slightly lower (43.0 %; SYRC 48 %, CWAF 52 %, LWC 35 %, DRBC 37 %). Relatively few animals (6.5 %) arrived at the sanctuary by other means (abandoned, zoo transfer, unknown).

There is great variation in where the chimpanzees were held before entering the sanctuary. About one third (30.4 %) were kept as pets in individual homes (SYRC 13 %, CWAF 44 %, LWC 35 %, DRBC 30 %). About one fifth (21.2 %) were held in villages (SYRC 39 %, CWAF 14 %, LWC 17 %, DRBC 15 %). Another fifth (22.4 %) were displayed as attractions for clients and tourists in public establishments such as restaurants, bars, hotels, zoos or amusement parks (SYRC 35 %, CWAF 11 %, LWC 19 %, DRBC 26 %).

The nationality of the previous owners was known for about two-thirds of all chimpanzees (64 %). Of these, close to two-thirds (68 %) were Nigerians or Cameroonians whereas around one third were ex-patriots (35 %).

It was difficult to collate data on the origins of the chimpanzees, but reasonable information is available about the place where they were last held. In Cameroon, most were found or brought from the province where the sanctuary was located, with provinces East (21 %), Central (20 %), and South-West (19 %) being the most common. In Nigeria, about half of all chimpanzees were supplied from the site of the sanctuary, Cross River State (48 %), while the place of origin of most other arrivals (33 %) was scattered throughout states with major towns (Abuja, Kano, Lagos, Warri).

Loss to the Wild Population

The average annual intake of 6 chimpanzees corresponds to different numbers of *P. t. vellerosus* taken from the wild, depending on how many dead chimpanzees one orphan is thought to represent.

- Assuming that 1 in 5 reaches the final destination, $6 \times 5 = 30$ would be taken from the wild each year.
- This figure doubles to a minimum of $30 \times 2 = 60$, because the mother has to be killed to obtain the infant.
- The loss increases to $30 \times 5 = 150$, if (as it is likely) at least 3 group members are killed in addition to the mother.
- The annual loss may well reach $30 \times 10 = 300$, if 8 group members are killed in addition to the removal of mother and infant.

Therefore, 6 infant chimpanzee arrivals per year over 13 years equate to a minimum of 780 and a maximum of 3900 *P. t. vellerosus* removed from the wild.

Is the Loss Unsustainable?

The intrinsic rate of population increase r can be calculated from equation (2), where α is the age at first reproduction (13.3 years), b is the birth rate of female offspring (0.101 / year), and ω is the age at last reproduction (40 years). This equation thus becomes:

$$1 = e^{-r} + 0.101\, e^{-r(13.3)} - 0.101\, e^{-r(40+1)} \qquad (14.5)$$

The solution $r = 0.041$ is obtained by numerical methods.

Population density at carrying capacity (*K*) (0.42 / km^2) and r_{max} (0.041) are needed to calculate maximum production P_{max} as given in equation (3):

$$P_{max} = 0.6\,K(R-1)f = 0.6(0.42)(e^{0.041}-1)\times 0.2 = 0.002$$

The maximum harvest is therefore equal to 0.002 chimpanzees / km^2.

For a population of 5000 *P. t. vellerosus*, which is thought to occur over an area of 12,160 km^2, we thus calculate annual harvest / km^2 as 0.005 (60 / 12160 km^2; for the minimum of 60 removed chimpanzees / year); 0.012 (for the more likely number of 150 / year); and 0.025 (for the higher estimate of 300 / year). A population of 8000 is thought to occupy an area of 19760 km^2. Here, annual harvests / km^2 are calculated to be 0.003 (i.e., 60 / 19,760 km^2), 0.008 (150 / 19760 km^2), and 0.015 (200 / 19760 km^2), respectively.

When compared to the maximum harvest rate of 0.002 / km^2, it is clear that current rates are unsustainable. For the population of 5000, they are greater than the sustainable rate by a factor 2.5, 6.0 or 12.5, and for the population of 8000 by a factor 1.5, 4.0 or 7.5.

Do Hunting Rates Exceed Growth Rates?

A simplified procedure would work under the assumption that hunting is unsustainable if hunting rates exceed annual growth rates (*Tab. 14.4*).

Table 14.4 Is the hunting of *P. t. vellerosus* unsustainable? A comparison of different calculations. B = borderline = ± 10%

			Chimpanzess annually removed from the wild		
		Population size	60	150	300
		5000	1.2%	3.0%	6.0%
		8000	0.8%	1.9%	3.8%
Source of formula	Annual growth (%)		Hunting unsustainable? Y = yes, N = no, B = borderline		
Cole (1954)	4.2	8000	N	N	B
		5000	N	N	Y
Slade *et al.* (1998), Marshall *et al.* (2000)	2.2	8000	N	N	Y
		5000	N	Y	Y
Robinson & Redford (1991)	0.8	8000	N	Y	Y
		5000	Y	Y	Y

The classic equation by Cole (1954) assumes an annual growth of 4.2 % per year. In this case, hunting would only be unsustainable if at least 300 infants were removed from a population of 5000. However, Cole's formula leads to a gross overestimation of growth as no mortality is included.

Slade *et al.* (1998) control for this by suggesting that survivorship data be incorporated. Marshall *et al.* (2000) tested this method with 2 % juvenile and 2 % adult mortality. This yields a prediction of 2.2 % growth. Therefore, whilst removal rates of 60 chimpanzees per year are still not unsustainable, removing 150 is not sustainable at population level 5000 and bordering on unsustainability at 8000. At both levels, 300 removed chimpanzees would be unsustainable.

Robinson & Redford's (1991) original formula is the most pessimistic, as it assumes that maximum production occurs at 60 % of carrying capacity, and that from this only 20 % of the maximum production can be removed. This yields a prediction of just 0.8 % annual growth. Harvesting therefore becomes unsustainable if more than 41 chimpanzees are removed from a population of 5000 or 64 from a population of 8000. Thus, only the most optimistic scenario – annual removal of 60 chimpanzees from a pool of 8000 – equates to "not unsustainable", i.e., a situation in which hunting *might* be sustainable.

Thus, predictions of unsustainability become more likely, the lower the predicted growth rate is.

Extinction Risk

As harvest models can only make inferences about *un*sustainability, we ran VORTEX simulations for a more fine-tuned analysis. All calculations work on the assumption that only one in five poached ape infants makes it to a sanctuary. Simulations were run for a lower population limit of 5000 and an upper limit of 8000 as well as 3 different numbers of annual removal of apes from the wild (the minimum of 60 / year; the more likely number of 150 / year, and the higher estimate of 300 / year; see above).

We ran VORTEX simulations to predict the timeframe of extinction for *P. t. vellerosus*. Here, we applied four basic scenarios, i.e., that no hunting takes place, or that 60, 150, or 300 chimpanzees were removed from the wild annually. The simulations for the population life cycles produce both sample path plots which show the evolution of each of the 100 individual runs (*Fig.* 4.5) as well as summary statistics (*Tab.* 14.5). The mean time to extinction in years (i.e., the probabilistic average of the path times over all the generated paths) is the best overall predictor of time to extinction.

The combined picture of these runs provide an intuitive summary of what is likely to happen in each scenario. With no harvesting at all, the probability of extinction within 100 years is 0 – although the population will still be declining. However, even at the lowest harvesting rate (60) and the higher population estimate (8000), P. t. vellerosus is predicted to go extinct in less than 50 years. In the worse case scenario (5000 / 300), the population becomes non-viable in all generated paths in just over 20 years.

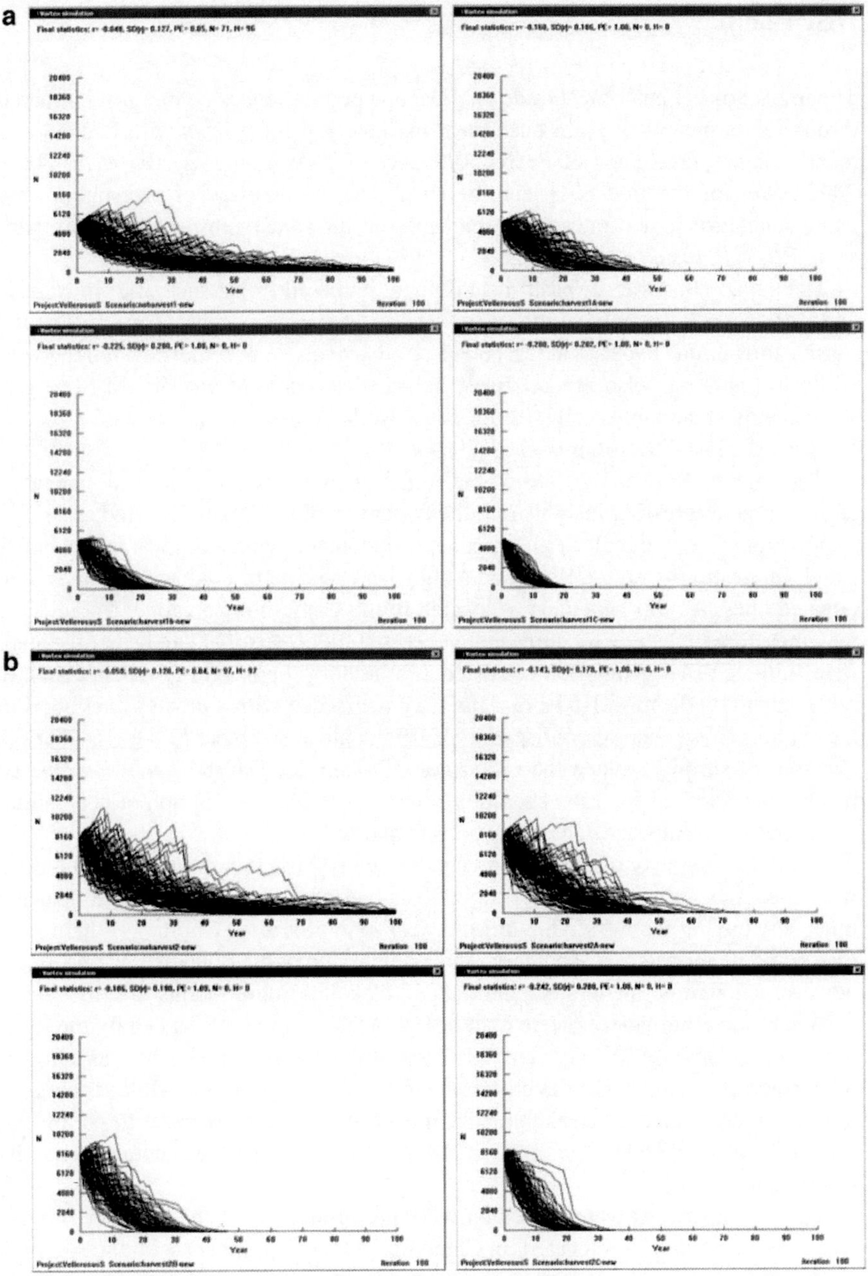

Fig. 14.5 Simulation runs of population viability analysis tool VORTEX for population life cycles of *P. t. vellerosus*. Calculations for two different estimates of initial population sizes: (**a**) minimum estimate of 5000, (**b**) maximum of 8000. For both population sizes, simulations were run for four different figures of chimpanzees annually removed from the wild (no hunting; 60; 150; 300). See text for further details

Discussion

The impact of the bush-meat trade on great ape populations is commonly estimated through data generated from bush-meat markets and interviews with hunters and local villagers (Ammann & Pearce 1995, Kemf & Wilson 1997, Fa *et al.* 2000, 2002). We, for the first time, employ an alternative pathway of investigation by using sanctuary intake rates while focusing on the rarest subspecies of chimpanzees, *P. t. vellerosus*.

The data on histories of bush-meat orphans gathered from sanctuaries in Nigeria and Cameroon – the only countries, where *P. t. vellerosus* occurs – are well suited for the aims of our paper as at least 80 % of all arrivals were removed from the wild while still nursing. Also, the relatively balanced sex ratio of arrivals (M : F = 1.1) corresponds reasonably well with the only available assessment for wild *P. t. vellerosus* (M : F = 0.9; Sommer *et al.* 2004).

A reconstruction of the fate of the infants before they arrived at a sanctuary allows some interesting insights into the dynamics of the bush-meat trade.

Slightly more than half of all orphans were donated, whereas 43 % were confiscated. Interestingly, only CWAF had a higher proportion of confiscated rather than donated apes. As this sanctuary is located in the Cameroonian capital Yaoundé, it may therefore have a larger surrounding population of pets that can be confiscated. Indeed, the CWAF population consisted of a higher proportion of ex-pets than any other sanctuary. It may also be that the law is enforced with a greater likelihood in the capital. However, the confiscation of pets is always subject to the risk that the owner may simply replace the infant that is taken (L. Gadsby *pers. comm.*). To break this cycle, confiscation should be a last resort when the option of persuading the owner to voluntarily donate has been exhausted.

While the majority of previous owners are West African nationals, about a third were non-African ex-pats. These individuals had often "rescued" the chimpanzee infants by buying them from hunters or traders and providing a home for them until they could be moved to a sanctuary. This compassion, whilst understandable, does, however, encourage the hunters and traders to acquire more infants to sell.

Where the chimpanzees were held before arrival is also influenced by the location of the sanctuary. SYRC, which has been established "in the bush", has a higher percentage of chimpanzees directly from hunters in local villages, while sanctuaries in towns receive chimpanzees that have moved from hunter to trader to pet owner. This reflects the different stages of the bush-meat trade that the sanctuaries can be involved in.

Most chimpanzees were picked up within or brought from the province or state where the sanctuary was located. In Cameroon, this is contingent with the southern half of the country, i.e., the region where most chimpanzees survive. However, in Nigeria, although most chimpanzees survive in Taraba, this state is underrepresented as an area that supplies the apes to sanctuaries. It is therefore likely that most chimpanzees orphaned in Nigeria will never become part of a sanctuary record, although we can neither exclude that hunting pressures are lower in Taraba or that infants from this region are brought to the south by an intensive trade.

It was rarely possible to pin-point the exact location from where the orphans originated, which would have allowed us to build up a picture of the intensity of hunting. However, a current genetics project (Leslie Knapp of Cambridge University; Volker Sommer of UCL) aims to fill in this gap, trying to determine genetic markers of orphans, which are typical for certain areas. One of the goals is to determine whether or not ape populations in certain regions lose genetic diversity due to hunting pressure.

In terms of overall numbers, chimpanzee arrival rates at sanctuaries in Cameroon and Nigeria have increased dramatically over the last two decades. The figures do not bode well for the survival of *P. t. vellerosus* who appears to be under the same hunting pressure as *P. t. troglodytes*, but has a much smaller wild population. Our models based on average annual intake rates into sanctuaries strongly indicate that *P. t. vellerosus* is being hunted unsustainably (cf. *Tab. 14.4*). The long inter-birth intervals and low birth rates of great apes suggest that sustainable harvesting could be impossible for these species (Milner-Gulland & Mace 1998). Moreover, population viability analysis with VORTEX estimates that extinction may occur in as little as 20 years (*Fig. 14.5*).

Admittedly, our calculations are based on various estimated parameters. However, we used a range of thresholds and methods, pessimistic as well as overtly optimistic ones, and all give essentially the same result. Several factors indicate that our results do not overestimate the risk posed by the bush-meat trade to wild populations but may in fact underestimate its severity:

- Estimates of 5000 – 8000 surviving individuals of *P. t. vellerosus* on which the calculations are based are – as explained above – almost certainly too high.
- Sanctuary populations are only the tip of the iceberg of the bush-meat trade. Sanctuary directors in Cameroon estimate that around 500 chimpanzees are held illegally throughout the country. This is almost five times as many as the number of sanctuary inhabitants reported here. The assumption that one sanctuary

Table 14.5 Time to extinction of *P. t. vellerosus* population. Mean, maximum, summary statistics as calculated by VORTEX (r = intrinsic population increase, SD = standard deviation, PE = probability of extinction; *n* = simulation runs; *H* = proportional gene diversity relative to starting population ["expected heterozygosity"]). Values for *Fig. 14.5*

Chimpanzees annually removed from the wild		Years to extinction		Summary statistics				
		Mean	Maximum	*r*	SD	PE	*n*	*H*
Population 8000	0 (no hunting)	> 100		–0.05	0.128	0.04	97	97
	60	49.1	80	–0.143	0.178	1	0	0
	150	34.3	50	–0.186	0.19	1	0	0
	300	25	31	–0.242	0.208	1	0	0
Population 5000	0 (no hunting)	> 100		–0.048	0.127	0.05	71	96
	60	41.3	54	–0.16	0.186	1	0	0
	150	27.6	33	–0.225	0.2	1	0	0
	300	20.1	22	–0.28	0.202	1	0	0

orphan represents 5 infants removed from the wild would therefore be an under-estimate, given considerable infant mortality associated with the hunting event.

- Population density at carrying capacity is the biggest variable in the calculation of production and harvest and is notoriously unreliable. Estimates vary widely due to the type of habitat (secondary versus primary forest), level of disturbance and methodology (i.e., nest counting or regression from body mass). However, even using a high density for primates of 1.8 / km^2 (Fa & Purvis 1997) changes the maximum harvest from 0.002 to just 0.009 / km^2. This would still imply unsustainability of current hunting levels for most imputed levels of harvesting.
- We have chosen to model only one population of *P. t. vellerosus* in VORTEX, due to the fact that there is little data available on the size and number of *P. t. vellerosus* communities. This creates the assumption that all members of the population live in one large supergroup and can interbreed. In reality, *P. t. vellerosus* exist in numerous, often small and independent communities. Despite the potential existence of corridors, many of these populations will not have reproductive contact. As such, the time to extinction will be over-estimated, as smaller groups are more vulnerable to stochastic environmental change and thus disappear faster (Cowlishaw & Dunbar 2000).
- Carrying capacity will likely reduce year after year due to the destruction of available habitat (see Adanu *et al.* this volume [Ch. 3]). Nevertheless, for the VORTEX simulations we assumed no habitat destruction at all. In reality, as the natural living space for chimpanzees becomes reduced, time to extinction will quicken. For example, a catastrophic 90 % decline of Ivory Coast's chimpanzee population over the last 17 years is related not only to increased hunting pressure due to a 50 % increase in human population, but also to reduction in forest cover, which ran as high as 93 % over a 7 year period for even a national park (Campbell *et al.* 2008).

In contrast, only a few factors in our analysis may have produced overestimates of the threat to wild populations:

- Yearly mortality rates are based on long-term data for habituated chimpanzee communities (Hill *et al.* 2001; cf. *Tab. 14.2*). These are comprised of deaths from disease, warfare, infanticide and predation and also include hunting. Harvesting values calculated from sanctuary arrivals thus elevate the hunting rates already accounted for by Hill *et al.* (2001). However, the effect is probably small because habituated chimpanzees – who do typically live in protected areas, which are closely monitored by researchers – are likely to be facing less pressure from hunting and predation (Boesch & Boesch-Achermann 2000: 33) than unprotected populations. Nevertheless, they are perhaps more susceptible to disease transmitted by humans (ibid., 35ff).
- Changing life-history variables also affects the time to extinction. For example, it is possible that chimpanzees could breed up to the age of 45 rather than 40. Changing this parameter at the most extreme level of harvesting (300 / year) where $K = 15{,}960$ and $P = 8000$ lengthens the estimated time to extinction by about 3 years. Changing the age at first breeding of males from 15 to 13 (Goodall 1986, Edroma *et al.* 1997) also prolonges the time to extinction by about 2.5 years.

- Similarly, previous calculations of maximum rate of population increase have used longevity in place of age at last reproduction (e.g., Ross 1992) as some females in the wild may live to > 50 years. However, these older females have not been observed to have given birth, and data from known study sites indicate that the age of last reproduction is closer to 40 years. As such, r_{max} is calculated using 40 and not 50 years as age of last reproduction. In any case, using 50 years as the age at last reproduction, Ross (1992) calculates r_{max} at 0.05. This value does not differ much from that computed here (0.041) or that proposed by Marshall *et al.* (2000) of 0.044.
- There is evidence for some habituated chimpanzee populations that the secondary sex ratio may include 55 % newborn females and only 45 % males (Boesch & Boesch-Achermann 2000). Using this ratio at again the most extreme level of harvesting (300 / year), where $K = 15960$ and $P = 8000$, prolonged the time to extinction by 6 years.

Our findings present a rather bleak outlook for *P. t. vellerosus*, suggesting that these chimpanzees could become extinct in two decades. This prediction is made solely on the basis of hunting pressure. However, the pessimistic picture is exacerbated by two main additional threats: habitat destruction and disease. Hunting pressure on the largest contiguous population of *P. t. vellerosus*, that at Gashaka Gumti National Park / Nigeria, currently exists around the edges of the park. However, vast swathes inside the park are denuded through cattle grazing and burning of savannah-woodland (Chapman *et al.* 2004. Adanu *et al.* this volume [Ch. 3]). Habitat destruction can promote the spread of infectious diseases (Walsh *et al.* 2003). For example, the bacterial disease anthrax, which killed chimpanzees of the Taï National Park / Ivory Coast has been linked to deforestation that enabled livestock migrations. The apes possibly caught the disease from passing cattle (Leendertz *et al.* 2004). It is conceivable that similar transmission threatens the chimpanzee population of Gashaka-Gumti where ungulates have in the past been decimated by rinderpest (Dunn 1999). Like other African apes, *P. t. vellerosus* is also threatened by zoonotic diseases such as monkeypox and Ebola, as well as by infections caught from humans such as respiratory disease, influenza viruses and measles. Contact with humans exists through farmers, hunters, researchers, conservationists, photographers, film crews and tourists. Clearly, locals pose less of a threat than those visiting from afar, in particular from outside Africa, as the apes can be exposed to diseases they have never encountered before and against which they have no natural immunity (e.g., Butynski & Kalina 1998).

With the ultimate demise of the Nigeria-Cameroon chimpanzee, one fifth of the sub-generic taxa of *Pan* would have vanished from our planet. It has to be emphasised that this not only impoverishes biodiversity but also cultural diversity, as different chimpanzee populations maintain varied portfolios of customs and traditions which depend on unbroken chains of inter-generational social transmission (McGrew 2004).

Long-term field studies have proven to be an important conservation tool, given that research is often an effective deterrent for poaching. Moreover, research can focus attention on troubled hot-spots through the generation of knowledge and the production of media-friendly information, particularly if areas harbour "charismatic megafauna" such as apes (Goodall 1986, Boesch & Boesch-Achermann 2000). It is therefore regrettable that currently only a single long-term field study pays attention to *P. t. vellerosus*, i.e., the *Gashaka Primate Project* in eastern Nigeria (Sommer *et al.* 2004), although other studies have been initiated recently (Chapman *et al.* 2004, Morgan & Abwe 2006).

The bush-meat trade is an extremely serious threat not only to *P. t. vellerosus*, but also to the other three subspecies of chimpanzees as well as to the other African apes; gorillas and bonobos. The 22 institutions of the Pan African Sanctuary Alliance (PASA) currently hold close to 1000 apes. Intake rates as re-constructed from individual records (Rosen *et al.*, in prep.) accelerated dramatically from the mid 1990s onwards (*Fig. 14.6*). The arrival of orphans in sanctuaries does therefore not reflect an accumulation of anecdotal events but clearly increasing levels of hunting (Ammann 2001, Peterson 2003).

This crisis can be averted only if bush-meat consumption is reduced. However, to design effective measures will be extremely difficult, as the crisis is brought about by current social, cultural and economic circumstances of millions of people. Some of the factors that need to be considered include providing alternative sources of protein; changing consumer preferences; providing alternative sources of income; improving enforcement of hunting laws and controlling logging and other heavy industries opening up the forests (Bowen-Jones 1998, Bowen-Jones & Pendry 1999, Wilkie 2001, Peterson 2003). However, it is probably fair to say that zero progress has been made to halt the trend since wider awareness about the bush-meat crisis began with the photographic campaigns of Karl Ammann about 15 years ago (Peterson 2003). The "Slaughter of the Apes" (Ammann & Pearce 1995) is likely to continue until only a few populations will remain in what are hopefully effectively protected areas.

Given this uphill battle, sanctuaries are quickly becoming important conservation tools in their own right. All sanctuaries visited are members of PASA that was formed in 2000 "to support, assist and encourage member sanctuaries in their efforts to save Africa's great apes and other primates" (Rosen *et al.* 2001:13). Nevertheless, sanctuaries are sometimes seen as an expensive way of focusing on a few individuals using money that could be spent on preserving remaining wild populations (MacKinnon 1977, Soave 1982). On the other hand, infant apes are increasingly regarded as persons who should have a basic right to care, similar to that of human orphans (Cavalieri & Singer 1993). Apart from this animal-rights perspective, sanctuaries can also raise awareness of conservation issues through education and tourism programs, which bring the plight of apes to a wider local, national and international audience. Sanctuaries also support the local economy through employment opportunities and purchase of supplies such as animal food. Moreover, sanctuaries encourage law enforcement, because "where no sanctuaries exist, no confiscation of apes occur" (Teleki 2001:144). They may also complement or support wild populations if reintroduction can be successful (Tutin *et al.* 2001). In addition, sanctuaries could become an increasingly vital source of information

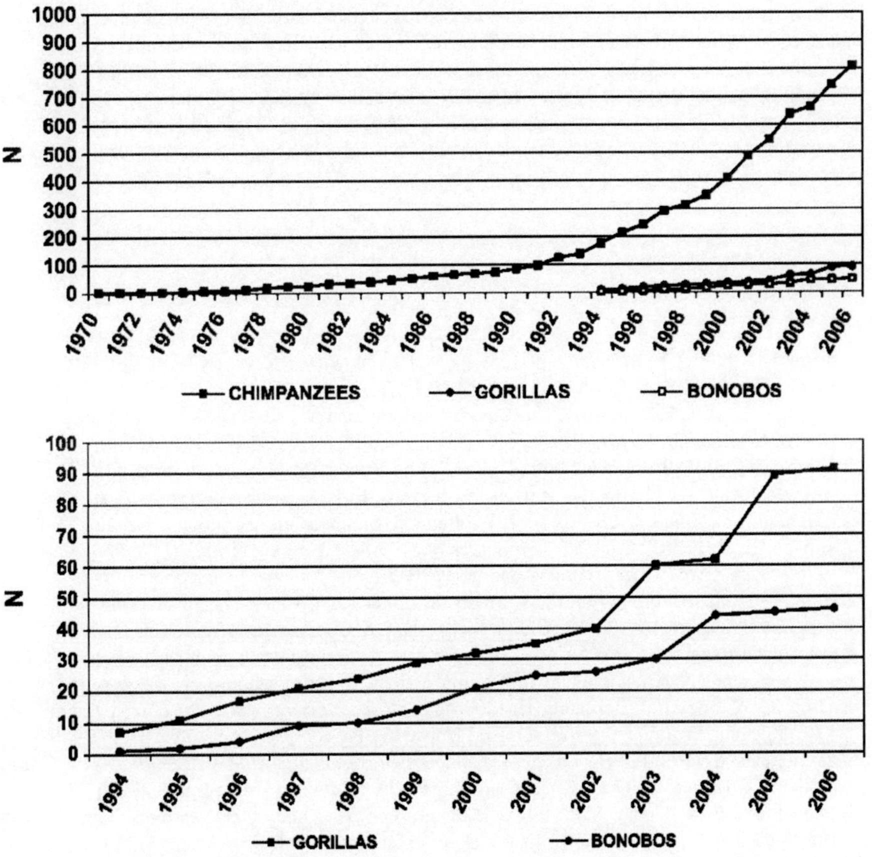

Fig. 14.6 Cumulative overall intake of African apes into 22 PASA sanctuaries across the whole continent (taking into account mortality and transfers out of sanctuaries; data from Rosen, Hughes, Alberts & Sommer, unpublished). Upper graph: gorillas, chimpanzees, bonobos, 1970 – 2006. Lower graph: amplified 1994 – 2006 intake of gorillas and bonobos

for science in fields such as pathenogenesis, zoonosis, genetics, taxonomy and – as illustrated by the present study – population viability analysis.

However, our study reveals that all these factors are less important than the fact that sanctuaries may, in a couple of decades, hold the only viable breeding populations of African apes, including *P. t. vellerosus*. This should be borne in mind when formulating policies, such as subspecies segregation in the light of the possibility of re-introduction, the pros and cons of birth control (castration, contraception vs. enabling the apes to breed), degree and type of human contact and allocation of funds for ape conservation.

We propose to apply our method of measuring extinction risk via intake rates into sanctuaries to other taxa of apes across Africa and Asia – particularly gorillas and orang-utans – and to continue to re-evaluate the role of sanctuaries as conservation tools.

Acknowledgements We thank the directors at the sanctuaries visited for their time, hospitality, and access to records: Sheri Speede (Sanaga-Yong Chimpanzee Rescue Centre), Colonel Avi Sivan and Talila Sivan (CWAF), Liza Gadsby and Peter Jenkins (Limbe Wildlife Centre, Drill Rehabilitation and Breeding Centre). In addition, the following sanctuary managers and personnel kindly provided assistance and hospitality: Saira Ndi, Edmund Stone, Livia Wittiger and Tony Chasar. Nicola Hughes was partly funded by the UCL Graduate School and the North of England Zoological Society.

Literature

Adanu, J. (1997). Determination of Life-span of Chimpanzee Nests (Nest-aging) in Gashaka-Gumti National Park. Report to NCF & WWF-UK.

Akçakaya, H. R. (1998). RAMAS metapop: Viability analysis for stage-structured metapopulations, Version 3.0. *Software Program.*

Ammann, K. (2001). Bushmeat hunting and the great apes. In Beck, B. B., Stoinski, T. S., Hutchins, M., Maple, T. L., Norton, B., Rowan, A., Stevens, E. F. & Arluke, A. (eds.), *Great Apes and Humans: the Ethics of Co-existence.* Washington: Smithsonian Institution Press, pp. 71–85.

Ammann, K. & Pearce, J. (1995). *Slaughter of the Apes: How the Tropical Timber Industry is Devouring Africa's Great Apes.* London: WSPA.

Baillie, J. E. M., Hilton-Taylor, C. & Stuart, S. (2004). *IUCN Red List of Threatened Species: A Global Species Assessment.* Gland, Switzerland & Cambridge / UK: IUCN.

Beck, J. & Chapman, H. (2008). A population estimate of the endangered chimpanzee *Pan troglodytes vellerosus* in a Nigerian montane forest: implications for conservation. *Oryx* 42: 448–451.

Boesch, C. & Boesch-Achermann, H. (2000). *The Chimpanzees of the Taï Forest. Behavioural Ecology and Evolution.* Oxford: Oxford University Press.

Bowen-Jones, E. (1998). *A Review of the Commercial Bushmeat Trade with Emphasis on Central / West Africa and the Great Apes.* Cambridge / UK: Flora and Fauna International.

Bowen-Jones, E. & Pendry, S. (1999). The threat to primates and other mammals from the bushmeat trade in Africa, and how this threat could be diminished. *Oryx* 33: 233–246.

Butynski, T. M. (2001). Africa's great apes. In Beck, B. B., Stoinski, T. S., Hutchins, M., Maple, T. L., Norton, B., Rowan, A., Stevens, E. F. & Arluke, A. (eds.), *Great Apes and Humans: the Ethics of Co-Existence.* Washington: Smithsonian Institution Press, pp. 3–56.

Butynski, T. M. & Kalina, J. (1998). Gorilla tourism: a critical look. In Milner-Gullard, E. J. & Mace, R. (eds.), *Conservation of Biological Resources.* Oxford: Blackwell, pp. 280–300.

Caldecott, J. & Miles, L. (ed.) (2005). *World Atlas of Great Apes and Their Conservation.* Los Angeles: University of California Press.

Campbell, G., Kuehl, H., N'Goran Kuamé, P. & Boesch, C. (2008). Alarming decline of West African chimpanzees. *Current Biology* 18: R903–R904.

Cavalieri, P. & Singer, P. (ed.) (1993). *The Great Ape Project. Equality Beyond Humanity.* New York: St. Martin's Press.

Chapman, C. A. & Peres, C. A. (2001). Primate conservation in the new millennium: the role of scientists. *Evolutionary Anthropology* 10: 16–33.

Chapman, C. A., Wrangham, R. & Chapman, L. J. (1994). Indices of habitat-wide fruit abundance in tropical forests. *Biotropica* 26: 160–171.

Cowlishaw, G. & Dunbar, R. (2000). *Primate Conservation Biology.* Chicago: The University of Chicago Press.

Cole, L. C. (1954). The population consequences of life history phenomena. *The Quarterly Review of Biology* 29: 103–137.

De Merode, E., Homewood, K. & Cowlishaw, G. (2004). The value of bushmeat and other wild foods to rural households living in extreme poverty in Democratic Republic of Congo. *Biological Conservation* 118: 573–581.

Dunn, A. (1993). *The Large Mammals of Gashaka-Gumti National Park, Nigeria: Line Transect Surveys from Forest and Savannah*. Report to WWF-UK, NCF & Federal Ministry of Agriculture, Water Resources and Rural Development / Nigeria.

Dunn, A. (1999). *Gashaka Gumti National Park: a Guidebook*. Lagos: Gashaka Gumti National Park, NCF / WWF-UK.

Edroma, E. L., Rosen, N. & Miller, P. S. (1997). *Conserving the Chimpanzees of Uganda: Population and Habitat Viability Assessment for Pan troglodytes schweinfurthii*. Apple Valley, Minnesota: IUCN / SSC Conservation Breeding Specialist Group.

Fa, J. E. & Purvis, A. (1997). Body size, diet and population density in Afrotropical forest mammals: a comparison with neotropical species. *Journal of Animal Ecology* 66: 98–112.

Fa, J. E., Currie, D. & Meeuwig, J. (2003). Bushmeat and food security in the Congo Basin: linkages between wildlife and people's future. *Environmental Conservation* 30: 71–78.

Fa, J. E., Garcia Yuste, J. E. & Castelo, R. (2000). Bushmeat markets on Bioko island as a measure of hunting pressure. *Conservation Biology* 14: 1602–1613.

Fa, J. E., Peres, C. A. & Meeeuwig, J. (2002). Bushmeat exploitation in tropical forests: an intercontinental comparison. *Conservation Biology* 16: 232–237.

Foster, K. L. (1998). *Censusing Chimpanzees in Gashaka-Gumti National Park, Taraba and Adamawa States, Nigeria*. Report to WWF-UK, NCF & Federal Ministry of Agriculture, Water Resources and Rural Development / Nigeria.

Fowler, A. (2006). *Socio-ecology of Nigerian Chimpanzees at Gashaka*. PhD thesis, University of London.

Fowler, A & Sommer, V. (2007). Subsistence technology in Nigerian chimpanzees. *International Journal of Primatology* 28: 997–1023.

Gonder, M. K., Oates, J. F., Disotell, T. R., Forstner, M. R. J., Morales, J. C. & Melnick, D. J. (1997). A new West African chimpanzee subspecies? *Nature* 338: 337.

Goodall, J. (1986). *The Chimpanzees of Gombe: Patterns of Behaviour*. Cambridge, MA: Harvard University Press.

Hill, K., Boesch, C., Goodall, J., Pusey, A., Williams, J. & Wrangham, R. W. (2001). Mortality rates among wild chimpanzees. *Journal of Human Evolution* 40: 437–450.

Hogarth, S. A. (1997). *The Distribution and Abundance of the Chimpanzees (Pan troglodytes troglodytes) of the Lowland Gallery Forest of Gashaka-Gumti National Park Nigeria: A Preliminary Report to the Gashaka Sector*. Report to WWF-UK, NCF & Federal Ministry of Agriculture, Water Resources and Rural Development / Nigeria.

Hughes, N. (2003). *Great Ape Sanctuaries in Nigeria and Cameroon: a Window into the Bushmeat-Trade*. MSc in Human Evolution and Behaviour thesis, University College London.

Inskipp, T. (2005). Chimpanzee (*Pan troglodytes*). In Caldecott, J. & Miles, L. (eds.), *World Atlas of Great Apes and Their Conservation*. Los Angeles: University of California Press, pp. 53–81.

IUCN (2007). *IUCN Red List of Threatened Species*. Gland / Switzerland & Cambridge / UK: IUCN / SSC.

Kemf, E. & Wilson, A. (1997). *Great Apes in the Wild*. Gland: WWF.

Kormos, R., Boesch, C., Bakarr, M. I. & Butynski, T. M. (eds.) (2003). *West African Chimpanzees. Status Survey and Conservation Action Plan*. Gland & Cambridge UK: IUCN / SSC Primate Specialist Group.

Lacy, R., M. Borbat & Pollack, J. (2003). *VORTEX: A Stochastic Simulation of the Extinction Process*. Version 9 User's Manual. Apple Valley / Minnesota: Conservation Breeding Specialist Group (SSC / IUCN).

Lameed, G. A. (2002). Chimpanzee (*Pan troglodytes*) population density and structure in Gashaka Gumti National Park. *ROAN: The Journal of Conservation* 1: 88–93.

Leendertz, F., Ellerbock, H. Boesch, C., Couacy-Hymann, E., Mätz-Rensing, K., Hakenback, R., Bergmann, C., Abaza, P. Junglen, S., Moebius, Y., Vigilant, L., Formenty, P. Pauli, G. (2004). Anthrax kills wild chimpanzees in a tropical rainforest. *Nature* 430: 451–452.

MacKinnon, J. 1977. The future of orang-utans. *New Scientist* 74: 697–699.

Matthews, A. & Matthews, A. (2004). Survey of Gorillas (*Gorilla gorilla gorilla*) and chimpanzees (*Pan troglodytes troglodytes*) in Southwestern Cameroon. *Primates* 45: 15–24.

McManus, E. (2005). Federal Republic of Nigeria. In Caldecott, J. & Miles, L. (eds.), *World Atlas of Great Apes and Their Conservation*. Los Angeles: University of California Press, pp. 378–383.

Miller, P. S. & Lacey, R. C. (1997). *VORTEX: A Stochastic Simulation of the Extinction Process*. Version 8 User's Manual. Apple Valley / Minnesota: Conservation Breeding Specialist Group (SSC / IUCN).

Milner-Gullard, E. J. & Akçakaya, H. R. 2001. Sustainability indices for exploited populations. *Trends in Ecology and Evolution* 18: 686–692.

Mittermeier, R. & Cheney, D. (1987). Conservation of primates and their habitats. In Smuts, B. B., Cheney, D. L., Seyfarth, R. M., Wrangham, R. W. & Struhsaker, T. T. (eds.), *Primate Societies*. Chicago: The University of Chicago Press, 477–490.

Morgan, B. J & Abwe, E. E. (2006). Chimpanzees use stone hammers in Cameroon. *Current Biology* 16: R632–R633.

Morin, P. A. (1993). Reproductive strategies in chimpanzees. *Yearbook of Physical Anthropology* 36: 179–212.

Ngalla, P. T., Miles, L. & Caldecott, J. (2005). Republic of Cameroon. In Caldecott, J. & Miles, L. (eds.), *World Atlas of Great Apes and Their Conservation*. Los Angeles: University of California Press, pp. 305–314.

Oates, J. F. (2006) Is the chimpanzee, *Pan troglodytes*, an endangered species? It depends on what endangered means. *Primates* 47: 102–112.

Oates, J. F., Groves, C. & Jenkins, P. D. (2008). The type locality of *Pan troglodytes vellerosus* (Gray, 1862), and implications for the nomenclature of West African chimpanzees. *Primates* 50: 78–80.

Peres, C. A. (1991). Humboldt's woolly monkeys decimated by hunting in Amazonia. *Oryx* 25: 89–95.

Peterson, D. & Ammann, K. (2003). *Eating Apes*. California: University of California Press.

Peterson, D. & Goodall, J. (1993). *Visions of Caliban: on chimpanzees and people*. Boston: Houghton Mifflin Company.

Robinson, J. G. & Redford, K. H. (1991). *Neotropical wildlife use and conservation*. Chicago: The University of Chicago Press.

Rose, A. L. (1998). Finding paradise in a hunting camp: turning poachers to protectors. *Journal of the Southwestern Anthropological Association* 38: 4–11.

Rose, A. L., Mittermeier, R. A., Langrand, O., Ampadu-Agyei, O., Butynski, T. M. & Ammann, K. (2003). *Consuming Nature. A photo essay on African rain forest exploitation*. Palos Verdes, CA: Altisima.

Rosen, N., Cox, D., Montgomery, C. & Cress, D. (2002). *Pan African Sanctuaries Alliance (PASA) Workshop Report*. Apple Valley / Minnesota: Conservation Breeding Specialist Group (SSC / IUCN).

Rosen, N., Seal, U. S., Cox, D., Montgomery, C. Boardman, W. (2001). *Pan African Sanctuary Alliance (PASA) Workshop Report*. Apple Valley / Minnesota: Conservation Breeding Specialist Group (SSC / IUCN).

Ross, C. (1992). Environmental correlates of the intrinsic rate of natural increase in primates. *Oecologica* 90: 383–390.

Slade, N. A., Gomulkiewicz, R. & Alexander, H. M. (1998). Alternatives to Robinson and Redford's method for assessing over harvest from incomplete demographic data. *Conservation Biology* 12: 148–155.

Soave, O. (1982). The rehabilitation of chimpanzees and other apes. *Laboratory Primate Newsletter* 21: 3–8.

Sommer, V. (2008). *Schimpansenland. Wildes Leben in Afrika*. Munich: C. H. Beck.

Sommer, V. & Ammann, K. (1998). *Die Großen Menschenaffen: Orang-utan, Gorilla, Schimpanse, Bonobo*. Munich: BLV.

Sommer, V., Adanu, J., Faucher, I. & Fowler, A. (2004). The Nigerian Chimpanzee (*Pan troglodytes vellerosus*) at Gashaka: Two years of habituation efforts. *Folia Primatologica* 75: 295–316.

Teleki, G. (1980). *Hunting and Trapping Wildlife in Sierra Leone: Aspects of Exploitation and Exportation*. Washington: WWF.

Teleki, G. (1989). Population status of wild chimpanzees (*Pan troglodytes*) and threats to survival. In Heltne, P. G. & Marquardt, L. A. (eds.), *Understanding Chimpanzees*. Cambridge: Harvard University Press, pp. 312–353.

Teleki, G. (2001). Sanctuaries for great ape refugees. In Beck, B. B., Stoinski, T. S., Hutchins, M., Maple, T. L., Norton, B., Rowan, A., Stevens, E. F. & Arluke, A. (eds.), *Great Apes and Humans: The Ethics of Co-Existence*. Washington: Smithsonian Institution Press, pp. 133–149.

The Sunday Times (1999). *Chimps on the menu in Brussels restaurants*. 16th May.

Tutin, C. E. G., Ancrenaz, M., Paredes, J., Vacher-Vallas, M., Vidal, C., Goossens, B., Bruford, M. W. & Jamart, A. (2001). Conservation biology framework for the release of wild-born orphaned chimpanzees into the Conkouati Reserve, Congo. *Conservation Biology* 15: 1247–1257.

Walsh, P. D., Abernathy, K. A., Bermejo, M., Beyers, R., De Wachter, P., Akou, E. M., Huijbregts, B., Mambounga, DI, Toham, A. K., Kilbourn, A. M., Lahm, S. A., Latour, S., Maisels, F., Mbina, C., Mihindou, Y., Obiang, S. N., Effa E. N., Starkey, M. P., Telfer, P., Thibault, M., Tutin, C. E. G., White, L. J. T. & Wilkie, D. S. (2003). Catastrophic ape decline in western equatorial Africa. *Nature* 422: 611–614.

Werikhe, S., Macfie, L., Rosen, N. & Miller, P. (1998). *Can the Mountain Gorilla Survive? Population and Habitat Viability Assessment Workshop for Gorilla gorilla beringei*. Conservation Breeding Specialist Group. Minnesota: Apple Valley.

Whiten A., Goodall, J., McGrew, W. C., Nishida, T., Reynolds, V., Sugiyama, Y., Tutin, C. E. G, Wrangham, R. W. & Boesch, C. (1999). Cultures in chimpanzees. *Nature* 399: 682–685.

Wilkie, D. S. & Carpenter, J. F. (1999). Bushmeat hunting in the Congo Basin: an assessment of impacts and options for mitigation. *Biodiversity and Conservation* 8: 927–955.

Wilkie, D. S. (2001). Bushmeat trade in the Congo Basin. In Beck, B. B., Stoinski, T. S., Hutchins, M., Maple, T. L., Norton, B., Rowan, A., Stevens, E. F. & Arluke, A. (eds.), *Great Apes and Humans: the Ethics of Co-Existence*. Washington: Smithsonian Institution Press, pp. 86–109.

Wilkie, D. S. & Godoy, R. A. (2001). Income and price elasticities of bushmeat demand in lowland Amerindian societies. *Conservation Biology* 15: 761–769.

Williams, J. M., Liu, H. Y. & Pusey, A. E. (2002). Costs and benefits of grouping for female chimpanzees at Gombe. In Boesch, C., Hohmann, G. & Marchant, L. F. (eds.), *Behavioural Diversity in Chimpanzees and Bonobos*. Cambridge: Cambridge University Press, pp. 192–203.

World Bank (2002). *World Development Indicators 2002*.

Wrangham, R. W. (2002). The cost of sexual attraction: is there a trade-off in female *Pan* between sex appeal and received coercion? In Boesch, C., Hohmann, G. & Marchant, L. F. (eds.), *Behavioural Diversity in Chimpanzees and Bonobos*. Cambridge: Cambridge University Press, pp. 204–215.

Index

V. Sommer and C. Ross (eds.), *Primates of Gashaka*, Developments in Primatology: Progress and Prospects 35, DOI 10.1007/978-1-4419-7403-7,